Comprehensive and Molecular Phytopathology

Comprehensive and Molecular Phytopathology

Editors

Yu. T. Dyakov
M.V. Lomonosov State University
Moscow, Russia

V. G. Dzhavakhiya
Russian Research Institute of Plant Pathology
Moscow reg., Russia

T. Korpela
JBL, Turku University, Turku, Finland

ELSEVIER

Amsterdam • Boston • Heidelberg • London • New York • Oxford
Paris • San Diego • San Francisco • Singapore • Sydney • Tokyo

ELSEVIER
Radarweg 29, PO Box 211, 1000 AE Amsterdam, The Netherlands
The Boulevard, Langford Lane, Kidlington, Oxford OX5 1GB, UK

First edition 2007

ISBN-13: 978-0-444-52132-3
ISBN-10: 0-444-52132-1

For information on all Elsevier publications
visit our website at books.elsevier.com

Transferred to digital print 2007
Printed and bound by CPI Antony Rowe, Eastbourne

Contents

Part 1 – Introduction to Molecular Phytopathology   **1**

Chapter 0   Overview on Parasitism   **3**
Yu. T. Dyakov

Types of host–parasite relationship   3
Features of parasites   4
Selected literature   17

Chapter 1   Plant Parasite Microorganisms   **19**
Yu. T. Dyakov, S. V. Zinovyeva

Fungal plant pathogens   Yu. T. Dyakov   19
Bacterial plant pathogens   Yu. T. Dyakov   25
Plant viruses and viral plant diseases   Yu. T. Dyakov   28
Parasitic nematodes as plant pathogens   S. V. Zinovyeva   39
Selected literature   47

Chapter 2   Structural Basis of Plant–Pathogen Interactions   **49**
J. Plotnikova, F. M. Ausubel

Alterations of plant and pathogen structures during
    infection process   49
Structural features of the plant cell response to infections   59
Biotrophic infections   59
Necrotrophic infections   64
Conclusion   71
Selected literature   72

Chapter 3   Advanced Methods of Plant Pathogen Diagnostics   **75**
L. A. Shcherbakova

Antigen–antibody interaction-based diagnostics   76
Diagnostics based on nucleic acid analysis   95
Conclusion   109
Using plant pathogen metabolites in diagnostics and research of
    plant–pathogen relationships   110
Selected literature   116

Chapter 4   Phenomenology of Plant–Parasite Relations   **117**
Yu. T. Dyakov

Types of resistance   117
Genetics of resistance   118
Phenotypic manifestation of resistance   122
Relations with parasites   127
Selected literature   135

Part 2 – Biochemistry and Molecular Biology of Plant–Parasite Relations   **137**

Chapter 5   Horizontal Pathosystem: Parasite Attack Factors   139
Yu. T. Dyakov, S. V. Zinovyeva

Biosynthetic processes during fungal differentiation
   and penetration   **Yu. T. Dyakov**   139
   Plant–nematode interrelationships   **S. V. Zinovyeva**   147
Suppression of plant defenses   **Yu. T. Dyakov**   150
Changes in metabolism of infected cells and tissues   **Yu. T. Dyakov**   155
Transport in the plant   **Yu. T. Dyakov**   157
Selected literature   159

Chapter 6   Horizontal Pathosystem: Resistance Factors   **161**
Yu. T. Dyakov, V. G. Dzhavakhiya

Anatomical–morphological resistance factors   **Yu. T. Dyakov**   161
Regulation of ontogenesis and damage reparation   **Yu. T. Dyakov**   162
Nutrient value of infected organs and tissues   **Yu. T. Dyakov**   163
Biochemical resistance factors   **Yu. T. Dyakov, V. G. Dzhavakhiya**   164
Selected literature   179

Chapter 7   Vertical Pathosystem: Avirulence Genes and Their Products   **181**
Yu. T. Dyakov, O. L. Ozeretskovskaya

Molecular interpretation of genetic data   **Yu. T. Dyakov**   181
Abiogenic elicitors   **O. L. Ozeretskovskaya**   187
Nonspecific biogenic elicitors   **O. L. Ozeretskovskaya**   187
Endogenous plant or secondary elicitors   **O. L. Ozeretskovskaya**   193
Avirulence genes and specific elicitors   **Yu. T. Dyakov**   200
Selected literature   215

Chapter 8   Vertical Pathosystem: Resistance Genes and Their Products. Signal Transduction   **217**
Yu. T. Dyakov, O. L. Ozeretskovskaya

Resistance genes and susceptibility genes   **Yu. T. Dyakov**   217
Signal transduction   **Yu. T. Dyakov, O. L. Ozeretskovskaya**   227
Conclusion   243
Selected literature   245

Chapter 9   Hypersensitivity   **247**
S. F. Bagirova

Morphological alterations in cells during apoptosis and necrosis   248
Genes involved in plant cell death – paranoid mutants   251
Signalling   260
Conclusion   262
Selected literature   262

Chapter 10   Immune Response   **265**
V. G. Dzhavakhiya, O. L. Ozeretskovskaya, S. V. Zinovyeva

Phytoalexins   **O. L. Ozeretskovskaya**   265
PR-proteins   **O. L. Ozeretskovskaya**   273
Antiviral proteins   **V. G. Dzhavakhiya**   276
Inhibitor of viral replication   277
Proteinase inhibitors   **O. L. Ozeretskovskaya**   279
Phenylpropanoids and lignin   **O. L. Ozeretskovskaya**   283
Oxyproline-rich glycoproteins   **O. L. Ozeretskovskaya**   287
Modification of plant gene expression at nematode invasion   **S. V. Zinovyeva**   290
Systemic acquired resistance   **O. L. Ozeretskovskaya**   298
Selected literature   314

**Chapter 11   RNA Silencing as a General Defence
                Mechanism against Pathogens**   **315**
P. Susi

Introduction   315
Coexistence and discovery of RNA silencing   315
Key characteristics and triggers of RNA silencing   316
Why RNA silencing mechanism exists?   320
Suppression of RNA silencing by plant virus-encoded proteins   320
RNA silencing in plant breeding   321
Techniques used to analyze viral suppressor activity   323
Selected literature   325

Chapter 12   Virulence Genes and Their Products   **327**
Yu. T. Dyakov, O. L. Ozeretskovskaya

Suppressors (impedins)   **O. L. Ozeretskovskaya**   328
Pathotoxins   **Yu. T. Dyakov**   332
Enzymes degrading antimicrobial plant compounds   **Yu. T. Dyakov**   343
Selected literature   348

Chapter 13   General and Specific Aspects of Plant and Animal Immunity   351
Yu. T. Dyakov

Immunomodulators of pathogenic microorganisms   353
Host immune molecules   357
Signal transduction and immune response   359
Conclusion   362
Selected literature   364

Chapter 14   Beneficial Plant–Microbe Interactions   365
I. A. Tikhonovich, N. A. Provorov

Nitrogen-fixing symbioses   365
Mycorrhizae   398
Defensive symbioses   409
Ecological and agricultural impacts of beneficial plant–microbe interactions   413
Selected literature   419

Part 3 – Practical Use of Molecular Studies   421

Chapter 15   Molecular Basis of Plant Immunization   423
Yu. T. Dyakov, V. G. Dzhavakhiya, T. Korpela

Vaccination   423
Immunization   426
Selected literature   437

Chapter 16   Creation of Disease-resistant Plants by Gene Engineering   439
V. G. Dzhavakhiya, L. A. Shcherbakova

Methods of transgenic plant creation   V. G. Dzhavakhiya   440
Development of transgenic plants resistant to viral disease   446
  Resistance of transgenic plants induced by expression   450
    of antiviral antibodies   L. A. Shcherbakova
Development of transgenic plants resistant to fungal diseases   V. G. Dzhavakhiya   452
Enhancement of plant resistance to bacterial diseases by
  genetic engineering   V. G. Dzhavakhiya   458
Possible limitations of commercial use of bioengineering methods in creation
  of plants resistant against bacterial diseases   V. G. Dzhavakhiya   464
Selected literature   465

Conclusion   467

Index   471

Preface

This book is a textbook, rather than a monograph. We see university students and post-graduates majoring in phytopathology, as well as researchers working in related areas of science as its readers. Therefore, the main task of this book is not a comprehensive review of molecular phytopathology studies, but rather a certain philosophy of the course, a system of presentation of the material that will allow understanding the internal logic of plant–parasite interaction, the consecutive stages of molecular dialogue between partners – what the Russian phytopathologist M.S. Dunin termed "immunogenesis" (Dunin, 1946). This task can be accomplished only by a small team of authors who write their sections of the book with a general plan and correct each other's manuscripts during the course of the work. This plan was based on the arrangement that is shown in Chapter 5 of this book (Figure 5.0). Following this arrangement could not but result – along with unquestionable benefits – in some shortcomings in presentation of the material, since nature is not a desk with separate drawers for "vertical resistance" and "horizontal resistance" folders. Therefore, phytoanticipins (Chapter 6), phytoalexins (Chapter 10), non-specific vivotoxins (Chapter 5), and specific pathotoxins (Chapter 12) come in different sections of the book. The same reasons also explain some repetitions (for instance, description of the chemical composition of the plant cell wall in different chapters, though from different angles). These inevitable shortcomings were allowed in full awareness, to ensure showing the *dynamics* of plant–parasite interaction processes. Moreover, repetitions in a textbook can be considered an advantage and not a shortcoming, since they help to show different dimensions of the same concept.

The first few chapters of this book explain the elements of general phytopathology. We thought it necessary to first introduce the basic concepts related to parasitism: trophics, specialization, and pathogenicity (Chapter 0); the basic groups of microorganisms, plant parasites, their biology, taxonomy, and symptoms of the diseases they cause (Chapter 1); and results of the classical genetic analysis of resistance and pathogenicity (Chapter 4). Therefore, that section of the book is entitled "Introduction to Molecular Phytopathology". The main body of the book is contained in Chapters 5–14. Chapters 5 and 6 describe the basic pathogenicity and resistance factors, and the remaining chapters discuss the factors responsible for specific parasite virulence and plant resistance. The last two chapters explain the principles and approaches to practical application of molecular studies in phytopathology.

We think that such a system of presentation will be interesting to a broader audience. Molecular studies in phytopathology provide an opportunity, unlike other approaches, to consider the abundant material regarding host–parasite interactions within a uniform system. While geneticists consider Flor's "gene-for-gene" concept as the basis of plant–parasite relationship, numerous biochemical studies of resistance and virulence carried out in 1950–1980s failed to provide a unique answer. It was found that the molecular dialogue between plants and parasites involves too many factors, the investigation of which blurred out the wood behind the trees. Also, a "multicomponent" hypothesis of plant immunity has emerged (Heitefuss, 1982), which is not a hypothesis but just an acknowledgement of the researchers' helplessness in the face of the multitude of facts found. It would be enough to look through the materials of the numerous conferences held in those years to see the intense debates on the nature of plant–parasite interactions between the supporters of the hypotheses of induction and suppression, elicitor specificity and non-specificity, etc. The suggested models of interaction, often the original ones (Bushnell, Rowell, 1981; Heath, 1981) were too speculative. However, historically, these discussions turned out to be extremely important, as "... a spark of new knowledge could be kindled only in collision of pieces of steel. A collision of dusty bags produces only a cloud of dust" (Meyen et al., 1977). A decisive step was necessary for dialectic synthesis of the opposite views. This step was taken by N. Keen, who bridged the gap between phytochemistry and classical genetics, on the one hand, and molecular biology and genomics, on the other. Keen was the first to experimentally confirm specificity of induction, working with recombinant DNA (Keen, Staskawicz, 1984). Since then, there has been an avalanche of molecular studies in phytopathology.

Molecular phytopathology developed as a compromise-making science. It enabled reconciliation of the apparently opposite theories and concepts:

1. Suppression and induction of plant defense responses by a parasite. It was found that the same parasite metabolites may play the role of suppressors or inductors (elicitors).
2. Specific and non-specific elicitors. The "guard" model of interaction brings together the processes generated by the action of non-specific elicitors, suppressors, and specific elicitors.
3. Host and non-host resistance. It is shown that they are based on the same factors.
4. Plant immunity and animal immunity. For many years, the difference between them seemed so obvious that a well-known Russian immunochemist observed: "there is as much in common between the immunity of plants and animals as between them and diplomatic immunity." Molecular studies have allowed it to be demonstrated that this is not so, and that the diversity of nature is built through a combination of the same blocks.

It is this concept of unity as against "the multicomponent hypothesis" that we have tried to stress in this book.

References

Bushnell WR, Rowell JB. Supressors of defense reactions. A model for roles in specificity. Phytopathology 1981; 71:1012–1014.

Dunin MS. Immunogenesis and its practical use (in Russian). Moscow: 1946.

Heath MC. A generalized concept of host-parasite specificity. Phytopathology 1981; 71:1121–1123.

Heitefuss R. General review of active defense mechanisms in plants against pathogens. In: Wood RKS, (ed.) Active Defense Mechanisms in Plants. Plenum Press; 1982:2–18.

Keen NT, Staskawicz BJ. Gene cloning as approach to understanding specificity in plant-pathogen systems. In: Struct, Funct, Biosynth Plant Cell Walls. Proc Annu Symp. Bot Dep Plant Pathol. Riverside: Univ of California; 1984:344–358.

Meyen SV, Sokolov BS, Schreider JA. Lyubitshev's Phenomenon. Vestnik Acad Sci USSR 1977, p.121.

Contributors

Frederick M. Ausubel, Department of Molecular Biology, Massachusetts General Hospital Boston, MA, USA.

Svetlana F. Bagirova, The Scottish Crop Research Institute, Invergowrie, Dundee, DD2 5DA, Scotland, UK.

Yuri T. Dyakov, M. V. Lomonosov State University, Moscow, Russia.

Vitaly G. Dzhavakhiya, Russian Research Institute of Plant Pathology, Moscow reg., Russia.

Timo Korpela, JBL, Turku University, Turku, Finland.

Olga L. Ozeretskovskaya, A. N. Bach Biochemistry Institute of the Russian Academy of Sciences, Moscow, Russia.

Julia M. Plotnikova, Department of Molecular Biology, Massachusetts General Hospital Boston, MA, USA.

Nikolai A. Provorov, Institute of Agricultural Microbiology, St. Petersburg, Russia.

Larissa A. Shcherbakova, Russian Research Institute of Plant Pathology, Moscow reg., Russia.

Petri Susi, Department of Virology, Turku University, Turku, Finland.

Igor A. Tikhonovich, Institute of Agricultural Microbiology, St. Petersburg, Russia.

Svetlana V. Zinovyeva, Institute of Parasitology of the Russian Academy of Sciences, Moscow, Russia.

Part 1

Introduction to molecular phytopathology

Phytopathology is a science on plant diseases. Like any other science it consists of several subtopics:

– studies of plant disease symptoms;
– etiology of pathogenic organisms that cause diseases;
– pathomorphology, anatomy, physiology, and biochemistry;
– epidemiology, a science covering large-scale plant diseases, epidemics;
– plant immunity;
– hygiene, prophylaxis, and therapy.

Phytopathology belongs to biosciences, such as botany, mycology, micro-biology, and virology, as it studies the infected plant as well as the pathogen. On the other hand, phytopathology belongs to agricultural sciences, such as crop growing, selection, and agricultural chemistry, as it assists creation of the conditions to minimize crop losses caused by diseases.

The above-mentioned sections of phytopathology can be combined into three blocks of questions:

1. Questions connected to diseases of an individual plant: pathogens, their growth in the plant, and response of a susceptible or resistant plant to an infection.
2. Plant diseases in populations and the factors that influence epidemics in natural phytocenoses and agrocenoses.
3. Research connected to diagnosis, accounting, forecast, and plant protection against diseases.

All these aspects are integrated in the focus on studying interaction between a higher and lower organism, host plant and parasite. Without knowledge, analysis, and practical use of this interaction, even the most resistant cultivars will rapidly lose resistance, and the most active fungicides will become inefficient.

Part 1
Introduction to molecular phytopathology

Chapter 0

Overview on parasitism

Yu. T. Dyakov

Types of host–parasite relationship

Organisms in ecosystems can interact with each other in different ways as shown in Table 0.1. Apparently, two organisms in direct contact, or in chemical contact through extracellular metabolites cannot avoid the interaction. Zeroes in Table 0.1 only show that the interaction is so weak that it cannot be detected by the usual methods. Among the types of interactions, a plant pathologist would be mostly interested in parasitism and less in commensalism and mutualism.

Parasitism is the ability of an organism to develop in/on another organism while consuming the host's nutrients for its own growth and reproduction. We will make a small digression before discussing parasites in detail.

Plants can host various fungal, bacterial, viral, etc. species of parasites. They can be isolated and identified. If a plant is pulled out of the soil and abandoned, some time later it will be decomposed by fungi and bacteria that can be also isolated and identified. Comparison of these two lists of microorganisms will show almost no species in common. This observation leads to two important conclusions:

1. Some microorganisms (saprotrophs) can only live on dead plant tissues. Thus, a living plant, though it has all the necessary nutrients, is not available to saprotrophs because of some defenses that would be lost after the plant death.
2. Many parasitic fungi that can overcome living plant's defenses either die or turn to a dormant state after the plant death.

We will first discuss the second conclusion, while the first one will make an important subject later in this book.

A. de Bary, a great botanist of the 19th century, divided all microorganisms into four groups:

1. obligate saprophytes: organisms feeding on dead plant remains or soil humus but unable to grow on plants;
2. facultative parasites: saprotroph organisms capable of attacking weakened plants or their parts;

Table 0.1.   Types of relationship of organisms

Types of relationship	Effect on each other	
	First on second	Second on first
Neutralism	0	0
Amensalism	–	0
Commensalism	+	0
Competition	–	–
Parasitism predatorism	+	–
Mutualism	+	+

"+" is positive effect; "–" is negative effect; "0" is no effect.

3. facultative saprophytes: parasites capable of continuing vegetative growth and reproduction after the host plant death;
4. obligate parasites: organisms that can only take nutrients from the cells of a living plant and turn to a dormant state or die after the plant death.

As can be seen, this classification is based on the balance of saprotroph and parasite phases in the life cycle of a microorganism. In a living plant, we observe obligate parasites, facultative saprotrophs, and much less frequently, facultative parasites. In a dead plant, we can find obligate saprotrophs and facultative parasites, and less frequently, facultative saprotrophs. To explain these facts, we discuss essential features of parasites in the next section.

Features of parasites

Trophicity

According to the type of feeding, microorganisms are classified into saprotrophs, necrotrophs, and biotrophs. Saprotrophs extract nutrients from dead tissues, i.e. they are saprophytes, while necrotrophs and biotrophs are parasites. However, nectrotrophs kill a plant or its part with toxic secretion before occupying it, which actually means that they feed on dead cells, similar to saprophytes. Biotrophs extract nutrients directly from living cells. The difference between necrotrophs and biotrophs is the rate of death (necrosis) of the affected tissue and the rate of the parasite growth in a plant. If the necrosis takes the lead over the parasite spread, consequently, it is the necrotrophic type of nutrition, and if the spread of the parasite is faster, it is biotrophic nutrition.

There are transitional forms between necrotrophs and biotrophs, hemibiotrophs, with a combined type of nutrition. First they feed biotrophically, and after the death of the infested tissue they continue to develop in it with necrotrophic feeding. The apple scab pathogen *Venturia inaequalis* forms intra-tissue (endophytic) mycelium between the mesophyll and epidermis without cell damage (biotroph), and after the cell death it spreads in the cells as a necrotroph; finally, when the leaves die-off and fall down the pathogen continues its growth in the leaves as a saprotroph.

Necrotrophic parasitism renders a rougher effect on host cells compared with biotrophic. Necrotrophic nutrition is less specific and seems to be the primary feeding. The evolution of nutrition types can be traced in soil fungi, where various transitional types can be observed.

Plant roots release to the environment many metabolites, both beneficial (amino acids, sugars, etc.) and toxic for soil microorganisms (phytoncides); the metabolite concentration decreases according to the distance from the roots. The organisms that have acquired resistance to harmful metabolites are able to live near roots using useful metabolites. Thus a specific microflora of rhizosphere (around roots) and rhizoplane (on root surface) has developed. The relationships between rhizospheric microorganisms and plants are close to commensalism, as the plant has a pronounced positive effect on the microorganisms, while the microorganisms have only a slight effect on the plant. Rhizospheric microorganisms possess certain specialization to plant species.

Rhizospheric microorganisms grow towards the concentration gradient of metabolites diffusing from the root. The diffusion considerably increases in case of root surface wounding. The least specific and most primitive group of necrotrophic parasites, wound parasites, apparently have evolved from rhizospheric microorganisms. For instance, zoospores of the fungal wound parasite *Pythium ultimum* were found to be concentrated around the wounded part of the root. An optimum development of another wound parasite *Rhizoctonia solani* occurs in artificial medium at 25°C, while it causes the greatest damage to cotton at 17°C. This difference in temperature optimums can be explained by the fact that at 17°C cotton roots release seven times more amino acids and sugars than at 30°C.

Wound parasites can saprotrophically grow in soil for an indefinably long period (facultative parasites), and they possess various antibiotic substances necessary for competition for substrate with other microorganisms; they cannot, however, penetrate through intact plant tissues. When they get into wounds, they release toxic products (phytotoxins) that kill host plant cells, thus increasing necrosis, the parasite's life basis. They often cause death of the entire plant. Their next stage is saprotrophic growth on the plant remains. The next phase of evolution is further development of necrotrophic parasitism. It is interconnected with the following features of necrotrophic organisms.

Ability to penetrate into intact plant parts

Penetration into a plant cell can be achieved through perforation of the cuticle and upper wall of epidermal cells, or through the stomata. Cell wall perforation is carried out by mechanical pressure and enzymatic degradation of the cell wall polymers.

Mechanical perforation. In many phytopathogenic fungi the spore germ ends with a thicker termination, appressorium. Development of appressoria is induced by cuticular wax; therefore, spore germs of phytopathogenic fungi do not form appressoria on the leaves with removed cuticle, or in vitro. The foot of the appressorium attaches to the substrate with extracellular carbo-hydrates and/or low-molecular hydrophobic proteins, hydrophobins, acting as vacuum lubricant. The foot cell wall has two layers: a thin outer membrane and a thicker inner membrane. In the layer above the substrate, many fungi, deposit melanin, a dark pigment consisting of condensed aro-matic rings and working similarly to a molecular trap: allowing solutions to pass to the cell but not out of the cell. Therefore, a high turgor pressure develops in the appressorium, resulting in breaking of the thin outer layer of the foot and perforation of the cell covers with the down-growing infectious hypha which is covered with the internal cover of the appressorium. For instance, the appressorium of the rice blast disease pathogen *Magnoporthe grisea* creates a pressure of 8 MPa; to develop such a pressure, the cell needs a solution concentration over 3 M.

Chemical degradation of cell covers.   Phytopathogenic fungi and bacteria possess a wide range of enzymes destroying the carbohydrate polymers, which constitute the building materials of the cell walls. First, the parasite, using these enzymes, penetrates into the cell and feeds on its nutrients. Second, in some necrotrophic parasites these enzymes are so active that they degrade a larger part of the cell wall; with this strong frame removed, the osmotic processes destroy the protoplast, and the cell dies and loses the immune properties of a living being. Third, the parasites destroy the septum (intercellular cement that glues cells together), which enables them to move further in the infected plant tissues. And fourth, parasites feed on the cell wall polymer degradation products, monosugars.

Longer residence of the parasite in plant tissues due to a gentler or local toxic effect

It is obvious that the motive force in the evolution of parasitic microbes is an attempt to overcome the competition with saprotrophs. By penetrating inside the plant, fungi can inhabit a separate niche, i.e. a living plant. However, it is not possible for most microorganisms due to the immune properties of the living cells. The easiest way to overcome cell immunity is to kill the host cells. But the death of the host means a return to competition with other microbes. Thus, the evolution of parasitism is a way to biotrophic nutrition, which means replacement of the rough ways of breaking the host immunity – necrotrophic nutrition – by gentler ways ensuring live conditions of the host cells for a longer time. This is achieved by particular features of the metabolism of biotrophic parasites, which are different from necrotrophs.

Decreasing diversity and concentration of plant-toxic hydrolytic enzymes.   The barley powdery mildew pathogen *Erysiphe graminis* secretes hydrolytic enzymes only from the tip of the infectious hypha, and they are active only as near as 0.1 m from the hyphea. A list of hydrolytic enzymes is shown in Table 0.2.

Table 0.2. Activity of extracellular enzymes destroying plant cell wall carbohydrates (mmol/min × cm²) compared for two phytopathogenic fungi: a necrotroph and a biotroph (Cooper RH, 1983)

Enzyme	Necrotroph *Verticillium albo-atrium*	Biotroph *Uromyces fabae*
Endopolygalacturonase	200	0
Exopolygalacturonase	0	5
Endopectatlyase	150	0
Cellulase Cx	775	0
Arabinase	61	10
Galactanase	4	14
Xylanase	62	9

No mechanical damage of membranes due to apoplastic growth (outside plasmalemma). Many phytopathogenic bacteria and some fungi grow only in intracellular space and secrete enzymes that loosen cell walls and help nutrients to diffuse into the intracellular space. Biotrophic fungi cause local cell wall breaking and penetrate into the cell through the resulting hole. They do not destroy the plasmalemma, but invaginate it and form a widened or even branching structure called the haustorium (Figure 0.1). The haustorium is surrounded by the plasmalemma extension, the extrahaustorial membrane. The latter slightly differs from a normal plasmalemma in chemical composition (different coloring, no intramembrane inclusions, sterols, or adenosine triphosphate (ATP) activity) as well as in multiple invaginations indicative of an active nutrient transport through the membrane. Between the haustorium cell wall and extrahaustorial membrane, there is an extrahaustorial matrix for metabolite exchange and prevention of mutual toxicity of the partners by their toxic metabolites. Thus, in addition to absorption of nutrients by the entire mycelial surface, typical for the fungi, biotrophs possess a special trophic organ, the haustorium.

Secretion of biologically active substances activating plant metabolism and enhancing parasite's growth conditions. Biotrophic parasites, unlike necrotrophs, do not secrete toxic products into the infected tissues but rather secrete hormonal substances regulating host metabolism. Rust, smut, Taphrinales, and many other fungi, as well as many phytopathogenic bacteria release indoleacetic acid, gibberellins, cytokines, and other substances into plants, which causes considerable morpho-physiological changes in the host plant, including increased inflow of photosynthetic products to the affected plant organs. This has been shown using radio autography of the wheat leaves infected by the rust pathogen. Wheat was grown in an atmosphere of radioactive labeled carbon dioxide ($^{14}CO_2$). Figure 0.2 shows the radioactive labeled concentrates in the affected area. One of the mechanisms to increase the transport of photosynthetic products into the affected area is their

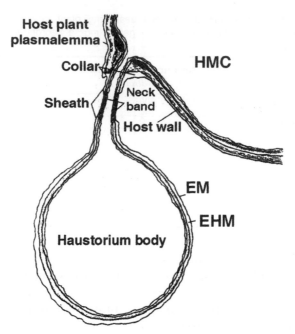

Figure 0.1.   Structure of the haustorium of *P. graminis* (Andreev, Plotnikova, 1989). HMC – haustorium mother cell, EM – extrahaustorial membrane (invaginated portion of plasmalemma), EHM – extrahaustorial matrix.

interception by the parasite and inclusion in the compounds absent in the plant. For example, the primary product of photosynthesis, glucose, is used by the plant for energy processes and synthesis of structural and reserve products. In fungal cells, glucose is quickly transformed into the hexatomic alcohol mannitol and disaccharide trehalose, which results in a sharp drop of glucose in the affected area and its transportation along the concentration gradient.

Parasitic processes are frequently accompanied by proliferation of the affected tissue, formation of tumors, galls, and other neoplasms. The affected parts of the plant, age slower than non-affected, and maintain juvenile condition for a longer time. If leaves are removed, they become chlorotic due to chloroplast destruction. However, green "islands" will remain around the rust pustules.

Biotrophs may increase non-specific resistance of the protoplast to damaging factors. For example, Yarwood, who made a big contribution in revealing the nature of biotrophy, inoculated one side of a bean leaf with a biotrophic parasite (rust or mildew fungi), with the other side left intact. Then the whole leaf was exposed to the factors causing cell damage and death (low and high temperatures, poisoning pollutants, etc.). In all the cases the inoculated leaf sides were more resistant and died only upon more intensive exposure. As the causes of cell death from different exposures are also different, a conclusion

Figure 0.2. Concentration of photosynthesis products in the zone of stem rust pustules (Shaw M, Samborski DJ, 1956).

could be made on the general improvement of cell protoplast resistance in the inoculated plants.

Thus, biotrophic parasites integrate into host cells, in various degrees. The greatest integration is observed for phytopathogenic viruses, and among eukaryotes, for the parasitic red algae. More than 15% of the red algae are obligate parasites. They have tiny reduced colorless thallomes (lost photosynthetic pigments). Most species are adelphoparasites that live on relative algal species of the same family. Some species (allo-parasites) are specialized to live on non-relative species. When a parasite cell contacts a host cell, it splits off a small conjunctive cell with a nucleus, and that cell retains connection with the parasite thallome through a glycoprotein plug (Figure 0.3). The conjunctive cell merges with the host cell and transfers the nucleus into the host cell. In some species it also transfers the mitochondria and protoplastids, which replicate in host cells and migrate to the adjacent cells through intercellular ties, replacing the nuclei and mitochondria of the host cells. After a certain period of parasitism, the thallomes of some parasites gain pigmentation, probably, due to the host plastid transfusion, along with nutrients. In some species of red algae, the information transferred by the parasite causes intense cell fission and canker tumor formation around the affected area.

The genome of many viruses includes only those genes necessary for the synthesis of own components – protein and nucleic acid. Viruses do not have energy exchange or protein-synthesis enzymes. Synthesis of the required enzymic and structural proteins occurs on the host cell ribosomes, using the host cell energy-reserving molecules, and in some cases also using the enzymes that catalyze RNA synthesis.

Thus, obligate parasitism can occur at the organelle or even molecular level. The biotrophic type of nutrition allows the parasite to stay in the host

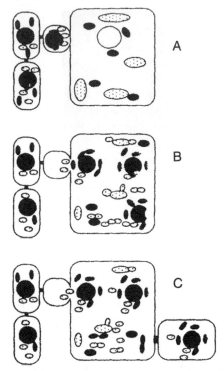

Figure 0.3.   Transfer of parasitic nuclei, mitochondria, and protoplasts into a host cell (Goff LJ, Coleman AW,1995).
(A) The parasite cells (left) have formed a conjunctor cell containing a parasite nucleus (black), mitochondria (gray), and protoplastids (white with dots). This conjunctor cells fuses with the host cell (host nuclei are white, host plastids are dotted, and hosts mitochondria are black ovals) delivering the parasite organelles into the host's cytoplasm. (B) The parasite nucleus and mitochondria replicate in the host cell, and the host plastids divide to form numerous protoplastids. The host nuclei and mitochondria replicate in the host cell. The host nucleus may disappear or persist. (C) Ultimately, a cell is cut-off from the heterokaryotic host plus parasite cell. This cell contains a parasite nucleus, parasite mitochondria, and protoplastids derived from the host plastid.

tissues for a long time. However, this advantage is acquired at high costs: (a) loss of antibiotic substances and reduced activity of enzymes degrading plant polymers; (b) modification of the membrane structure that enables biotrophs to release organic substances to the environment; (c) a high degree of adjustment of the parasite's own metabolism to host metabolism; specialization to attack only certain host species. Therefore, after the plant's death, biotrophs cannot return to saprophytic nutrition, they fail in the competition with necrotrophs and saprotrophs (environmentally obligate parasites), they cannot attack non-host-relative plants, and some biotrophs cannot be even cultured in artificial nutrient media (physiologically obligate parasites).

Biotrophic parasitism, at least in the first phase, represents a form of symbiotrophism, as the parasite stays in living host cells or tissues, and even stimulates their metabolism. Apparently, biotrophic parasitism and symbiosis of fungi and bacteria represent two branches of phytopathogen evolution from necrotrophy to biotrophy.

The type of nutrition causes considerable changes in parasite's metabolism and morphogenesis. Therefore, trophicity is an important taxonomical property for dividing some bacterial and fungal species. For instance, among the fungus species of the genus *Cladosporium*, there are obligate saprotrophs growing on plant and animal substrates, various other materials, and even on liquid fuel. However, there are very weak parasites attacking dying plant tissues, there are species with predominant necrotrophic nutrition; and there are biotrophs (Table 0.3).

Biotrophs have bigger conidia than saprotrophs, they show less enzymatic activity for degradation of amine nitrogen and carbohydrate polymers they are very susceptible to antibiotic substances produced by saprotrophs, hence, biotrophs are less competitive in comparison to saprotrophs. Necrotrophs keep an intermediate position.

The *Phytophthora* fungi also include the saproptrophs that grow on plant remains in water bodies and humid soil (*Phytophthora gonapodyides*), slightly specialized necrotrophs (*P. cactorium*), and highly specialized biotrophic parasites (*P. infestans*).

The genus *Pseudomonas* also includes saprotrophic bacteria, for instance, *P. aeruginosa* that lives in soil and can cause wound rot (pyocyanic rod), saprotrophs that can infect weakened and aging plants (some strains of *P. fluorescens*), and active biotrophic parasites (*P. syringae*).

Table 0.3.   Characteristics of the fungi from the genus *Cladosporium* (Levkina LM, 1974)

| Species | Conidia length (mm) | Amine nitrogen consumption (mg/ml) | Enzyme activity, standard units | | Lethal concentration of trichotecin (µg/ml) |
			Polygalactu-ronases	Amylases	
Biotrophs					
fulvum	23.0	2.5	22	8	0.12
Necrotrophs					
cucumerinum	11.4	6.4	34		0.46
paeoniae	7.8	7.0	33	40	0.97
Average	9.6	6.7	33	40	0.71
Saprotrophs					
macrocarpum	6.0	11.7	80		3.9
cladosporioides	7.6	8.7	92	59	15.6
sphaerospermum	9.3	9.6	65	42	3.9
herbarum	6.8	8.4	71	47	7.8
Average	7.4	9.6	77	49	7.8

Specialization
Phylogenetic specialization

It is specialization for the host plant taxons. On this basis, parasites will be divided into: monophages that attack plants within the same genus (or several close genera); oligophages specializing within the host plant family; and polyphages that attack plants from a variety of families, orders, and even classes. Biotrophs are mostly concentrated among monophages or oligophage necrotrophs, among oligophages and polyphages. This is understandable as biotrophy supposes good fitting of the parasite and plant metabolism. For instance, the fungus *P. infestans* attacks many species of the subgenus *Eusolanum* of the complex genus *Solanum*, but it does not attack the species of the subgenus *Stelatipillium* of the same genus. This feature of biotrophs is used in plant taxonomy to determine the level of relationship between plants with unclear taxonomic status.

In the process of experimental research, the concept of oligophagy and polyphagy of many parasites have been revised. The first revision of the scope of specialization was made a hundred years ago by the Swedish plant pathologist Eriksson. He showed that spores of the cereal stem rust pathogen *Puccinia graminis*, taken from wheat, could infect wheat only but not other cereals. Therefore, morphologically indistinguishable strains were found to be physiologically different, and a species of the oligophage *P. graminis* includes several specialized forms. In this connection, for many phytopathogenic fungi their specialized form is indicated after the species name, for instance, *P. graminis* f. sp. (formae speciales) tritici (wheat form of stem rust pathogen); *E. graminis* f. sp. *hordei* (barley form of cereal mildew pathogen). Thus, trophicity is used for differentiation of species, and specialization is used for intra-species differentiation, division of species into specialized forms.

The concept of polyphagy suffered one more blow in the last quarter of the 20th century when morphological species of many fungi were found to be complexes consisting of a large number of genetically isolated biological twin-species. For instance, the fungus *Rhizoctonia solani* that lives in soil as sterile mycelium, can sometimes develop the basidial stage (theleomorph) *Thanotephorus cucumeris* – a polyphage affecting shoots of a great number of gymnospermous and angiospermous species. However, the hyphae of certain strains were found to merge only with the hyphae of some particular strains. As for this fungus, the mutual nuclei migration through the hyphae anastomoses is the only way to exchange genetic information, the non-anastomosis strains are genetically separated biologic species. They are referred to as anastomosis groups (AGs) and ordinal number. More than ten AGs have been described, with four of them being most common.

AG1. It occurs in surface soil layer and parasitizes on a wide variety of plants (mostly legumes and cereals). It causes rot of seeds, shoots, and mature plants. Three subgroups with different specializations have been described: AG1-1A causing rot of rice ear scales; AG1-1B causing cobweb-like rot of vegetative parts in various plants; and AG1-1C attacking sugar beet and buckwheat.

AG2. The AG2 isolates are divided into two subgroups according to the frequency of mutual anastomoses: AG2-1 (isolated mostly from crucifers) and AG2-2 (isolated from goosefoot and other plants). Subgroup AG2-2 is further subdivided into two groups: isolates that cause fire blight of rush sheaths (IIIB) and root rot of sugar beet (IYB).

AG3. It attacks mostly *Solanacea* and causes such common potato diseases as potato black scab of potato tubers, shoot rot, and white leg.

AG4. This is another highly specialized group that causes rot of seeds and shoots in the families *Chenopodiaceae, Solanaceae, Fabaceae, Rosaceae*, etc. AG4 comprises several subgroups differing in a number of traits, including specialization: AG4-1 attacks sugar beet, AG4-2 attacks flax, and AG4-3, potato.

Apparently, the ancestor strains of *R. solani* lived in soil as saprotrophs or weak non-specialized parasites on weakened plants. A group of strains was isolated from virgin soil and called AGB1. Those strains are saprotrophic and develop anastomoses with the strains of other groups. In the agrocenoses, where large areas are covered by the plants with similar susceptibility (same cultivars) new strains emerged, capable of attacking these plants. New genetic barriers developed between these new strains and saprotrophs in order not to undermine the parasitic properties of the newly emerged strains by the genes of the saprotrophs through anastomoses. A hypothetical evolution of the anastomosis groups of *R. solani* is shown in Figure 0.4.

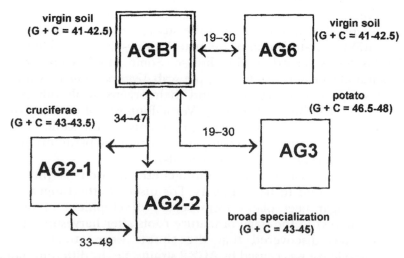

Figure 0.4. Hypothetic evolution of the *R. solani* anastomose groups in agrocenoses. AG – anastomose groups; G+C – percent of DNA guanine and cytosine bases; digits above the arrows show percent of DNA homology by hybridization results.

A similar situation as with AGs was described for another phytopathogenic fungus, *P. megasperma*. This species is divided into two subspecies *P. megasperma* and *P. sojae*, according to the morphological (spore size), physiological (optimum growth temperature), molecular (structure of some proteins and DNA) features and parasitic specialization. The former is a polyphage attacking different plants (Coniferous, Rosaceae, etc). The latter is divided, in turn, into the strains specialized for soybean, alfalfa, peas, etc., and they are assigned a status of species, sub-species, or specialized forms.

Even in such a non-specialized parasite as *Botrytis cinerea* that causes gray rot in wild strawberry, cabbage, sunflower, and other plants, evolution seems to be directed towards a narrower specialization. This assumption is supported by the presence of the genus *Botrytis* along with a widely specialized species *B. cinerea*, of more narrow specialized species *B. allii* (onion parasite), *B. tulipae* (tulip parasite), and *B. fabae* (horse bean parasite).

New molecular genetic data have necessitated grouping of some narrow specialized species into one species. For instance, several dozens of species of phytopathogenic bacteria of the genus *Pseudomonas*, divided by specialization in tobacco, beans, soybean, and other host plants, were found to be so negligibly different in terms of DNA structure that they were grouped into one species with narrow specific pathological variants. A similar grouping was conducted for the fungi of the genus *Alternaria* attacking pear tree, apple tree, wild strawberry, etc., into the species *A. alternata*.

Ontogenetic specialization

Ontogenic specialization is manifested through confinement of a parasite to particular phases of host plant ontogenesis. A Russian plant pathologist M. S. Dunin made special investigations on ontogenetic specialization of parasites for regulation of the rate of the susceptible phase of ontogenesis. He used various agro-technical or other techniques, and was able to reduce considerably losses caused by the disease. Some fungi (causal agents of covered smut of wheat *Tilletia tritici*, apple scab *Venturia ineaqualis*, and shoot root rots) attack only young plants and their organs, while others (causal agents of potato late blight *P. infestans*, Verticillium wilt of cotton *Verticillium dahliae*, etc.), on the contrary, attack aging plants. At the same time, some widely specialized parasites can accompany a susceptible plant throughout its lifetime (for instance, fusarial wilts of cereals).

In a number of cases a wide ontogenetic as well as phylogenetic specialization is found to be imaginary. For instance, the fungus *R. solani* accompanies sugar beet plants from seeds (development of mould on sprouts and shoots) to the stage of mature roots after harvesting (pile rot). However, it was discovered that the shoots were attacked by AG4 strains, while pile rot was caused by AG2-2 strains, i.e. by different biological species.

Table 0.4.   Affected and disease areas in plants

Affected area	Disease area	Examples
Local	Local	Cereal rust
General	Local	Cereal smut
Local	General	Root rots
General	General	Viral diseases

Tissue (histotropic) and organotropic specialization

Some parasites can grow only in certain organs or tissues of the host plant. For instance, some viruses attack phloem, others, parenchyma. Fungi and bacteria that cause infectious wilt are located in plant xylem and restrict water transport from roots to above-ground parts. The fungus *Synchytrium endobioticum* a potato wart pathogen, affects stolons and tubers but not roots. Many fungi and bacteria that cause leaf spots attack only or mostly on leaves.

When studying the organotropic specialization, it is necessary to distinguish between the infected area (plant area where the parasite is located) and the disease area (plant parts where visual disease symptoms are manifested). These areas can or cannot overlap (Table 0.4).

Spores of rust fungi penetrate into cereal leaves, mycelium spreads inside the leaf for a short distance, then spore-bearing develops. The spores look like rust powder and expose through breaches in epidermis. Similar local symptoms are also caused by other fungi and bacteria attacking leaves. Smut fungi infect developing seeds and young cereal shoots and systemically grow inside the plant.

After the formation of the generative organs (ears, panicles), the mycelium migrates into those parts, grows there intensely and splits into dark-colored spores, which makes the ears look charred. Consequently, the entire plant gets infected, though the symptoms are seen only in the generative organs.

Many root rot pathogens are confined to the roots, however, they undermine the plant growth, cause yellowing and even death of the plant.

Majority of viruses are spread over the entire plant and cause its general disease.

Pathogenicity

Pathogenicity is the main feature of pathogenic organisms. This term can be considered in a broad or narrow sense.

In the narrow sense, pathogenicity is the ability to cause damage in the infected plant. Necrotrophs that cause rapid death of the plant or its parts are more pathogenic than biotrophs with their gentle effect on plants.

In the broad sense, pathogenicity is viewed as a complex of traits distinguishing parasites from non-parasites. However, the data in Table 0.3 shows that parasites differ from saprotrophs in a large number of traits (and, consequently, also in genes): morphological, physiological, and biochemical.

The complex term of pathogenicity must therefore be discussed for its qualitative and quantitative components.

Virulence is the qualitative component and describes the ability of a genetically homogeneous strain to grow on a genetically homogeneous host plant. According to their virulence, species of phytopathogenic organisms are classified into specialized forms and races. A physiological race, or race by host, is a group of strains virulent to the plants with identical resistance genes. Since virulence is a qualitative feature, one cannot refer to its higher or lower level, but the races can be characterized as those with more or less virulence, meaning the number of genetically diverse plants that can be attacked by the races being compared.

Aggressivity is the quantitative component of pathogenicity. Aggressivity is interpreted as a level of pathogenicity of the race, i.e. ability to cause a large-scale disease in susceptible plants, epiphytoty. The concept of aggressivity includes:

1. A minimum infectious efficiency. Many biotrophic fungi can infect the host plant with one spore. Thus, in this respect they are more aggressive than necrotrophs. In necrotrophs, spore germs first need to produce an anastomosis network wherein the gene product amplification will result in accumulation of metabolites in amounts necessary for the infection (enzymes, toxins, etc.).
2. A short latent period (from infection to reproduction) that allows production of several successive generations during one season (polycyclic growth).
3. Numerous progeny.

Therefore, aggressivity can be discussed in terms of a higher or lower level, unlike virulence. The relationship between aggressiveness and virulence is illustrated in Table 0.5. Strain "a" has a wide virulence (attacks all the three cultivars) but low aggressivity. Strain "b" has high aggressivity but narrow virulence; and strain "c" has wide virulence and high aggressivity. The severity of the attack is shown by numbers 0–4(0 is none, 4 is the most severe). Thus, the strains avirulent to a particular cultivar are not aggressive towards it, while the virulent strains can have different levels of aggressivity.

Table 0.5. Severity of plant infestation with parasites of different virulence

Parasite strain	Plant cultivar		
	A	B	C
a	1	2	1
b	0	0	4
c	4	4	4

Selected Literature

Cooper RH. Biochem. Plant Pathol. Ed. Callow JA, 1983; 101–135.

Goff LJ, Coleman AW. Plant Cell. 1995; 7:1908 (fig 10).

Hansen EM. Speciation in plant pathogenic fungi: the influence of agricultural practice. Canad J Plant Pathol 1987; 9:403–410.

Hardham AR. Cell biology of pathogenesis. Ann Rev Plant Physiol Plant Molec Biol 1992; 43:261–278.

Levkina LM. Problems of phylogeny of lower plant (In Russian) 1974; 99-101.

Lewis DH. Concepts in fungal nutrition and the origin of biotrophy. Biol Rev 1973; 48:261–278.

Scott KJ. Obligate parasitism by phytopathogenic fungi. Biol Rev 1972; 4: 537–572.

Shaner G, Stromerg EL, Lacy GH, et al. 1992; 30:47–66.

Shaw M, Samborski DJ. Canad. J. Botany. 1956; 34:389–407.

Stall RE, Roberts DA. Plant pathology and control of plant diseases. In: Plant Pest Control. San Francisco. 1978; 89–121.

Chapter 1

Plant parasite microorganisms

Yu. T. Dyakov, S. V. Zinovyeva

Fungal plant pathogens

General characteristics

Fungi are classified into a separate group of organisms differing from both plants and animals, primarily by the type of nutrition. Fungi are not autotrophs, they have no chloroplasts, they can only use the energy stored in organic compounds. This distinguishes fungi from plants. As against animals, fungi are osmotrophic: they obtain food by absorbing nutrients from the environment. These feeding features correlate with fungal morphology and physiology.

1. The body of most fungi is made of mycelium consisting of very branched hyphae. Such a structure allows a maximum occupation of the substrate, whether it is soil or plant, to extract nutrients. Fungi absorb nutrients by the entire body.
2. The osmotrophic type of feeding makes a vegetative body plunge fully into the substrate, which impedes its propagation and occupation of new substrates. Therefore, in most fungi, spores are brought out above the substrate in special structures, which in many cases have a complex arrangement (sporangiophores, conidiophores, and fruit bodies). Sporiferous structures of endophytic fungi (those developing inside plants) are released through stomata or breaches in epidermis.
3. Fungi need to use, as energy sources, complex organic compounds that cannot pass to the cell through cellular covers because of large molecular weight. Therefore, fungi release the enzymes depolymerases to the environment that cause degradation of polymers. Degradation products enter cells in a dissolved form. Fungi are sources of highly active depolymerases.
4. Fungi need to develop high turgor pressure in the cells to provide entrainment of nutrient solutions from the substrate to the mycelium.

Both saprotrophic and parasitic fungi feed mostly on plant tissues. Apparently, the association of fungi and plants developed at the very early stages of their evolution. The most primitive fungi Chytridiomycetes and Oomycetes parasitize on the most primitive plants, algae. Some mycologists believe that fungi came to live on the land under the cover of plants that had come to live on the land, as their parasites and symbiotes. Symbiotic fungi are also believed to have provided adaptation of green plants to life on land. There are almost no fungi living in symbiosis with animals, while a huge number of fungi live in continuous symbiotic relations with plants. The enzymatic system of fungi is designed to decompose carbohydrates – structural materials and reserve nutrients of plants. It is not only parasitic fungi that mainly attack plants, but also saprotrophic fungi feeding on dead plants, leaving dead animals to bacteria. Dead wood is almost entirely decomposed by fungi.

System of fungi and fungal plant diseases

The above listed properties of fungi characterize them as an ecotrophic group, an ecomorph. Phylogenetically, fungi belong to several independent kingdoms. The differences between them are shown in Table 1.1.

It can be seen from the table that myxomycetes are not fungi by definition, as they have mixotrophic (mixed) feeding: osmotrophic (as fungi) and zootrophic (as animals). Their classification among fungi is rather a traditional approach.

1. Kingdom Myxobiontes includes organisms whose life cycle consists of several phases. The first phase, *amoeboid*, is represented by amoebiform cells, without cell wall, and feeding both osmotrophically on dissolved organic matter and on separate cells (bacterial, yeast) using endocytosis. In the second phase, *plasmodial* ameboids form slimy mass by growth, coalescence (plasmodium), or coordinated association of individual cells

Table 1.1. Properties of phylogenetically independent taxons including fungi and fungaceous organisms

Property	Eumycetes	Oomycetes	Myxomycetes
Type of feeding	Osmotrophic	Osmotrophic	Osmotrophic, zootrophic
Structural carbohydrates	Glucan (mannan), chitin	Glucan, cellulose	Cellulose, chitin
Reserve carbohydrates	Glycogen, trehalose, polyols	β-glucan (laminarin)	Glycogen
Lysine synthesis	Through α-aminiadipate	Through diaminopimelate	Through α-aminiadipate
Mitochondrial crista	Teniform	Tubular	Teniform
Sexual process	Somatogamy, gametangiogamy	Oogamy	Somatogamy Plasmodium, pseudoplasmodium
Dominating type of thallus	Mycelium	Mycelium,	
Molecular phylogenies	Kingdom Mycobiontes	Kingdom Stramenopiles	Kingdom Myxobiontes
Dominating mode of life	Terrestrial	Aqueous	Terrestrial

(pseudoplasmodium). At the third stage plasmodia and pseudoplasmodia are transformed into spore-bearing organs, on the surface or inside of which non-motile spores are developing. Stepping up to each subsequent stage is controlled by external factors (starvation, illumination, etc.) and is accompanied by cardinal changes in genetic regulation of intracellular biochemical processes, which makes this group of fungi an extremely interesting subject of experimental biology.

Myxomycetes from <u>division</u> <u>Plasmodiophoromycota</u> are obligate parasites that grow on underground organs of plants and cause their proliferation (clubroot) or cracking (powdery scab of potato).

2. Kingdom Stramenopila. They are rather algae than fungi, as they possess heterocont spores (with two morphologically different flagella), cellulose and glucan in the stem wall (most fungi contain chitin and glucan), oogamic sexual process, and metabolism similar to algae. Plant parasites are present in two divisions.

<u>Division</u> <u>Labyrinthulomycota</u> are aqueous organisms developing plasmodia as an anastomosis net in the body of large algae (macrophytes) or the sea weed eelgrass. They frequently cause large-scale lesions and destruction of the infected plants.

<u>Division</u> <u>Oomycota</u> are aqueous and terrestrial organisms; most species form mycelium without septa. They reproduce by zoospores with two flagella unequal in length and shape. The sexual process is oogamy. They feed saprotrophically on plant or animal remains, or parasitize on algae, plant roots, or plant parts above ground. Many oomycetes cause damaging diseases of useful plants (root rots, mildews).

3. Kingdom Fungi (true fungi)

<u>Division</u> <u>Chytridiomycota</u> are organisms living in water, wet soil, or stomach of ruminants. Different species form plasmodium, rhizomycelium (central "collective" cells and branched mycelial processes without nuclei), or true mycelium with nuclei and without septa. They reproduce by motile zoospores that have one smooth flagellum turned backward. Many species parasitize on algae or terrestrial plants.

<u>Division</u> <u>Zygomycota</u> are terrestrial organisms; they form mycelium without septa, have non-motile spores, which in most species develop endogenously (in sporangia), and a peculiar sexual process similar to conjugation (zygogamy). They feed saprotrophically or symbiotrophically (develop endotrophic mycorhiza with plant roots) in soil, but they can cause rot of ripe succulent organs of plants (strawberry, onions); many species parasitize on insects.

<u>Division</u> <u>Ascomycota</u> form mycelium with septa, with voids in the center. They reproduce by non-motile spores developed exogenously (conidia). The zygote formed in the sexual process immediately divides in a reduction way, producing eight (less often four) endogenous haploid meiospores (ascospores) in asca (sacs). Many species have lost the sexual process and reproduce entirely by conidia (they were earlier classified separately as

imperfect fungi, or Deutoromycetes). Saprotrophic species live in soil, dead plants, or manure of herbivores. The parasites cause large-scale plant diseases.

Division Basidiomycota form mycelium with complex doliporous septa. They reproduce by conidia and meiospores (basidispores), which, unlike ascomycetes, develop as four pieces exogenously in outgrowths (sterigmas) of special cells, basidia. They cause large-scale diseases of crops and wild-growing herbaceous and woody plants, and decomposition of dead wood; many species live in symbiotic relations with plant roots (ectrotrophic mycorhiza).

There are plant parasites of various trophic groups almost in all the listed groups of fungi.

Necrotrophs cause the following major types of diseases:

Root rots are caused by many soil fungi, frequently wound parasites. The major species are: *Olpidium brassicae* (division Chytridiomycota) causes black-leg of cabbage shoots; *Pythium ultimum* (division Oomycota) causes root rot of beet, cotton, wood crops, etc.; *Thielaviopsis (Thielavia) basicola* (division Ascomycota) causes black root rot of tobacco, peas, and cotton; *Rhizoctonia solani (Thanatephorus cucmeris)* from division Basidiomycota causes root rots of beet, cotton, wood plants, and many others.

Canker lesions of tree bark are frequently caused by wound parasites that settle on dead, frozen, or otherwise damaged bark areas, secrete toxins to surrounding tissues and occupy them after their degradation. Such fungi include *Phytophthora cactorum* (oomycete) attacking fruit species from the family Rosaceae, sac fungi from the genera *Valsa, Nectria,* and *Criphonectria* that cause degradation of trunks and branches of foliage trees.

Tracheomycosis affects xylemic vessels. It is typically caused by numerous sac and imperfect fungi developing in vessels and destroying xylemic transport due to release of hydrolytic enzymes and toxins. The external sign of disease is wilt due to lack of moisture. Fungi from the genera *Verticillium* and *Fusarium* can saprotrophically live in soil and infect plants from many orders by damaging the rootage. Fungi from the genus *Ophiostoma* cause wilt of broad-leaved plants, infecting them through insects.

Trunk rots are usually caused by Polyporales or Agaricales fungi from the division Basidiomycota. The most damaging are *Heterobasidium annosum* (root fomes), *Armillaria mellea* (honey agaric) which infect trees through roots, *Phellinus igniarius* (false tinder fungus), *Fomitopsis pinicola* (fringed tinder fungus) and others, the spores of which penetrate tree tissues through damaged trunks and branches.

Postharvest rots are one of the most damaging groups of diseases for the following reasons: (1) usually the produce in storage has high water content in cells (juicy fruits, root crops, etc.); (2) organs separated from the plant have lower resistance compared with those within the plant; (3) loss of immunity during fruit ripening developed due to their need to rot for seed release. Therefore, most storage diseases are caused by non-specialized necrotrophic parasites, as can be seen in Table 1.2, while biotrophs and hemibiotrophs attack vegetating plants.

Table 1.2. Ratio of major necrotrophic and biotrophic fungal diseases of vegetable crops

| Crop | Biotrophs and hemibiotrophs | Including, % | | | Necrotrophs | Including, % | | |
		In field	In field and in storage	In storage		In field	In field and in storage	In storage
Potatoes	3	67	33	0	10	20	30	50
Cabbage	2	100	0	0	8	37	50	13
Beets	4	75	25	0	5	20	20	60
Carrots	0	0	0	0	4	0	50	50
Onions	3	100	0	0	1	0	0	100

Table 1.3.   Occurrence of biotrophs among fungi (Lewis DH, 1973)

Groups	Species, total	Including biotrophs
Protistoid fungi*	1600	500 (32%)
Zygomycetes	600	150 (25%)
Ascomycetes		
Not lichens	15000	4000 (27%)
Lichens	18000	18000 (100%)
Basidiomycetes	12000	7000 (58%)
Deuteromycetes	15000	150 (1%)
All	62200	29800 (48%)

* Fungi forming zoospores (oomycetes, chitridiomycetes).

Hemibiotrophs mostly cause local leaf damage, spot diseases. The causal agents of these diseases are usually sac and imperfect fungi from the genera *Mycosphaerella* (white spots), *Alternaria*, *Septoria*, *Venturia* (apple and pear scab), etc. When fungi from the genera *Colletotrichum* and *Gloeosporium* affect succulent parts of plants (valves of legumes, fruits of cucurbits) they produce deeper ulcers called anthracnoses.

Biotrophs together with *symbiotrophs* (including lichens) cover about half of the fungal species described (Table 1.3).

The basic diseases caused by biotrophic fungi are:

Bloom of spore-bearing parasite on the surface of affected organs (leaves, fruits, stems). Very common are white or brownish blooms called powdery and downy mildew. The former are caused by ascomycetes from the order Erysiphales. Mycelium of most mildew fungi spreads over the surface of an affected organ, and only haustoria are submersed in epidermal cells. Conidiophores with chains of conidia also develop on the surface and make the bloom look powdery. Downy mildew is caused by oomycetes from the order Peronosporales, the mycelium of which develops endophytically, and the bloom is formed by sporangiophores with sporangia coming out through the stomata. Since most stomata are on the leaf underside, downy mildew bloom also forms on the leaf underside.

Pustules formed by breaches in epidermis, which uncover spore-bearing of a parasite, are characteristic of basidiomycetes from the order Uredinales (rusts) and oomycetes from the family Albuginaceae (white rust).

Smut looks like dark, as if charred, areas of affected tissue, typical of destruction by basidial fungi from the order *Ustilaginales*/(smuts). These fungi attack only herbaceous monocotyledons and dicotyledons, causing signs of smut in generative organs or on stems and leaves.

Hyperplasia and *hypertrophy*. An attack of biotrophic parasites frequently causes a change in the hormonal balance of a plant and intensive proliferation of the affected organs or tissues due to intensified cell fission or their distention. Gall formation and deformations are caused by the fungi from different divisions: myxomycetes *Plasmodiophora brassicae* (clubroot), oomycetes *Albugo candida* (white rust), chitridiomycetes *Synchytrium endoboiticum* (potato wart), ascomycetes from the order Taphrinales (deformation of various organs), basidiomycetes *Ustilago maydis* (corn smut), etc.

Bacterial plant pathogens

General information on phytopathogenic bacteria

Bacteria are the smallest cellular organisms with a prokaryotic cell structure. Their size varies from 0.5 to 4.5 μ long and from 0.3 to 0.6 μ wide. The size of some species is almost the size of large viruses. In dense nutrient media, phytopathogenic bacteria form colonies: bright or opaque, transparent or turbid, sometimes colored yellow. Some bacteria form capsules around cells, produced due to release of polysaccharides. One of the functions of the slimy capsules is protection of the cells from drying and exposure to sunshine. In addition, the capsules protect phytopathogenic bacteria from binding of their cells, contained endophytically in intercellular space of plants, to the glycoproteins (lectins) of plant cells. Therefore, the capsular forms of bacteria are more virulent than non-capsular forms. The plant organs affected by bacteria frequently bear exudates, which are slimy excretion of the encapsulated bacteria.

According to the shape, there are several types of bacterial cells and their aggregates; however, most phytopathogenic species occur as straight or slightly curved rods. Sometimes rods form easily breaking chains (for instance, causal agents of bacterial blight of oat *Pseudomonas syringae* pv. *coronafaciens*). In Corynebacteria such chains can even branch, and in actinomycetes the branching chains form long branched filaments (mycelium).

Some bacteria (genus *Bacillus*) form endospores inside the cells, which are covered with a thick wall and can withstand harsh environment without losing viability.

Most phytopathogenic bacteria are motile. The flagella used for motion have a polar (at the cell ends) and peritrichal arrangement (over entire surface). The polar flagella can be single (in monotriches) or bundled (in lophotriches). Flagella proteins, flagellins, can play an important role in relation with plant cells.

A system of phytopathogenic bacteria and bacterial plant diseases

For a long time the system of bacteria was based on such attributes as the cell wall structure (determined by Gram staining), morphology, ability for sporulation, and enzymatic activity (determined by fermentation of various proteins, carbohydrates, etc.). For the system of phytopathogenic bacteria, the signs of disease caused were very important (for instance, the bacteria causing soft rot due to pectin decomposition were classified into the genus *Pectobacterium*) as well as the scope of plants attacked; therefore, the specific name of the bacteria is frequently correlated with the host plant name (for instance, *Pseudomonas tabaci* or *Xanthomonas malvacearum*). Use of the molecular methods (structural analysis of DNA, rDNA gene sequencing) showed that the genes controlling such seemingly essential features as production of enzymes or specialization towards hosts, occupy such small areas of the genome that they cannot possess a great diagnostic value. Hybridization of DNA of various strains paved the way for the concept of genomic species, which include the strains with more than 70% of hybridizing sites and for which the difference in the DNA melting point is within 5°C. The molecular classification which includes sequencing of ribosomal genes (determination of the sequence of nucleotides), DNA–DNA and DNA–RNA hybridization, and other approaches, allowed avoiding the contradictions of the phenetic classification. Today, comprehensive approaches are used, combining molecular and phenetic data at hierarchically different taxonomic levels of the system. Under such classification, phytopathogenic bacteria are included in the following taxons.

DIVISION FIMICUTES *(Gram-positive)*

Genus <u>*Bacillus*</u> are motile spore-forming peritrichous bacteria.

B. megathericum causes rot in vegetable marrow, cucurbit, potato, corn, and other plants, mainly in southern regions. It can live in soil for a long time, with saprotrophic feeding.

Genus <u>*Clavibacter*</u> *(=Corynebacterium)* are non-motile, chain-forming bacteria. *C. michiganense* are the bacteria living in the xylem of the Solanaceae and causing infectious wilt, with two subgenera: subsp. *Michiganense,* the causal agent of bacterial wart of tomato, and subsp. *Sepedonicum,* the causal agent of the potato ring rot.

DIVISION GRACILICUTES (*Gram-negative*)

Family <u>Enterobacteriaceae</u>

The motile peritrichous bacteria, classified into the genus *Erwinia,* cause two types of diseases:

Fire blight of the above-ground parts of a plant, caused by infiltration of bacterial toxins in infected tissues. *E. amylovora* causes darkening of blossom and leaves of apple and pear trees.

Soft rot of tubers, stems, and fruits of various plants due to decay of infected tissues degraded by pectolytic enzymes. The causal agents are *E. carotovora, E. Chrysanthemi, E. artroseptica,* etc.

Family <u>Pseudomonadaceae</u>

Motile monotriches and lothotriches. The family includes the genera *Pseudomonas, Burkholderia, Ralstonia,* and *Xanthomonas.* Members of the latter genus produce pigment, hence the yellow color of their colonies.

Bacteria from the genus *Pseudomonas* cause several types of diseases:

1. Tumors on branches of oleander and olive trees. The agent is *P. savaastanoi* pv. *savastanoi.*
2. Spots on leaves surrounded with a light halo (due to infiltration of the bacterial toxin causing decomposition of chloroplasts). Bacterial blight agents were classified on the basis of the host specialization: *P. tabaci* – agent of

tobacco bacterial wild fire, *P. phaseolicola* – agent of halo blight of beans, etc. The use of the methods of genetic systematics showed that the differences between them are within the intra-species variations; therefore, 45 species causing leaf blight were united in two species and the names were given according to the priority rule: *P. syringae* and *P. savastanoi*, with a large number of the pathology variants (*P. syringae* pv. *tabaci*, *P. savastanoi* pv. *phaseolicola*, etc.).

3. Trachebacterioses. The bacterium Ralstonia (=Pseudomonas) solanacearum causes the damaging disease southern wilt of Solanaceae.

Bacteria from the genus *Xanthomonas* cause blight of leaves and other affected organs (cotton hommosis, etc.), affect vessels (vascular bacteriosis of cabbage), and cause fruit rot (cucumber bacteriosis). Like in the previous example, many species were assigned a status of pathology variants within the species *X. campestris*, *X. axonopodis*, *X. oryzae*, or *X. vesicatoria*.

Family Rhizobacteriacea

They are motile bacteria with polar flagella. They cause tumors of the affected parts of plants.

Genus *Rhizobium* includes nodule bacteria that penetrate the root cells of legumes and cause their nodulation. The bacteria in the nodules, using the enzyme nitrogenase, involve molecular nitrogen of the air in chemical compounds that are digested by plants. Therefore, despite of some harm caused by the bacteria (outflow of nutrients to the infested parts of the root to form nodules and feed the bacteria), the presence of *Rhizobia* in the roots is very beneficial for the plant (mutualistic symbiosis).

The most common species in the genus *Agrobacterium* is *A. tumefaciens*. This bacterium comes to plant tissues through wounds in the root and causes proliferation of the infested tissue and gall formation. Most frequently, galls develop on the radical part of the stem (crown gall), but can also develop on roots, stems, leaves, and other organs. Bacterial canker affects plants of more than 40 dicotyledon families, including such important plants as grapes, apple, sunflower, tomato, etc. Another common species *A. rhizogenes* causes intense proliferation of roots ("beard") in the affected plants.

Tumor growth is induced by a large Ti (tumor inducing) plasmid which leaves the bacterial cells and goes to the infected plant cells. A part of it integrates into the plant chromosome, replicates and expresses in it as additional genes. The cell starts synthesis of additional substances, plant hormones, which cause gall formation, and special amino acids opins, unavailable for plant digestion, but serving as nutrition for bacteria.

After the specific features of agrobacteria and plant cell relations have been revealed, not only plant pathologists but also molecular biologists paid close attention to those bacteria. It was found that the Ti plasmid is a convenient vector to be loaded with the useful genes to transfer them to a plant from the donor separated with the acceptor species by a reproductive non-cross barrier. *A.tumefaciens* became a favorite subject of gene manipulations with plants (see the final chapter).

DIVISION TENERICUTES (bacteria without cell wall, or mycoplasmas)

Mycoplasmas are very small bacteria that have no cell wall and are consequently capable of changing the cell shape. Many mycoplasmas have an ovoid shape, turning into pear shape, dumbbell shape, etc. Others are stretched, helical cells (spiroplasmas). The mycoplasmas attacking plants are called phytoplasmas. They are biotrophic parasites living in vessels of affected plants (most of them in the phloem). The infection destroys the phloem transport of metabolites and regulation of growth processes, which results in small leaves, bushiness, and chlorosis. The organism is transmitted from plant to plant by sucking insects. The obligate intracellular parasitism, the way of transmission, and the signs of the disease caused by phytopathogenic mycoplasmas approach those of viruses.

GROUP ACTINOMYCETES *(actinomycetes)*

Actinomycetes form a thin-branched mycelium with the spores forming at the ends of hyphae, separately or in sporangia. They have typical prokaryotic cells, however, their taxonomic status among the bacteria is not clear. Actinomycetes are soil saprotrophs developing in soil abundant in organic matter; however, some species of the genus *Streptomyces* can infest potato tubers, causing their surface damage (common potato scab).

Plant viruses and viral plant diseases

Structure of plant viruses

Viruses include non-cellular prokaryotes with the following common features: (1) only one type of nucleic acid (RNA or DNA); (2) lack of own protein-synthesizing system (ribosomes); (3) lack of energy metabolism enzymes. These features make viruses fully dependent of the host: all viruses are obligate parasites on bacteria, plants, or animals.

Chemical composition of viruses is considerably simpler than composition of the smallest bacteria. All viruses contain one or several molecules of a nucleic acid coated with protein molecules of the same type (some viruses have several types of coat proteins) that form crystalline structures. In the viruses with most complex arrangement, these structures are covered with an outer membrane made of proteins, carbohydrates, and lipids.

Viral genome and its functioning. In cellular organisms, the major functions of the nucleic acids – replication, transcription, and translation – are distributed between two types of molecules: double-stranded DNA and single-stranded RNA. In replication, the DNA molecule is unzipped and on each strand a complementary second strand is constructed using the enzyme DNA-polymerase. In transcription, the enzyme RNA-polymerase builds an RNA strand (minus strand) complementary to the DNA strand (plus strand) in a definite direction (from 5′ end of the molecule towards 3′ end). In translation, a protein molecule is synthesized from individual amino acids, using ribosomes, on the RNA molecule as a matrix, in the opposite direction (from 3′end towards 5′ end).

Table 1.4.   Structure of plant viral genomes (Harrison BD, 1982)

Genome	Number of viruses	Percent
Single-stranded DNA	26	4
Double-stranded DNA	13	2
Single-stranded DNA (plus strand)	470	76
Single-stranded DNA (minus strand)	85	14
Double-stranded RNA	26 + all fungal viruses	4

In viruses, the same type of molecule (RNA in most phytopathogenic viruses) performs all the three functions: replication, transcription, and translation. Besides, in most viruses the nucleic acids are single-stranded, and can fulfill both (+) and (−) functions (Table. 1.4).

As can be seen, only a small number of viruses contain double-stranded information molecules, reading from which is done similar to cellular organisms. Most phytopathogenic viruses have one multifunctional RNA (+) strand. Its functions are storage and realization of information. In addition, the RNA (+) strand is an infectious molecule. After arrival of the virus particle containing an RNA (+) strand to the cell, the first stage is decapsulation, i.e. release of the RNA molecule from its protein coat. This process is carried out on the cell receptors by the proteases of the host plant. The next process is replication, catalyzed by the enzyme RNA-dependent RNA polymerase (replicase), which in most viruses is encoded by the own genome. The RNA of the tobacco mosaic virus (TMV), typical of this group, encodes four proteins (Figure 1.1).

Translation begins from the 5′ end of the molecule – synthesis of 126 kDa and 183 kDa proteins, replicase components. These proteins occur in the cell at early stages of the infection process; hence they are called early proteins. Another virus – turnip yellow mosaic virus – develops hybrid replicase in the infected cell: one of its components (115 kDa) is encoded by the viral genome, the other – a 45 kDa protein – by the host cell genome. Apparently, the host enzyme system is routinely used for replication of viral RNA, as for many plants infected with viruses a substantial increase in the synthesis of RNA-dependent RNA polymerase is typical.

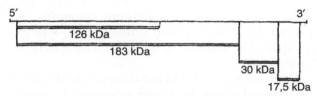

Figure 1.1.   TMV genome and TMV genome-encoded proteins.

RNA polymerase builds an RNA (–) strand complementary to the (+) strand of the viral RNA. This process results in formation of a replicative form of the viral RNA in the cell, represented partially and completely by a double-stranded structure. The (–) strand serves as a matrix for synthesis of new molecules of the viral (+) RNA, which functions as mRNA in translation of the late viral protein cells in ribosomes. In TMV they consist of a 30 kDa transport protein and a 17.5 kDa structural protein of the coat (Figure 1.1). The final process, encapsulation, consists of maturation of the whole particle – self-assembly of the structural protein molecules on the surface of the RNA molecule.

Thus, in viruses with the RNA (+) strand, the viral particle disappears after arrival to the cell, and disjunctive (separated) replication of a new generation of viruses occurs, similar rather to the factory conveyor than reproduction of cellular organisms (synthesis of individual components, sometimes occurring in different compartments of the cell, and assembly of the whole particles).

In viruses with the RNA (–) strand, this strand is non-infectious, because it cannot function as a matrix RNA. Therefore, in addition to protein-coated RNA molecules, the particle also contains enzymes, in particular, RNA transcriptase, and all these are covered with an additional coat containing lipids. Such a particle enters the cell like a "Noah's Ark" with its own enzymes. Further synthesis of the new generation of viruses is carried out not separately, but in one compartment.

Unlike cellular organisms, the viral genome experiences a deficit of information, as it can encode only several proteins. In some viruses, the molecule of nucleic acid contains not four reading frames, like the TMV RNA, but a larger number; however, the increase in the amount of information needs to be accompanied by an increase in the length of the information molecule, while the single-stranded molecules, typical of most viruses, have no structural rigidity of the double-stranded molecules, and with an increased length they lose the structure necessary for recognition by enzymes. Different viruses approach the shortage of information problem in different ways.

1. Multifunctionality of the viral proteins. In POTY viruses (Y virus of potato and related viruses) the ca. 10 kb genome contains a single long open reading frame (ORF) translated into a large polyprotein (340–370 kDa), which is divided into ten viral proteins in co-translation and post-translation, using own proteases. Almost all the proteins are multifunctional, i.e. contain several domains providing various functions. For instance, the capsid protein (CP) is responsible for aphid transmission, cell-to-cell and systemic transport, and virus assembly; protein HC-Pro for aphid transmission, systemic movement, papain-like cysteine proteinase, and synergism in combined infections.

2. Fragmentation of the genome. On centrifugation of some viruses isolated from infested plants, in density gradient of cesium or sucrose,

it was found that they constitute a mixture of larger and smaller particles (solutions of cesium chloride or sucrose are layered in a centrifuge test tube from more concentrated to less concentrated solutions, the specimen being studied is put on the top and is centrifuged; the mixture of particles in the specimen, differing by molecular weight, is easily separated in gradients into separate fractions). Each separate fraction is not infectious or is slightly infectious, while the mixture is highly infectious. The ratio of larger and smaller particles in a plant is usually constant. It was found that though the particles are coated with the similar protein molecules, their RNA differ in the structure and encoded proteins. For instance, the RNA contained in the long particles encodes early proteins, enzymes, while the RNA in the short particles encodes the structural proteins of the coat. A fraction consisting of the long particles can infect plants and form a new RNA generation, but it is unstable and cannot survive outside a plant cell; the short particles are stable, they cannot infect plants and reproduce in them. Two components were found in the viruses of tobacco rattle, ring spot of raspberry, black ring spot of tomato, cucumber mosaic, etc., three in the cow pea mosaic virus, and four in the alfalfa mosaic virus.

3. Use of a helper. It was written earlier that shortage of information made some viruses use the host cell enzymes during replication and translation. There are viruses (suggested to be called virusoids) that have an RNA molecule consisting of several hundred nucleotides and are incapable of encoding more than one protein (for instance, structural protein of the coat). The virus receives the other proteins, necessary for intracellular maturation, from another virus, the helper; therefore, it cannot live in the cells not infected with the helper such as the satellite virus (SV), satellite of tobacco necrosis virus (TNV). It never occurs in TNV-free plants and though it is covered with its own coat, it uses the early proteins encoded by the TNV genome.

4. Helper-independent replication of a short ring RNA molecule that contains no information on the structural protein. Such molecules (viroids) can self-replicate in a plant and cause serious diseases (spindle tuber disease, etc.).

Structural viral proteins. Molecules of structural viral proteins of plants combine with each other in a particular way to form crystalline structures called capsids. One or more very close proteins of the molecules of which the capsid is made up are called a morphological unit, or capsomere. In different viruses, capsomeres are arranged helically (helically symmetric viruses) or polygonally (viruses with icosahedron symmetry). A capsid containing the nucleic acid of the virus is called nucleocapsid.

In an electron microscope, viruses with a helical structure look like rods or filaments. Inside the helically arranged protein molecules (capsomeres) there is an RNA strand, also helically arranged, the pitch of which repeats that of the capsid helix. The shape of the particle depends on the length to width ratio. If the ratio is less than 50, the virus particles look like rods (for instance, tobacco rattle virus with the particle dimensions $130 \times 320 \times 20$ nm or tobacco mosaic virus with the particles 300×18 nm). With the length to width ratio of 50 or more, the particles look like flexible filaments. The examples are the potato X viruses ($480-580 \times 10$ nm) and potato Y viruses (750×15 nm). Incidentally, the length of beet yellows virus is as much as 2000 nm with the width only 10 nm.

Viruses with a cubic symmetry of the capsid have the shape of icosahedrons, polygons with 12 apexes and 20 facets. The polygon apexes are made by the capsomers consisting of five aggregate protein molecules (pentamers), and the facets and edges, from a large number of capsomers (multiple of 20) consisting of six molecules (hexamers). Most plant viruses with cubic symmetry contain 180 protein molecules (12 pentamers and 20 hexamers) in the capsid.

Finally, in some plant viruses (bronzing of tomato, yellow dwarf of potato, etc.) a long nucleocapsid with helical symmetry or several nucleocapsids (if the virus has a fragmented genome) are enclosed in an outer membrane consisting of proteins densely packed in a hexagonal structure, and lipids.

Thus, the viral particles observed in an electron microscope are immobile structures (virions). They are absent from the vegetative phase during intracellular reproduction of virions. Mature viral particles of the next generation produce, in the infected cells, viral inclusions in the form of various crystals or amorphous x bodies, where the viral particles are mixed with the host cell cytoplasm. The inclusions can form in the cytoplasm (TMV, potato virus Y (PVY), etc.), nuclei (beet yellows virus, barley yellow mosaic), chloroplasts (virus of turnip yellow mosaic, barley streak), mitochondria (tobacco rattle virus), or vacuoles (cucumber mosaic virus, tobacco necrosis virus, potato virus X (PVX)).

Virions of some viruses (TMV, PVX) are accumulated in infected cells at a very high concentration, and they are extremely resistant to external effects (heating, drying). Other viruses (bronzing of tomato, PVY) are non-resistant and are inactivated at much milder exposure.

Symptoms of viral diseases

According to the tissue specialization, viruses are classified as *parenchymal* and *phloemic*. The former occur in cells of the leaf parenchyma and cause its various damage. The most common sign of the diseases caused by parenchymal viruses is alternation of dark green and lighter colored areas of a leaf, mosaic. The light areas may be light green (green mosaic), yellow (yellow mosaic), or white (white mosaic). On infection with tobacco necrosis virus (TNV) and some other mosaic viruses, the mosaic coloring is observed only if the leaves have been infected at a young age; in mature leaves, the infection develops asymptomatically. The light areas contain 250 times more viral particles and four times less chlorophyll than the dark green areas. The dark green, slightly infected leaf areas are called "green islands."

Growth retardation of the light areas results in deformation of the infected leaves: blisters in some leaf areas, wrinkles caused by a deeper location of the veins, and in some cases destruction of the leaf parenchyma that survives only next to the veins. In some viral diseases, the light areas are not distributed on a leaf randomly but form patterns as rings (ring mosaics), wavy strips (etch), etc. A disease is frequently accompanied by necrotic lesions of the leaf veins (streak mosaic) passing on to stem petioles (tomato streak caused by TMV).

Phloem viruses cause diseases which were earlier called "yellows" due to resemblance to the disease "aster yellows." Today, however, it is shown that the causal agent of aster yellows is not a virus but a bacterium without cell wall, a mycoplasma. Localization of viruses in the phloem frequently results in interruption of the phloem flow and rolling of the leaves overflowed with the products of photosynthesis. Leaf roll attacks potato and cotton. Many phloem viruses interfere with phytohormone metabolism, which causes various malformations (teratomorphs): transformation of vegetative organs to generative organs, and vice versa, proliferation of flowers, excessive bushiness, and dwarf.

Survival and propagation of phytopathogenic viruses

Survival. Viruses cannot reproduce and most of them cannot survive for a long time outside living plant cells. The best source of their survival in winter is vegetative parts of infested plants. Therefore, viruses inflict the greatest loss to the plants reproduced by vegetative organs: potatoes, fruits and berries, and flower crops. Accumulating year after year, viruses gradually decrease the yield and change the characteristic attributes of a cultivar, which is called cultivar degeneration. Most viruses have a broad host range, for agricultural crops infection of weeds and wild plants is very dangerous, in wintering organs of which potentially hazardous viruses can survive.

Survival of viruses in seeds is a more complex aspect. Only 18% of plant viruses were found to be able to survive in seeds, and the number of infected seeds developed in infested plants can vary from fractions of percent to 100%. However, even a low infection rate of seeds threatens the future yield; for instance, 0.1% of chick pea seeds infected with the mottling virus has been estimated to result in 20,000 infested plants per acre. Most often the viruses of alfalfa mosaic, barley striate mosaic, ordinary and southern bean mosaic, soybean mosaic, chick pea mottling, and raspberry ring spot survive in seeds. Apparently, many viruses survive in seeds of legumes, which is related to the anatomical features of formation of their blastemas. The most resistant viruses, such as TMV, can survive on seed surface. The probability of survival of this virus on seeds depends on the fruit structure of the infested plant. The succulent pulp of a tomato fruit forms a drying film on the seed surface, where the virus particles can survive till planting; in dry tobacco pods the viral particles are unlikely to stick to the seed surface, therefore, tobacco seeds taken from infected plants are usually not infected.

Some viruses survive in the bodies of their vectors.

Transmission of viruses can be vertical (from parents to progeny) or horizontal (from infected to healthy plants). Vertical transmission by vegetative parts of plants and by seeds has been discussed earlier. Horizontal transmission can proceed in different ways.

Contact transmission occurs at touch of leaves or roots of infected and healthy plants. In contact of plants in rows, rubbing leaf surfaces produce small wounds (for instance, hair breaking) through which virus-bearing sap passes from one plant to another. Therefore, contact transmission is possible only for the viruses accumulating in epidermal cells (phloem viruses cannot be transmitted through contact) and able to survive in sap without inactivation. The contact way is the predominant way of transmission for such viruses as TMV, PVX, and turnip yellow mosaic virus. Contact transmission of viruses poses the greatest threat at the time of plant care involving their damage: cultivation, removal of side-shoots of tomatoes, or breaking of tobacco leaves.

Engrafting transmits all known viruses. This way of transmission is important in orcharding where plants are reproduced by engrafting on stocks. Engrafting transmits such destructive diseases as apple-tree mosaic, peach rosette, or hop curly top.

Some viruses (plum dwarf, tobacco streak, cherry leaf roll) can adsorb on pollen in exine folds and spread from sick to healthy plants by wind or pollinating insects.

The most common way of virus transmission in nature is vectors. Most viral vectors are arthropods (insects, mites) with piercing and sucking mouth parts. All of them eat liquid feed; therefore, the primary function of the mouth parts is the sucking of sap. To reach plant cells, they have to pierce the integument; therefore, the second function of the mouth parts is piercing. These animals secrete salivary enzymes in plant cells to hydrolyze plant polymers to simpler derivatives. Hence the third function of the mouth parts: salivation. The mouth parts are evolved into bristles, some piercing integument and others forcing salivary enzyme and sucking sap to the internal cavity, in a syringe-like way. Feeding habits of piercing and sucking arthropods create perfect conditions for transmission of viruses. For phloem viruses, transmission by vectors is the only natural way to arrive in susceptible tissue (their vectors feed on phloem sap).

According to virus–vector relation, viruses are classified as non-persistent (stylet) and persistent (circulating). Vectors of stylet viruses obtain them from the epidermal cells very promptly, within 30 seconds to 2 minutes of feeding (insect stylet cannot penetrate deeper cell layers in such a short time); they can transmit viruses to healthy plants immediately after feeding and quickly lose the transmission capability. Obviously, this works only for parenchymal viruses, and the basic mechanism of transmission is adsorption of particles on the stylet. An evidence is loss of viruses by insects after molting, when all chitinous parts are shed, including the stylet. However, transmission of viruses cannot be explained by mechanical adsorption of particles on a stylet, as some viruses, including TMV, are not transmitted by insects despite a high concentration of particles in the sap. A certain mutual specialization is known to exist between viruses and vectors, and even presence of helper viruses in transmission by insects. For instance, aphids do not transmit some potato viruses (aucuba mosaic, etc.); however, if the potato is infected with a mix of aucuba mosaic viruses and PVY, then both viruses are transmitted by aphids. Consequently, PVY serves as a helper to aucuba mosaic virus in transmission by aphids. The PVY genome contains a gene encoding, a 58 kDa protein called helper component (HP) and promoting connection of viral particles with particular stylet areas. Mutations of this gene make the virus incapable of transmission by aphids.

Persistent viruses have more complex relations with vectors. The vectors get the viruses after a longer feeding time (more than 30 minutes). Hence, this group includes all phloem viruses. The vector needs a certain latent period after feeding (from several days to several weeks) during which the vector is unable to transmit the viruses; however, having acquired such ability, the vector will maintain it for a long time, in many cases the whole life. The viruses are absorbed with sap, come to intestine through the wall from which they arrive to hemolymph, are carried over the body, get into salivary glands, and finally are delivered to a new plant through the hole in the stylet together with saliva. Such circulation in the vector body requires time defined as a latent period. Many persistent viruses replicate in the vector body and precipitate as crystalline or amorphous inclusions.

Stylet transmission is typical of aphids, while the circulatory way is used by many animals: aphids, cicadas, trips, mites, etc.

Soil-habitant viruses are transmitted by nematodes and fungi.

Nematodes transmit over 20 viruses, many of which cause ring spot and survive in seeds. Most important among them are vine roncet virus, raspberry ring spot, and tobacco rattle virus. Their vectors are free-living nematodes from the genera *Xiphinema*, *Trichodorus*, and *Longodorus*. They also have a chitinous stylet and feed by sucking cell juice. Different viruses are transmitted by the stylet or circulatory way. All these viruses have a fragmented genome consisting of two types of particles; the shorter RNA strand contains the gene determining specificity to a vector (capsid protein).

The fungi transmitting plant viruses fall into division Chytridiomycota and Myxomycota. Tobacco necrosis viruses and lettuce big vein viruses are transferred by the fungus *Olpidium brassicae* that frequently occurs on roots of many plants. Tobacco necrosis virus is on the surface of zoospores and comes under the envelope only after incystment of zoospores on root hairs, and lettuce big vein virus is always inside zoospores. Some cereal viruses (barley yellow mosaic, American wheat mosaic, oat mosaic) are transmitted by the myxomycete *Polymyxa graminis*, parasite on cereal roots.

Interaction of viruses and mixed infections

Interactions between different viruses in an infested plant can be antagonistic, synergistic, or neutral.

Antagonism between viruses is manifested in the phenomenon of interference or cross protection. This phenomenon can be defined as follows: a plant infested by a virus is more resistant to the subsequent infection by other virus. Interference most frequently occurs between the relative viruses from the same group (see below). As interference shows at primary plant inoculation not only with whole virions but also with capsid protein or viral RNA, apparently, there are several mechanisms of intracellular interference, in particular:

1. Competition for specific deproteinization receptors;
2. Loss of replication and translation ability of the second virus RNA due to its covering with the capsid protein of the first virus, accumulated by the moment;
3. Inhibition of reproduction of RNA that has homology areas with the RNA earlier arrived to the cell.

Cross protection has a wide practical use (artificial inoculation of plants with a low pathogenic strain yielded good results in reducing the damage caused by TMV, tomato bronzing virus, apple mosaic virus, and citrous tristeza virus). The molecular mechanisms of interference are used in gene engineering to assure resistance of transgenic plants to viruses (Chapter 13).

Synergism. Combined plant inoculation with non-relative viruses often results in an increased reproduction rate of the virus particles, intensification of the symptoms of disease and the losses it causes. For instance, cucumber mosaic virus and black eye mosaic virus reduce cow pea yield by 14 and 2.5%, respectively. Combined infetion causes dwarfing and yield reduction by more than 85%. The PXV pathogenicity sharply increases in the presence of PYV, in particular, the concentration of PXV particles in potato cells infected with PYV grows 3–5 times compared with a single infection.

Virus nomenclature

The attempts to develop a hierarchical system of viruses, similar to the Linne system of cellular organisms (division – class – order – family – genus – species), have failed due to polyphyletic origin of viruses. Therefore, non-hierarchical systems were developed, where the viruses with similar properties were grouped together, without attempts to find any hierarchical relations between the groups. Today, a compromise system has been suggested with the hierarchy of the lower taxons (family – genus – species). The division into families is based on the fundamental attributes: type and structure of the virus nucleic acid (DNA or RNA, number of strands, (+) or (–) form of RNA, fragmentation and length of the genome, number of transcripts, etc.). Viruses are grouped into genera on the basis of phenotypical attributes, such as particle shape (rod-shaped, filamentous, isometrics, bacilliform), way of transmission and conservation (aphid vector, nematode, fungi, seeds, pollen, mechanical transmission, persistent, non-persistent). Species are made of the traditional groups of viral strains, called after the host plant and signs of disease (Table 1.5).

Table 1.5. A classification of plant viruses (as approved by International Committee on Taxonomy of Viruses in May 2000)

DNA	Caulimo-viridea	Caulimovirus	Cauliflower mosaic virus
		SoyCMV-like	Soybean chlorotic mottle virus
		CVMC-like	Cassava vein mosaic
		PVCV-like	Petunia vein cleaning virus
		Badnavirus	Commilina yellow mottle virus
		RTBV-like	Rice tungro bacilliform virus
	Gemini-viridae	Mastrevirus	Maize streak virus
		Curtovirus	Beet curly top virus
		Begomovirus	Bean golden mosaic virus
	–	Topocuvirus	Tomato pseudo-curly top virus
	Reo-viridae	Nanovirus	Subterranean curly top virus
dsRNA		Phytoreovirus	Wound tumor virus
	Partiti-virivae	Fijivirus	Fiji disease virus
	–	Oryzavirus	Rice ragged stunt virus
	Rhabdo-viridae	Alphacryptovirus	White clover cryptic virus 1
	Bunya-viridae	Betacryptovirus	White clover cryptic virus 2
	–	Varicosavirus	Lettuce big-vein virus
(–)ss RNA	–	Cytorhabdovirus	Lettuce necrotic yellows virus
	Bromo-viridae	Nucleorhabdovirus	Potato yellow dwarf virus
		Tospovirus	Tomato spotted wilt virus
		Tenuivirus	Rice stripe virus
(+)ss RNA	Clostero-viridae	Ophiovirus	Citrus porosis virus
	Como-viridae	Bromovirus	Brome mosaic virus
	Luteo-viridea	Cucumovirus	Cucumber mosaic virus
		Alfamovirus	Alfalfa mosaic virus
	Poty-viridae	Ilarivirus	Prunus necrotic ringspot virus
		Oleavirus	Olive latent virus 2
		Closterovirus	Beet yellows virus
		Crinivirus	Lettuce infectious yellows virus
		Comovirus	Cowpea mosaic virus
	Sequi-viridae	Nepovirus	Tobacco ringspot virus
	Tombus-viridae	Luteovirus	Barley yellow dwarf virus
		Polerovirus	Potato leaf roll virus
		Enamovirus	Pea enation mosaic virus
		Potyvirus	Potato virus Y
		Rymovirus	Ryegrass mosaic virus

Continued

Table 1.5. A classification of plant viruses (as approved by International Committee on Taxonomy of Viruses in May 2000)—cont'd

Bymovirus	Barley yellow mosaic virus
Macluravirus	Maclura mosaic virus
Ipomovirus	Sweet potato mild mottle virus
Tritimovirus	Wheat streak mosaic virus
Sequivirus	Parsnip yellow fleck virus
Waikavirus	Rice tungro spherical virus
Tombusvirus	Tomato bushy stunt virus
Carmovirus	Carnation mottle virus
Necrovirus	Tobacco necrosis virus A
Machlomovirus	Maize chlorotic mosaic virus
Dianthovirus	Carnation ringspot virus
Avenavirus	Oat chlorotic stunt virus
Aureusvirus	Pothos latent virus
Tobravirus	Tobacco rattle virus
Tobamovirus	Tobacco mosaic virus
Hordeivirus	Barley stripe mosaic virus
Furovirus	Soil-borne wheat mosaic virus
Pomovirus	Potato mop-top virus
Pecluvirus	Peanut clump virus
Benyvirus	Beet necrotic yellow vein virus
Sobemovirus	Southern bean mosaic virus
Marafivirus	Maize rayo findo virus
Umbravirus	Carrot mottle virus
Tymovirus	Turnip yellow mosaic virus
Ideovirus	Raspberry bushy dwarf virus
Ourmiavirus	Ourmia melon virus
Potexvirus	Potato virus X
Carlavirus	Carnation latent virus
Foveavirus	Apple stem pitting virus
Alexivirus	Shallot virus X
Capillovirus	Apple stem grooving virus
Trichovirus	Apple chlorotic leaf spot virus
Vitivirus	Grapevine virus A

Parasitic nematodes as plant pathogens

Nematodes, or ringworms, are numerous and ubiquitous organisms. Out of 20,000 described species of nematodes approximately 20%, or ca. 4000 species, are connected with plants. Phytoparasitic nematodes (phytohelminths) belong to four taxonomic orders, two in each of the two classes of nematodes: orders Dorylaimida and Triplonchida of the class Adenophorea, and orders Aphelenchida and Tylenchida of the class Secernentea.

Phytonematodes are microscopic organisms, their size varying from 300 μ to 8 mm. The size of most phytohelminths does not exceed 2 mm. Their body is always round in cross section, non-segmented, threadlike or spindle-shaped (Figure 1.2); less often, in sedentary or inactive species of nematodes, the body is pear-shaped or ball-shaped, (root-knot and cyst nematodes). They are diecious animals, with distinct sexual dimorphism manifested not only as difference in the structure of sexual glands but also as presence of the secondary sexual characters. Sexual dimorphism is especially distinct in sedentary nematodes. Females of cyst and root-knot nematodes are pear-shaped and sit motionless on plant roots. Males, as against females, retain a threadlike body typical of nematodes and are capable of independent motion. Hermaphroditism is rare and a secondary character.

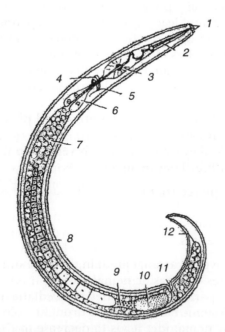

Figure 1.2. Diagram of a typical plant–parasitic nematode.
(1) Lip region; (2) stylet; (3) muscular pumping esophageal bulb; (4) nerve ring; (5) excretory pore; (6) digestive gland; (7) intestine; (8) ovary; (9) quadricolumella; (10) undeveloped egg; (11) vulva; (12) rectum.

For nematodes, only sexual reproduction is known. Reproduction is by eggs. In most cases the eggs are fertilized, though parthenogenesis is also possible and known to occur in cyst and root-knot nematodes. In root-knot nematodes, enough males usually occur at deterioration of life conditions, for example, at decay of damaged plant roots. Fertility varies in different species; in some most specialized phytohelminths, such as root-knot nematodes of the genus *Meloidogyne*, a female lays up to 2500 eggs. Phases of nematode development include eggs, larvae, and adults. Postembryonic development is accompanied by structural changes similar to metamorphosis. Constitution of a mature larva is generally similar to the structure of adult worms, but differs primarily in the underdeveloped sexual system. A strong external skeleton in the form of cuticle determines the type of growth: by molts. Postembryonic development of phytonematodes includes four larval ages and the fifth age as the adult, is distinctly differentiated by molts. Separate age categories clearly differ in dimensions. Besides, each age is characterized by certain phases of development of sexual and digestive system. Parasitizing is possible due to the presence of certain structures. The key structure is the stylet (or spear) developed to penetrate plant cell wall, and also secretory glands of the esophagus.

All parasitic phytonematodes are obligate parasites feeding exclusively on the contents of living cells of plants (phytohelminths) or fungi (mycohelminths). Many of them are polyphages not possessing specificity to particular plant species, but oligophages and monophages also occur. They are divided into root, stem, and leaf phytonematodes depending on which part of plant they prefer for feeding. Less often nematodes occur in flowers and fruits. They mostly attack roots, less often, tissues of other organs. Annual damages from these parasites are estimated as more than 100 billion US dollars. There are no species of plants, cultivated or wild-growing, not to be known as a host of one or more species of parasitic nematodes.

The diseases caused by these parasites are called phytohelminthoses. The external signs of the disease (signs of lesion) caused by nematodes are often discernible to the naked eye and are directly related to the type of nematode feeding and mode of life. They include:

1. Ectoparasites that pierce root tissues by the stylet or only dip the top end in the root;
2. Migratory endoparasites; and
3. Sedentary endoparasites.

Figure 1.3 shows a typical arrangement in the root and typical shape of the body of the key species of parasitic nematodes.

Plants invaded by parasitic nematodes showed the signs similar to the effect of extreme abiogenic factors, such as drought, cold, mineral starvation, etc. Plant invasion by nematodes leads to decrease in the number of leaves, weight of the plant, stem and tubers, size of the blade, and drop in the leaf water potential. Inhibition of photosynthesis is observed and aqueous and respiratory metabolism of the plants change.

Figure 1.3. Localization of different nematode groups in plant tissues (Siddiqi MR, 1986).

Ectoparasitic nematodes of the genera *Helicotylenchus, Rotylenchus, Tylenchulus,* etc., feeding on plant roots, can cause root tissue necrosis, expressed as browning of the necrotized spots. Invasion of the root system by nematodes of the genus *Trichodorus* shows as thickened and shortened ends of the roots. Deeper and broader necrosis, the so-called "wounds," are caused by endoparasitic nematodes of the genus *Pratylenchus* that are able to penetrate, feed, and reproduce inside a root. Dark brown, almost black spots with a diameter from several millimeters to 5 cm develop on the root surface. The wound depth varies from 2–10 mm. Slightly smaller lesions are caused by annelid nematodes of the family *Criconematidae.* Of special importance are the ectoparasitic nematodes of the orders Dorylaimida and Triplonchida: longidorids, xyphynemes, and trichodorids. These nematodes are virus vectors, the key viruses being grapevine fan leaf, raspberry ring spot virus, peach rosette virus, tobacco ring spot virus, and tobacco rattle virus.

Most parasitic nematodes, more than 20 genera, belong to the order Tylenchida. Among them, sedentary endoparasitic nematodes are of greatest economic importance. They include cyst nematodes of the genera *Heterodera* and *Globodera* (most important of them are the potato cyst nematodes *Globodera rostochiensis* and *G. pallida*, beet cyst nematode *Heterodera schachtii*, and soybean cyst nematode *H. glycines*), as well as root-knot nematodes of the genus *Meloidogyne* (*Meloidogyne incognita*, *M. javanica*, *M. arenaria*) causing characteristic formations on roots, root-knots. Cyst nematodes have a small host range, whereas root-knot nematodes are polyphages. The characteristic features of ontogenesis and morphogenesis of cyst nematodes allow regarding them as evolutionarily advanced parasites. The signs of plant diseases caused by these parasites include growth inhibition, leaf chlorosis, and fading.

Sedentary parasitic nematodes are in focus of close investigation by nematologists and plant pathologists worldwide, and in the recent years considerable progress has been made in investigation of the subtle mechanisms of the relations of these parasites with plants.

Characteristic features of development of sedentary nematodes

Relationship between sedentary nematodes and plants constitute a complex process that develops in accordance with development of the life cycle of the parasites. A more detailed description of the life cycles of sedentary nematodes, provided below, is necessary to understand the processes underlying the interaction of these pathogens with plants. Life cycle of sedentary nematodes is shown in Figure 1.4.

The cycle of development of nematodes begins with the egg where the development of the first larval phase is completed. After molting, a second age larva with a well-developed stylet leaves the egg. The nematode can infest the plant root system only at this phase. Larvae from soil penetrate the root tissue growth area directly in front of the root cap. To penetrate the root, the larvae pierce a cell with the stylet and secrete the enzymes that cause cell wall degradation. These enzymes are produced in the esophageal glands (Figure 1.5). After penetrating the root, the larvae move towards the central cylinder (stele). Larvae of root-knot and cyst nematodes migrate by different pathways: the root-knot nematode moves in the intercellular space, whereas the cyst nematode moves intracellularly. Upon arrival to the place of final localization, the larvae position themselves parallel to the central cylinder (stele), become motionless, and develop a feeding site.

Irrespective of the nematode–plant combination, the mechanism of feeding site induction is similar in the cyst and root-knot nematodes. Secretory discharge of nematodes induces formation of the special structures which ensure parasite feeding (Figure 1.6). In the root-knot nematode parasitism these are several huge cells the structure of which develops during hypertrophy and anomalies of cell division. The general characteristics of the huge cells are: lobed and swollen nuclei, can be numerous (up to 100); the nucleoli are enlarged;

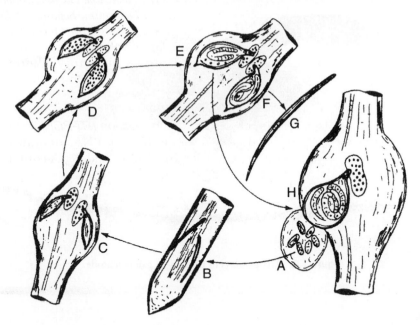

Figure 1.4. The life cycle of the root-knot nematode.
(A) 1st egg stage; (B) larval stage within the egg; (C) 2nd larval infections stage; (D) 3rd larval plant feeding stages; (E) 3rd molt (male and female); (F) 4th larval stage (male and female); (G) adult male; and (H) adult female.

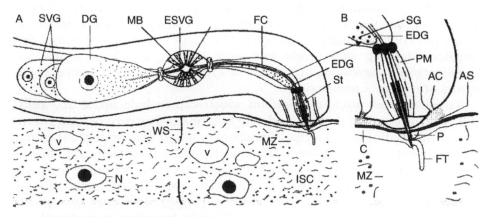

Figure 1.5. Schematic presentation of a *Heterodera schachtii* J2 feeding from an initial syncytial cell (ISC), about 36 hours after induction (Jung C, Wyss U, 1999).
AC – amphidial canal; AS – amphidial secretions; C – callose; DG – dorsal gland; EDG – exit dorsal gland; ESVG – exit subventral gland; FC – food canal; FT – feeding tube; ISC – initial syncytial cell; MB – muscular bulb; MZ – modified cytoplasmic zone; N – nucleus; P – feeding plug; PC – pump chamber; PM – stylet protractor muscles; SG – secretory granule; St – stylet; SVG – subventral glands; V – vacuole; and WS – wall stubs.

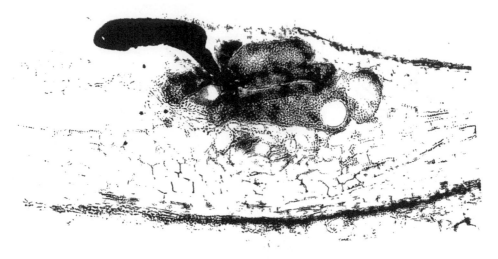

Figure 1.6. The giants cells formed by gall nematode.

the cytoplasm is grained and electron dense; the vacuoles are small or missing; the walls are irregularly thickened with ingrowth inside the cell; the number of mitochondria and plastids is highly elevated. The cause of multinucleosis is abnormal division after invasion resulting from the numerous cytokinesis-free mitoses. The cytoplasm is dense and grainy; it contains lipids, nucleic acids, and proteins, with the protein content 10 times more than normal. Cell membranes of the huge cells were found to contain cellulose and pectin, but they contain no lignin and suberin; the cells contain no starch and contain numerous vacuoles. The plant cells which surround the nematode feeding sites divide and swell causing root-knot formation. The emergence, development, and life of a root-knot is closely related to life activity of root-knot nematodes.

Cyst nematodes produce syncytium which can include up to 200 cells. The cells surrounding the nematode are necrotized. Presence of local necrosis near the feeding sites is a characteristic sign of cyst nematode parasitizing. Formation of local necrosis results from mechanical damage of the cells due to larvae motion in the plant tissue. Parenchymal root cells adjacent to the parasite localization places are hypertrophied. Their size is larger than that of normal parenchymal cells. Alongside the hypertrophy of the crust cells, proliferation is observed. Pericycle cells also proliferate. Syncytium is produced.

The main purpose of both the huge cells and syncytium is to provide the optimum conditions of feeding and development. Duration of development of a post-invaded second age larva coincides with duration of development of the feeding structures, i.e. the second phase larvae are completely responsible for their formation. Little is known about the mechanisms underlying the induction of the feeding sites, though there are some assumptions today regarding the effect of some larvae-secreted signal molecules as mediators in the development of huge cells and syncytium.

Upon the formation of the feeding site (huge cells or syncytium), the larvae molt three times. Development of both root-knot and cyst nematodes includes a series of morphological changes: they grow in size and change the body shape: from vermiform they become lemon-shaped or pear-shaped. Internal rearrangement also takes place: the intestine is reduced, and ovary with eggs develops in the body cavity. About one week after the last molting a female of the root-knot nematode starts to secrete gelatinous matrix where it lays eggs. Reproduction of cyst nematodes always requires males; after fertilization the eggs remain inside the female, and their number grows with age, gradually filling the entire body of the female. After death, a female turns into a cyst which constitutes a bag with eggs.

Role of nematode discharge in plant–nematode relationship

A clue to understanding of the signals that trigger a complex response of plants to invasion is the nematode biology. It was already mentioned that parasitic nematodes possess certain structures and organs enabling parasitizing on plants. These structures can secrete particular molecules involved in plant–nematode interaction. Nematodes possess the central nervous system and sense organs. Chemical sense organs, amphids, are at the sides of the head (in the subclass of aphasmidia), or on the lips (in the subclass of phasmidia) (Figure 1.7). Amphids are recesses in the cuticle from which large nerves come up. Chemoreceptor signals seem to be important for attraction of nematodes by a plant and for recognition of the appropriate places to attack the plant and begin feeding. Amphids are especially well developed in males, and, probably, they help the males in search of a female. Amphid secreta were found to contain some glycoproteins which can be involved in perception of signals from the environment. The tactility organs, tangoreceptors, are papilla-shaped or bristle-shaped. They are located mostly on the head part of the body (less often on the body itself), and in males they are also located near the cloaca. An important role is played by rectal glands, which in meloidogins secrete gelatinous mass (egg pouch) that protects the eggs from predators and dehydration. In addition, cuticle of the nematode, with its selective permeability, can be a source of the compounds perceived by the plant as signal molecules.

However, the most characteristic features distinguishing the phytoparasitic nematodes from other plant and soil-related nematodes are the presence of the special structures: the stylet and esophageal secretory glands which are necessary attributes for parasitizing (Figure 1.7). The stylet is a very thin structure 20–23 μ long (hollow in sedentary nematodes). The stylet lumen is very narrow, literally capillary, and it can deliver only liquid-feed to the nematode esophagus. The stylet is retracted due to elasticity of the esophagus tissues. The capillary lumen of the stylet ensures a sufficient suction force of the organ, which works like a syringe. The key organs used by the nematode for feeding on the cell contents of plant tissues are three secretory glands: the

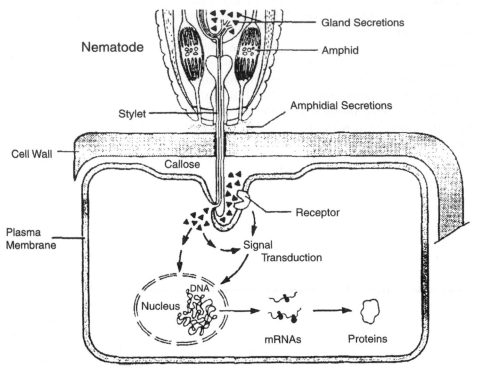

Figure 1.7.   Schematic model interactions of sedentary nematode with plant cell (Moroz VM, Hussey RS, 1996).

dorsal gland (DG) located in the spinal part of the esophagus, and two sub-ventral glands (SVG) at its sides. Each secretory gland is a single cell containing a large nucleus, the golgi system, the endoplasmic reticulum, secretory granules, and other inclusions typical of secretory cells. It has been recently found that the secreta of SVG and DG of sedentary nematodes can perform different functions: SVG secreta are important in nematode penetration and movement over the plant cells, whereas the DG secreta induce formation of the syncytium or huge cells. As the nematodes develop in plant tissues, activity of the enzymes being secreted changes. During feeding, DG secreta of sedentary endoparasitic nematodes enter, through the stylet, the unique formations called feeding tubes, then to cytoplasm of the feeding cells. These tubes develop within the plant cells from the compounds secreted by DG, and can work as a molecular sieve to limit entering from the modified plant cells of large feed molecules which can clog the stylet hole. Studying these formations using fluorescent inclusions showed that they pass feed particles sized within 20–40 kDa. Plasmalemma remains intact at the initial stages of the nematode feeding process. The nematode takes nutrients from the cytosol through small pores in the plasmalemma produced in the place of its contact with the stylet hole. A plug grows where the edge of the stylet enters the syncytium cell, from the protein products secreted by amphids, and the parasite entry place is "sealed up" (Figure 1.7). Investigation of the feeding processes of the cyst nematode *H. schachtii* showed that the secretory proteins released by amphids in intercellular space between the nematode head and cell wall fix (glue) the nematode head to the syncytium wall.

Selected Literature

Abad P, Favery B, Rosso M-N, et al. Root-knot nematode parasitism and host response: molecular basis of a sophisticated interaction. Molecular Plant Pathology 2003; 4(4):217–224.

Barr DJS. Evolution and kingdoms of organisms from the perspective of a mycologist. Mycologia 1992; 84:1–11.

Frasier RSS, ed. Recognition and Responses in Plant-Virus Interactions. Springer-Verlag: 1990.

Froissari R, Michalakis Y, Blanc S. Helper component – transcomplementation in the vector transmission of plant viruses. Phytopathology 2002; 92:576–579.

Fundamentals of Bacterial Plant Pathology. Academic Press; 1992.

Harrison BD. In "Active Defense in Plants". Ed. Wood RKS. Plenum Press; 1982. 193–210.

Hibbett DS, Donoghue MJ. Integrating phylogenetic analysis and classification in fungi. Mycologia 1998; 90:347–356.

Kohn LM. Developing new characters for fungal systematics: an experimental approach for determining the rank of resolution. Mycologia 1992; 84:139–153.

Lewis DH. Biol Rev 1973; 48:261–278.

Milne RG. Plant virus taxonomy and nomenclature: what's new? Revista de investigaciones Agropecularias 1990; 22:206–224.

Mink GI. Pollen and seed-transmitted viruses and viroids. Ann Rev Phytopathol 1993; 31:375–402.

Jung C, Wyss U. Appl Microbiol Biotechno 1999; 51:439–446.

Moroz VM, Hurrey RS. The Plant Cell 1996; 8:1735–1745.

Siddiqi MR. Tylenchida: Parasites of plants and insects. CAB common wealth. Institute of Parasitology. St. Albans, UK. 1986.

Symons RA. The intriguing viroids and virusoids: what is their information content and how did they evolve? Molec Plant-Microbe Interact 1991; 4:111–121.

Young JM, Takikawa Y, Gadau L, et al. Changing concepts in the taxonomy of plant pathogenic bacteria. Ann Rev Phytopathol 1992; 30:67–105.

Young JM, Sadder GS, et al. Names of plant pathogenic bacteria 1864–1995. Rev Plant Pathol 1996; 75:721–763.

Chapter 2

Structural basis of plant–pathogen interactions

J. Plotnikova, F. M. Ausubel

The strategies that biotrophic and necrotrophic pathogens employ to attack their plant hosts are different. Necrotrophs (from the Greek words *nek* =death and *trophy* =feeding) kill plant cells in order to obtain their nutrition, whereas biotrophs (from the Greek words *bios* =life and *trophy* =feeding) have a long symbiotic phase and even stimulate their hosts' metabolism at the initial stages of the infection process. Necrosis of plant tissue at the host–pathogen boundary region stimulates the growth of necrotrophs, but inhibits or suppresses that of biotrophs. Alterations of host-plant structure and ultrastructure under the influence of biotrophs and necrotrophs are different and are good markers of the type of infection strategy that the pathogens employ. Hemi-biotrophic and hemi-necrotrophic pathogens cause less dramatic alterations in their host-plants' ultrastructure than necrotrophic pathogens, but more than the biotrophic pathogens. Although there is a large diversity of plant pathogens, we will consider some of the well-studied members among the biotrophs and necrotrophs.

Alterations of plant and pathogen structures during infection process

Phytopathogenic viruses invade plant cells and propagate in their cytoplasm as biotrophic pathogens. Viral structures, invasion and propagation in plant cells are described in previous Chapter 1. Phytopathogenic bacteria can be either necrotrophs or biotrophs. Many necrotrophic bacteria cause soft rot, spots and blights, vascular wilt, cankers and scabs, whereas some biotrophic bacteria produce galls. The structure and taxonomy of phytopathogenic bacteria, as well as the disease symptoms they cause in their host plants are discussed in Chapter 1. The structural basis of plant–bacterial interactions will be described in this chapter. Taxonomy of phytopathogenic fungi is also described in Chapter 1. We will consider the typical biotrophic fungal pathogens, such as the rust fungi, the powdery mildew fungi as well as the necrotrophic fungi and bacteria in this chapter.

Biotrophic fungi can grow on plant surfaces like the powdery mildew fungi (Ascomycota) or inside host tissues like the rust and smut fungi (Basidiomycota). Biotrophic fungi typically invade only a few plant cells to produce nutrient-absorbing structures termed haustoria. Biotrophs cannot survive without a host plant. In contrast, fungal necrotrophs require a host plant only for a part of their life cycle and can grow in culture or survive on dead organic matter.

Fungal infection starts with the imbibition of spores, their attachment to plant surfaces and germination (Figures 2.1–2.3). Figure 2.1 shows the diagram of *Puccinia graminis* urediospore germination and infection structure differentiation. An appressorium is formed from the swollen apical portion of the germ tube delimited by septa from its subapical portion. Thin appressorial outgrowth becomes a penetration peg which gives rise to a substomatal vesicle. Growth of intercellular hyphae starts from the ends of the substomatal vesicle. The diagram shows that a germ tube wall is an extension of the inner layer of a urediospore wall. The thicker wall of an appressorium is due to the appearance of a new additional layer. The substomatal vesicle wall is an extension of the inner layer of appressorial wall. It is thus apparent that the walls of infection peg, substomatal vesicle and intercellular hyphae are not continuous with that of the germ tube.

Temperature shock or chemical induction at the appressorial stage can induce complete *P. graminis* infection structure formation consisting of sequentially an appressorium, infection peg, substomatal vesicle, infection hyphae and sometimes the first haustorial mother cell (HMC). During the differentiation of infection structures a nuclear transformation takes place (Figure 2.1). Initially, two nuclei move from a spore into the germ tube and later they migrate into the appressorium where mitosis occurs. A mature rust

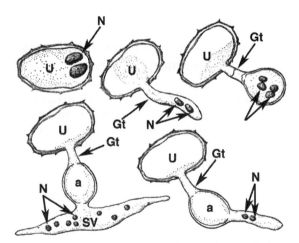

Figure 2.1.   Diagram of *P. graminis* urediospore germination and infection structure differentiation. Nuclear transformation occurs during this process whereby two urediospore nuclei migrate into the germ tube and later into the appressorium. Once inside the young appressorium, they undergo the first round of mitosis, and later during substomatal vesicle formation, the second round of mitosis. U – urediospore, Gt – germ tube, a – appressorium, SV – substomatal vesicle, N – nucleus (From Andreev, Plotnikova, 1989).

appressorium contains four nuclei. A second mitosis occurs during substomatal vesicle formation. A substomatal vesicle may contain 8 or 16 nuclei. One to eight infection hyphae can grow from the ends of the substomatal vesicle. Two nuclei move to each newly formed infection hypha.

A typical feature of filamentous fungi is their apical growth: their hyphal tips contain a population of numerous membrane-bound vesicles that contribute membrane and cell-wall precursors to the fast growing apices. The cytoplasmic volume of growing hyphae does not increase proportionally to their whole volume and concentrates only in their apical portion. Vacuoles or one large vacuole develop in the vacated portion of the fungus as its cytoplasm moves into newly formed apices. The vacated portions of fungal hyphae are outlined by a plasmalemma and a tonoplast. The cytoplasm of both germ tubes and the infection hyphae of various pathogenic and saprophytic fungi contains nuclei, mitochondria with plate-like cristae, tubular and vesicular endoplasmic reticulum (ER), ribosomes, microbodies, lipid droplets, vacuoles, microtubules and microfilaments. Microtubules support a framework of rigid cytoskeleton. Actin microfilaments create a much more extensive system stabilized by their association with microtubules. Membrane-bound vesicles and various types of vacuoles are common in intercellular hyphae. Lipid droplets are often associated with microbodies in rust urediospores. These microbodies might function as glyoxysomes and participate in lipid breakdown during spore germination.

Scanning electron micrographs (Figures 2.2A and B) show the germination and further development of representative biotrophic pathogens on plant leaf epidermis. Urediospore of *P. striiformis*, causal agent of yellow rust, germinates on the wheat leaf surface; germ tube grows perpendicularly to the longitudinal axis of the wheat leaf and penetrates into stoma (arrow) without forming an appressorium (Figure 2.2A). The directional growth of rust fungi towards the stomata may be a thigmotropic response to epicuticular

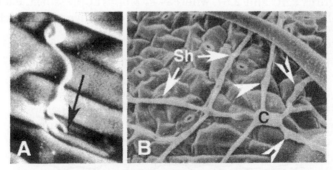

Figure 2.2. Scaning electron micrographs of the germination and further development of biotrophic pathogens on plant leaf epidermis.
(A) *Puccinia striiformis* germ tube penetration into wheat leaf stoma (arrow) without appressorium formation (From Andreev, Plotnikova, 1989);(B) germination and a colony formation of *Erysiphe orontii* on *Arabidopsis thaliana* leaf 5 dpi. Superficial hyphae (Sh) give rise to appressoria with nipple-shaped protuberances (arrows heads) (Plotnikova, unpublished).

wax platelet distribution and plant surface topography. If the wax is removed with organic solvents, germ tubes will not adhere to the surface. Germ tubes of rust fungi grow predominantly perpendicular to the longitudinal axis of the plant leaf and at right angles to the parallel ridges of the host cuticle (Figures 2.2A and 2.4). Figure 2.2B shows a conidium germination and a colony formation of the powdery mildew pathogen *E. orontii* on *A. thaliana* leaf 5 dpi. Each conidium produces 3–5 germ tubes which give rise to super-ficial hyphae and appressoria with nipple-shaped protuberances.

Many leaf pathogens, both bio- and necrotrophs, invade their hosts by piercing the outer walls of the epidermal cells. For example, the biotrophs *Erysiphe* sp. and the necrotrophs *Botrytis* sp. both penetrate directly through the outer epidermal wall rather than via stomata (Figures 2.3 and 2.5A). Figure 2.3 is a diagram detailing *E. orontii* development on *Arabidopsis* leaf epidermis. *E. orontii* conidium produces a germ tube that is attached to the surface of the upper leaf epidermis. Appressorium is formed from the swollen apical portion of a germ tube. The formation of the infection peg to pierce the thick outer epidermal wall is initiated from the basal site of an appressorium. *E. orontii* haustorium is formed inside the epidermal cell volume from the apex of the infection peg.

Figure 2.4 is a scanning electron micrographs of *P. graminis* infection structure differentiation *in vitro* and *in vivo*. Urediospore germinated *in vitro* produces a short germ tube and an appressorium (Figures 2.2A and B). Complete infection structures differentiated *in vitro* consist of germ tube, appressorium, substomatal vesicle and infection hypha (Figure 2.4B). *P. graminis* appressoria can be formed *in vitro* after heat shock or other stimuli (Figure 2.2A and B) and *in vivo* after fungal contact with stomata (Figure 2.4C), or with the grooves above anticlinal wall junctions. If plant cells lose their turgor or die, no appressoria develop on their surface. The spores, germ tubes and appressoria stick to the plant's cell surface by their mucilaginous sheath. *P. graminis* appressoria are formed primarily above stomata (Figures 2.4C and 2.6), whereas *E. orontii* appressoria, each having a small lateral protuberance, are usually located above the periclinal walls,

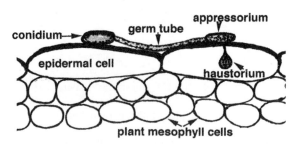

Figure 2.3.   Diagram detailing the powdery mildew pathogen *E. orontii* development on Arabidopsis leaf epidermis. *E. orontii* conidium produces a germ tube that is attached to the surface of the upper leaf epidermis. (Plotnikova, unpublished).

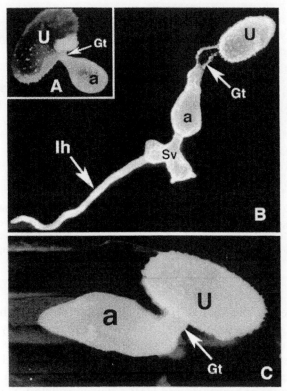

Figure 2.4.   Scanning electron micrographs of *P. graminis* infection structure differentiation *in vitro* and *in vivo*. (From Andreev, Plotnikova, 1989.)
(A) Urediospore germinated *in vitro* produces a short germ tube and an appressorium; (B) Complete infection structure formation *in vitro*; (C) Appressorium above stoma on the wheat leaf. U – urediospore, Gt – germ tube, a – appressorium, Sv – sub-stomatal vesicle, Ih – infection hypha.

and a thin penetration peg emerges from below the protuberance and pierces the outer wall of an epidermal cell (Figure 2.3). Rust and powdery mildew appressoria can be formed in response to purely thigmotropic stimuli even above the scratches on artificial membranes. Appressoria can be ellipsoidal in shape like those of *Puccinia* sp. and *Erysiphe* sp. (Figures 2.2 and 2.4).

The conidia of the fungal necrotroph *Botrytis cinerea* form extensive germ tubes that bear large, apical, digitate appressoria or holdfasts (Figure 2.5A). The appressoria are formed commonly by the repeated branching of hyphae that come in contact with the host surface. In the case of *Botrytis*, appressorial outgrowths penetrate into host cells by piercing the outer epidermal cell wall. *Botrytis* overwintering organs sclerotia have pseudoparenchymatous structure consisting of closely packed mycelial threads. The several outer layers of dark-walled cells give the external black colour to the sclerotium; the remaining interior is made up of hyaline cells. Sclerotia give rise to bunches of conidiophores bearing numerous conidia (Figure 2.5B).

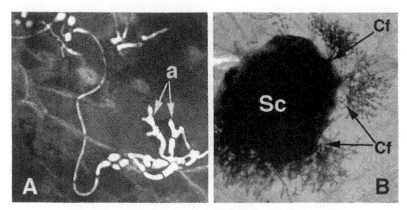

Figure 2.5. *Botrytis cinerea* spore and sclerotium germination (Plotnikova, unpublished).
(A) *B. cinerea* spore germination on *Arabidopsis* leaf 2 dpi. Germinating spores give rise to superficial, intra- and intercellular hyphae. Superficial hyphae produce large branched appressoria (a); (B) Germinating sclerotium (Sc) produces bunches of conidiophores (Cf) bearing numerous spores.

When intercellular hyphae of rusts establish contact with the primary wall of a mesophyll cell, the hyphal apical portion is delimited by a septum and forms a terminal HMC. Fungal nuclei, mitochondria, ribosomes, ER and glycogen bodies move from the HMC into the haustorial body, and only the fungal plasmalemma, tonoplast and a large vacuole remain in the HMC. The plasma membranes of haustoria and HMCs stay connected. Young *P. graminis*, *Uromyces fabae*, and *E. orontii* haustoria have an ellipsoidal shape, but when they mature, they become spherical. *E. graminis* haustoria are highly distinctive, consisting of a rounded body with finger-like projections in a wheat epidermal cell. Haustorial cytoplasm contains the same basic components found in other stages of the fungal life cycle. Glycogen particles and lipid droplets contain storage products. They vary in abundance and distribution in various fungi. The volume of lipid droplets increases in aged haustoria. Whorled membrane aggregates usually appear at the periphery of mature and older haustoria. Storage lipids and glycogen are accumulated during spore formation and are utilized during the spore germination.

Biotrophic pathogen propagules form profuse superficial (powdery mildew) or intercellular (rust, smut) mycelium, and many haustoria. Figure 2.6 is a diagram of *P. graminis* development inside a wheat leaf. A urediospore that lands on the wheat leaf surface produces a germ tube and an appressorium. The outgrowth of the appressorium pushes aside the closed guard cells and gives rise to a substomatal vesicle in the leaf substomatal cavity. Two to four intercellular hyphae originate from the ends of a substomatal vesicle. When the apex of the intercellular hyphae contacts a plant mesophyll cell, it stops growing, attaches itself to the plant wall and forms HMC. The basal outgrowth of the rust's HMC can pierce the thin host primary mesophyll cell wall by the enzymatic degradation of host-wall matrix and mechanical pressure.

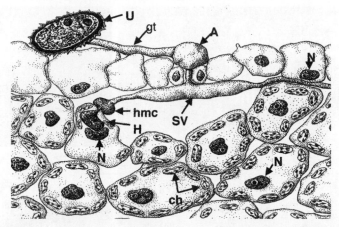

Figure 2.6. Diagram of *P. graminis* development inside a wheat leaf. A urediospore (U) that has landed on the wheat leaf surface produces a germ tube (gt), and an appressorium (A). The outgrowth of the appressorium pushes aside the closed guard cells and gives rise to a substomatal vesicle (SV) in the leaf substomatal cavity. Two to four intercellular hyphae originate from the ends of a substomatal vesicle. When the apex of an intercellular hyphae contacts a plant mesophyll cell, it stops growing, attaches itself to the plant wall, and forms a haustorial mother cell (hmc). Infection peg protruding from the bottom portion of the hmc penetrates through plant cell wall and gives rise to a haustorium (H). Host nucleus (N) moves towards the haustorium and partly surrounds it. Chloroplasts (Ch) and other plant organelles which are normally located at the cell periphery in uninfected plant cells move towards a haustorium. Host cytoplasm surrounds the whole haustorial body. (From Andreev, Plotnikova, 1989.)

From there, it then grows into the host-cell lumen and forms a tubular haustorial neck. Apical portion of the haustorial neck expands to form a haustorial body. The host nucleus, chloroplasts and other plant organelles which are normally located at the cell periphery in uninfected plant cells move towards haustoria. Haustoria are localized in the periplasmic space between the cell wall and the invaginated portion of the plasmalemma (Figures 2.6 and 2.7B).

Figure 2.7 shows scanning electron microscopical images of cross-fractured wheat leaves infected with *P. graminis*. The wheat mesophyll cells are separated from each other by rust intercellular mycelium (Figure 2.7A). The HMCs are seen attached to plant walls. Figure 2.7B shows the interior of an infected wheat mesophyll cell and the surrounding intercellular space that contains intercellular mycelium. The HMC is separated from the rest of the hyphae by a septum (arrow). It triggers the growth of a lateral hyphal branch. The HMC is shrunken due to the migration of its cytoplasm into the growing haustorium, which is seen in the periplasmic space between the plant mesophyll wall and plasmalemma. Figure 2.7C shows *P. graminis* haustorium inside a wheat mesophyll cell. Spherical haustorial body and thin cylindrical haustorial neck are surrounded by the cytoplasm containing chloroplasts and thin plant mesophyll wall. A haustorial body and neck are seen inside the wheat epidermal cell (Figure 2.7D). A haustorial band is seen in the middle portion of the neck.

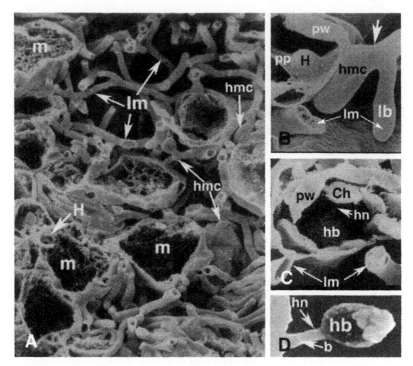

Figure 2.7. *P. graminis* intercellular mycelium and haustoria.
(A) Cross section of a wheat leaf infected with *P. graminis*. Wheat mesophyll cells (m) are separated from each other by rust intercellular mycelium (Im). Haustorial mother cells (hmc) are attached to plant walls. The cross section through the collar of a haustorium (H) is seen from inside the plant cell volume (Plotnikova, unpublished); (B) Cross section of an infected wheat mesophyll cell and the surrounding intercellular space that contains intercellular mycelium (Im). The HMC is the apical portion of an intercellular hyphae, which is enlarged and separated from the rest of the hyphae by a septum (arrow). The HMC formation triggers the growth of a lateral hyphal branch (lb). The HMC is shrunken because some hmc cytoplasm had already migrated into the growing haustorium (H), which is seen in the periplasmic space between the plant mesophyll wall (pw) and plant plasmalemma (pp) (From Andreev, Plotnikova, 1989); (C) *P. graminis* haustorium inside a wheat mesophyll cell. Spherical haustorial body (hb) and thin cylindrical haustorial neck (hn) are surrounded by the cytoplasm containing chloroplasts (Ch) and thin plant mesophyll wall (pw) (From Andreev, Plotnikova, 1989); (D) A haustorial body (hb) and neck (hn) are seen inside the wheat epidermal cell. A haustorial band (b) is located in the middle portion of the neck. Freeze-fracturing (From Andreev, Plotnikova, 1989).

Fungi propagate primarily by spores. There is great variation in the types and functions of fungal spores. Some pathogens, such as *P. graminis*, produce several types of spores (Figures 2.8 and 2.9). Massive propagation of this pathogen occurs by urediospores, which are formed in structures termed urediopustules. *P. graminis* can produce up to seven generations of urediospores during its vegetative period. When a plant senesces, teliospores are formed at the periphery of urediopustules. In the following spring, the over-wintered teliospores germinate and form structures termed basidia, which produce basidiospores. Germinating basidiospores give rise to heterothallic picnidia,

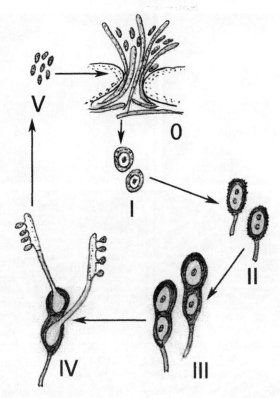

Figure 2.8. Various types of *P. graminis* spores: 0 – a pycnium with flexious hyphae and pycnidiospores; I – aeciospores, II – urediospores, III – teliospores, IV – germinating teliospores producing basidia with basidiospores, V – basidiospores (After Andreev, Plotnikova, 1989).

which form picnidiospores. The fusion of + and − picnidiospores results in the formation of dikaryotic mycelium and aeciospores. Thus, *P. graminis*, the causal agent of stem rust, can form five types of spores: uredio-, telio-, basidio-, picnio- and aeciospores (Figure 2.8).

A new generation of fungal spores is formed from various types of conidiophores (Figure 2.9B). Powdery mildews do not form pseudoparenchyma, its superficial mycelium produces numerous conidiophores which grow vertically up from the hyphae. Mature short-lived conidia detach from conidiophore apices (Figure 2.9C). Scanning electron micrographs detailing the sporulation of *B. cinerea*, P. graminis and *E. orontii* are shown in Figure 2.9. Large number of pigmented conidiophores bearing numerous one-celled, ovoid conidia are produced on *Botrytis* infected plant tissues (Figure 2.9A). Rust uredio- and teliospores are formed on the vertical outgrowths of pseudoparenchyma. Figure 2.9B shows a small urediopustule of *P. graminis*. Urediospores appear through the epidermal cracks of the leaf surface. Young urediospores have smooth outer surface, while the mature urediospores are echinulate, bearing on their surface 200–300 conical spines. These spines are initially formed inside the walls of young urediospores as small bumps, which increase in size

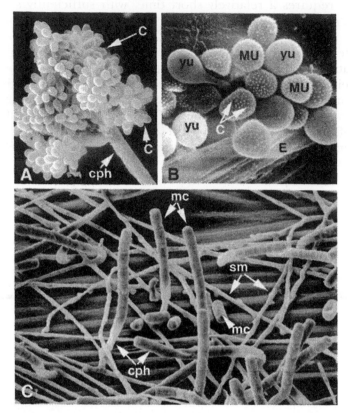

Figure 2.9. Scanning electron micrographs detailing the sporulation of *B. cinerea, P. graminis* and *E. orontii* (Plotnikova, unpublished).

(A) *B. cinerea* conidiophore (cph) bearing conidia (c); (B) Small urediopustule of *P. graminis*. Urediospores appear inside the epidermal cracks (E) of the leaf surface. Young urediospores (yu) have smooth outer surface, mature urediospores (MU) are echinulate, bearing on their surface 200–300 conical spines (C). These spines were initially formed inside the wall of young urediospores as small bumps, which increased in size as the primary wall dissolved and the material for the secondary wall centripetal to the primary wall was being deposited (Plotnikova, unpublished); (C) *E. orontii* growth and sporulation on *Arabidopsis* (Ws-O ecotype) stem. Superficial mycelium (sm) gives rise to vertical conidiophores (cph), the apical portion of which mature conidia (mc) arise and detached from. They can drop on the host - plant surface, or, are spread by air to neighboring leaves or plants (From Plotnikova et al, 1998).

as the primary wall dissolves and the material for the secondary wall centripetal to the primary wall is being deposited.

E. orontii growth and sporulation on *Arabidopsis* Wassilevskaya (Ws-O) ecotype) stem is seen in Figure 2.9C. Superficial mycelium gives rise to vertical conidiophores, from the apical portion of which mature conidia arise and detach. They can drop on the host-plant surface, or are spread by air to neighboring leaves or plants.

Structural features of the plant cell response to infections

Plants are resistant to most phytopathogenic fungi, and only a few have acquired the capability to invade the tissue of a particular host plant and to suppress its resistance. The first lines of plant defense against phytopathogenic fungi are the external wax layer, cuticle and polysaccharide-rich cell wall. Pathogens breach these barriers to penetrate into the plant's intercellular space. The effectiveness of the host plant's defense response is determined by both its speed and magnitude. The triggering of host defense is preceded by the recognition of the pathogen by the host cells.

How plants recognize pathogens and activate defense is still not completely clear. All multicellular organisms have developed mechanisms to ward off potential microbial invaders. In their natural environment, plants encounter a vast array of pathogens, such as bacteria, fungi and viruses. Successful development of infection or its inhibition depends greatly on the events taking place at the host–pathogen interface during the establishment of infection. An active local plant cell response at the host–pathogen interface can result in resistance to a pathogen. Slow and passive response can lead to fungal proliferation. Host expression of pathogen resistance genes in the very proximity of growing fungal hyphal tips can stop or inhibit fungal development at the interface rather than at a distance from the pathogen.

Biotrophic infections

Obligate biotrophic pathogens, which include viruses, viroids, some fastidious bacteria, nematodes, protozoa, and fungi and oomycetes that cause downy mildews, powdery mildews, smuts and the rusts, establish long-term relationships with the living cells of their hosts, resulting in a continuous flow of plant metabolites to the pathogen. At the initial stages of infection process, biotrophs stimulate plant metabolic processes such as photosynthesis, respiration, nutrient absorption and create a sink to the infection site. These pathogens keep the host alive as a long-term source of food up to the time of mass sporulation.

In compatible biotrophic combinations, fungal spore germination and infection structure formation do not cause any degenerative changes in the host's ultrastructure. First visible under the transmission electron microscope, alterations of the host cell wall and cytoplasm occur at the site of the rust HMC attachment to the host cell wall and at the haustorial neck intrusion into the plant cell lumen. Similar host cytoplasm alterations also occur at the sites of *Erysiphe* sp. appressorium attachment and infection peg penetration through the host cell wall (Figure 2.3). Around the infection peg, the plant plasmalemma invaginates and produces numerous vesicles, whose appearance and behavior in the host cytoplasm at the sites of attempted penetration are controlled genetically. Localization of some proteins to fungal entry sites was studied recently. It was shown that the PENETRATION2 (PEN2) protein, which is a glycosyl hydrolase, localizes to peroxisomes and acts as a component of a pathogen-inducible pre-invasion resistance mechanism.

It was hypothesized that at this stage, powdery mildews block the soluble NSF (N-ethylmaleimide-sensitive factor) attachment protein receptor (SNARE) domain containing syntaxin, which is found in vesicles at fungal entry sites as part of a host response. SNARE proteins are targeted to vesicle trafficking in eukaryotes, leading to the proposed vesicle-associated defense response.

Young metabolically active haustoria contain nuclei, well-developed mitochondria, ribosomes, ER, small vacuoles, whereas older haustoria contain fewer mitochondria, ribosomes and ER cisternae, enlarged vacuoles, lipid globules and peripheral membrane whorls. Although haustoria of biotrophic fungi are usually referred to as intracellular structures, they do not penetrate into the host cytoplasm. Instead, during penetration, they invaginate the plasmalemma of the host cell and gradually produce a local enlargement of the host periplasmic space. This haustorium–host cytoplasm interface was named the extra-haustorial matrix (EHM) by Bushnell (1972). The EHM is surrounded by an invaginated portion of host plasmalemma called the extra-haustorial membrane (EM) on one side and the outer surface of the haustorial wall on the other. The haustorium together with the EHM and the EM is referred to as the haustorial apparatus. Host plasmalemma invaginated by a haustorium differs from non-invaginated portion of the plasmalemma by its higher electron density and increased thickness, increased polysaccharide staining, and higher number of multienzyme complexes participating in the synthesis of plant wall microfibrils. Whereas both the fungal haustorial membrane and non-invaginated plant membranes have ATPase (adenosine triphosphatase) activity for driving nutrient uptake, the EM lacks ATPase activity and thus the ability to control the movement of nutrients across this membrane.

The development of a haustorial body is accompanied by a proliferation of plant ER around haustorium. Furthermore, the ER cisternae are arranged parallel to the EM and in a polar fashion: cisternae closest to the haustorium are agranular and those more distant are granular with multiple ribosomes attached to their surface. Uninfected mesophyll cells usually contain granular ER. The content of ER cisternae around a haustorium has similar electron density to the EHM. This observation suggests that ER-derived substances are being released into the EHM. Another characteristic of fungal biotrophic infections is a close association between the fungal haustorium and the host nucleus. This association cannot even be broken during the procedure of haustorial isolation such that many isolated haustoria remain still attached to the host nucleus. The density of interchromatin regions in the host nuclei associated with the haustorium increases and is coincident with an increase in nuclear RNA. Haustorium invasion also causes reorganization of the host cell. An uninfected mesophyll cell contains a large central vacuole surrounded by a thin peripheral layer of cytoplasm. In contrast, in an infected cell, the cytoplasm, which contains the ER, mitochondria, chloroplasts and other organelles, moves towards the haustorium and surrounds it (Figure 2.10). The host organelles attached to the EM are also altered; the chloroplast granal structure is disorganized, and there is a reduction in the number of lamellae. These observations support the idea that host photosynthates are being released into the EHM. During the later stages of haustorium invasion, alterations in the endomembrane system become more pronounced and result in plant organelle degradation. For example, the volume of lipid globules in the chloroplasts gradually increases. Interestingly, haustorium-induced changes in the ultrastructure of host organelles are similar to those exhibited during the normal process of senescence, suggesting that the organelles around haustoria are undergoing premature senescence. During the process of haustorial maturation, the tightly packed membrane complexes in the host cytoplasm surrounding the haustorium, consisting of parallel-oriented agranular tubular ER cisternae along with an association of ER and Golgi membranes, are found around rust and downy mildew haustoria but are absent in uninfected plant cells. Figure 2.10 shows the transmission electron micrographs of wheat organelle alteration around *P. graminis* haustoria.

Figure 2.10A shows the section through haustorial neck and body inside a wheat mesophyll cell as well as the portion of *P. graminis* hmc. Host cytoplasm containing ER, polyribosomes, Golgi apparatus and microbodies is concentrated around the haustorium. Figure 2.10B shows the electron-opaque haustorial body surrounded by a thin electron-lucent EHM and electron-opaque EM. A chloroplast and a mitochondrium are tightly attached to EM. Portion of the chloroplast adjacent to the EHM contains many vacuoles.

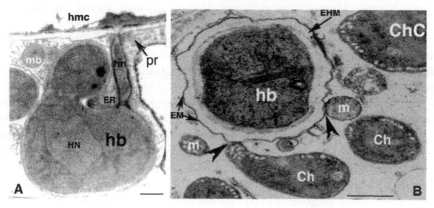

Figure 2.10. Transmission electron micrographs of wheat organelle alteration around P.graminis haustoria (From Andreev, Plotnikova, 1989).

(A) Median section through haustorial neck (hn) and body (hb) with electron-dense content inside a wheat mesophyll cell as well as the portion of P. graminis haustorial mother cell (hmc). Host cell organelles: chloroplasts (Ch), ER, polyribosomes (pr), and microbodies (mb) are concentrated around the haustorium. Bar =1 µm; (B) Cross section through *P. graminis* haustorial body and a portion of host cytoplasm. The electron-opaque haustorial body is surrounded by a thin electron-lucent extrahaustorial matrix (EHM) and electron-opaque extrahaustorial membrane (EM). A chloroplast (Ch) and a mitochondrium (m) are tightly attached to extrahaustorial membrane (EM) (arrows heads). Vacuolization along the chloroplast's periphery, especially portion adjacent to the EHM, is evident. Bar =1 µm.

The formation of haustoria is one of the hallmarks of the interactions between obligate biotrophic fungi and their host plants, and plays a major role in nutrient uptake of the pathogen. Normal haustoria development is necessary for establishing a biotrophic relationship, and signaling between haustoria and host is one of the most important determinants of biotrophic parasitism. For example, recent studies of the pathogen *Uromyces fabae* have demonstrated that the fungal Rust Transferred Protein 1 (*Uf*-RTP1p) and its homolog in *U. striatus*, *Us*-RTP1p, were both detected not only in the EHM, but also inside infected plant cells. These proteins were transported from the haustoria into infected plant cells and localized in the host cytoplasm and nuclei. The concentration of *Uf*-RTP1p in the host cytoplasm and nucleus increased with haustorium maturation. The localization of RTP1p suggests that it might play an important role in the maintenance of the biotrophic interaction.

Additionally, the hexose transporter HXT1 and a putative amino acid transporter were found to be highly expressed in rust and powdery mildew haustoria, but were not detected in earlier structures such as infection structures and intercellular mycelium. The HXT1 is localized in the haustorial plasma membrane and has substrate specificity for D-glucose and D-fructose. Its localization and biochemical requirements support the idea that the haustorial apparatus separated from the plant apoplast serves as a sink for host-derived nutrients and thus enables fungal biotrophs to compete with natural sink organs in the host plant.

Callose deposition is a non-specific response of higher plants to injury and to other various types of infection. The polysaccharide callose is typically electron-lucent. Most electron-lucent collars around young metabolically active haustoria contain this polysaccharide as a major component. Callose-containing papillae, collars and encasements are formed on the surface of an invaginated host plasmalemma as part of a non-specific host response to pathogen invasion. Attempted penetration of leaf surface results in the appearance of callose papilla surrounding the point of penetration. Papillae are fluorescent under ultraviolet light following 1% aniline blue staining. Figure 2.11A shows papilla formation by *Puccinia thlaspeos* on *Arabidopsis glauca* leaf. *P. thlaspeos* basidiospore forms a germ tube and a small appressorium. Large papilla is seen around the point of fungal penetration (dashed arrow). The wide band of callose deposition is seen on the inner side of the cell wall as part of the host's response to invasion by *Gymnosporangium juniperi-virginianae* (Figure 2.11B). Its aeciospores germinate on *Arabidopsis* Col-O leaf and produce appressoria. Invasion by appressorial infection peg induces papilla formation.

In compatible combinations, many sporelings enter plant organs and give rise to fungal colonies and a new generation of spores. In incompatible combinations, most sporelings fail to enter plant cells. In such cases, infection peg development can be stopped by an active host cell defense response involving the secretion of callose, Si-containing compounds, lignin and cell wall precursors. Incompatible combinations can result in papilla formation

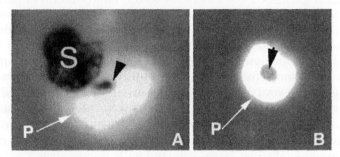

Figure 2.11. Papilla formation by rust fungi. Papilla are fluorescent under ultraviolet light following 1% aniline blue staining. Attempted penetration of leaf surface results in the appearance of callose papilla surrounding the point of penetration (Plotnikova, unpublished).
(A) *P. thlaspeos* basidiospore (S) with a short germ tube and a small appressorium on *A. glauca* leaf surface. Large papilla (P) is seen around the point of penetration (arrow head); (B) Aniline blue staining shows the wide band of callose deposition on the inner side of the cell wall in response to invasion by *G. juniperi-virginianae*. Its aeciospores germinate on *Arabidopsis* (Col-O ecotype) leaf and produce appressoria. Invasion by appressorial infection peg (arrow head) induces papilla (P) formation.

directly beneath penetration attempts (Figures 2.11A and B). In incompatible combinations, a very small percent of the attempted penetration sites showed a haustorium formation. Of those, most of the penetrated cells underwent a hypersensitive-response (HR)-like cell death, which can be visualized as whole-cell autofluorescence. Callose-containing deposits encase small haustoria if they are formed. Later, a network of cellulose microfibrils appears around infection pegs and/or haustoria. This newly synthesized wall material surrounding pegs and/or haustoria prevents metabolite exchange between plant and pathogen cells, and usually appears only around low metabolically active and necrotic haustoria.

Attempts to visualize any plasmodesmatal connections between the fungal haustorium and host cytoplasm have failed. The absence of plant–pathogen symplastic connection suggests that the bulk of host solutes absorbed by the pathogen cross the EHM via apoplastic transport. The tight association of the host and fungal membranes along the neckband (or the haustorial collar) seems to prevent the release of host nutrients from the extrahaustorial matrix into host apoplast and thus force the haustorium to rely solely on the uptake of host substances through the EM. In other words, the formation of the neckband most likely leads the haustorium to deploy specialized membrane transport proteins in the EM to regulate the flow of nutrients into and out of haustoria.

Intercellular mycelium obtains at least some nutrients directly from the host cells without passage through the haustorium, but the flow of sugars from host plants to the rust and powdery mildew pathogens occurs largely through the haustorial apparatus. As a consequence, the fungal gene expression involving nutrient uptake is largely specific to haustoria. For example, *U. fabae* haustoria isolated from broad bean leaves differentially expressed 31 *in planta* induced genes (PIGs) that were not expressed in germ tubes, appressoria, substomatal vesicles or HMC. Sequence analysis of PIG cDNAs revealed similarities to genes encoding proteins involved in amino acid transport, thiamine biosynthesis, short-chain dehydrogenases, metallothioneins, cytochrome P-450 monooxygenases and peptidyl–prolyl isomerases. Powdery mildews, rusts and viral infections result in a general increased levels of apoplastic monosaccharides due to the activity of cell wall invertases, which are enzymes that catalyze the cleavage of sucrose to glucose and fructose. The pathogen-specific induction of cell wall invertases has been reported in tobacco (*Nicotiana tabacum*) leaves infected with potato virus Y as well as the induction of the cell wall invertase Atfruct1 in *A. thaliana* leaves infected with white blister rust. Powdery mildew infection also results in an increase in the expression of the *Arabidopsis* sugar transport protein 4 (*At*STP4) in infected leaves. Reporter lines containing *At*STP4 promoter fusions to β-*glucuronidase* (GUS) gene showed promoter activity mainly in sink tissues such as roots and not in leaves. However, after infection with powdery mildew, promoter activity was induced in mature infected leaves and increased over time during the 6-day experiment. This changing expression pattern suggests that the pathogen acts as an additional sink, competing with host sinks resulting in a considerable modification of assimilate redistribution within the host plant.

Necrotrophic infections

The largest group of plant pathogens is the necrotrophic fungi and bacteria. Necrotrophs primarily penetrate through small wounds or cracks in the cuticle or enter through the stomata. They require more nutrients during propagule germination and penetration in comparison to biotrophic pathogens. The required supplemental nutrients are obtained by leaching them from surrounding host tissues. Successful necrotrophic infections are therefore greatly dependent on the condition of the infected organ. For example, senescing organs are more susceptible than young or mature undamaged ones. As a consequence, the susceptibility of plant organs (leaves, stems, petals, sepals, siliques, transmission tissues, pollen grains, stigmas) to necrotrophic infections can vary greatly from organ to organ within a single host plant.

In order to leach host nutrients, necrotrophs increase plant membrane permeability and produce copious amounts of active exoenzymes to breach host defenses. Unlike biotrophic pathogens, necrotrophic pathogens kill host cells after cell penetration and, in some cases, prior to penetration. How necrotrophs breach the cell wall barrier is of great interest. Because necrotrophs generally have low cellulase activity, the plant cell wall cellulose microfibrils remain intact. Instead, pectins in the plant wall appear to be an early target of fungal digestion, and as a consequence, numerous studies on necrotrophic pathogens have focused on the structural features of pectins. Pectolytic enzymes degrade pectins of middle lamellae, primary and secondary walls and create access for other enzymes like cellulases, proteases and phospholipases.

Fungal necrotrophic infections. Fungal necrotrophic pathogens usually survive on plant debris and cause infection when conditions are favorable for the pathogen but not favorable for the plant. While some biotrophic pathogens do not germinate if spore concentration in the inoculum is very high due to the presence of self-inhibitors, necrotrophic pathogens germinate well and penetrate into the plant tissues faster at higher spore concentrations. Concentration of *Botrytis* conidia in an infection drop has synergistic effect on lesion formation. Thus, the number of disease lesions on the broad bean is directly proportional to the conidial concentration in the inoculum, and lower conidial concentrations are correlated with small non-spreading lesions while higher concentrations are correlated with large fast-spreading lesions.

Our understanding of the mechanisms of general and specific virulence of necrotrophs lags behind our understanding of virulence in other groups of pathogens. Necrotrophic fungi have a diverse repertoire of virulence mechanisms, which include the detoxification of plant anti-microbial compounds (phytoalexins and preformed compounds) and the secretion of cell wall degrading enzymes, toxins, peptides and small proteins. Secreted enzymes as well as secondary fungal metabolites, such as fungal toxins, are particularly important to necrotrophic fungal pathogens as both general and specific agents of virulence. This is reflected in the large number of secondary metabolite genes present in the genome sequences of necrotrophic fungi such as *Botrytis, Fusarium, Cochliobolus* and *Magnaporthe.*

A typical facultative necrotrophic fungus, *B. cinerea* is the causal agent of gray mold in many plant species. In 1888, Marshall Ward demonstrated that host-plant cells collapsed prior to *B. cinerea* penetration. We now know that toxic fungal compounds play an essential role in *B. cinerea* pathogenesis, since host cells have to be killed before a parasitic relationship can be established. In particular, *B. cinerea* causes plasmolysis of and alterations in membrane structures in *Arabidopsis* cells around infection loci just beyond the reach of the growing hyphae. Pathogen-induced cell plasmolysis can also occur inside the leaf when the hyphae are still on the leaf surface. Local plasmolysis, changes in host membrane structures and later maceration and necrosis of invaded cells are particularly evident in *Arabidopsis* cells marked with green fluorescent protein (GFP)-tagged ER, mitochondria and cytoplasm. Alterations of *Arabidopsis* membrane structures in front of growing *B. cinerea* hyphae occur before its penetration inside the host-plant tissues. These marker lines clearly show that *B. cinerea* hypha invasion into the intercellular space leads to the degradation of plant plasmalemma, ER, mitochondria, chloroplasts, nuclei and cell walls. Figure 2.12 shows initial and advanced *B. cinerea* colonization of *Arabidopsis* LL-O leaves. Fungal pectolytic enzymes cause plasmolysis of surrounding *Arabidopsis* mesophyll cells: plant cytoplasm is separated from the wall and is concentrated into the center of the cell volume.

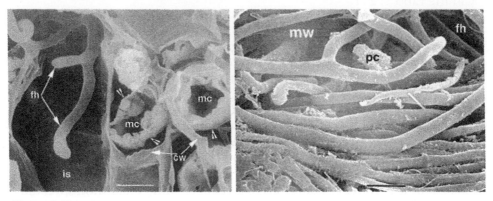

Figure 2.12.   Scanning electron micrographs of initial and advanced *B. cinerea* colonization of *Arabidopsis* (LL-O ecotype) leaves (Plotnikova, unpublished).
(A) Penetration of fungal hyphae (fh) into *Arabidopsis* intercellular space (is) *B. cinerea* hyphae degrade middle lamellae and increase greatly the intercellular space. Fungal pectolytic enzymes cause plasmolysis (arrowheads) of surrounding mesophyll cells (mc): plant cytoplasm is separated from the cell wall (cw) and concentrates into the center of the cell. Bar =10 µm; (B) Advanced *B. cinerea* colonization of *Arabidopsis* leaves results in maceration and degradation of plant cells. Fungal hyphae (fh) grow rapidly utilizing the plant nutrients, only mesophyll walls (mw) and remnants of plant cytoplasm (pc) are seen in this infection locus. Bar =10 µm.

Necrotrophic pathogens usually secrete cell wall degrading enzymes (CWDEs) during the early stages of infection. Because the penetrating hyphae cause a swelling of the epidermal outer cell wall, it is thought that CWDE production aids the fungus in the penetration of the epidermal outer cell wall with its wax and cuticle covering. Secretion of pectolytic enzymes by necrotrophic pathogens has a profound effect on infection ingress and spreading. They diffuse in advance of the pathogen and cause dissolution of the middle lamellae and thus the separation of plant cells from one another (Figures 2.12 and 2.13). Mesophyll cells of *Arabidopsis* leaves are macerated around growing *B. cinerea* hyphae (Figure 2.13).

The activity level of pectinases found in *B. cinerea* germinating spores is similar to that found in *Aspergillus niger* germination spores. *A. niger* is one of the highest fungal pectinase producers, and a well-known source of pectolytic enzymes used in biotechnology. It is notable that the pectinase activity of both *Botrytis* and *Aspergillus* germinating spores is much higher than that of their mycelia. Among the *B. cinerea* CWDE, endo-PG with basic isoelectric points (pI) is one of the first enzymes secreted. Activity of this enzyme denoted as PG-A is predominantly detected in the outer zone of the infected area while other endo-PGs are only detected in the central zone of the lesion. Because endo-PGs exhibit both macerating and killing activity, plant cell death may not be a mere consequence of maceration. *B. cinerea* hyphae degrade middle lamellae thereby increasing greatly the intercellular space (Figure 2.12A). Advanced *B. cinerea* colonization of *Arabidopsis* leaves results in maceration

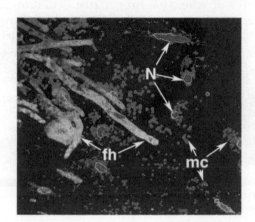

13. Tissue degradation in *Arabidopsis* leaves colonized by *B. cinerea*. Confocal microscopy. ophyll cells (mc) are macerated in front of growing fungal hyphae (fh). *Arabidopsis* nuclei are / GFP (Plotnikova, unpublished).

gradation of plant cells. Fungal hyphae grow rapidly utilizing plant ts such that only mesophyll walls and remnants of plant cytoplasm are this infection locus (Figure 2.12B).

iscussed above, successful development or inhibition of pathogen n depends greatly on the events taking place at the host–pathogen :e during the establishment of infection. An active localized plant cell .e can result in resistance to a pathogen. Slow and passive response d to fungal proliferation. Microarray analyses have revealed a large r of *Arabidopsis* genes that are differently regulated in regions sur- g *B. cinerea* infection loci. Many of those genes encoding MAP , WRKY transcription factors and pathogenesis-related (*PR*) proteins duced around lesions but not in remote leaf areas or at least at much xpression levels. These proteins have been implicated in many plant : responses. Global transcriptional profiling, measuring levels of messenger RNA and densitometry of transgenic plants bearing ic promoter–GUS fusion showed that induction of pathogenesis- genes such as *PR1*, *BGl2*, *PR5*, *PGIP2*, *GST1* and *PDF1.2* was higher infection loci in comparison with the remote leaf tissues.

e 2.14 shows strong *PR1*gene induction only around *B. cinerea* lesion s *GST1* gene was induced around the lesion as well as in remote areas leaf (Figure 2.14). These results and others indicate that the most lefense response occurs in cells directly surrounding infection loci t its primary role may be to contain the pathogen to the infection site. xpression levels of defense-related genes in remote leaf tissues may • limit a secondary infection. Apoptosis and necrosis of plant cells at t–pathogen interface stimulates the facultative pathogen *B. cinerea*

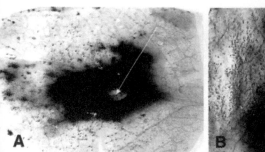
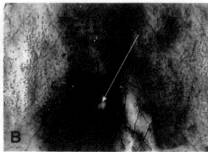

e 2.14. Pathogenesis-related (*PR*) gene expression around *B. cinerea*-induced lesions 4 dpi.
hed *Arabidopsis* (Col-O ecotype) leaves containing *GUS* reporter gene under the control of
PR1 or *GST1* promoter were inoculated with drops of *B. cinerea* spore suspension (105/ml) at
temperature and then stained for GUS activity.
he GUS staining shows strong *PR1* promoter activity only around *B. cinerea*-induced lesion;
ST1 promoter activity is seen around lesion as well as in remote areas of the leaf (Plotnikova,
blished).

ow zone defining the area of immediate contact between host and
ogen.

acterial necrotrophic infections. Many phytopathogenic bacteria are
otrophs. They usually enter the host tissues through stomata or wounds.
scanning electron micrograph shows the accumulation of *Pseudomonas
ginosa* PA14 bacteria above stoma on the *Arabidopsis* LL-O ecotype upper
epidermis (Figure 2.15).
ter penetration inside the plant tissues, bacteria concentrate in the
tomatal cavity and in intercellular spaces. At this early stage of the infec-
, most host cells are intact except for the ones bordering the substomatal

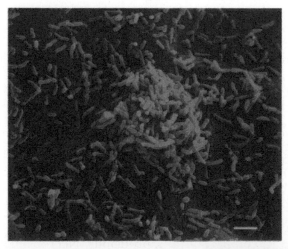

e 2.15. The scanning electron micrograph of *P. aeruginosa* PA14 bacteria accumulation above

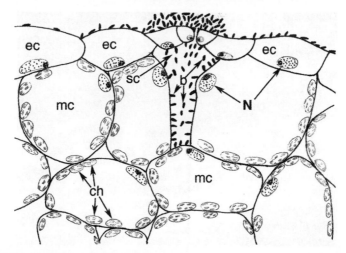

Figure 2.16.   Diagram of *P. aeruginosa* PA14 concentration above stoma and its penetration into sub-stomatal cavity. Bacterial cells are attached to the plant leaf surface as well as to the walls of mesophyll and epidermal cells inside the substomatal cavity: *b* – bacterium, *ec* – epidermal cells, *mc* – mesophyll cell, sc - substomatal cavity, *N* – plant nucleus, *ch* – chloroplast (After Plotnikova et al, 2000).

cavity and the wounded cells exhibiting plasmolysis and degradation of their organelles. Figure 2.16 shows a diagram of *P. aeruginosa* PA14 accumulation on the *Arabidopsis* epidermis and its penetration through stoma into substomatal cavity. Bacterial cells are attached to the plant leaf surface above stoma as well as to the walls of mesophyll and epidermal cells inside the enlarged substomatal cavity.

At later stages of infection, the bacterial cells proliferate in the intercellular spaces and spread along leaves in basipetal direction. Eventually, many mesophyll cells become separated because of the destruction of the middle lamella and subsequent increase in the intercellular space and concomitant decrease in the plant cell volume. This shrinkage of mesophyll cells is most likely a consequence of the absorption of host cell nutrients by the proliferating bacteria.

Bacterial infection has a dramatic effect on the structures of the host cell walls and cytoplasm and on the location and organization of host cell organelles. The first sign of host cell degeneration is slight plasmolysis and concentration of host membrane structures including chloroplasts, ER and dictyosomes at the site of bacterial contact. The proliferation of bacteria in intercellular space results in a redistribution, alteration and later degradation of host organelles. Later in the infection process, the outer chloroplast membranes become swollen and their thylakoids as well as host cell mitochondrial cristae are disrupted. In addition, the cytoskeleton becomes degraded and cell organelles including chloroplasts become redistributed in the cell volume. Host plasmalemma becomes highly undulated. Plant walls become thinner and highly convoluted as a result of their matrix being digested by bacterial pectolytic enzymes (Figure 2.17).

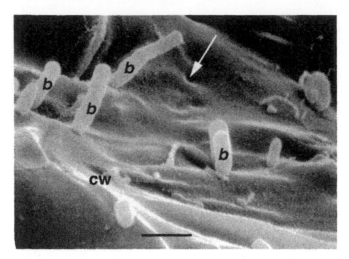

Figure 2.17. Scanning electron microscopy of *P. aeruginosa* PA14 penetration through mesophyll wall of *Arabidopsis* LL-O ecotype leaf. Bacterial cells (*b*) are seen on the surface of a highly convoluted plant cell wall (cw) of vessel parenchyma cells at various stages of penetration 96 hpi. Holes (arrow) in plant cell wall with a diameter similar to that of the bacteria are apparent. Bar = 1 μm. (From Hendrickson et al, 2001.)

Scanning electron microscopy of freeze-fractured infected *Arabidopsis* leaf tissues revealed that the bacteria *P. aeruginosa* when parasitizing in *Arabidopsis* intercellular space attach themselves perpendicularly to plant cell walls (Figures 2.17 and 2.18). At 96 hpi bacterial cells are seen on the surface of highly convoluted walls of vessel parenchyma cells at various stages of cell-wall penetration (Figure 2.17). Unusual convoluted or undulated surface of *Arabidopsis* cell walls in *P. aeruginosa* infected tissues suggest that the walls have been subjected to a significant degree of overall enzymatic digestion by hydrolytic bacterial enzymes. Furthermore, holes can be seen in the plant cell wall with a diameter similar to the diameter of *P. aeruginosa* cells (Figure 2.17). These holes are presumably formed as a consequence of bacterial attachment and/or penetration, and only appear to form on cell walls that have a convoluted appearance and thus are marked with a sign of wall matrix degeneration. In contrast, to the highly virulent *P. aeruginosa* PA14, the cells of the less virulent bacterium *P. syringae* pv. *maculicola* are attached lengthwise along the rigid plant cell wall, which, in turn, does not show any convolutions, undulations or holes. Perpendicular attachment of *Ralstonia solanacearum* has also been observed in tobacco tissue culture cells and in tomato xylem cell walls. Aside from these three, there is little data on the attachment of other bacterial pathogens to plant hosts. The human pathogen *P. aeruginosa* strain PAK A549 is also capable of attaching perpendicularly to the surface of human respiratory epithelial cells, indicating that *P. aeruginosa* may utilize a similar mode of attachment to both human and plant tissues, and may form the holes in order to gain easier access to host nutrients.

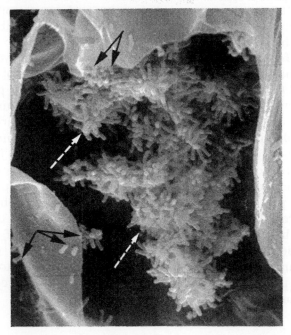

Figure 2.18.   The scanning electron micrograph of a cryo-fracture through *P. aeruginosa* PA14-infected Arabidopsis mesophyll cells show bacteria attached perpendicularly to the inner and outer surfaces of plant cell wall (solid arrows) and to the degrading host cytoplasm concentrated into the central portion of the cell volume (dashed arrows) (Plotnikova, unpublished).

During systemic bacterial infection of *Arabidopsis* leaves, most vessel parenchyma cells are invaded by a large number of bacterial cells. Figure 2.18 shows the internal structure of *Arabidopsis* mesophyll cells infected with *P. aeruginosa* PA14. Infected *Arabidopsis* leaves were frozen in liquid nitrogen, cryo-fractured and studied under scanning electron microscope. The micrograph shows bacteria attached perpendicularly to the inner and outer surfaces of plant cell wall and to the degrading host cytoplasm concentrated into the central portion of the cell volume (Figure 2.18). Systemic *P. aeruginosa* infection results in basipetal movement of bacterial threads along *Arabidopsis* veins, apparently from one vessel parenchyma cell to the next. *P. aeruginosa* parasitism in vessel parenchyma gains its access to photosynthates in the phloem vessels, whereas other bacterial necrotrophs such as *Xanthomonas phaseoli* and *X. campestris* are detected in xylem vessels, where they can travel large distances. Host cell collapse is the final step of the *P. aeruginosa* and many other bacterial infections.

Conclusion

Fungal, bacterial and viral pathogens greatly affect plant cell structure and ultrastructure. A diversity of plant pathogenic microorganisms can be classified into two major groups, biotrophs and necrotrophs, although many pathogens fall into the intermediate groups of hemibiotrophs and heminecrotrophs.

Biotrophs infect and colonize young plant cells with active metabolism. They have a long-lasting mutualistic stage where they do not destroy plant structures until the time of massive sporulation. During their parasitic stage, biotrophs can even stimulate the formation of new host structures in infected cells such as new membrane complexes, dictyosomes and polarized ER cisternae so as to exploit plant biosynthetic machinery by inducing infected cells to compete with the growing plant apices and developing flowers for assimilates.

In contrast, necrotrophs are adapted to the metabolism of older plants and/or their senescing parts with their activated catabolic pathways. With their arsenal of pectolytic and proteolytic enzymes, necrotrophs can increase greatly the permeability of plant membranes and stimulate their autolysis. This turns the host's own enzymatic machinery towards itself, churning out monosaccharides, amino acids, vitamins etc., which are then leached through the damaged, highly permeable plant membrane into the apoplast, where they are utilized by a fast-growing necrotrophic pathogen. During this process, the plant tissues are destroyed.

Selected Literature

Agrios GN. Plant pathology (5th edition). Acad. Press 2005, 952 p.

Alexander D, Lawton K, Uknes S, et al. Defense-related gene induction in plants. Genetic Engineering 1994; 16:195–212.

Allen PJ. Control of spore germination and infection structure formation in the fungi. Encycl Plant Physiol, NY 1976; 4:51-85.

Andreev LN and Plotnikova JM 'Wheat Rusts: Cytology and Physiology', Nauka, M, 302 p.1989 (in Russian).

Bracker CE, Littlefield LJ. Structural concepts of host-pathogen interfaces/ Fungal pathogenicity and the plants response. Ed RJW Byrde, Cutting CV. NY Acad Press 1973 pp.159-318.

Bushnell WR. Physiology of fungal haustoria. Annu Rev Phytopathol 1972; 10:151-176.

Coffey MD, Paleviz BA, Allen PJ. The fine structure of two rust fungi: *Puccinia helianthi* and *Melampsora lini.* Canad J Bot 1972; 50:231-240.

Colllins NC, Thordal-Christensen H, Lipka V, Bau S, Kombrink E, Qiu J-L, Huckelhoven R, Stein Monica, Freialdenhoven A, Somerville Shauna C and Shulze-Lefert P. 2003. Snare-protein-mediated disease resistance at the plant cell wall. Nature 425: 973-977. Hendrickson E L, Plotnikova JM, Mahajan-Miklos S, Rahme L. and Ausubel FM Differential Roles of the *Pseudomonas aeruginosa* Strain PA14 *rpoN* Gene in Pathogenicity in Mice, Nematodes, Insects and Plants. J. Bact., 2001 Dec;183(24):7126-34.

Littlefield LJ and Heath MC. Ultrastructure of Rust Fungi. N.Y.:Acad. Press 1979, 227 p.

Voegele RT, Struck Christine, Hahn M, and Mendgen K. 2001. The role of haustoria in sugar supply during infection of broad bean by the rust fungus *Uromyces fabae*, Proc Natl Acad Sci USA 98(14): 8133–8138. Plotnikova JM, Rahme Laurence G0 and Frederick M. Ausubel 2000. Pathogenesis of the human opportunistic pathogen *Pseudomonas aeruginosa* PA14 in *Arabidopsis thaliana*, Plant Physiology, 124, 1766-1774.

Plotnikova JM, Reuber TL, Ausubel FM, Pfister DH 1998. Powdery mildew pathogenesis of *Arabidopsis thaliana*, Mycologia, 90(6):1009-1016.

RahmeL.G., Frederick M. Ausubel, Hui Cao, Eliana Drenkard, Boyan Goumnerov, Gee W.Lau, Shalina Mahajan-Miklos, Julia Plotnikova, Man-Wan Tan, John Tsongalis, Cyntiha L. Walendziewicz and Ronald G.Tompkins 2000, Plants and animals share functionally common bacterial virulence factors, PNAS (Proceedings of the National Academy of Sciences of the United States of America), 97 (16): 8747-9348.

Reuber TL., Plotnikova JM, Julia Dewdney, Elizabeth E. Rogers, William Wood and Frederick M. Ausubel (1998) Correlation of Defense Gene Induction Defects with Powdery Mildew Susceptibility in *Arabidopsis* Enhanced Disease Susceptibility Mutants, The Plant Journal 16(4), 473-485.

Sukhorukov KT Physiology of plant immunity. 1952, M USSR Acad Sci 146 p. (in Russian).

Vavilov NI Plant immunity to infectious diseases M 1918, 244 p (in Russian).

Chapter 3

Advanced methods of plant pathogen diagnostics

L. A. Shcherbakova

Plant disease control can be successful only with timely detection of diseases and correct identification of their causative agents. However, diagnostics of diseases and their causative agents by symptoms is rather unreliable, as (i) the external symptoms of infections of plants are similar for many diseases or resemble the symptoms of physiological disorders caused by adverse external effects; (ii) plant pathogens can cause asymptomatic disease or disease with indistinct symptoms. Diagnosis using indicator plants or differential cultivars is not always reliable. Often disease control measures, especially, treatment with pesticides, are less efficient when applied at the stage of symptom manifestation. Identification of bacterial and fungal pathogens using isolation in pure culture with subsequent testing by microbiological methods, as well as diagnostics of plant viruses by means of electron microscopy is time-consuming and laborious. In addition, these techniques have low productivity, and they can hardly be automated. Moreover, there are many phytopathogens whose cultivation *in vitro* is either difficult (strict biotrophs) or impossible (viroids, viruses, and phytoplasmas). The need to overcome all these problems of conventional phytopathology diagnostics led to the development of newer methods of plant pathogen detection and identification. In many respects, these new methods fundamentally differ from the earlier techniques and can supplement or replace conventional techniques.

The range of modern technologies used in phytopathology for diagnostics of plant pathogens is extremely wide. Some of them are improved or modified conventional phytopathological methods (certain cultural, histochemical or biochemical techniques, and electron microscopy), while others have been adapted from other sciences such as molecular biology, biochemistry, or immunology.

Developing new methods for detection and identification of plant pathogens is a continuous active process. Some of the methods have limited application (e.g. pyrolysis gas liquid chromatography, precipitating serology, immunoelectrophoresis, isoenzyme assay and two-dimensional electrophoresis), while the utility of other methods is increasing. At present, more state-of-the-art and diverse techniques are being introduced in microbial diagnostics. The list of pathogens detected and identified by these methods is growing longer from year to year. In writing this section, the author found it was difficult to fit the very extensive and diverse material within a limited size of a general textbook. Rather than duplicating the theoretical aspects described in respective textbooks and monographs or turning this section into a list of examples of applications of the new diagnostic methods for particular pathogens, only the basic principles of those biomolecular, immunochemical, and biochemical methods that are most relevant for detection of plant viruses, bacteria, oomycetes, and fungi will be discussed.

This section is mainly focused on immunochemical assays and methods of molecular biology. The diagnostic potential of biochemical methods has been illustrated by chromatography of fungal metabolites. The main goal is to help the reader to navigate in a large volume of information and prepare for reading more specialized publications. The specific examples of application of the above methods for detection and identification of viruses, bacteria, fungi, and oomycetes are limited to the most important pathogens and are mostly mentioned without references to the original papers. References to the publications from which the examples were taken as well as the detailed protocols of assays can be found in the monographs and major reviews listed in the end of the chapter.

Antigen–antibody interaction-based diagnostics

Plant pathologists have seen the great potential of immunological methods in diagnostics and inter- or intraspecific differentiation of phytopathogens for a long time. These methods were primarily developed to diagnose viral diseases in plants, although attempts to adopt them for identification of other plant pathogens were also made. The first antisera against fungi were prepared as early as 1903, and already in 1937 an attempt was made to determine the species and geographic forms of head bunt of wheat using immunological reactions. However, more systematic use of immunochemical methods began in the 1960s, which was connected with the rapid development of immunochemistry techniques (of immunodiffusion, immuno-fluorescence, and immunoelectrophoresis). The first milestones of serological diagnostics of plant pathogenic microorganisms were reviewed by A.J. Crowle. Although at that time immunological methods in most cases confirmed the known taxonomic status of the microorganisms, they also allowed revealing intraspecific distinctions based on different serotypes of the phyto-pathogens. A little later, fluorescent antibodies were applied to detect and identify the bacteria and fungi that were difficult to isolate. For example, *Polyporus fomentarius* was detected in a mixed culture, soil, and infected pine roots. It became evident that immunochemical methods are suitable to diagnose plant pathogenic microorganisms in plants and other matrices in the environment. By the end of the 1960s, these methods, especially immunod-iffusion and immunofluorescence, were quite widely used for identification of practically important pathogens and more precise determination of their taxonomic status. A new form of immunoassay, enzyme linked immunosorbent assay (ELISA), evolving from the radioimmunoassay, was developed in the early 1970s and introduced into plant pathology laboratory practice in the 1970s–1980s. It significantly expanded the scope of plant disease diagnosis and phytopathogen detection.

All immunodiagnostic methods are based on the ability of antibodies to bind antigens. The primary goal of the assays is to detect or quantify the binding of a specific (diagnostic) antibody to a homological (target) antigen. Such binding can be displayed with a number of techniques. The antigen–antibody complexes can be visually detected due to their ability to form macro or microscopically visible precipitates in a liquid medium (agglutination and immunoprecipitation tests) or gel (single, radial, or double immunodiffusion tests). The antibodies can be tagged with a fluorescent dye, and the binding can be detected under a fluorescence microscope (direct and indirect immunofluorescence tests). It is possible to use radiolabelled antibodies and quantify the complexes with antigen (radioimmunoassay). Colloidal gold is used as an antibody marker in immunohistochemical analyses using electron microscopy or immunoblotting. However, more often, coupling the antibody to an enzyme is involved, and generation of a colour product upon addition of a substrate is used for detection of antigen–antibody binding. This technique is called the enzyme immunoassay (EIA).

Applicability of EIA in practical diagnostics is based on two fundamental scientific facts. The first fact is that antibodies, noncovalently bound to a solid carrier, and enzymes, covalently bound to antibodies, can retain their functional activity, i.e. bind the antigens (antibodies) or cleave the substrate (enzymes). The second fact is that the resulting antibody–enzyme complex, as conjugate, retains its biological activity in solution. Antibodies and their enzyme conjugates possess very high specificity and sensitivity. Antibodies recognize and selectively bind to the antigen against which they were produced. An antibody molecule tagged with one enzyme molecule can catalyze conversion of tens and hundreds of thousands of the substrate molecules to products. This results in an increase in the homologous antigen detection threshold, due to the "signal enhancement".

To date, dozens of reaction variants, based on the above facts, have been suggested, however, ELISA became the most common in diagnostics of plant diseases.

Many reviews have been published on the use of ELISA in phytopathology studies to diagnose the causative agents of plant diseases: viruses, phytoplasmas, bacteria, fungi, and oomycetes. Some of these references are listed below. A number of companies produce the diagnostic kits and equipment for ELISA, and commercial ELISA kits are available for many viruses, bacteria, oomycetes, and fungi (e.g. Agdia, Elkhart, IN and ADGEN, Ayr, Scotland).

Enzyme linked immunosorbent assay (ELISA)
Antibodies

Antibodies are the key component of ELISA and provide the basis for specificity and sensitivity of this technique. Antibodies are glycoproteins (immunoglobulins or γ-globulins) produced by a mammalian immunosystem in response to penetration by foreign organisms or injection of foreign macromolecules.

To detect a plant pathogen, antiserum (also called immune serum) containing antibodies against its antigens can be used directly in ELISA. However, a common practice for this assay is the use of antibodies previously isolated from the immune serum. To prepare conjugates with enzymes, the antibodies must be purified from the antisera. Regardless of whether ELISA is used for qualitative or quantitative diagnostics, it requires highly specific and sensitive antibodies.

Depending on the structure, immunoglobulins are subdivided into several classes. In ELISA, class G immunoglobulins (IgG) or their fragments (usually Fab) are commonly used. The IgGs have two light and two heavy chains connected by disulfide bonds (Figure 3.1). Both chains have constant (C_H1, C_H2, C_H3, C_L) and variable (V_H and V_L) regions. The Fab portion of the antibody has a hypervariable sequence with an antigen-binding site, and it controls specificity of the interaction with the antigen (Fab). Thus, an IgG molecule carries two antigen-binding loci. Cleavage of an antibody molecule by papain in the "hinge" area results in production of three fragments: one constant fragment (Fc), carrying constant anti-species determinants, and two antigen-binding fragments (Fab) retaining their antigen-binding ability. Because these fragments do not lose the functional properties possessed by the whole antibodies, they are also used in some ELISA variants (see "ELISA formats").

The main property of the specific antibodies that largely determines ELISA's success is their affinity. It characterizes the strength of binding of the active centres of the antibody molecule with the reactive (determinant) sites of the antigen. Affinity depends on the mutual steric complementarity of the

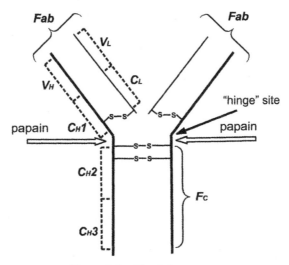

Figure 3.1.   The IgG structure.

antibody active centre and antigenic determinant (epitope): the higher the complementarity, the higher the affinity.

There are several ways to prepare antibodies suitable for immunodiagnostics of plant pathogens.

Polyclonal antibodies (pABs).   Ordinary sera from immunized animals are a mixture of similar but not identical antibodies produced by a large number of different lymphocytes. These antibodies are designated as polyclonal antibodies. Therefore, antisera obtained against an immunogen that may be an extract from pathogen or specially isolated individual compounds produced by it, contains a heterogeneous population of antibodies which may react with different determinants of the immunogen.

Polyclonal antibodies are routinely used in immunoassays of plant pathogenic viruses, bacteria, and fungi. The procedure of polyclonal antibody preparation is relatively easy and inexpensive, and can be accomplished in 3–4 months. In addition, the ability to bind several antigenic determinants creates advantages if it is necessary to detect a pathogen in natural population where its antigenic pattern may vary. In this case, pABs give a chance "to cover" the biological diversity of the given pathogen. However, pABs are not always sufficiently specific, which manifests as cross-reactivity with related organisms. This disadvantage partly impedes the practical application of pABs for detection of some viruses, bacteria, and especially fungi. Target pABs are generated in limited amounts. In any given antiserum, the antibodies specific to homologous antigen make from 5–10% of the total immunoglobulin. Their specificity and affinity can change and vary from batch to batch. One approach to overcome the undesired cross-reactivity is to partially purify pABs by either absorbing polyclonal antisera with heterologous antigens or diluting high titre antisera to minimize or eliminate cross-reactive components and decrease background reactions, while maintaining the specific reactivity at an appropriate level. Another possibility is to use monoclonal antibodies.

Monoclonal antibodies (mABs).   These are a homogeneous population of antibodies that have "narrow" specificity towards one epitope. Monoclonal ABs are produced by hybridomas. Hybridomas are hybrid cells that combine the ability to generate epitope-specific immunoglobulins with unlimited proliferation. The principles and technical procedures of mAB production are described in detail in a number of books and technical articles. In general, the mABs production technique consists of hybridization of lymphocytes of immunized mice with the tumour cells cultured in a medium (murine myeloblasts) followed by the selection of the clones, during the hybridoma passages, to produce antibodies with the required specificity. The standard specificity for a particular epitope, reproducible in passages, is the principal advantage of mABs. The extremely high degree of specificity allows detection of very subtle differences in closely related strains or serotypes of plant pathogenic microorganisms. Monoclonal ABs enable detection of both common and different antigens in various morphological structures of phytopathogens. For example, antigens in zoospores or cysts of pathogenic oomycetes can be detected. Monoclonal antibodies are homogeneous not only in specificity but also in other biological properties including avidity, stability, and affinity.

The first application of mABs in phytopathology was for virus diagnostics. The ELISA involving mABs is now used for diagnostics of bacteria, fungi, oomycetes, and phytoplasmas. In many cases, mABs were found to be more effective than pABs in differentiation of the viral strains, bacterial serotypes as well as fungal races and isolates. However, the use of mABs for phytopathogen detection and identification requires certain caution. Due to their high specificity, these antibodies may fail to recognize a natural strain or isolate, if the structure of the epitope responsible for the connection with this antibody has changed in mutation. On the other hand, in case of common antigens, for instance, in case of barley yellow dwarf virus (BYDV) serotypes, mABs do not possess enough specificity.

The hybridoma technology enables production of virtually unlimited amounts of mABs to various antigens, which, in turn, makes possible the production of commercial diagnostic products, including ones for the detection and quantitative analyses of plant pathogens by ELISA.

Phage displayed antibodies. A new method of producing antibodies without the need for animal immunization has been developed based on achievements of molecular biology. It is known as a phage display technique. The phage display technique makes it possible to produce libraries of functional fragments of immunoglobulins from various animals. The amplified immunoglobulin fragments are displayed on the phage surface. These fragments contain variable domains with hypervariable sequences which are responsible for the antibody specificity. For immunoassays, this technology creates an opportunity of using an extremely wide diversity of the specificity and opens unlimited cheap supply for production of recombinant antibodies. The antibody fragments specific for a wide range of antigens, including plant pathogens, were obtained by selection from the large phage display libraries. The first instances of using antibodies produced by phage display technique for immunodiagnostic assays of plant viruses, bacteria, and fungi have been described.

Antigens

Antigens are the molecules of foreign organisms against which antibodies are produced. Such biological molecules as proteins, polysaccharides, polynucleotides can be immunogens. In addition, almost any polymeric compound having functional groups which are recognized by antibodies can be an antigen. Certain sites of immunogenic macromolecules (for instance, 6–10 amino acid residues in a protein molecule), or functional groups of other immunogens that induce antibody production are called antigenic determinants or epitopes. If one antigen molecule has several of such determinants it is considered polyvalent. Haptens are small molecules that are not immunogenic, i.e. they cannot independently induce generation of antibodies. However, if they are covalently bound with a high-molecular compound (e.g. protein molecule) they become antigenic determinants. Antibodies produced in response to the antigen can bind free haptens. In application of polyclonal antibodies specific to haptens, it is advantageous to produce two conjugates with two different carrier proteins. One of them is utilized for the immunization to produce then enzyme-labelled antibodies whereas the other conjugate is used for coating in ELISA. This allows avoiding the absorption of the antiserum by the first protein, which otherwise will be necessary to remove the antibodies with specificity to its determinants.

Various types of immunogens are injected into animals (usually rabbits) to produce antibodies against plant pathogens, and different plant pathogenic antigens are used as analytes. Injection of viruses, whole cells, cell walls, soluble or solubilized high-molecular antigens, extracellular secreta, extracts from fungal or bacterial pathogen into animals result in production of polyclonal antisera. There is no direct indication as to what type of immunogen enhances specificity of the antiserum. In general, injection of the purified antigens results in production of more specific antibodies. Therefore, wherever possible, highly purified antigens with known steric and chemical structure should be preferred for immunization.

Conjugates and substrates

Conjugates of antibodies with enzymes are detection molecules in ELISA. Horseradish peroxidase (HRPO) and alkaline phosphatase (AP) are the most common enzymes for tagging antibodies. Covalent binding of the enzymes and immunoglobulin molecules are carried out with glutardialdehyde (alkaline phosphatase) or periodate oxidation (in case of HRPO). AP can be covalently linked to antibodies by one step glutardialdehyde procedure. The antibodies are labelled with HRPO in the reaction with sodium periodate in the carbohydrate moieties of the enzyme. The aldehyde groups of HRPO which form as a result of the oxidation, react with amino acid residues of immunoglobulins. The substrate for conjugates with horseradish peroxidase is *o*-phenylene diamine mixed with hydrogen peroxide. The coloured reaction product has the absorption maximum at 492 nm. The substrate for alkaline phosphatase is *p*-nitrophenyl phosphate that turns into *p*-nitrophenol measured at 405 nm.

Besides HRPO and AP, conjugates with other enzymes have also been applied in plant pathological research. For instance, detection of plant viruses was carried out with conjugates of inorganic pyrophosphatase and penicillinase.

To detect plant pathogens in plant tissues, conjugates with alkaline phosphatase are preferable, as this enzyme, in contrast to HRPO, does not occur in higher plants, which reduces probability of false positive results in the assay. Nevertheless HRPO is very often exploited as an enzyme label due to their higher specific enzyme activity, stability and small size reactivity. An important characteristic of HRPO–antibody conjugates is the RZ (Reinheitszahl, degree of purity) which is expressed as the ratio of the absorbance at 403 nm (heme group of HRPO) to that at 278 nm. RZ shows proportion of antibodies actually tagged with the enzyme and characterizes the purity of prepared conjugate. The absorbance at 403 nm should be 0.3–0.4 fold of that at 278 nm. Conjugates are stored conveniently by adding an equal volume in 50% glycerol at −20°C.

The sensitivity of ELISA can be enhanced by the streptavidin–biotin (or avidin–biotin) system or by the conjugates with luciferase and fluorescent substrate, luciferin (luminescent or fluorescent ELISA).

Diagnostic potential of ELISA has been enhanced by introduction of multiplex assay. Multiple ELISA is employed when it is necessary to identify more than one plant pathogen species in the same microtitre plate well. They can be assayed by means of different enzyme labels.

Solid phases

ELISA is often conducted by using the 96-well microtitre polystyrene plates. Polystyrene can easily adsorb antibodies and various antigens. There are also a number of chemical methods of treating the plate surface to provide reactive functional groups (e.g. hydrazine, maleimide) for covalent binding of antibodies and antigens. Treatment of microtitre plates with 1% nitrocellulose enhances the capacity of polystyrene to absorb proteins. Moreover, formation of a thin film on the surface of plate wells results in more uniform adsorption. Such treatment procedures can improve efficacy of ELISA. Nitrocellulose membranes are used as solid supports for immobilization of both the antigens and antibodies in so-called dot-ELISA. In this case, substrates yielding insoluble products of the enzymatic reaction, which precipitate onto the membrane, are preferred. Magnetic beads or particles can also serve as solid phases. The membranes or magnetic beads accelerate ELISA, which makes such assays more suitable for field use.

ELISA formats used in plant pathology

The ELISA is classified into **homogeneous** and **heterogeneous** depending on where the specific antigen–antibody interaction occurs. If all assay stages, including the enzymatic reaction, proceed in solution, the ELISA format is homogeneous. To determine the target antigen or measure its quantity in a homogeneous assay, there is no need to separate the reacted and unreacted components. The heterogeneous format assumes separation of free and immunochemically bound labelled reagent (antigen, antibody) using a solid carrier with immobilized complementary reagent (antibody, antigen). At the same time, unbound components and unwanted inhibitory substances are eliminated from the reaction system by washing. The homogeneous assay is usually applied for detection and quantification of low molecular weight compounds, e.g. haptens, such as toxins, hormones, drugs, etc. This format has relatively restricted application in plant pathogen diagnostics as compared to the heterogeneous type which is very suitable for the purpose of diagnostics. In addition, ELISA can be classified into **competitive** and **noncompetitive**, based on the type of reagents used at the first stage of the assay. The distinguishing features of a competitive format are simultaneous presence of an analyzed antigen from the sample and the labelled antigen at the first assay stage, and their competition for binding to a limited number of antibody binding sites. However, the competitive assays are less common in diagnostic of plant pathogens. One of the constraints is that plant pathologists are often dealing with the systems where the antigen cannot be easily labelled with an enzyme. The ELISA can be referred to as **direct** or **indirect** depending on what type of antibody is coupled to the enzyme. In the direct format, the target antigen is detected by the enzyme conjugate of homologous (antigen-specific) antibodies. In the indirect format, unlabelled homologous antibodies (ABs_1) are employed for trapping the target antigens, and then the antigen–antibody complexes are detected by the enzyme conjugates with other (the second) antibodies produced against ABs_1. These second enzyme-tagged antibodies specifically react with the constant domains in the Fc part of the antigen-specific IgG. Finally, it is possible to subdivide ELISA according to the method of measuring the end product of the enzyme reaction (colourimetric, fluoriometric, bioluminometric ELISA) or the way of the enzyme activity measurement (application based on principles of enzyme kinetic or end point).

Modifications of ELISA involve the combinations of the above-described basic formats, as well as with variations of the immunoreagents, immobilized on solid phase. In this section we focus only on the variants of the assays that possess advantages particularly for plant pathology.

ELISA is used to detect or quantify pathogens in plants or soil, diagnose whether the analyzed plant is infected by a pathogen to reveal infected plants before appearance of the disease symptoms, or to select healthy seeds or planting stock DAS-ELISA (double antibody sandwich-ELISA), where antibodies are immobilized onto a solid phase, is usually employed.

Basically, DAS-ELISA (Figure 3.2) on microtitre plates involves the following steps:

1. Addition of antibodies against plant pathogen antigen(s) in plate wells.
2. Incubation of the plate to immobilize antibodies followed by washing off the unbound antibodies.
3. Coating of the plate with a protein (e.g. bovine serum albumin) to block nonspecific binding sites.
4. Addition of samples to trap the target antigen (analyte) by the immobilized antibodies.
5. Incubation of the plate to accomplish the immune reaction and then washing the wells from nonspecific components of the sample.
6. Addition of specific antibodies labelled or unlabelled with enzyme. If enzyme-conjugated antibodies are used in step 6, step 7 will follow.
   If unlabelled antigen-specific IgGs are used at step 6, the next required step will be 6a.

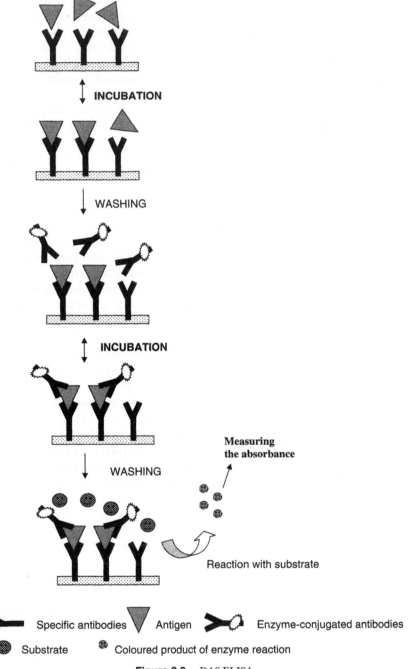

INCUBATION

WASHING

INCUBATION

WASHING

Measuring the absorbance

Reaction with substrate

Specific antibodies   Antigen   Enzyme-conjugated antibodies

Substrate   Coloured product of enzyme reaction

Figure 3.2.   DAS-ELISA.

6a. Addition of enzyme conjugate of anti-species antibodies, for example enzyme-labelled mouse antibodies against rabbit IgG specific to target antigen.
7. Incubation of the plate and washing to eliminate nonspecifically adsorbed conjugate.
8. Addition of enzyme substrate followed by incubation for development of the colour and termination of the enzyme reaction.
9. Recording the assay results by visual observation or measuring of the absorbance.

Actually, the above outline is a combined description of two conventional ELISA formats well-known as, "direct" and "indirect" DAS-ELISA. The indirect sandwich ELISA is also termed TAS-ELISA (triple antibody sandwich-ELISA). It involves antibodies from two different species of animals. Antibodies specific to the target antigen(s) are produced in one species, and in order to synthesize conjugates, antibodies prepared from antiserum of another species are used. These second antibodies, as was indicated earlier, possess specificity against IgG's Fc that contains the determinants constant for particular animal species. Instead of the second (anti-species) antibodies, staphylococcal protein A conjugated to enzyme can also be used. If necessary, immune serum from the same animal species can be used in the indirect assay. To do this, antigens of the analyzed pathogen are captured by the immobilized Fab antibody part containing the specific binding sites. The bound antigens are coated with the whole specific antibodies, and then the sandwich is detected with the enzyme conjugates that have been prepared against Fc portion of IgG.

Although DAS-ELISA can be used only for analysis of the antigens which contain at least two determinants, and though it consists of several stages and thus requires quite a long time, it possesses a number of benefits. The decisive benefit is high sensitivity, exceeding the potentials of the other ELISA formats. Assuming that the maximally high equilibrium constant of antigen–antibody interaction is $10^{12} \cdot M^{-1}$, the limit sensitivity of DAS-ELISA may theoretically reach 10^{-14} M. An advantage of TAS-ELISA is universality of the enzyme-labelled anti-immunoglobulins and protein A which are used for analysis of various antigens.

Following from the outlines above, the enzyme activity in DAS-ELISA and TAS-ELISA is proportional to the amount of antigen bound to the immobilized antibodies. Hence, the level of absorbance is directly proportional to the amount of antigen in the sample.

The DAS-ELISA can be simplified in terms of procedure by means of simultaneous addition of the detected antigen (the sample) and specific antibody conjugates preincubated with each other for a short time, to the immobilized antibodies. It is the so-called two-step DAS-ELISA, where a part of the immunocomplexes between antigen and conjugated antibodies can form in homogeneous conditions and are removed in the subsequent washing. The higher the antigen concentration in the sample, the fewer conjugate molecules will react with the antigen captured by the immobilized antibodies; therefore, the level of absorbance will be inversely proportional to the amount of antigen in the sample being analyzed.

Another format of ELISA is the so-called DAC form (direct antigen coating) of ELISA. The PTA-ELISA (plate-trapped antigen ELISA) is another name of the same format. In DAC-ELISA, as against DAS-ELISA, the antigens are immobilized onto solid phase. In general, this format is predominantly used in the analysis of the immune response to an antigen. It is not so commonly used for diagnostics of plant pathogens, as well-characterized antigens are not always available. In any case, it is necessary to take into consideration that polystyrene will adsorb various proteins in an increasing amount depending on their concentration; peptides of length 15–20 amino acid residues can adsorb on plate, but it is a very weak adsorption; bacteria and viruses displaying on their surface proteins can be directly adsorbed on plates; while carbohydrates and highly glycosylated glycoproteins due to their hydrophilic properties poorly bind to plastic.

Both the (above-described) ELISA formats (DAS and DAC) are the noncompetitive versions of the assays. They can be modified to become a competitive format of ELISA.

Some application aspects

Prior to the development of ELISA, relatively little work was done on serological detection of plant pathogenic bacteria, oomycetes, and fungi. Most applications for laboratory-based tests were aimed at virus detection with relatively less emphasis on fungal and bacterial pathogens.

The list of plant disease causative agents for which ELISA has been used is steadily growing. However, viruses still lead in this list, primarily due to the fact that, in addition to nucleic acid, only ELISA provides fast and reliable diagnostics of viral plant diseases. Simplicity and high throughput of ELISA makes it convenient for viral disease monitoring and studying the spread of plant viruses. Due to ELISA, many viruses were shown to have a wider range of host plants than it had been assumed earlier. Quite often it is found that symptoms of unknown etiology on crops are caused by the virus which was not previously found in these plants. ELISA allows updating information on occurrence of plant viruses, and a number of them were first detected in new geographic regions using this assay. ELISA enables a large-scale monitoring of plant viruses in seeds and planting stock, in particular, potato seed stocks originating from meristem tissue, or in bulbous plants (e.g. in Holland alone, hundreds of thousands of tulip, lily, and iris samples are tested annually). ELISA applications for detection and identification of viral plant diseases are too numerous for detailed description and full listing. However, to illustrate the accessibility of the practical applications of the assay, it should be noted that commercial diagnostic reagents and kits for identification of more than one hundred viruses in cultivated plants are available. Together with isolates and serotypes, they have been developed for 25 viruses of cereal crops (barley, maize, rice, sorghum, and wheat) and 3 viruses of cereal grasses (brome and Johnson's grass); 7 viruses of grain leguminous crops and forage grasses (bean, pea, soybean, cowpea, and red clover); 17 viruses of fruit crops (apple, banana, cherry, citrus, plum, papaya, peach, peanut, and prune) and 20 viruses of berries (blueberry, grapes, raspberry, and strawberry); 51 viruses of vegetables, melons, root and tuber crops (beet, cassava, cauliflower, celery, cucumber, hop, leek, lettuce, onion, pepper, potato, radish, shallot, squash, tomato, turnip, and zucchini), as well as 8 tobacco viruses, 2 groundnut viruses, and 2 sugarcane viruses. In addition, commercial kits are available for detection of at least 29 viruses on cultivated flowers and ornamental plants.

Bacterial cells are very complex and heterogeneous relative for their surface antigens and can vary with the environment where bacteria are growing. Taxon-specific polyclonal antibodies with acceptable specificity are used to detect plant pathogenic bacteria, but cross-reactions with unrelated species are common. For this reason, monoclonal antibodies are actively preferred in ELISA applications for numerous taxa of plant bacterial pathogens. On the other hand, most anti-bacterial mABs were initially screened with ELISA. Presently, this assay is used for detection and identification of pathovars, subspecies and species that belong to many economically important genera of phytopathogenic bacteria (*Acidovorax, Agrobacterium, Clavibacter, Corynebacterium, Erwinia, Pectobacterium, Pseudomonas, Ralstonia, Xanthomonas, Xylophilus*). Typical sensitivity of ELISA for bacteria is 10^5–10^6 CFU ml^{-1} (colony forming unit), which is sufficient for detection of bacterial pathogens in infected plants and colonies cultivated *in vitro*. However, the best diagnostic results are achieved if the analyzed samples consist of fresh lesions containing a high titre of the target pathogen.

ELISA is a routine laboratory technique for detection of various fungi and oomycetes in plants, soil, air, and other biological matrices or environmental samples. The Table 3.1, summarizing information on diagnostics of various fungi and oomycetes is by no means exhaustive, however, it provides an idea for what pathogens and for what purposes ELISA is suitable. This method is most often used to detect pathogens in tissues of symptomatic plants to diagnose the disease. In addition, ELISA enables finding latent infection and revealing fungal pathogens in asymptomatic plants at early stages of the disease, before manifestation of its signs. As ELISA is a quantitative method, it can be used to determine the content of pathogen biomass in tissues of the infected plants. This provides an idea of the degree of tissue colonization, which, in turn, allows studying the plant-pathogen interaction, evaluation of efficacy of fungicidal treatment, and monitoring of diseases. In a number of cases, ELISA is used to differentiate fungal and oomycete species, and, providing sufficient specificity of the antibodies, it can also be used to detect the intraspecific differences. ELISA is applied to find these pathogens in agricultural products and in quality control.

As for diagnostics of phytoplasmal diseases, ELISA is applicable for a number of them. Polyclonal and monoclonal antibodies have been prepared against phytoplasmas. However, because of difficulties in obtaining the

Table 3.1.   The instances of ELISA application for plant pathogenic fungi and oomycetes

Pathogen	Diseases and plants	Antibodies used	Application goals
Alternaria solani	Early blight of potato	pAB	Detection of the pathogen in plants.
Alternaria spp.	Alternaria diseases		Differentiation of 3 species.
Antroda carbonica	Brown rot of pine tree	pAB	Early detection of the pathogens and disease diagnosis.
Aphanomyces euteiches	Pea root rot	mAB	Quantification of the pathogen in resistant pea lines.
Botrytis allii	Grey mould neck rot	pAB	Detection of latent infection in storage onion.
B. cinerea	Vine grape grey mould (and must spoilage). Moulds of various vegetable, fruit and berry crops	mAB pAB	Pathogen detection in grape berries, must and vine, in various vegetables and fruits. Detection and quantification of the fungus in pear stems during cold storage.
Cladosporium fulvum	Cladosporium leaf and fruit mould	mAB	Detection and measuring of biomass in tomato leaves.
Colletotrichum lindemuthianum *C. orbiculare* *C. trifolii* *C. malvarum*	Anthracnose of legumes Anthracnose of forage grasses Jute anthracnose	mAB	Confirmation of relation of these species.
Coniophora puteana	Brown rot of pine tree	pAB	Early detection of the pathogens and disease diagnosis.
Erysiphe graminis	Powdery mildew of cereals		*In planta* quantification of biomass.
Fusarium graminearum	Head blight of cereals (scab)	pAB mAB	Detection of the pathogen and quantification of its antigens. Differentiation at the species level.
F. oxysporum f. sp.cubense	Banana wilt (Panama disease)	mAB	Race differentiation.
F. o. f. sp. lini	Flax wilt	pAB	The pathogen detection. Quantitative evaluation of flax root colonization.
F. o. f. sp. lycopersici	Tomato wilt	mAB	The disease diagnosis.
F. solani f. sp. pisi	Fusarium rot of pea	mAB pAB	Comparison of mAB and pAB. Demonstration of mAB suitability for the pathogen detection.

Gaeumannomyces graminis var. tritici	Wheat foot rot	pAB	Detection and quantification of initial infection.
Gliocladium roseum	Biocontrol fungus	mAB	Identification in wood.
Gloephyllum trabeum	Brown rot of pine tree	pAB	Early detection of the pathogens and disease diagnosis.
Glomus occultum	Mycorhizial fungus	mAB	Differentiation of species.
Humicola lanuginosa	Yellowing of harvested rice	mAB	Detection of the pathogen in plants and grains.
Lentinus lepideus	Brown rot of pine tree	pAB	Early detection of the pathogens and disease diagnosis.
Mycosphaerella pinodes	Pea ascochitosis	pAB	Detection and identification of the pathogen in seeds.
Ophiostoma ulmi	Dutch elm wilt disease	mAB	Disease diagnosis, detection of a fungal metabolite. Usage of mAB to determine the pathogen species, races, and aggressive groups.
Penicillium islandicum	Yellowing of harvested rice		Detection of the pathogen in grains.
Peronospora viciae	Pea downy mildew	pAB	Detection of seed infection.
Pestalotiopsis theae	Grey blight of tea plant	pAB	Detection of the pathogen in tea plant leaves.
Phoma exigua var. foveata	Potato gangrene		Detection of the pathogen in plants. Detection of small quantities of fungal spores.
Phoma tricheiphila	Citrus drywilt	pAB	Detection of the pathogen in lemon plants, the disease diagnostics.
Phomopsis longicolla	Phomopsis seed decay of soybean	pAB	Detection of latent infection in seeds and plants.
Phytophthora species including: - *cactorum*, - *cambivora*, - *capsici*, - *cinnamomi*, - *citrophthora*, - *cryptogea*, - *drechsleri*, - *megakaria*, - *megasperma*, - *nicotianae*, - *palmivora*, - *parasitica*, - *roseum*, - *syringae*	Phytophthora root rots (including watery root rots) of vegetable, melons, root and tuber crops, forage grasses, and other fodder plants. Collar, stem, and fruit rots of fruit trees, berry cultures, nut plants.	pAB mAB	Diagnostics of the species. Detection of the species in plants or soil. Research for field detection and study of distribution in soil.

continued

Table 3.1. The instances of ELISA application for plant pathogenic fungi and oomycetes—cont'd

Pathogen	Diseases and plants	Antibodies used	Application goals
P. fragare	Red core disease of strawberry	mAB	The pathogen detection in infected plants.
P. infestans	Potato late blight	pAB	Early detection in potato tubers and leaves. Study of host–pathogen relationships.
Plasmopara halstedii	Sunflower downy mildew		Detection of the pathogen in plants.
Polymyxa graminis	Obligate intracellular parasite of Graminaceae roots and vector of plant viruses	pAB	Detection of the pathogen in sorghum roots at various stages of life cycle under natural and controlled conditions.
Postia placenta	Brown rot of pine tree	pAB	Early detection of the pathogens and disease diagnosis.
Pseudocercosporella her potrichoides	Eyespot of cereals fodder	mAB	Detection of the pathogen in plants.
Puccinia graminis f. sp. tritici	Wheat stem rust	pAB	Detection of the pathogen and infected plants before appearance of the disease symptoms. Quantification of mycelium to study the pathogen–host-plant relationships.
Puccinia recondita	Wheat leaf rust	pAB	Detection of the pathogen in infected plants before appearance of the disease symptoms. Quantification of mycelium to study the pathogen–host-plant relationships.
Pyricularia grisea	Grey leaf spot of millet and rice	mAB	Detection of the pathogen and differentiation of its strains.
Pyricularia oryzae	Rice blast disease	pAB mAB	Detection and quantification of the fungus in plants, to study the pathogen-host-plant relationships. Study of antigenic characteristics of nine strains.
Pythium aphanidermatum	Pythium blight of many vegetable crops and fruit trees	mAB	Detection of the pathogen based on recognition of surface antigens of spores and cysts.
P. ultimum	Seedling blight and root rots of vegetables and root crops	mAB	Identification of the species among other Pythium isolated from sugar beet. Detection of the fungus in bean, cabbage, sugar beet roots. Quantification of fungus in plant tissues.

P. violae *P. sulcatum*	Carrot rot	pAB	Detection of the pathogens.
Rhizoctonia solani and other Rhizoctonia species	Rhizoctonia diseases of various crops	pAB mAB	Detection of live propagules in soil. Differentiation of anastomosis groups, quantification.
Rhynchsporium secalis	Rye and barley scald	pAB	Detection of latent seed infection.
Sclerospora graminicola	Downy mildew of cereal crops and grasses		Quantification of the pathogen in tissues of millet.
Septoria avenae f. sp. triticea	Septoria disease of oats	mAB	Identification of isolates.
S. nodorum *S. tritici*	Leaf spot of wheat	pAB mAB	Early detection of the pathogen, disease diagnosis, differentiation of the species, quantification in plants, evaluation of fungicidal treatment, the disease monitoring.
Serpula incrassata	Brown rot of pine tree	pAB	Early detection of the pathogens and disease diagnosis.
S. lacrimans	Dry rot of trees		In vitro determination of species-specific antigens to develop diagnostic method.
Sirococcus strobilinus	Shoot blight of conifers		Detection of the pathogen in seeds.
Spongospora subterranea	Potato powdery scab	mAB	Detection of the pathogen in plants.
Thielaviopsis basicola	Cotton black root rot	pAB	Detection of the pathogen at early stages of the disease.
Tilletia spp.	Wheat bunt		Species differentiation.
Trichoderma harzianum		mAB	Detection of the fungus in peat.
Uromyces appendiculatus	Bean rust		Detection of the pathogen in plants.
Verticillium dahliae	Rape wilt Verticillium wilt of potato	pAB	Detection of the pathogen in plants, determining the degree of cultivar resistance. Detection of the pathogen in stem tissues.
Verticillium lecanii	Hyperparasite (nematodes, aphids, insects, and plant pathogenic fungi)		Detection of the pathogen and quantification.

purified antigens in sufficient quantities, for many phytoplasmas the antisera have not been raised. Nevertheless, commercial kits for detection of a limited number of phytoplasmas are available.

Fast ELISA formats

Modern clinical studies make laboratories use widely ELISA tests that enable receiving diagnostic yield results in a very short time (several tens of minutes or even few minutes). In addition to the fastness, an advantage of fast ELISA formats is the possibility of their use *in situ* without involvement of highly skilled personnel, as well as reading the results without special instruments. Presently, rapid ELISA is used in diagnostics of plant diseases for detection of pathogens in plant and soil extracts.

One of the ELISA modifications suitable for fast diagnostics is the dot-blot immunoassay (DBIA) or dot-ELISA. The carrier to immobilize the sample (extract of homogenized plant tissue) is either nitrocellulose or polyvinylidene difluoride (PVDF) membrane. It is marked up into sections, and a few microlitres of the extract are added in each section. The nonspecific binding sites are blocked by application of foreign protein (for instance, solution of defatted dried milk) on the membrane. The membrane is first incubated with the antibody conjugates, then with the substrate. Both pathogen-specific antibody conjugates and enzyme-labelled anti-immunoglobulin antibodies can be employed in dot-ELISA. During the analysis, the membrane is washed like in plate-ELISA. The substrates to yield nonsoluble, stained reaction products are bromo-4-chloro-3-indolyl phosphate (BCIP) and nitro blue tetrazolium (NBT). A positive result of the assay looks like a spot on the membrane in the sample application place.

There also exists another format of the fast ELISA, the so-called dipstick ELISA. A strip with immobilized antibodies is dipped in the sample, and then the trapped antigen is detected by enzyme-labelled antibodies interacting with the antigen or antigen-specific antibodies.

Another fast and simple technique adopted by plant pathologists is known as lateral flow assay (LFA). The test is usually arranged as a flat support bar that houses, in sequence, the sample application area, a nitrocellulose membrane with a thin strip of immobilized specific antibodies, and an absorbent pad. The sample application area contains specific colloidal gold-bound antibodies that coat the latex beads. If the target antigen is present in the liquid sample, the antibodies link to it and the antigen–antibody complex is drawn with the fluid flow through the membrane. As soon as the complex reaches the antibody strip, it is captured by the antibody strip and concentrated, which results in generation of a colour line. To make sure that LFA is functioning, an additional thin strip of antibodies reacting with antigen-specific antibodies is applied in front of the absorbent wick. Thus, two coloured lines indicate a positive result whereas one line shows the absence of target antigen in the analyzed sample.

Commercial kits and immunoreagents for fast ELISAs of plant pathogens in plants and soil, including some bacteria (e.g. *Clavibacter michiganensis* subsp. *michiganensis, Xanthomonas horotum* pv. *pelargoni, Ralstonia solanacearum*), oomycetes (e.g. *Phytophthora cinnamomi*), and fungi (e.g. *Penicillium islandicum,* a saprophyte *Humicola lanuginose*) are available.

Other immunodiagnostics methods

Immunofluorescence

Strictly speaking, immunofluorescence, known ever since 1942, can hardly be referred to as a new method of immunodiagnostics. However, the method received a new birth after development of the hybridoma technology. The use of monoclonal antibodies in the immunofluorescence reaction enhanced its sensitivity and specificity.

To detect plant pathogenic microorganisms by immunofluorescence reaction, antibodies raised against their antigens are conjugated with fluorescent dye molecules such as fluorescein isothiocyanate or rhodamine isothiocyanate. In the direct immunofluorescence assay (IFA), fluorescing conjugates of antigen-specific antibodies are used. In case of indirect IFA, as well as in TAS-ELISA, the second (anti-immunoglobulin) antibodies tagged by fluorescent dyes are applied. The antigens present in the sample (spores or other pathogen propagules, sections of plant tissue, bacterial cells, etc.), which are applied on a slide, bond to the labelled antibody, and the fluorescent immune complexes are observed under a fluorescent microscope. The IFA results may be quantifiable, if the number of images can be quantified by counting the number of fluorescent units. Such a counting is a very laborious procedure, therefore, the number of samples analyzed is limited. The IFA throughput can be enhanced by an automated microscope system supplied with a computer and appropriate software. It is possible to perform the reaction in wells of microtitre plates and measure emission of the analyzed samples using special equipment.

As in the case of ELISA, sample components can interfere results of IFA. Firstly, plant tissues or soil samples possess their own fluorescence. Secondly, along with autofluorescence, nonspecific absorption of antibodies on sample particles and cross-reactions with other microorganisms can occur. These factors should be taken into account when plant–pathogen diagnostics is based on IFA.

IFA has been practically used to detect spores of soil-borne fungi (e.g. *Phytophthora, Fusarium*) and determine localization of fungal pathogens in plants, as well as to diagnose some bacterial diseases.

Immunoblotting

Immunoblotting (Western blotting) is a highly sensitive method for identification of proteins, including antigens of viruses and other plant pathogens. The method is based on combination of gel-electrophoresis and antigen–antibody interaction. High resolution of immunoblotting is provided by combination of electrophoretic separation of proteins, glycol- lipo- and nucleoproteins and the high specificity of antibodies or immune sera used for detection. Under optimized conditions, immunoblotting can detect less than 1 ng of target antigen in a sample.

Immunoblotting is carried out in three stages: (a) separation of the proteins to be analyzed in SDS-polyacrylamide gel (SDS – sodium dodecyl sulfate); the separated proteins can be visualized after staining and comparing with the reference samples; (b) a nitrocellulose membrane is placed on the gel, the protein bands are transferred (blotted) onto the membrane by electrophoresis and fixed on it; (c) detecting pABs or mABs carrying the enzyme label are applied on the membrane (to detect the bound antibodies, anti-species labelled serum is also used). In other words, the final stage of blotting is similar to the solid phase immunological tests.

Western blot analysis can detect one protein in a mixture of any number of proteins. This allows identification of viruses in plant extracts and vector cells. For example, immunoblotting was effective in revealing and identifying both individual and mixed viral infection in symptomatic wheat, corn, grapevine, and garlic plants. However, it is necessary to take into account that after SDS-PAGE (SDS-polyacrylamide gel electrophoresis) the proteins are in denatured condition, therefore, they may not be visualized by the antibodies specific to the native protein. If sera to the component peptides are available, the antigenic range of the tested protein can be revealed, which can be used to establish antigenic relationship between individual viral strains.

Tissue blot immunoassay

The tissue blot immunoassay (TBI) technique allows exact localization of plant pathogen antigens (predominantly those of viruses, and, to a lesser degree, fungi and phytoplasmas) in plant tissues. Direct tissue blot immunoassay is "transfer of antigens from the specimens onto a nitrocellulose membrane support by means of blotting a freshly cut tissue surface onto the supporting substrate" followed by detection of antigens immobilized on the membrane by the enzyme-tagged antibodies. This blotting relies on extremely high binding capacity of nitrocellulose towards macromolecules or viral particles. It is similar to Western blotting for protein transfer after electrophoresis from gel to nitrocellulose membranes but the transfer process of TBI is much simpler. To obtain a tissue imprint on nitrocellulose matrix, the surface of freshly cut plant tissue is slightly pressed to the dry membrane for less than a second. Powerful contact with the membrane is not recommended as higher and longer pressure can destroy the surface and distort the tissue image. TBI may be direct or indirect depending on the use of viral-specific or anti-immunoglobulin enzyme-tagged antibodies (or protein A) for antigen localization.

In contrast to ELISA, horseradish peroxidase is not a fully suitable enzyme for TBI, since many plant tissues contain highly active peroxidases, which can interfere with the analysis. Therefore, alkaline phosphatase is preferred in this assay. Furthermore, the final coloured product of enzymatic reaction must be insoluble and precipitate on the membrane at the reaction site. To localize plant viruses in tissues that contain endogenous pigments impeding the assay, chemiluminescent substrates and X-ray film should be used.

Diagnostics based on nucleic acid analysis

The progress of molecular biology in the 1980s–1990s allowed the development of specific, sensitive, and rapid methods for plant disease diagnostics. Perhaps, the most important advantage of these, so-called biomolecular methods in detection and identification of plant pathogens, compared with immunochemical assays, is that they can be used for analysis of genomic information, i.e. the most conservative properties of a pathogen.

As it was demonstrated in the previous section, we now have at our disposal a variety of immunochemical diagnostic methods providing good detection and identification of disease causative agents. However, the constraints of antigen–antibody diagnostics grow as target agents become more complex. For instance, bacteria and especially fungi present a challenge for accurate immunodiagnostics as compared to viruses. The antigenic pattern of bacteria and fungi can change depending on the developmental stage, and the antigens present at one stage may differ or be absent at another stage. Besides, it may be difficult to obtain antibodies with the required degree of specificity to distinguish between races or closely related species. In diagnostics of plant viruses, such a problem as low immunogenicity can take place because of spatial structure features of coat proteins, that sometimes create certain complication in production of specific antibodies. In addition, it may be difficult to distinguish between closely related strains of some viruses. Viroids are RNA molecules, they contain no proteins, and antibodies are difficult to produce against them.

Implementation of nucleic by techniques allows researchers to overcome a number of disadvantages encountered acid detection-based antigen–antibody techniques, in particular, insufficient specificity (showing as cross-reactivity and leading to false positive results) and insufficient sensitivity (when the target pathogen is present in too low quantities, e.g. in case of initial latent infection in seeds or other planted material). It is important that samples (microlitres) can be analyzed by biomolecular methods with high accuracy. An important advantage of some nucleic acid methods is the combination of fast detection and identification of plant pathogens.

There is a range of biomolecular methods adopted in various areas of plant pathology, e.g. for development of genetic characteristic of species, *formae speciales*, stains, races, and isolates, studying the population structure of pathogenic species, identification of pathogenicity genes, investigation of plant resistance to diseases, toxinogenesis, early stages of host–parasite interactions, etc. In the recent decade, the number of phytopathology studies with these techniques has been growing in an avalanche-like way. Many facts discussed in the chapters of Part 2 of this book were obtained due to application of these methods.

This section is focused on the methods that are most important for plant pathogen detection and identification and currently used for this purpose. They may all be divided into two groups: the probing methods based on nucleic acid hybridization with DNA/RNA probe sequences and those based on polymerase chain reaction (PCR) amplification of nucleic acid sequences. The techniques of the first group were developed before the PCR methods and were alternatives to PCR for pathogen identification till middle of 1990s. However, now the probing methods are used in conjunction with PCR. In some cases, PCR combined with probe enhances the diagnostic potential of PCR. Depending on whether the target nucleotide sequence is known or partially or fully unknown, a particular version of a method within each group may be employed. Plant pathogen diagnostics involving restriction fragment length polymorphism (RFLP) will also be briefly discussed. The PCR, nucleic acid hybridization, and RFLP are the subjects covered in huge and rapidly growing number of publications, including molecular biology textbooks, explaining the theoretical aspects of these techniques. Therefore, the sections below discuss only general aspects of their application. The specific examples of application of the above methods for detection and identification of viruses, bacteria, fungi, and oomycetes are limited to the most important pathogens. References to the publications from which the examples were taken can be found in the reviews used in the writing of this section and listed at the end of the chapter.

Nucleic acid hybridization methods

Application of hybridization of nucleic acids with specific probes was first reported in the middle of 1970s. A dot-blot assay, involving hybridization of immobilized DNA with labelled DNA or RNA probes, was developed in the 1980s and opened new opportunities for plant pathogen detection. In this assay, DNA of the analyzed samples is immobilized on a nitrocellulose (more seldom on nylon) membrane. To convert a double-stranded DNA (dsDNA) into a single-stranded (ss) molecule, the membrane with immobilized target DNA is treated with alkali or heated. The free binding sites on the membrane are blocked by DNA from other organisms (e.g. from salmon sperm) or protein (e.g. bovine serum albumin or dried skimmed milk). Next, the membrane is bathed in a probe-containing solution at a specific temperature (usually at 65 °C) and specific salt concentrations. The probe is a small single-stranded DNA or RNA molecule carrying radioactive or nonradioactive label. If DNA of the analyzed sample contains complementary sequences, the probe hybridizes to them. The greater the number of sequences complementary to the probe are present in the target DNA the higher the binding degree of the probe. The unbound probe molecules are removed by washing.

Results of hybridization can be detected a number of ways depending on the type of label (reporter molecule). Initially, a phosphorus isotope (P^{32}) was used as the label, and the results were displayed using autoradiography, as darkening of the X-ray film. Radioactivity of the label significantly constrained application of the hybridization technology in plant disease diagnostics. This limitation was overcome very quickly, instead of introducing the radioactive label into the probe the latter was biotinylated. After hybridization with such a probe, the membrane is incubated with streptavidin/avidin conjugated to HRPO or AP, then the substrate is added, and the colour change is detected. In addition, chemiluminescent and fluorescent probes are available. In this case, darkening of the X-ray film is caused by emitted light. The use of probes labelled with fluorogenic label will be discussed below in the subsection on PCR.

The above-described stages of the hybridization analyses, especially the enzyme-mediated variant, are very similar to the dot-ELISA procedure. However, it should be noted that these two methods have entirely different principles of binding. In ELISA antibodies capture the antigens, whereas hybridization results from annealing of the complementary strands of the nucleic acids.

Cloned or uncloned probes can be prepared from double-stranded nucleic acids of organisms and viruses or synthesized chemically. Synthetic probes are short single-stranded oligonucleotides which usually consist of 20–40 bases. Synthetic oligonucleotides as probes, especially in combination with PCR, open great prospects for phytopathogen diagnostics, as they allow development of race-, strain-, and isolate-specific tests for detection of closely related races and strains, as well as tests with wide specificity range for determining whether a phytopathogen belongs to a particular taxon. Sequencing the genomes of an increasing number of phytopathogens expands the scope of synthesis of highly specific tests, which improves the accuracy of identification of pathogens both on species and subspecies level.

The diagnostics has considerable limitations in terms of sensitivity as well as duration and labour input required. Its sensitivity is comparable to that of ELISA. Its implementation requires isolation of sufficiently intact DNA from a relatively large amount of the analyzed sample (e.g. at least, from 10,000 bacterial cells), and the hybridization process cannot be fully automated.

One of the present day tasks of plant pathogen diagnostics is to detect one or more target pathogens among other pathogenic or nonpathogenic microorganisms in microbial communities and biological matrices. From this point of view, the reverse probing technique is very interesting. The procedure is as follows: a panel of reference DNAs specific for the pathogen(s) being detected is fixed to the support. Total DNA preparation isolated from a sample (plant tissues, soil extracts, or other matrices), usually previously amplified by PCR, is labelled and used as a probe. Hybridization with reference DNAs results in a pattern of positive signals indicating what target organisms are contained in the sample. Thus, several microorganisms are detected and identified in a single experiment. For instance, a number of oomycete species (*Phytophthora* and *Pythium*) as well as bacteria were detected by this way.

Plant pathogen diagnostics based on reverse probing should further evolve towards increasing the scale of detection. The number of simultaneously detected microorganisms could be brought up to hundreds or thousands. Such large-scale diagnostics will have quite a realistic future if the approach of the **microarray technology** (other names are biochip, DNA chip, gene array) is applied, with hundreds (macroarrays) and thousands of (microarrays) genes simultaneously analyzed in the same experiment.

Restriction fragment length polymorphism (RFLP)

To analyze RFLP, DNA isolated from plant tissues is "cut" by specific bacterial enzymes, restrictases, which recognize characteristic nucleotide combinations. The restriction fragments are separated by electrophoresis in agarose gel, transferred to nitrocellulose membrane with the Southern blotting, hybridized with probes (the probes are usually DNA fragments of the size 500–2000 bases), then analyzed according to the band position in the gel or with the autoradiography. Random products of restriction of genomic DNA can serve as probes. However, more often specially selected clones of unique and rarely repeating DNA sequences, used for drawing up molecular maps of the genome, are exploited as probes.

The major advantages of RFLP-analysis are its universality and good reproducibility. The RFLP-analysis requires considerable amounts of DNA. Depending on the genome dimension, electrophoresis is carried out with 2–15 µg of the mixture of restriction fragments per one well in agarose gel. Therefore, it cannot be used when the amount of material in analysis is limited (for example, for detection of infection in individual plantlets). In addition, the method requires thorough DNA purification from protein, RNA, carbohydrates, and other substances which can interfere with restriction or have an adverse effect on electrophoretic separation of the restriction fragments. Maintenance of a large RFLP-clone library and procurement of reagents and materials require considerable expenses; in addition, safety measures in case of handling radioactive materials are also expensive. In view of all these reasons, RFLP-analysis is practiced in diagnostic laboratories less often than PCR, regardless of all its indisputable advantages.

PCR methods

Polymerase chain reaction (PCR) was developed in 1983 by Kerry Mullis. The American journal "Science" called PCR as one of the most remarkable discoveries of the last decade. Practical applications of the method were first reported in 1985. PCR spread worldwide at a lightning speed. The number of publications on the studies where PCR is used as a method of scientific and applied research is recorded high. However, the primary application of the method was in microbiological diagnostics and became a preferred tool for detection and identification of plant pathogens. The importance of PCR for diagnostic tasks follows from the fact that it enables repeated finding and amplification *in vitro* of specific DNA sequences (or cDNA transcripts) present in trace amounts among a huge number of other polynucleotides. This provides extremely sensitive detection and identification of a target pathogenic microorganism in complex biological or environmental matrix. For instance, PCR allows detection of about 10 bacterial cells in a sample.

PCR is based on the process of natural DNA replication which includes untwining of the DNA double helix, separation ("unzipping") of the DNA strands, and complementary assembling of both strands by the cell enzyme DNA-dependent DNA polymerase. The DNA replication cannot begin at any point but only in fixed starting blocks, short double-stranded areas. Marking the DNA site specific only for the given species (but not for other species) with such blocks allows repeated reproduction (amplification) of this site.

To carry out a PCR process in a test tube, two oligonucleotide probes, called primers, are added to the sampled DNA melted by temperature. The primers are short oligonucleotides, each approximately 16–35 bases in length, and they serve as triggers for synthesis of the chosen DNA site. They are complementary to the sequences on the left and right ends of the fragment and are oriented so as to constrain the completion of the new DNA strand. The forward and reverse primers, added in the sample being analyzed, bind to the DNA sequence to which they are complementary. After binding (annealing) of two primers, the second strand of the specific DNA fragment is generated, extending from the primers, with the help of a thermostable DNA polymerase, Taq polymerase (isolated from the thermophilic bacterium *Thermus aquaticus* and having the activity optimum at 70–72 °C). The newly synthesized DNA fragments serve as matrix for synthesis of the new strands in the next amplification cycle, i.e. chain reaction takes place. The number of copies grows in geometric series, and after 25 amplification cycles (from 30 seconds to a few minutes each) about 100 fragment copies are produced. In an automated thermal cycler, 2–3 hours are sufficient for 30–40 cycles, and they generate more than a million copies of DNA or complementry DNA (cDNA) fragments, from the length as short as 50 bp to over 10,000 bp. This results in production of enough DNA for visual registration of the reaction results after electrophoresis.

Classical PCR-analysis consists of three basic stages: sample preparation which is nothing but a fairly simple procedure of DNA or RNA isolation, PCR itself, and detection of the reaction product (amplified DNA fragment). The reaction mixture consists of the following four main components: (1) DNA containing the fragment to be amplified; (2) a pair of primers; (3) Taq DNA polymerase; and (4) four types of deoxyribonucleotide triphosphate (dNTP) to build the complementary strand on the amplified site. In addition, Mg^{2+} ions must be present in reaction buffer.

The synthesized product of polymerase reaction (amplicon) can be detected by various ways. One of the simplest, efficient, and common way is electrophoretic separation of the amplified DNA with subsequent detection of the fragments with intercalating fluorescent probes, such as ethidium bromide. In addition to ethidium bromide, PCR products can be detected in gel by staining with silver. The difference of several nucleotides between the amplified fragments is found with electrophoresis in agarose gel. However, one-nucleotide repeat difference requires separation in polyacrylamide gel. Prior to electrophoresis, the amplified DNA may be "cut" with a restriction endonuclease to study RFLP of the amplified product. Detection of the amplified DNA by hybridization with different labelled probes is also used. For example, it is possible to measure fluorescence of an oligonucleotide probe (directly or indirectly) after its hybridization with the reaction products in liquid phase, or on a special support.

Detection methods using microtitre plate format where the reaction products are revealed by ELISA (PCR–ELISA) have become common. Combination of PCR and ELISA has been used for analysis of plant pathogens. According to this technique the target fragment is amplified with biotinylated primer, and a DNA sequence is detected by a DNA-binding protein. Resulting hybrid of the amplified DNA and biotin-labelled probe is then captured by the DNA-binding protein-coated microtitre plate, and bound via the biotin moiety to HRPO conjugated with avidin. Upon addition of HRPO substrate, the increase of coloured product is measured with usual ELISA.

There are also other methods to analyze the amplification reaction product. They include high performance liquid chromatography (HPLC), capillary electrophoresis, and mass spectroscopy. Development of these approaches would be promoted by automation of obtaining the results. It should be especially noticed that similar techniques are now also used in biochips, most of them being nothing else but miniature devices for detection of amplification products.

Speaking about PCR and hybridization techniques fundamentally new trend in development of registration systems should be mentioned: the optical biosensors. Diagnostics of pathogens by these devices does not require any modification of PCR-products or probes. They allow real time determination of picogram amounts of the nucleic acid fragments immobilized on the surface of the optical cell. The cell design allows several hundred repeated uses of one optical cell with the appropriate regeneration cycles after each determination. Such devices appear promising for quantitative detection of plant pathogens.

In modern plant pathology, PCR is applied not only for plant pathogen diagnostics but also in taxonomy and phylogeny of viroids, viruses, phytoplasmas, bacteria, fungi, oomycetes, and nematodes. This technique is widely used for monitoring plant diseases, and also for detection of their causative agents in organs of vegetating plants and seeds and fruits in storage. It has been used for detection of many viruses of economically important crops and other cultivated plants. The list of species, races, and isolates of oomycetes and fungi has already included the representatives of the most damaging taxa (e.g. *Phytophthora, Pythium, Aspergillus, Colletotrichum, Fusarium, Gaeumannomyces, Helmintosporium, Mycosphaerella, Phoma, Puccinia, Rhizoctonia, Septoria, Tilletia, Ustilago, Verticillium*, etc.). The PCR protocols have been elaborated for many important plant pathogenic bacteria, and a number of appropriate primers for their identification are described in the Laboratory Guide for Identification of Plant Pathogenic Bacteria.

Modifications of PCR method
Reverse transcription-polymerase chain reaction (RT-PCR)
The ordinary polymerase reaction is not suitable for RNA identification, therefore, it is unsuitable for diagnostics of viroids and RNA-containing

viruses, as Taq polymerase cannot catalyze DNA synthesis on the RNA template. In practice, the task of detection and identification of these pathogens is accomplished by an additional enzyme, RNA-dependent DNA polymerase (reverse transcriptase). The reaction catalyzed by this enzyme leads to formation of single-stranded cDNA fragments that are further amplified by Taq polymerase.

RT-PCR is successfully used for detection of viral and viroid pathogens as well as viruliferous vectors (aphids). It is well known that detection of viroids, present in infected plants in minute amounts, is especially difficult. These causative agents are diagnosed by hybridization analysis or return gel electrophoresis. These two methods are 100 and 1000 times less sensitive, respectively, as compared to RT-PCR for which isolation of 1–100 pg of total nucleic acid from the plant tissue will suffice.

An important advantage of RT-PCR in diagnostics of bacteria, fungi, or oomycetes is its ability not only to detect pathogens but also to confirm their viability.

The reverse transcription can be also accomplished by, namely, DNA polymerase of the other thermophile, *Thermus thermophilus*. This enzyme (Tth polymerase) catalyzes synthesis of a single-strand complementary DNA on RNA in the presence of manganese ions. Interestingly, after addition of magnesium ions and manganese ion chelators in the reaction medium, the enzyme is able to catalyze an ordinary DNA-dependent DNA polymerase reaction. This enzyme can be used in diagnostics of viroids and RNA-containing viruses to synthesize simultaneously a single-stranded DNA, complementary to RNA of the target pathogen, and amplify the DNA fragments.

Nested PCR

One of the widespread PCR modifications involves two pairs of primers, one of them capable of amplifying the internal part of the fragment resulting from the completion of the cycles of the first amplification with the external pair of primers. There are two options of nested PCR application in diagnostics. Under the first option, amplification (15–30 cycles) is carried out with the external pair of primers, and a basic DNA fragment is amplified. Next, part of the contents is taken to another test tube containing the reaction mix which includes a pair of primers detecting the sequences in the first product. The second stage of amplification (reamplification) also consists of 15–30 cycles. Under the second option, the annealing temperature for the internal pair of primers is chosen so that they are not involved in the reaction at the first round of amplification. In this variant of nested PCR, the primer annealing temperature is usually taken 10–15 °C lower than for the external pair of primers. After amplification with the external primers, the temperature for annealing the second pair of primers is set ten degrees lower in the programmed amplifier, which provides effective involvement of the internal pair of primers in reamplification. Nested PCR has both advantages and disadvantages. The advantage is firstly high sensitivity of the technique. Secondly, nested amplification makes unnecessary other methods of confirming specificity of the reaction. Besides, the transfer of reaction products from one test tube to another considerably reduces the concentration of inhibitors that may occur in the preparation of nucleic acids. However, on the other hand, transfer of the amplified fragment from one test tube to another can lead to false positive results due to cross contamination of the samples with the amplification products. To prevent it, the second variant of the assay is used, and both rounds of amplifications are carried out in the same test tube. An example is a sensitive, specific, fast, and simple RT-PCR developed for *Erwinia amylovora* in naturally infected asymptomatic plants.

Multiplex PCR

Multiplex, or multiprimer PCR is usually interpreted as process of coamplification of several DNA matrices in the same reaction medium with several pairs of primers. This allows simultaneous screening of several pathogens in one experiment. It is important to avoid the primers complementary to each other and select conditions for effective annealing of all participating primers to assure the equal yield of different amplified products. In addition, to ensure the distinguishability, amplified products generated for different target pathogen, must have different sizes. Multiplex PCR have been applied for detection of phytoplasmas, nematodes as well as multiple viruses or viroids in a single sample.

Immunocapture PCR (IC-PCR)

PCR can be combined with ELISA at the final detection stage. There is also another combination of PCR and immunoenzyme assay termed as immunocapture PCR. In this, antigen–antibody interaction is used at the initial stage of the analysis. For instance, sample being investigated (e.g. an extract containing target virus) is added into centrifuge microtuber or microtitre plate well, the surfaces of which are coated with specific antibodies, and they capture viral particles. DNA or RNA is then released in alkaline conditions in the presence of a detergent and analyzed by PCR or RT-RCR, respectively. Usually, IC-PCR improves the specificity and enhances the sensitivity of plant pathogen detection and can reduce an undesirable influence of inhibitors present in the sample. The IC-PCR has been used for some viroids, viruses, and bacteria.

Quantification of results in conventional PCR

Conventional PCR is highly sensitive and specific analysis. However, quantitative measurement of the amplification products and, especially, the initial template, in standard PCR is a challenge. In calculation of standard curves, it is practically impossible to take into account all the factors of the *in vitro* biochemical reaction. Therefore, application of the measures, that allow taking into account numerous parameters during PCR, are necessary. The test specimens can produce different product yields under initially identical conditions in terms of reagents and temperature in the amplifier, and it is not always possible to calculate accurately the number of initial amplification targets (i.e. ultimately, the amount of initial pathogen DNA/RNA in the sample). Hence, adjustments are made in the procedure to record the course of the reaction or amendments to take into account possible deviations for each sample. One of the most common approaches to quantification is application of internal standards introduced into the PCR tube together with the sample. The internal standards are DNA fragments (so-called competitor DNA) which, like DNA, have primer annealing sites, but yield products of different sizes. The internal standards with different number of copies per sample allows determination of the number of copies of the initial nucleic matrix in the reaction.

A practical quantitative PCR technique is "the method of final dilutions". The key point of the method is determination of the final dilution of a DNA or RNA sample to yield a positive PCR result. A panel of standards to reveal sensitivity of the test system is used in each assay. This method does not provide a real opportunity to control the amplification process with all its deviations, however, the procedure is very simple.

Another available way of the quantitative analysis is measuring the product during the reaction. It is with real-time PCR.

Real-time PCR

Real-time PCR allows monitoring (registration) of the PCR product accumulation in real time and to plot calibration curves for actual processes in each particular test tube during each cycle throughout the reaction. This method does not require the time-consuming stage of electrophoretic separation of the PCR products. As compared to the classical PCR technique, it makes plant pathogen detection faster, more specific, and sensitive. As real-time PCR is a quantitative method, it produces immediate data for the initial amount of the target sequence immediately after amplification, as well as calculation of the content of pathogen gene material in the analyzed samples. Sometimes, specificity of real-time PCR can be better than in conventional PCR, since specific reporter probes complementary to a sequence inside an amplicon are added to the specific primers. This enhancement of specificity allows distinguishing between the sequences differing in one nucleotide base pair, e.g. in order to differentiate fungicide resistant strains from sensitive ones. Detection is carried out in a closed system (without opening the test tube) which is less susceptible to contamination.

Amplification proceeds in a special instrument equipped with a device to illuminate each amplification cell with excitation light, and with a multichannel fluorescent detector with the spectral resolution to record amplicons.

To detect real-time PCR product, fluorescent probes are used. Detection of the amplified product is based on measurement of the fluorescence emitted by a probe incorporated into newly formed amplicon or, on the contrary, by a probe that releases into buffer solution during the amplification process.

For simultaneous detection of several targets it is possible to use probes labelled with different fluorescent reporter dyes. Detection can be carried out by one of the following ways.

The first way (Figure 3.3-1) involves an intercalating agent (SYBRgreen® that has replaced ethidium bromide used in the past). This is the simplest way of detection, however, it has an important disadvantage: the fluorescent stain equally intercalates the target and nonspecific amplicon, if the latter is produced in the reaction.

The second way involves several types of fluorogenic probes, including TaqMan® which is currently used in most applications (Figure 3.4). The probe is a sequence of 25–30 oligonucleotides labelled with fluorescein derivative (usually 6-FAM) at the 5'-end and with fluorescence quencher (usually TAMRA) at the 3'-end. This probe hybridizes with analyzed DNA. Taq polymerase that catalyzes the build-up of a new DNA strand from 3'-end to 5'-end possesses 5'–3'-endonuclease activity: if in synthesis of the new strand the enzyme encounters a double-stranded DNA fragment, polymerase will destroy it by elimination of one or more nucleotides. As Taq polymerase extends the primer during the amplification cycles, it cleaves the probe and releases the fluorescent chromophore into the reaction solution. Fluorescence of the released probe can be measured.

The principle of spatial separation of the fluorophore and quencher is applied in Molecular Beacons and in Scorpion™ probe (Figure 3.3-2). They are designed so as to produce a structure reminding a hairpin, where the quencher and fluorophore are attached to the pin end nucleotides. Since the quencher and chromophore are positioned in immediate proximity, the probes do not fluoresce. As the temperature rises, the probes acquire an ability to hybridize to the target amplified product. After hybridization, the fluorophore moves aside from its quencher, and fluorescence is observable.

A new type of PCR probe is the fluorescent resonance energy transfer (FRET) probe (Figure 3.3-3). Detection is carried out with two different probes labelled with two different fluorogenic dyes which light at different wavelengths. The first probe is labelled at 3'-end, while the second at its 5'-end. The probes are designed in such a way that when they hybridize to an amplicon, they are ranked in a "head-to-tail" way. The first probe emitting higher wavelength light (green) excites the second probe whenever set close to it. The second probe then emits lower wavelength light (red). The red fluorescence is measured at 640 nm.

For the last few years, real-time PCR has been actively penetrated in applied phytopathology. A number of present-day reviews have been fully or partially given over to this method and its application in detection and identification of plant pathogens. It is used for field diagnostics of the plant diseases including those caused by the most hazardous and quarantine pathogens, when promptness of the exact diagnosis and identification of the pathogen is especially important (e.g. tomato spotted wilt virus, plum pox potyvirus *Ralstonia solanacearum, Pseudomonas syringae, Clavibacter michiganensis* subsp.

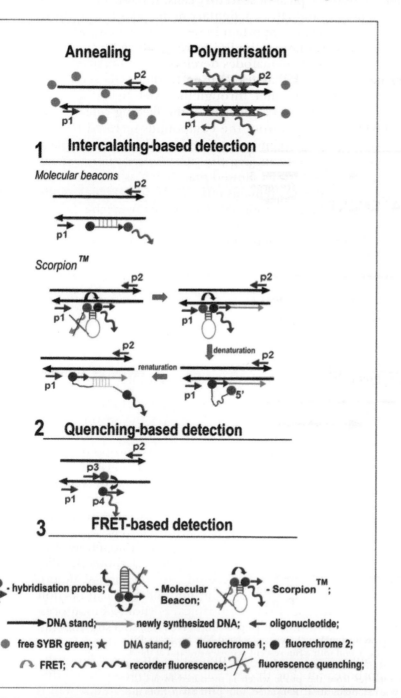

3. (1) Detection of real-time PCR product using SYBRgreen®; (2) Molecular Beacons and
M; (3) FRET. Is modified from Gachon C, Mingan A, Charrier B, 2004.

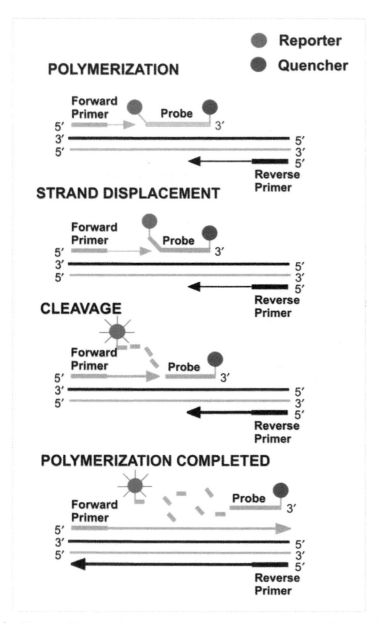

e 3.4. The principles of real-time PCR using TaqMan chemistry (Schaad NW, Frederick RD, , et al, 2003). The probe is labelled at the 5′ end with a fluorescent reporter dye and at the 3′ ith a quencher dye. During the strand displacement step, the 5′–3′-endonuclease activity of olymerase releases the fluorescent reporter from the probe which is measured by a CCD ge-coupled device) camera at each cycle of amplification. Reprinted, with permission, from the l *Review of Phytopathology*, Volume 41 © 2003 by Annual Reviews www.annualreviews.org

sepedonicus, Phytophthora infestans, P. citricola, Septoria tritici, Puccinia recondita, P. striiformis, Tilletia indica, Phomopsis longicola, Rhizoctonia solani, etc.). Real-time PCR protocols have also been developed for analysis of the genes responsible for toxinogenesis in trichothecene-producing Fusaria and aflatoxin-producing *Aspergillus* species.

Alongside with the above listed advantages, an important benefit of real-time PCR is its user-friendliness and low labour intensity. The process of analysis is fully automated, and many samples can be analyzed in the same experiment. Today, a range of instruments for rapid-cycle real-time PCR is available, including portable modifications (Light Cycler™ from Roche Diagnostic Corp., RAPID from Idaho Technologies, Smart Cycler® TD from Cephid, Inc., etc.).

Real-time PCR has overcome many shortcomings of conventional PCR, though the technique certainly has limitations, both common with the standard PCR and its own. Inhibitors originating from plant or soil sample as well as simple analytic procedures excluding DNA isolation along with very small analyzed volumes (1–10 µl) can lead to false negative results at low content of target DNA in a sample. Both PCRs can detect not only living but also dead cells or free DNA, which can result in false positive diagnosis. The way to overcome the above disadvantages in diagnostics of plant pathogenic bacteria has been developed and termed as BIO-PCR.

BIO-PCR

BIO-PCR allows detection of viable plant pathogens. It is a combination of cultivation of bacteria on selective media with PCR or real-time PCR. The sample to be analyzed is plated onto agar or added in liquid medium and cultivated for enriching the target bacterium. The agar surface is washed off or the medium is centrifuged, and 1 µl or 10 µl of the sample is taken to PCR. This method is recommended when the sensitivity is critical, and it is especially suitable for species growing fast (pinpoint-size colonies form during 15–48 hours). The number of bacterial units (over 1000 cells/ml) normally generated by one bacterial colony of such size during this time is a threshold for PCR. The sensitivity of BIO-PCR can be increased if the sample is filtered through a surfactant-free membrane. The filter is then placed right-side up onto a semiselective agar medium and incubated till the appearance of colonies of target organism. A single cell trapped on the membrane may be sufficient for thousand-fold increase in the number of cells after enrichment by cultivation.

RAPD-PCR

The modern plant pathology possesses a variety of PCR-based diagnostic tools. A researcher should have an idea of the target sequences that could be selected for DNA assay for a plant pathogen to be detected and identified. Therefore, availability of information derived on phytopathogen genome sequencing is essential. As the specificity of PCR is assured by the primers, a correct choice and design of the primers is a requirement for success of any PCR analysis. The primer choice is the first stage of PCR diagnostics.

As genomes of viruses and viroids are relatively small, full data on their sequences are available in databases and appropriate primers can be easily found. Less information has been accumulated about the genomes of bacteria, oomycetes, and fungi, although the volume of data is growing. For these pathogens, general approaches to selection of particular known DNA target fragments are available, and the techniques based on screening of random regions of DNA have been developed.

For bacteria, oomycetes, and fungi, DNA encoding ribosomal RNA (rDNA) is generally used as a target sequence. Several facts make rDNA suitable for diagnostic purpose. Many copies of rDNA are present in each cell, thus enhancing the sensitivity of detection. The genes are present in all organisms and contain highly conserved 5.8 S region that gives rDNA universal applicability. At the same time, there are highly variable regions, such as internal transcribed spacer (ITS) regions. The conserved regions can be used to design universal primers for the group detection of microorganisms within a taxon (for all oomycetes, fungi, or bacteria), while the presence of variable regions allows finding distinctions between races, strain, and isolates. Other target sequences which are used for detection of fungi are beta-tubulin genes which are connected with resistance to fungicides. DNAs contained in bacterial plasmids and pathogenicity-associated genes usually serve as the sources of the target fragments.

One of the techniques used if the target nucleotide sequence is unknown is random amplified polymorphic DNA PCR (RAPD-PCR), or arbitrary primed polymerase chain reaction (AP-PCR). The RAPD-PCR is usually applied, alone or together with RFLP, in studying DNA polymorphism, in gene mapping, and in population and evolutionary biology. The RAPD-PCR is important for plant pathogen diagnostics as it enables screening the sequences specific for closely related species, strains, races, and isolates, and differentiate them.

In contrast to the described PCR analyses, where two primers restricting the amplified sequence are used, RAPD-PCR involves annealing of single primers. The primer binds to the random complementary sequences of the genomic DNA, and after amplification, RAPD-PCR product of arbitrary length, which is partially or completely homologous to the arbitrarily primed sequence at both ends, is generated. The DNA polymorphism, resulting from insertions, deletions, and base substitutions, influences generation of the RAPD-PCR product, which ultimately shows as presence or absence of bands in gel after RAPD-PCR. With this method, it is possible to amplify gene products from many organisms analyzed, and the pattern of the bands after electrophoresis will be specific for a particular organism. Many different primers have to be tested to identify a band that is specific for a target. Specific bands can be used for synthesis of highly specific primers.

Finishing, it should be noted that PCR is not the only amplification diagnostic technique. For instance, some plant pathogens have been detected with ligase chain reaction (LCR). It is based on the ability of DNA-dependent DNA-ligase to ligate a DNA strand in the presence of adenosine triphosphate (ATP) and Mg^{2+} ions, at rupture of the phosphodiesteric bond. This method was suggested by Wu and Wallace in 1989. A characteristic feature of DNA ligase work is high specific activity in ligation of single-stranded ruptures at the template which constitutes the second complementary strand, and low specific activity in simultaneous ligation of two ruptures in both strands or rupture in single-stranded DNA. Implementation of LCR requires finding two pairs of primers complementary to each other and to the initially chosen fragment of the matrix (for instance, DNA of some causative agent), as "head to tail" arrangement in direction from 5′ to 3′ end. As early as after the second LCR cycle, the reaction mix accumulates the product which is a ligated double-stranded DNA fragment, structurally identical to the four primers used. It is characteristic that even a one-nucleotide error in the place of annealing leads to a negative result. Therefore, LCR is promising for enhanced detection of plant pathogens and revealing the point mutations in the wild types of causative agents. The LCR has been adapted in a PCR format and modified to detect the potato viruses A and Y in tubers, identify *Erwinia stewartii*, and to distinguish *Phytophthora infestans, P. mirabilis*, and *P. phaseoli* from other *Phytophthora* species. In the latter case LCR was combined with ELISA.

Conclusion

Nucleic acid and immunochemical techniques occupy the leading positions as methods of diagnostics in the present plant pathology. Among them, PCR and its modifications as well as various ELISA formats are the most popular. They are highly sensitive, specific, rapid, user-friendly, and generally excel the conventional methods in these parameters. Combinations of PCR and ELISA further increase their thresholds, enhance the specificity, and improve results of the analyses.

PCR and ELISA cover the whole range of plant pathogens (viroids, viruses, phytoplasmas, bacteria, fungi, and oomycetes). They are applied for pathogen detection in plant, soil, and other complex biological and environmental matrices, they can detect a target pathogen among many other microorganisms. These methods have been adapted to laboratory and field applications and can be used to distinguish and identify closely related species, strains, races, and isolates.

A number of precision instruments and facilities are developed for PCR and ELISA tests; commercial kits, master mixes, additional reagents. Public databases are available for application of the described technologies. This makes many of them as routine ways of plant pathogen detection.

Considerable progress has been achieved in plant disease diagnostics due to introduction of biomolecular and immunochemical techniques. To date, the methods elaborated on the basis of these technologies, alone or in conjunction with traditional phytopathological methods, allow reliable diagnostics of plant pathogens as well as more successful control and proper management of many economically important diseases. For the last few years, a number of promising approaches have been developed. Further progress in plant disease diagnostics should be expected, as novel methods and new applications of detecting modes are introduced. Much diagnostic technologies as microarray and biosensors seem to be good candidates for leadership in this field.

Using plant pathogen metabolites in diagnostics and research of plant–pathogen relationships

Some characteristic metabolites of plant pathogens can be used for diagnostic goals and to study disease development. For instance, metabolic differences between fungi and plants are used in plant pathology to address many practical and research problems. The comprehensive potential of biochemical methods in revealing such differences will be viewed here with the example of chromatographic analysis of some secondary metabolites produced by cereal pathogens, which is applied to determine pathogenic biomass in host plants. In this section, analysis of bacterial fatty acids is very briefly discussed as one of the examples showing applicability of biochemical methods for diagnostics of plant pathogenic cultures.

Analysis of specific fungal metabolites in infected plant tissues

Plant pathogenic fungi synthesize many secondary metabolites that do not occur in plants or are present in plant tissues in threshold amounts. Accumulation of any of these metabolites in infected plants can be interpreted as evidence of plant pathogen expansion. In this sense, the metabolites may be considered as markers of the infection, and indicative to the degree of development. ELISA provides information on accumulation of pathogenic biomass by measuring the content of antigen, whereas the operational activity of the pathogen is provided by analysis of biochemical markers. Although the marker metabolites cannot be used to discriminate between different fungal species, biochemical ways of determination of the degree of plant tissue colonization by fungal pathogens are efficient enough in characterization of cultivar resistance, in fungicide selection, as well as in investigation of plant diseases with complex signs, such as head blight.

Out of a large number of marker metabolites indicating presence of entophytic or pathogenic fungal biomass in plant tissues, this section discusses ergosterol, polyols (arabitol and mannitol), and chitin.

Ergosterol is a polycyclic alcohol (sterol) belonging to steroids. Sterols are very common in nature, but ergosterol is a dominant cell membrane sterol found almost exclusively in fungi. Sterols are integral constituents in the lipid layers of membranes of living organisms, they are involved in packing of membrane phospholipids and regulation of conformation changes in membranes,

ding to formation of a stronger membrane framework. Sterols play a
role in protection of microorganisms from adverse environmental
ns and stress. Oomycetes (*Phytophthora* and *Pythium*) are unable to
ze sterols and incorporate host-plant sterols into their membranes.
ls (acyclic polyhydroxy alcohols) are secondary metabolites of fungi.
ɟh they have been found in some plant species, the major producers
e metabolites are fungi. Fungi generally accumulate polyols such as
, erythritol, arabitol, or mannitol. Very high level, up to several hun-
illimoles per liter, can be accumulated in the cells. Polyols are also
l to the environment. Fungal polyols fulfil different physiological func-
hey are involved in osmoprotection of fungi. These metabolites are
tly found in spores where they probably act as storage carbohydrates.
ì is a biopolymer consisting of N-acetyl-glucosamine. Chitin's deacety-
roduct is chitosan. In fungi, chitin is a cell wall component of spores
ohae. Together with cell wall glucan, chitin forms microfibrils and
s the framework in cell wall morphology. A characteristic feature of
ɔhitin is its species diversity, and its physicochemical properties differ
rustacea chitin. Cell wall of oomycetes contains no chitin.
ɔility of ergosterol, arabitol, mannitol, and chitin as biochemical
s for studying plant tissue colonization by pathogens can be convinc-
emonstrated for wheat plants infected with rust fungi (*Puccinia*
s f. sp. *tritici* and *P. recondita*), fusarial root rot, scab pathogens (in par-
Fusarium culmorum, *F. sambucinum*, and *F. graminearum*), and Septoria
(*Septoria nodorum*), as well as rice plants infected with rice blast
(*Pyricularia oryzae*).
sess fungal expansion in wheat and rice plants, the marker metabo-
e isolated in different ways from the infected plant tissues, and,
or after derivatization, analyzed by gas liquid chromatography (GLC)
performance liquid chromatography (HPLC). The isolation proce-
cludes plant tissue homogenization, co-extraction, purification from
ctive substances, and concentration of the resulting marker-containing
. It is important to optimize all steps of analyte preparation in order
de adequate recovery of the target metabolite. For instance, direct
cation of the samples to be analyzed for ergosterol is preferred to
on with methanol with subsequent hydrolysis. Extractivity of *Fusarium*
m ergosterol depends on the properties of fungal species and its cell
tmentation. Such facts should be taken into account in development
rtical procedures for this metabolite. All preparatory and analytical
ɔns should be standardized as much as possible.

ımics of fungal metabolite accumulation in process of pathogenesis
marker metabolites, absent from healthy leaves of cereals can be
ɪ them at different times since the fungal inoculation. Although bio-
ɪl tests are much less sensitive than compared to PCR and ELISA,

:tion process, and in many cases they show the real course of disease. For
.nce, a noticeable increase in chitin and polyols in the first leaf of wheat
d be observed 5–7 days after inoculation with the spores of *P. recondita* or
uminis f. sp. *tritici*. The amount of fungal metabolites grew rapidly during
e five days up to the end of the experiment, the duration of which was con-
ned by the duration of life of the first leaf. While accumulation of patho-
c biomass in plant tissues starts earlier (Figure 3.5), antigens of fungal
lium is detected in wheat leaves 2.5–3 days after infection by
aminis f. sp. *tritici*. Active accumulation of polyols in wheat tissues could
been caused by the generation of spore-forming structures. The exper-
its with rice blast pathogens showed the presence of polyols in conidia
almost total absence in vegetative mycelium. Polyol concentration in
lospores was virtually constant for each of the two investigated species of
inia. An increase in ergosterol in the wheat plants inoculated both with
al rust and *S. nodorum* became noticeable earlier than the increase in
itol and mannitol, and the dynamics of ergosterol accumulation coin-
d with the dynamics of accumulation of vegetative mycelium antigens of
e fungi.

ost probably, the dynamics of polyol content in the inoculated leaves is
:ative of the rate of generation of spore-forming structures in the phy-
thogenic fungi, whereas ergosterol accumulation is mostly indicative of
growth rate of the vegetative mycelium of the pathogen in host-plant
es. These facts should be taken into account in selection of a marker for
al biomass in an infected plant.

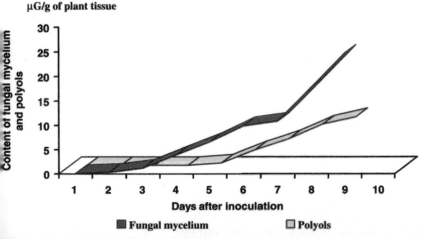

e 3.5. Accumulation dynamics of *P. graminis* f.sp. *tritici* biomass and polyols in wheat leaves.

The use of fungal metabolites to assess plant–pathogen relationship

In standardized plant growth and inoculation conditions, the dynamics of fungal metabolite accumulation reflects the rate of accumulation of pathogenic biomass which well corresponds to the degree of compatibility of the pathogens with wheat and rice cultivars. In other words, the fungal mass accumulated during a certain time is less, the larger the number the higher the efficiency of the barriers that impede pathogen growth and development in plant. Leaves of a susceptible cultivar are colonized by a phytopathogenic fungus more intensely than those of a resistant cultivar. This was shown by many experiments on determination of the metabolites in the wheat cultivars both susceptible and resistant to stem and leaf rust or Septoria blight as well as in the rice cultivars which unequally respond to inoculation with blast disease pathogen. Generally, determination of fungal biomass by accumulation of the marker metabolites in infected wheat and rice seedlings provided quite complete and correct idea of the level of resistance of a particular cultivar to the disease of interest.

There is no great problem in distinguishing resistant from susceptible cultivars in the field by characteristic external signs of infection. It is somewhat more difficult to determine the level of resistance of the tolerant cultivars which cannot be definitely referred to a particular group, or cultivars with the so-called partial resistance. At a high compatibility of the partners, low receptivity of the plants with such type of resistance assures a slow course of disease. The external signs of disease on cultivars with partial resistance can be indistinguishable from those in susceptible cultivars, which may become a reason for the culling of promising prototypes. In this case, determination of dynamics of fungal biomass accumulation can considerably promote solving the problem.

The level of resistance or susceptibility of a cultivar can be assessed in laboratory by accumulation of biochemical markers, which provides an opportunity to grow plants in controlled conditions of vegetation, experiment and standardize infection pressure, and inoculation procedure. If in laboratory conditions inoculation of a cereal crop assures the pathogen reproduction, seedlings should be preferred, as this will provide considerable saving of time and funds. However, for some pathogens more mature plants have to be used, for example, inoculation of wheat with *S. nodorum* is carried out after the complete formation of the third leaf. A laboratory experiment requires a relatively short time, with sufficiently small labour input, and, most important, its results are reproducible, as it is conducted in controlled conditions. It is important to note that many experiments have shown quite close relationship between the results of laboratory experiments and the data characterizing the disease development in field conditions at multiple reinfection of plants. For instance, it was shown that the data on polyol accumulation in the first leaves of seedlings of eight wheat cultivars inoculated with a stem rust pathogen well correlated with the results of determination of the disease development in several field experiments during the entire vegetation period. Similarly, accumulation of leaf rust mycelium in the first leaves of four wheat cultivars had the same dynamics as the disease development in field conditions during the entire vegetation period.

Screening of cultivars with race-nonspecific resistance (e.g. slow-rusting cereal cultivars) that can be durable due to independence from variation of the race composition in the pathogen population can be used as a basis for the analysis of marker fungal metabolites. It is not easy to find cultivars with potentially race-nonspecific resistance in a field experiment but for no other reason that several strains or races have to be used for plant inoculation, while the plots must be securely isolated from each other. At the same time, a laboratory experiment requires much less effort. To determine the degree of race-nonspecificity of resistance, seedlings of the cultivars being tested are inoculated with fungal races/strains/biotypes possessing different sets of virulence genes. The growth rate of fungal biomass in the plants of the cultivars being tested is calculated by the content of marker metabolites in the infected tissue which is determined by chromatographic analysis. A cultivar universally susceptible to this pathogen serves as the reference. An equal inhibition of expansion of all biotypes in the tissues of the tested cultivar is indicative of race-nonspecific resistance of the given cultivar.

Investigation of dynamics of accumulation of phytopathogenic fungal mass in infected cereal seedlings can provide very interesting and useful information. The results of such a study provide an idea of the degree of compatibility of the partners, hence, of the level of plant resistance to a particular pathogen. If the seedlings are inoculated with subspecies forms of pathogens with a different range of virulence, this yields an opportunity to determine the type of resistance, too. Especially valuable is the capability to reveal the cultivars (or germplasms) with race-nonspecific resistance. Biochemical testing of fungal mass was found to be useful in studying diseases with complex signs. A representative example of such a disease can be head blight, where visual observation of the development of disease is fairly difficult, whereas the data on ergosterol localization and dynamics of its accumulation are very informative. Quantitative determination of the marker metabolites in plant tissue also allows obtaining information on fungicide efficacy. Simultaneously, activity of a number of compounds can be compared, and an idea of their mechanisms of action can also be gained, as analysis of some fungal metabolites can show the effect of fungicides on mycelium growth, while testing others can help to assess the effect on sporogenesis.

Fatty acid analysis for identification of plant pathogenic bacteria

Diagnostics based on the biochemical properties of pathogens may be straightforwardly illustrated by fatty acid fingerprinting used for identification and differentiation of cultured bacteria. Analysis of cellular fatty acids has been routinely used to characterize, differentiate, and identify genera, species, and strains of bacteria, including plant pathogenic ones. It can also be applied for oomycetes and some fungi. A number of universally recognized laboratory guides for *in vitro* identification of bacteria are available.

Over 300 fatty acids and related compounds have been found in bacteria. The types of fatty acids produced and the relative concentrations of individual fatty acids strongly depend on the genotype, and they represent the characteristic traits of the taxa. Environmental factors also have an effect on the fatty acid composition. The growth substrate, incubation temperature, and incubation time are among the determining factors. However, if these factors are maintained constant, the qualitative and quantitative composition of fatty acids is highly reproducible, consistent, and conserved among different taxa. Simple mutations, or modifications of cell genetic material as a result of plasmid activity, do not change the composition of microbial fatty acids.

For identification of bacteria, automated gas chromatographic determination of fatty acid methyl esters (FAME) on capillary columns is used. Preparation of a sample intended for GC analysis include harvesting bacterial cells from the plate surface, saponification of the bacterial material, methylation of the fatty acids, and FAME extraction. To characterize and identify bacteria and compare quantitative and qualitative differences of the constituent bacterial FAMEs, a computer software package (Microbial Identification System; Microbial ID, Inc., [MIDI], Newark, DE) is used. The determination is rapid, sensitive, and reproducible. The chromatographic system automatically determines the composition of the complex of fatty acids in the cells of bacteria or fungi, then identifies it by comparison with the complexes of fatty acids of the known microorganisms stored in the extensive database of MIDI Sherlock (Figure 3.6). This allows classification of the species of the tested microorganism promptly with maximum accuracy and minimal costs. Standardized conditions for culturing of microorganism are described in the microorganisms' identification procedure by MIDI Sherlock. The Sherlock database contains the data on the most pathogenic

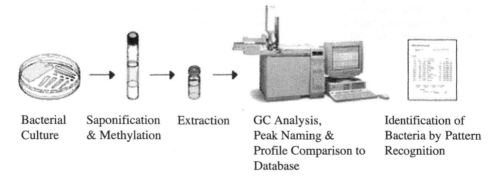

| Bacterial Culture | Saponification & Methylation | Extraction | GC Analysis, Peak Naming & Profile Comparison to Database | Identification of Bacteria by Pattern Recognition |

Figure 3.6. Diagram of MIDI System for FAME-based identification of bacteria by gas chromatography (from Myron Sasser. "Identification of bacteria by gas chromatography of cellular fatty acids" 1990, revised 2001. www.midi-inc.com that is publicly accessible at the address http://www.midi-inc.com).

species of *Xanthomonas and Pseudomonas*, as well as on other groups of phytopathogenic organisms, including fungi.

Although fungi and oomycetes generally produce fewer types of fatty acids and in lower amounts, compared with bacteria, the amount of data is growing to show that the fatty acid pattern may also be useful for identification and differentiation of plant pathogenic fungi (*Penicillium, Rhizoctonia*). The oomycete genera *Phytophthora* and *Pythium* produce several fatty acids not usually produced by fungi. The profiles of fatty acids of these groups that are under investigation can yield valuable information for their biochemical identification.

Selected Literature

Alvarez AM. Integrated approaches for detection of plant pathogenic bacteria and diagnosis of bacterial diseases. Annu Rev Phytopathol 2004; 42:339–366.

Clark MF. Immunosorbent assay in plant pathology. Annu Rev Phytopathol, 19:83-106.

Gachon C, Mingan A, Charrier B. Real-time PCR: what relevance to plant studies? Journal of Experimental Botany 2004; 55(402):1445–1454.

Halk EL, DeBoer SH. Monoclonal antibodies in plant disease research. Annu Rev Phytopathol 1985; 23:321–350.

Llop P, Bonaterra A, Penalver J, Lopez MM. Development of highly sensitive nested-PCR procedure using a single closed tuber for detection of *Erwinia amylovora* in asymptomatic plant material. Appl Environ Microbiol 2000; 66:2071–2078.

Martin RR. Impact of molecular diagnostic technologies on plant disease management. Annu Rev Phytopathol 2000; 38:207–239.

Miller SA, Martin RR. Molecular diagnosis of plant diseases. Annu Rev Phytopathol 1988; 26:409–432

Singh RP, Singh US (eds). Molecular methods in plant pathology. Boca Raton: CRC Press Inc.; 1995.

Schaad NW, Frederick RD. Real-time PCR and its application for rapid plant disease diagnostics. Can J Plant Pathol 2000; 24:250–257.

Schaad NW, Frederick RD, Shaw J, et al. Advances in molecular-based diagnostics in meeting crop biosecurity and phytosanitary issuses. Annu Rev Phytopathol 2003; 41:305–324.

Ward E, Foster SJ, Fraaije BA, McCartney HA. Plant pathogen diagnostics: immunological and nucleic acid-based approaches. Ann Appl Biol 2004; 145:1–16.

Ziegel A, Torrance L. Applications of recombinant antibodies in plant pathology. Molecular Plant Pathology 2002; 3:401–407.

Chapter 4

Phenomenology of plant–parasite relations

Yu. T. Dyakov

Types of resistance

From the phenomenological point of view, plant pathologists discovered at least two types of resistance to many diseases, a long time ago. Van der Plank (1963) called them vertical and horizontal resistance. According to his definition, "If a cultivar is resistant to several strains of the pathogen, we call such resistance vertical, or perpendicular. If the resistance is evenly distributed over all strains of the pathogen, we call such resistance horizontal, or lateral". The difference between them can be explained with a drawing that shows the response of four potato cultivars to the races of the late-blight pathogen *Phytophthora infestans* (Figure 4.1). The cultivars Severnaya Roza and Voltman have no vertical, but differ in the level of horizontal resistance. Therefore the former cultivar is attacked by all the races equally severely, while the latter equally slightly. The cultivars Domodedovsky and Olev also differ in the level of horizontal but they have vertical resistance, too. Therefore, they are resistant to different races (Domodedovsky is resistant to races 0 and 1.4, but it is attacked by race 3, while Olev is resistant to races 0 and 3, but is attacked

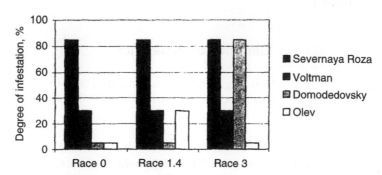

Figure 4.1. Infestation of four potato cultivars by three races of *P. infestans*. The cultivar Severnaya Roza is susceptible to all races (no resistance); the cultivar Voltman is moderately resistant against all races (horizontal resistance); the cultivar Domodedovsky is absolutely resistant against races 0 and 1.4 but is strongly affected by race 3 (vertical stability to particular races); the cultivar Olev is absolutely resistant against races 0 and 3 and moderately resistant against race 1.4 (combination of vertical stability to particular races with horizontal resistance to all races).

Table 4.1.   Differences between vertical and horizontal resistance

Trait	Vertical	Horizontal
Genetic control	Oligogenic	Polygenic
Phenotypic expression	Qualitative (alternative)	Quantitative (non-alternative)
Modification by external conditions	Weak	Strong
Relations with parasites	Race-specific	Non-specific
Mechanisms of resistance	Active defense reactions	Various mechanisms

by race 1.4). As these cultivars also differ in the horizontal resistance level, they are affected by virulent races in a different degree: Domodedovsky is strongly affected, and Olev is moderately affected. Thus, the basic difference between vertical and horizontal resistance is specific and non-specific response to infection by different strains of a pathogen. However, these properties are also linked with the other ones, which can be seen from the example below.

If the potato cultivar Domodedovsky (resistant to particular races of Phytophthora) is crossed with the cultivar Severnaya Rosa (susceptible to the same races), the resulting hybrid progeny F1 will be as resistant as the resistant parent. The F2 plant population will be divided in the ratio 3 resistant: 1 susceptible, which is an evidence for monogenic dominant inheritance of resistance. Hybrids F1 and F2 from crossing the cultivars Voltman and Severnaya Rosa (differ in horizontal resistance level) will produce infestation distribution as a bell-shaped curve with the edges in the high resistance and high susceptibility areas and the maximum in the affection area, intermediate between the parent parameters. That is, vertical resistance and horizontal resistance also differ in inheritance pattern.

Some differentiating features of horizontal and vertical resistance are shown in Table 4.1.

Genetics of resistance

Variations of any trait of an organism, including its response to infection, depend both on its genotype and conditions of its life. This double relation is described by the equation:

$$VP = VG + VE$$

where VP is phenotypic variance, VG is genetic variance, and VE is the variance caused by the environmental effect (environmental or ecological variance).

The contribution of the environmental variance is estimated from the behavior of the same genotype in different life conditions. The genetic variance is more complex, as the manifestation of the same trait is often controlled by several genes that can, in addition, interact with others. Assume that locus A with two alleles A1 and A2 influences manifestation of some trait, and homozygotes A1A1 and A2A2 provide the greatest and the smallest expression of the trait, respectively:

A2A2 _____ A1A1

Designate the genetic variance caused by homozygote as $y11 = m + a$, the variance of homozygote A2A2 as $y22 = m - a$, and the variance of heterozygote A1A2 as $y12 = m \cdot d$, where m is the mean of two homozygotes; $a - 1/2$ differences between the homozygotes, and d is the deviation of a heterozygote from the mean of two homozygotes. It is often not one but several genes (B, C, etc.) that influence manifestation of a trait. Therefore, they should also undergo similar procedures.

If there is no interaction between alleles A1 and A2, B1 and B2, etc., then $d = 0$, and $y12 = 1/2 \, (y11 + y22) = m$. In this case, the total genetic variance is governed only by the number of loci determining the given trait, and is called non-interacting, or additive (A). However, in addition to the number of genes, their interaction can influence the manifestation of a trait. Interaction of allelic genes is called dominance (D). It indicates the deviation from the mean of heterozygotes and has a magnitude d (at $d = a$ the dominance of allele A1 over A2 is complete, i.e. the heterozygote does not phenotypically differ from the dominant homozygote). In addition to the interaction of allelic genes, non-allele interaction, or epistasis (I) is possible, where some genes influence manifestation of a trait controlled by other genes, often unlinked with them.

Thus, the total genetic variance is described by the equation:

$$VG = VA + VD + VI$$

and the total phenotypic variance:

$$VP = VA + VD + VI + VE$$

Consider the contribution of individual components of the phenotypic variation in horizontal and vertical resistance manifestation.

Dominance

The dominance has a key influence on manifestation of vertical resistance of plants (the first generation shows the same resistance as the resistant parent). Resistance to fungi is usually dominant resistance, while resistance to viruses is often recessive. The same resistance gene can be dominant regarding one race of pathogen and recessive regarding another race. Probably, it is connected with the gene dosage, one dosage (heterozygosis) is sufficient to enhance resistance of a "weaker" race, while a "stronger" race will require two dosages. For example, corn gene Rp3 is dominant regarding race 901 of the rust pathogen *Puccinia sorgi* (both homozygotes and heterozygotes are resistant); however, it is recessive regarding race 933 (only homozygotes are resistant). The effect of the gene dose may depend not only on the parasite genotype, as was the case in the previous example, but also conditions of inoculation. For instance, the pea plants, heterozygous in resistance genes to viruses of bean yellow mosaic and horse bean yellow mosaic, are resistant at a temperature of 18°C and less (dominant resistance), but susceptible at 27°C and more (recessive resistance), while the homozygotes do not change a response to inoculation at the temperature change.

Epistasis

Phenotypic interaction of non-allelic genes, epistasis, can render a stronger influence on vertical resistance than on horizontal, since probability of one gene modification is higher than that of many. Genetic analysis of resistance does not show many cases of epistasis, and epistatic segregation of progeny for resistance can be caused not only by the plant gene interaction, i.e. true genetic epistasis, but also by external conditions, in particular, by mixed infection, i.e. environmental epistasis, which was analyzed by G. Sidhu. For instance, two plants with resistance genes Aabb and aaBB, where gene A controls resistance to parasite X, and gene B resistance to parasite Y, are crossed. In the absence of gene linking, there will be dihybrid segregation of the second generation: 9AB: 3Ab: 3aB: 1ab. Hence, in the absence of relations between parasites, segregation for each one will be 3 resistant to 1 susceptible, and for both, 9 resistant to X and to Y, 3 resistant to X, but not to Y, 3 resistant to Y, but not to X, and 1 susceptible to both parasites. However, cases are known when plant infection by one parasite enhances or weakens its resistance to the subsequent infection by another; at the same time, the ratio of segregating classes in F2 will differ from the above described and will resemble epistatic segregation. For example, inoculation of tomatoes resistant to Fusarium wilt *Fusarium oxysporum* f. sp. *lycopersici* with the pathogen makes the plants also resistant to Verticillium wilt (causative agent *Verticillium alboatrum*), i.e. the plants infected by parasite X become resistant to parasite Y (phenomenon of plant immunization by parasite X). In this case, the dihybrid segregation in the second generation 9AB: 3Ab: 3aB: 1ab will change as follows: 3: 1 to parasite X (*Fusarium*), 15: 1 to parasite Y (*Verticillium*) (two-locus dominant epistasis), X + Y - 12: 3: 1 to the combination of parasites (the first two classes are resistant to both parasites, the third only to Verticillium, and the fourth is susceptible to both); such segregation is typical of a single-locus dominant epistasis.

A reverse situation is possible: plants infected by parasite X lose their resistance to the subsequent infection by parasite Y (suppression of resistance to parasite Y by parasite X). Such event is observed at a combined inoculation of tomato with *Fusarium* and gall nematode. The second generation will have the following pattern of infection: 3: 1 for parasite X, 9: 7 for parasite Y (two-locus recessive epistasis), 9: 3: 4 for the combination X + Y (single-locus recessive epistasis).

These facts give ground to two practically important conclusions:

1. To avoid artifacts at genetic analysis, cleanliness of the infectious background must be thoroughly monitored; inoculation should be done using genetically uniform parasite strain.
2. At detection of epistatic relations in the segregating progeny, a thorough phytopathologic study of the relations of the two parasites within the plant and with the plant is necessary, as the presence of the above-described interactions may require adjustment of the crossing program for the development of resistant cultivars and of the disease control plan of actions.

Additivity

At vertical resistance, one gene can provide to a plant such a high degree of resistance that addition of one or more genes will not have a significant effect on the phenotype. Contribution of the additive variance to the total genetic variance is then negligible. Horizontal resistance is different because each gene brings a small, phenotypically undistinguished contribution, and the total resistance is determined by the addition of the genes, i.e. additivity.

The principal genetic difference between horizontal and vertical resistance does not contain a number of genes that control resistance but the ratio of interacting and non-interacting effects. Prevalence of interacting effects (dominance and epistasis) results in discrete segregation typical of vertical resistance. Prevalence of additivity results in continuous segregation, characteristic of horizontal resistance. These differences can be illustrated by the binomial segregation (Van der Plank, 1982).

Expansion of the binomial $(r + s)^n$, where r is the alleles controlling resistance, s is the alleles controlling susceptibility, and n is the number of loci, will result in the following ratios of alleles r and s:

1. Locus $(r + s)^2 = r^2 + 2rs + s^2$. If r completely dominates s, homozygotes and heterozygotes will be equally resistant, and expansion of the binomial will correspond to the monohybrid segregation of progeny F2, i.e. 3: 1. In full absence of dominance, we will observe not two but three segregating classes: highly resistant, intermediate, and highly susceptible, in the ratio 1: 2: 1.
2. Loci $(r + s)^4 = r^4 + 4r^3s + 6r^2s^2 + 4rs^3 + s^4$. If r fully dominates s, then again homozygotes and heterozygotes with allele r will make one class of resistant progeny, and only the homozygote in allele s in both genes will make a class of susceptible progeny, and segregation will correspond to a dihybrid case: 15r: 1s. If dominance is absent and the level of resistance is determined only by the gene dose (number of r alleles), variation of the level of resistance will result in five segregating classes the ratio of which corresponds to the Gaussian curve.
3. Loci $(r + s)^6 = r^6 + 6r^5s + 15r^4s^2 + 20r^3s^3 + 15r^2s^4 + 6rs^5 + s^6$. At complete dominance of resistance, out of seven genetically different classes,

six classes (possessing allele r) will be phenotypically equally resistant, and only one will be susceptible, their ratio corresponding to trihybrid segregation in F2: 63: 1. In the absence of dominance, the level of resistance is proportional to the number of r alleles, and the binomial will expand into seven phenotypically different classes (actually continuous segregation corresponding to the normal curve).

Thus, in the absence of gene interaction, i.e. in the presence of only the additive variance, even heterozygosis in one resistance locus can result in a continuous series of phenotypic segregation corresponding to horizontal resistance.

Modification by environmental conditions

Plants manifest horizontal resistance in the field and can be weakened or even lost with artificial inoculation in conditions promoting development of the disease. Vertical resistance is displayed in a broader range of environmental conditions (temperature, humidity, illumination), with more concentrated inoculum, and in a longer part of a plant life cycle than horizontal resistance. For instance, plant selection for vertical resistance to cereal leaf parasites (rust and mildew fungi) is usually conducted in the phase of sprouts which have no horizontal resistance.

Vertical resistance genes can also differ in response to the inoculation conditions. For instance, flax has several loci where rust resistance genes are mapped. In locus L, resistance genes L1, L2 and L3 provide only two alternatives depending on the race used for inoculation, complete immunity or complete susceptibility, and response of gene N2 (in locus N) varies depending on the illumination. Since at vertical resistance the greatest contribution in manifestation of a trait is made by interacting genes, while the role of the additive and environmental variances is negligible, the phenotypic variance could be shown as follows, with certain simplification:

$$VP = VD + VI$$

In manifestation of horizontal resistance, the additive effects and the external environment play the principal role, i.e.:

$$VP = VA + VE$$

Hence, the inheritance coefficient ($h = VG/VE$) is much higher for vertical than for horizontal resistance, therefore, it is much easier to use the former than the latter in selection programs for development of disease-resistant crops.

Phenotypic manifestation of resistance

Qualitative and quantitative manifestation of vertical and horizontal resistance influences the signs of disease in the plants possessing these types of resistance. Vertical resistance is mostly manifested as hypersensibility response, i.e. the active response of inoculated cells expressed as their death and the concurrent death or termination of development of the parasite in tissues. The hypersensibility response is most frequently displayed as a small necrosis developing soon after the inoculation at the spot of the parasite penetration. For instance, in potato late blight, any resistance gene responds in such a way to inoculation of a leaf with an avirulent race, with spore-bearing of the parasite being completely suppressed. Vertical resistance of wheat inoculated with rust fungi is manifested in more diverse ways. The scale of accounting of the types of wheat response to inoculation with the stem rust pathogen *Puccinia graminis* f. sp. tritici (Stakman scale) is based on the ratio of two parameters: size of the pustule and intensity of the defense response (necrosis or chlorosis) (Figure 4.2). This scale indicates five basic types of response: 0–4 (from chlorosis without a pustule to a large pustule without chlorosis), one combined type (X) where different leaf areas develop different types of response, and many additional subtypes (0; 2 + 3 =, etc.). All types and subtypes

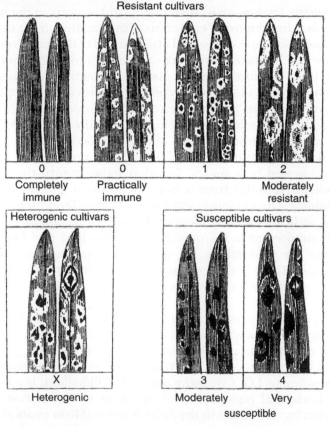

Figure 4.2.   The scale of the types of wheat response to inoculation with the *P. graminis* races (Stakman scale).

that develop defense response (necrosis or chlorosis) around the spot of inoculation or the small pustule, are often combined into one indicator "resistance" (R); the responses where no protective necrosis develops around a large or medium pustule (types 3 and 4) are combined into "susceptibility response" (S).

Unlike late blight of potato, in resistance responses of wheat to rust fungi, controlled by some vertical genes, interstitial development and even sporulation of the parasite is possible, though the number of produced spores is much less than in susceptible plants. Apparently, the rate of development of defense responses and the intensity of response processes (expression) is different for different resistance genes.

For instance, the rice cultivars resistant to the bacteria *Xanthomonas oryzae* show three phenotypes: asymptomatic response (resistance genes Xa1, Xa7, Xa10, and Xa12), browning (gene Xa2), and small yellow spots (Xa3, Xa4, and Xa5).

Diseases caused by mildew fungi are very suitable for studying the level of expressivity of different vertical genes. Their mycelium extends over the leaf surface. After formation of the primary haustorium in the epidermal cell and feeding, secondary hyphae are formed that form haustoria in the neighboring epidermal cells. After removing epidermis with mycelium from the leaf using a transparent adhesive tape, cytological observations of the fungus development can be conducted. The taken mycelium can be used to study accumulation of the nutrients it has obtained from the plant. A. Ellingboe showed that the vertical gene of wheat Pm1 almost completely suppresses the development of the secondary hyphae of the parasite; gene Pm2 activates the defense mechanisms later and has no effect on development of the secondary hyphae but causes necrosis or chlorosis 3–4 days after the inoculation. Gene Pm3 suppresses the development of only 60–70% of infections, while 30–40% continue to develop and bear spores. In other experiments, the wheat leaves obtained radioactive sulfur from a nutrient solution and then were inoculated with mildew. The mycelium was taken off at different intervals and the accumulated radioactivity was measured. The entrainment of sulfur had been sharply suppressed in the presence of genes Pm1 and Pm4 as soon as 16 hours after the penetration, i.e. these genes promptly engaged in the defense reactions. Gene Pm3 suppressed the entrainment of sulfur only after 20 hours and not so sharply as the previous genes, and gene Pm2 did it after 24 hours.

Horizontal resistance provides a pure form of a susceptible type of response (3 or 4 points by the Stakman scale). The level of resistance of a cultivar is expressed as the reduction of the number of lesions on a plant and the number of spores in one lesion. The product of these two parameters, called as resistance index by I.G. Odintsova and L.A. Mikhailova, is a good expression of the horizontal resistance level of wheat to brown rust. Therefore, observing behavior of cultivars in the field, it is possible to establish their form of resistance through the manifestation of disease. If there is a hypersensibility response, the cultivar possesses vertical resistance to the pathogen races widely spread in the field; if two cultivars with the same type of spots have reliable differences in similar conditions, one of the cultivars has a more horizontal resistance than the other. However, such conclusion can only be very preliminary, as the number of spots is not always indicative of the type of resistance. At vertical resistance hypersensibility response often proceeds so promptly that no necrosis is observed (resistance is expressed as absence of the signs of disease). If the field population consists of a combination of virulent and avirulent races, the former will cause formation of disease lesions, and the latter will be manifested as reduction of the number of infectious spots, i.e. as typical horizontal resistance. This event was observed by the Japanese plant pathologists in studying the resistance of rice cultivars to rice blast disease. The fungal pathogen *Magnaporthe grisea* has a high spontaneous variability; therefore, its field populations usually consist of a combination of races, and the number of spots on leaves is often not determined by the level of horizontal resistance so much but rather the number of genes of vertical resistance (Table 4.2).

The cultivars 1 and 2 behave like having horizontal resistance (few infection spots), although they actually contain many effective vertical resistance genes. Their presence inhibits formation of the signs of infection in most spores occurring on the leaves. These data show that assessment of the level of horizontal resistance in the presence of vertical resistance requires the use of the races virulent for the given cultivar.

To study horizontal resistance, the Dutch plant pathologist K. Zadoks suggested a method of analysis of the infection cycle i.e. studying all its stages from arrival of spores to the plant surface to the formation of new spores. Table 4.3 shows that the population density drops at each stage of the infection cycle. The increase in the number of pustules in comparison with the number of interstitial colonies may be due to the presence of two pustules in the same colony, in the upper side of the leaf and in the underside.

The share of infectious units at each stage of the cycle, i.e. their ratio to the number of infectious units, is expressed by the index l_i; the share of infectious units destroyed at each stage is expressed by the index g_i; and the share of the survivors $1 - g_i$, which is calculated by division of the number of infectious

Table 4.2. Disease severity of Pyricularia leaf-blight on the different rice cultivars (according to Ahn SW, Ou SH, 1982)

Cultivar	Number of spots on the plant	Number of avirulent races
1	2	64
2	14	55
3	39	40
4	66	19
5	97	4

Table 4.3. Component analysis of brown rust infection cycle (isolate 1073) on leaves of the wheat cultivar Rubis (Zadoks JC, Shein RD, 1979)

Time (days)	State	Infectious units per 10^3 stomata (N)	g_t	$1 - g_t$	Ratio	l_t
0	SPD	294	0.07	0.93	GTR	1.00
1	GED	273	0.11	0.89	AGR	0.93
1	APD	243	0.55	0.45	VAR	0.83
3	VED	110	0.16	0.84	HVR	0.37
3	HYD	92	0.33	0.67	CHR	0.31
9	COD	61	-	1.53	PCR	0.21
9	PUD	94	0.53	0.00	SPR	0.32
25	SUD	0	1.00			

SPD, total spore density; GED, germinated spore density; APD, appressorium density; VED, substomatal vesicle density; HYD, hyphal density; COD, colony density; PUD, pustule density; SUD, surviving pustule density. Ratios: GTR, germinated/total spore ratio; AGR, appressoria/germinated spore ratio; VAR, vesicles/appressoria ratio; HVR, hyphae/vesicles ratio; CHR, colonies/hyphae ratio; PCR, pustules/colonies ratio; SPR, surviving pustule ratio.

units at a subsequent stage of the cycle by the number of units at a previous stage (GED, APD, VED, etc.).

Comparison of g_t of the cultivars with different levels of horizontal resistance can provide important information for revealing the stage of the infection cycle where the effect of the horizontal resistance genes of these cultivars is pronounced. The higher the g_t, the more pronounced is the horizontal resistance. The best index (0.55) of the cultivar Rubis (Table 4.3) falls at the stage between the formation of appressoria and substomatal vesicles. Crossing of cultivars, the resistance of which is manifested at different stages of the infection cycle, can result in combination of different horizontal genes and thus receiving hybrids transgressive in resistance.

Horizontal and vertical resistance assessment for systemic diseases is even more complex. In field trials, an absolute non-infestation, or, vice versa, 100% infestation of a cultivar can be rarely observed, therefore, even in the presence of vertical resistance the differences between resistant and susceptible cultivars are quantitative. For instance, the boundary between cereal cultivars resistant and susceptible to smut fungi with a 10% infestation was determined entirely by an empirical way. The cultivars affected by less than 10% are regarded as possessing vertical resistance, and stronger affected cultivars as possessing some degree of horizontal resistance.

The degree of infestation often has to be taken into account, too, as different cultivars with the same percent of affected plants can considerably differ in intensity of the course of disease, for instance, at such systemic disease as Verticillum cotton wilt (pathogen *Verticillium dahliae*). Apparently, in such cases infestation is determined by expressivity and penetrance of the resistance genes. The number of affected plants depends on the gene penetrance, and the level of infestation depends on their expressivity. Such approaches allow controlled breeding to improve expressivity and penetrance of resistance genes. However, the relation of these gene properties with vertical or horizontal resistance is not clear.

Relations with parasites

Phenomenology of relations

By definition, vertical resistance is race-specific, while horizontal resistance is not. Race specificity is manifested in the ability of resistance genes to lose their efficiency after plant inoculation with virulent pathogen races. The race specificity is not so simple as it seems at first glance. For instance, see the data on interaction of cotton cultivars and strains of the cotton-wilt pathogens in Table 4.4. It shows the differences in the cotton cultivar resistance and pathogenicity of the parasite strains. There are no intersecting parameters (one cultivar is resistant to a first race and susceptible to a second race, whereas the other cultivar is susceptible to the first race and resistant to the second), i.e. there are specific interactions between a strain and a cultivar. According to Van der Plank's definition, only the presence of intersecting interactions between cultivars and races, like for late blight of potato, can indicate vertical relations: the cultivar Lyubimets is susceptible to race 1 and resistant to race 2, whereas the cultivar Veselovsky is susceptible to race 2 and resistant to race 1.

The differences between these two examples above are revealed in ranging host plant cultivars and pathogen races. For cotton wilt, the range of the cultivar resistance at inoculation with different races and the range of race pathogenicity at inoculation of different cultivars do not change: in all cases cultivar 8763 is least resistant, and the cultivar Tashkent-1 is the most resistant. In all events, race 1 is the least pathogenic and race 3 is the most pathogenic (Table 4.4). In late blight of potato, we see a change in resistance range of two cultivars at trial with different races and change of the pathogenicity range of two races at inoculation of different cultivars. Only the second case

Table 4.4. Interaction of cotton cultivars with the strains *Fusarium oxysporum* f. sp. *vasinfectum* (Sidorova SF, Akmuradov B, 1983)

Physiological race	Response of cotton cultivars		
	8763	9647	Tashkent-1
1	Susceptible	Resistant	Resistant
2	Susceptible	Susceptible	Resistant
3	Susceptible	Susceptible	Susceptible

is indicative of differential relations, i.e. presence of vertical resistance in plants and presence of virulence in their parasites. The first shows a general variation in the resistance and pathogenicity properties characteristic of plants and parasites, i.e. horizontal resistance and aggressivity. Host-plant resistance genes and parasite aggressivity genes work in a similar and additive way. Different levels of infestation can be expected depending on gene numbers in the host–parasite system. A cotton cultivar and its parasite strains have a growing number of additive susceptibility and aggressivity genes, and if the total number of the host susceptibility genes and pathogen aggressivity genes is less than four, resistance response will develop, in case of four or more genes, the plant is susceptible. Then Table 4.4 will assume the form shown in Table 4.5.

In conclusion, ranging is a good tool to identify the type of relations between host cultivars and parasite strains (horizontal relations are additive, vertical are non-additive). Statistical methods allow estimation of reliability of empirical data deviation from an established range.

A method, widely used in determination of specificity of host–parasite relations, is the two-factor variance analysis. The first factor is the assessment of differences in cultivar resistance, and the second is the assessment of differences in parasite strain pathogenicity. The absence of a reliable interacting variance indicates the absence of vertical relations. At well-manifested vertical resistance, the interacting variance is highly reliable. Variance analysis allows assessment of interaction intensity between cultivars. The strains of honey mushroom *Armillaria mellea sensu lato* are usually believed to have no specialization for attacking various tree species. However, processing of experimental data showed a weak interaction indicative of a small variance component responsible for differential relations between the honey mushroom strains and wood varieties. Presence or absence of interaction is not assessed in absolute terms but statistically by an arbitrary criterion (5%, 1%, or 0.1% level of significance). Therefore, an interacting variance, reliable at one level, can be unreliable within a more stringent assessment or within

Table 4.5.   Differential relations between cotton cultivars and *F. oxysporum*, strains based on additive-interacting genes of aggressivity and susceptibility

Parasite races	Number of aggressivity genes	Cotton cultivar		
		8763	9647	Tashkent-1
		Number of susceptibility genes		
		3	2	1
1	1	Susceptible	Resistant	Resistant
2	2	Susceptible	Susceptible	Resistant
3	3	Susceptible	Susceptible	Susceptible

Table 4.6. Percentage of barley leaf surface covered with pustules of rust pathogen *Puccinia dispersa* (Parlevliet JE, 1978)

Cultivars	Races				
	11-1	18	1-2	22	24
Berak	8.1	6.7	3.1	5.0	0.9
Julia	4.5	12.1	1.8	1.1	0.6

assessment of different size of samplings (variation of the degrees of freedom). Hence, there are numerous transitions between a high level of differential interactions characteristic of vertical resistance–virulence, and absence of differential interactions characteristic of horizontal resistance–aggressivity. The presence of an interacting variance and deviation from a range at horizontal resistance is reported in scientific papers.

As can be seen from Table 4.6, the range of resistance and aggressivity was upset in two cases (in the investigated cultivars, the susceptible type of infection to all races and polygenic inheritance are indicative of typically horizontal relations): 1. the cultivar Julia is more resistant to all races, except for race 18, than the cultivar Berak; 2. races 22 and 1.2 change ranks in the same cultivars.

Another example is reliable interactions of the strains of brown rust pathogen (*Puccinia recondita*) and the wheat cultivars Derbent Black Spike and Saratovskaya 29 which have no pronounced vertical resistance and differ only in the number of pustules (Table 4.7). Such data questioned a complete independence of the host–parasite systems at horizontal resistance postulated by Van der Plank. The Dutch plant pathologists Zadoks and Parlevliet believe that specificity also occurs at horizontal resistance, although the interacting effect is small and approaches the magnitude of experimental errors. Therefore, the effect cannot be detected especially in the presence of considerable environmental variance, typical of horizontal resistance.

In the analysis of plant–parasite relations as well as analysis of genetic monitoring of plant resistance, the major difference between vertical and horizontal resistance consists in the ratio of interacting and non-interacting effects. In the presence of interaction only (or almost only), vertical resistance

Table 4.7. Variance analysis of the number of pustules of three strains of brown rust pathogen on the leaves of two wheat cultivars differing in the level of horizontal resistance (Odintsova IG, Shelomova LF, 1983)

Sources of variation	Degrees of freedom	Sum of squares	Variance
Cultivars	1	20.0	20.7
Strains	2	14.4	7.2
Cultivars × strains	2	16.2	8.1

Table 4.8.   Types of plant–parasite relations (Carson SD, Carson MJ, 1989)

1	Monogenic	Qualitative	Present
2	Monogenic	Quantitative	Present
3	Polygenic	Qualitative	Present
4	Polygenic	Quantitative	Present
5	Monogenic	Qualitative	Absent
6	Monogenic	Quantitative	Absent
7	Polygenic	Qualitative	Absent
8	Polygenic	Quantitative	Absent

is observed, and in complete (or nearly complete) absence of interaction horizontal resistance is observed. Intermediate types are indicative of the absence of hiatus between them. Almost all of the possible types of relations can be expressed as a table with three varying traits and two categories of variation of each trait (Table 4.8). Four among the eight states (states 1–4) correspond to vertical resistance according to Van der Plank (occurrence of specificity in plant–parasite interaction), and the other four (states 5–8) correspond to horizontal resistance. However, only state 1 complies with the set of vertical resistance properties shown in Table 4.8, and state 8 complies with all horizontal resistance properties. Many transient states are well-known. For instance, polygenic, quantitative but specific resistance of wheat to brown rust (state 4), or monogenic, qualitative but non-specific resistance of corn to *Cochliobolus carbonum* (state 5) have been reported.

Genetics of plant–parasite relations

Investigation of the genetics of plant–parasite relations includes a parallel genetic analysis of the plant resistance and virulence of its parasite. The latter is possible for phytopathogenic fungi with the sexual process: Ascomycetes, Basidiomycetes, and Oomycetes. An important role for studying the nature of host–parasite relations was played by the investigations of cereal rust fungi. The methods of hybridological analysis of the stem rust pathogen *P. graminis* was developed in 1940s by the Canadian plant pathologists M.Newton and T.Johnson. The sexual process of this fungus occurs on berberry leaves, where haploid spermatia fertilize haploid hyphae of spermagonia. The resulting two-nuclear (dikaryotic) aecidiospores infect wheat, on stems and flag leaves of which urediniospores develop in summer, and their shoots can reinfect the wheat. By autumn, the resting teliospores form in the infested stems. After wintering, karyogamy and meiosis occur with the development of a tetrade of haploid basidiospores that are responsible for the primary infection of berberry (Figure 4.3). For hybridization, the berberry bushes in flowerpots are covered with lantern glass, a petri dish is put on its top, and wheat straw with teliospores previously exposed to negative temperatures for

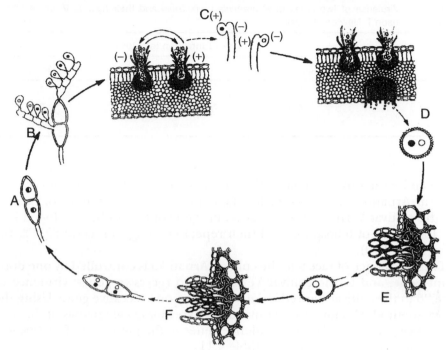

Figure 4.3.   Life cycle of *P. graminis*.
(A) Karyogamy in dormant teliospores; (B) meiosis and formation of promycelium with four haploid basidiospores that infect berberry; (C) berberry leaf with spermagonia and spermatia transferred by insects from one spermagonium to another; (D) upon the fusion of heteroallelic cells, aecium with two-nuclear aeciospores forms on the leaf underside; (E) infection of the wheat stem by an aeciospore leads to the formation of a pustule with two-nuclear urediniospores; (F) upon the stem die-off, thick-walled wintering two-nuclear teliospores develop in the pustules.

stimulation of germination is fixed on the internal surface of the dish. The forming basidiospores occur on the berberry leaves and infect them. Spermatia from the spermagonia developed in the inoculation spots are transferred to the other spermogonia by a sterile platinum needle. For inbreeding, the spermagonia resulting from the inoculation are fertilized with the basidiospores of the same race; for hybridization, the spermatia are transferred to the spermagonia resulting from the inoculation of berberry with another race. Since *P. graminis* is heterothallic and the probability of fertilization is 50%, approximately half of the fertilized spermagonia develop aecia on the leaf underside. Aeciospores are transferred to the wheat shoots that do not have efficient genes of vertical resistance, in order not to lose slightly virulent segregants among the hybrid spores. The resulting urediniospores are used to inoculate the wheat cultivars, whose resistance contrasts that of the parent strains, and a judgment of dominance or recessivity of the virulence genes can be made on the basis of the response to the inoculation (Table 4.9).

Table 4.9. Virulence of two races of *P. graminis* f. sp. *tritici* and their hybrids to some wheat cultivars (Johnson T, Newton M, 1946)

| Races | Type of infection in cultivars | | |
	Kanred	Arnautka	Vernal
9	0	4	4
3	4	1	0
69 × 36	0	4	0

It can be seen from the table that virulence of race 9 is dominant for the cultivar Arnautka and recessive for the cultivar Vernal; virulence of race 36 for the cultivar Vernal is also recessive. Progeny of F1 can be developed into the formation of teliospores and then repeat crossing to receive F2, F3, B1, etc. (Table 4.10).

Thus, virulence of race 9 to the cultivar Arnautka is controlled by one dominant gene, and to the cultivar Vernal by two recessive genes; virulence of race 26 to the cultivar Kanred is determined by one recessive gene. Using the above method, H. Flor carried out a cross hybridological analysis of flax-rust resistance genes and flax-rust virulence genes of the pathogen of *Melampsora lini* regarding the flax cultivars (Table 4.11).

Since heterozygotes of the host and parasite (F1) are resistant to both races or avirulent to both cultivars, respectively, resistance and avirulence are controlled by the dominant genes of the host and parasite, and susceptibility and virulence show in the presence of homozygous recessive alleles of the dominant resistance and avirulence genes. In F2, the ratio 3:1 of the plant hybrids resistant and susceptible to each race and parasite hybrids avirulent and virulent to each cultivar points to the presence of one dominant gene. Occurrence of four classes of responses in F2 shows that the host and parasite populations contain two independently segregating genes each. Segregation of the host or parasite populations F2 interacting with the partner populations F1 responds to inoculation as 15:1, which points to the presence of interacting pairs of the host and parasite genes. These data led Flor to a

Table 4.10. Virulence segregation of races 9 and 36 of *P. graminis* f. sp. *tritici* to wheat cultivar hybrids F2 (Johnson, Newton, 1946)

| Cultivar | Number of hybrids causing responses | | Expected segregation |
	0–1	4	
Kanred	227	98	243.7: 81.3 (3:1)
Arnautka	83	242	81.3: 243.7 (1: 3)
Vernal	307	18	304.7: 20.3 (15: 1)

Table 4.11. Chart of inheritance of resistance and virulence in the flax-rust system (from H.D.Flor)

Parasite race	Host cultivar		F1	F2				Ratio
	Ottawa	Bombay	R1r1R2r2	11	32	43	9	
	R1-r2r2	r1r1R2-		(9)	(3)	(3)	(1)	R:S
22 (a1a1A2–)	+	–	–	+	+	–	–	3:1
24 (A1-a2a2)	–	+	–	+	–	+	–	3:1
F1 (A1a1A2a2)	–	–	–	+	–	–	–	15:1
F2 (A1-A2–)	– 78 (9)	–	–	+	–	–	–	15:1
(A1-a2a2)	– 23 (3)	+	–	–	+	–	–	3:1
(a1a1A2–)	+ 27 (3)	–	–	–	–	+	–	3:1
(a1a1a2a2)	+ 5 (1)	+	+	+	+	+	+	–

The sign (–) designates the resistance/avirulence response (unambiguous); (+) stands for the susceptibility/virulence response.

conclusion about the presence of interacting genes (or, rather, gene products) of the host and parasite, formulated as the gene-for-gene conception. As can be seen in the Table 4.11, if a plant possesses any dominant resistance gene (R1 in the cultivar Ottawa or R2 in the cultivar Bombay), it can only be attacked by a pathogen race that has a recessive avirulence gene complementary to it (A1 in race 22 or A2 in race 24). An incompatible response (resistance/avirulence) occurs only if both the interacting genes (of host and parasite) are dominant.

It follows from Flor's theory that the presence of a dominant resistance gene in the host-plant genome is not observed in the absence of the corresponding dominant avirulence gene (the cultivar Ottawa is attacked by race 22, while the cultivar Bombay by race 24, though they possess resistance genes R1 and R2, respectively; Table 4.11). In addition, the dominant avirulence allele will be observed in the parasite genome only in the absence of a corresponding resistance gene. Moreover, in the absence of any host dominant resistance genes, the corresponding parasite virulence genes (recessive alleles of the avirulence genes) will not be observed, either. Genetic investigations of numerous plant diseases that followed Flor's studies have basically confirmed his conclusions. It was shown that: a. resistance genes (R stands for "resistance") are more frequently dominant, while virulence genes are more frequently recessive (P stands for "pathogenicity" or a stands for "avirulence"); b. R-genes are frequently allelic or closely linked with each other, and P-genes usually segregate independently; c. R- genes and P-genes usually interact according to the "gene-for-gene" theory. The first conclusion shows that irrespective of the interacting host and parasite genotypes, there can be only two resulting phenotypes: R or S, "minus" or "plus", or 0 or 1. This statement is well illustrated by the presentation of host–parasite relations as a diallellic system (square net) (Table 4.12).

Table 4.12.   Diallelle host–parasite relations in the "gene-for-gene" system

Host-plant genotype	Parasite genotype	
	aa	AA (or Aa)
RR (or Rr)	+ (S)	– (R)
Rr	+ (S)	+ (S)

R is resistance (incompatibility); S is susceptibility (compatibility).

However, cases of incomplete dominance of resistance or avirulence genes were reported. In such a situation not two but three phenotypes can be observed in the "square net": minus, plus, or intermediate. For instance, D. Samborski found a brown rust strain which caused a response 1+ in the wheat cultivar transfer, homozygous in gene resistance Lr9, instead of response 0 common for this gene. The inbreeding, carried out according to the above described Johnson and Newton method, showed that the avirulence gene is heterozygous, and allowed receiving isolates homozygous in this gene. Table 4.13 shows the relation between the resulting fungal cultures and wheat cultivars.

A feature of the organization of resistance and avirulence genes is allelity, or close linking of the former but not the latter, and it also influences the analysis of the effects of host–parasite relations. Since a diploid organism can have only two allelic states of genes, the number of possible combinations of the resistance genes is less than that of the avirulence genes. For instance, a system of two allelic resistance genes and two corresponding unlinked avirulence genes can have four gene combinations in the host (R1R1, R1r2, r1R2, r1r2) and nine gene combinations in the parasite (A1A1A2A2, A1A1A2a2, A1A1a2a2, A1a1A2A2, A1a1A2a2, a1a1A2A2, a1a1A2a2, A1a1a2a2, a1a1a2a2). A system of three allelic host genes and the corresponding unlinked parasite genes can have 10 combinations in the host and 27 in the parasite.

Codominance of resistance genes (phenotypic occurrence of both alleles in the heterozygote) imposes some limitations on the use of the "square net" for the analysis of the effects of host–parasite relations. For instance, there is a heterozygous plant with two allelic codominant resistance genes of R1/R2.

Table 4.13.   Pathogenicity (type of response to inoculation) of segregants of the strain *P. recondita* avirulent to the cultivar transfer (Samborski DJ, 1963)

Host genotype	Pathogen genotype		
	A9A9	A9a9	A9a9
Lr9Lr9	0	1+	4
Lr9lr9	0	3	4
Lr9lr9	4	4	4

Gene R1 interacts with the genes located in one locus of the parasite (A1A1, A1a1, and A1a1), and gene R2 interacts with a homozygous gene a2a2 located in the other locus (allele A2 is absent). In F2, the heterozygote will segregate into R1R1, R1R2, and R2R2 in the ratio 1: 2: 1. However, if the parasite strains being tested have no allele A2, all of them will be virulent to the plants possessing gene R2. Therefore, gene R2 will not be observed, its resistance-imparting properties will not be used, as it behaves as a recessive susceptibility allele in the segregating progeny, which is shown in the diagram suggested by Person and Mayo:

	Actual situation			Observed situation	
Relative frequency	¼	½	¼	¾	¼
Host genotypes in F2	R1R1	R1R2	R2R2	R–	rr
Parasite strains:					
A1– a2a2	–	–	+	–	+
a1a1 a2a2	+	+	+	+	+

Thus, actually digenic host segregation is phenotypically manifested as monogenic with the ratios typical of the "square net" (see Table 4.12).

The loss of phenotypic manifestation of the resistance gene in a parasite in the absence of the testing dominant avirulence allele will be observed even at independent inheritance of the resistance genes. For instance, the flax cultivar Bison is attacked by all North American strains of rust pathogen and it was used by Flor in hybridological analysis as a universal susceptible control without vertical genes; however, in Australia a resistance gene to some Australian races was found in the cultivar Bison and called L9. Apparently, all North American races have a recessive allele complementary to it.

Linking between avirulence genes is found much less frequently than between resistance genes. Close linking for virulence genes aL8 and aM and group of genes aP (a-P, a-P1, a-P2, a-P3) of *Melampsora lini* has been reported.

The most important conclusion from the parallel studies of resistance and virulence genes is that the interaction occurs between the partners on the genomic but not gene level. Hence, there is the necessity to study the genetics of the partners only in interaction, i.e. interorganismal genetics (Loegering, 1978). The emergence of interorganismal genetics as a separate discipline allowed linking biochemical studies of plant resistance factors and parasite virulence factors with genetic studies, i.e. making a step towards analysis of molecular genetic mechanisms of resistance and virulence.

Selected Literature

Ahn SW, Ou SH. Phytopathology 1982; 72:282–284.

Browder LE. Parasite: host: environment specificity in the cereal rust. Ann Rev Phytopathol 1985; 23:201–222.

Carson SD, Carson MJ. Ann Rev Phytopathol 1989; 27:373–395.

Crute IR, Norwood JM. Gene-dosage effects on the relationship between *Bremia lactucae* (downy mildew) and *Lactuca sative* (lettuce): the relevance to a mechanistic understanding of host-parasite specificity. Physiol Molec Plant Pathol 1986; 29:133–145.

Johnson T, Newton M. Bot Review 1946; 12:337–392.

Loegering WQ. Current concepts in interorganismal genetics. Ann Rev Phytopathol 1978; 16:309–320.

Newton AC, Andrivon D. Assumptions and implications of current gene-for-gene hypothesis. Plant Pathol 1995; 44:607–618.

Odintsova IG, Shelomova LF. In Variability of Plant-Pathogenic Microorganisms. M. "Kolos" (in Russian) 1983; 51–60.

Parlevliet JE. Neth J Plant Pathol 1978: 84:121–126.

Parlevliet JE, Zadoks JC. The integrated concept of disease resistance; a new view including horizontal resistance and vertical resistance in plants. Euphytica 1977; 26:5–21.

Samborski DJ. Canad J Bot 1963; 41:475–479.

Sidorova SF, Akmuradov B. In Variability of Plant-Pathogenic Microorganisms. M. "Kolos" 1983; 221–225 (In Russian).

Zadoks JC, Shein RD. Epidemiology and plant disease management. Oxford Univ. Press 1979.

Part 2

Biochemistry and molecular biology of plant–parasite relations

The attack weapons of parasites and defense strategies of plants can be specific or non-specific, or, in the terms of the famous plant pathologist, Van der Plank, vertical or horizontal. As host and pathogen develop a close relationship during their contact, one can speak about the host and parasite as a system of interacting organisms, a *pathosystem* (R. Robinson). The interactions usually result from specific metabolites produced by the parasite that induce a response chemical reaction only in particular plant genotypes. This is called a *vertical pathosystem*. The relationship induced by the factors inherent in a parasite, regardless of their phylogenetic or tissue specialization, on the one hand, and constitutional (present before the infection) plant resistance factors, on the other hand, can be referred to as a *horizontal pathosystem*. The most important components of the vertical and horizontal pathosystems are schematically shown in Figure 5.0 over page.

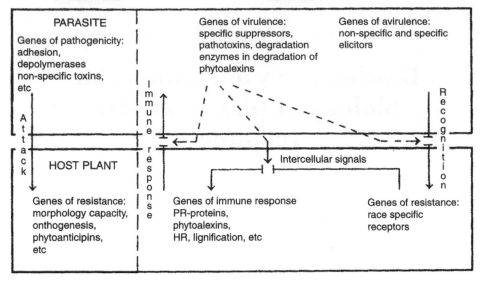

Figure 5.0. Modes of interrelations between host plants and parasites

Chapter 5

Horizontal pathosystem: parasite attack factors

Yu. T. Dyakov, S. V. Zinovyeva

Different groups of parasites (f = fungi, b = bacteria, vr = viruses, pv = fungal, bacterial or viral vectors) have both similar and different stages of pathogenesis, therefore, similar and different weapons of attacking the host:

– germination on a plant surface (f);
– development of infection structures (f);
– mechanical perforation of plant covers (f, pv);
– secretion of depolymerases and chemical destruction of plant covers (f, b);
– passive penetration into plant cells through wounds (f, b, vr);
– formation of haustoria (f);
– release of vivotoxins (f, b);
– secretion of phytohormones (f, b);
– spread of the parasite throughout the plant (f, b);
– production of progeny and their release (f, b).

It is obvious that fungi involve the most diverse chemical and structural factors of pathogenicity. Fungal pathogens develop various morphological structures and produce metabolites providing infection process. Bacteria are abundant with metabolites necessary for parasiting but only in rare cases they produce special morphological structures related to pathogenesis. For example, root nodule bacteria that have lost their cell walls can combine into infectious structures, bacteroids. The only enzymes possessed by viruses are those necessary for their reproduction, therefore, they penetrate into plants in a passive way (mostly via arthropod vectors).

Biosynthetic processes during fungal differentiation and penetration

Mode of fungal attachment

The fungal hypha closely attaches to a hydrophobic surface using various compounds available on its surface.

1. *Integrin*, as interaction mediator. Integrins are a family of animal proteins that permeate the cell membrane and ensure cell adhesion through interaction with integrin receptors of the adjacent cells. As the cytoplasmic "tail" of integrin molecules is short and is not subject to

enzymatic recognition, they transduce signals through the adapter protein that connects integrin with cytoskeleton, protein kinases and transmembrane growth factor receptors. This pathway of transduction is called the integrin cluster. Integrins have been found in fungi and oomycetes. They are necessary for apical growth of the hypha, orientation of cytoskeletal microfilaments and hypha attachment to the surface, and they also provide thigmotropism. Hypha grow strictly perpendicular to anticlinal cell walls of epidermal cells (such growth facilitates finding stomata).

2. *Hydrophobins.* Among the early expressed genes of *Magnaporthe grisea* the gene *MPG1* has been isolated. Its product is a low-molecular protein, hydrophobin, which is necessary for hypha attachment to hydrophobic leaf surface and development of infection structures (appressoria).

Pathogenesis-related type of morphogenesis occurs on the surface and inside the infected plants, but some of its phases can be induced *in vitro*, by culturing on colloid membranes, by cyclic mononucleotides, the plant maturation hormone ethylene, and by other exposures. Several morphogenesis-regulating genes – *INF (infection)* or *PIG* (in *planta induced genes*) – have been isolated in this way. The first differentiation-related protein was **calmodulin**, the protein regulating the **cytoskeletal organization**. Calmodulin is necessary to change orientation of microtubules and microfilaments in transition from a surface-growing hypha to the appressorium.

Many fungal appressoria initiate biosynthesis of **melanin**, a dark-colored pigment synthesized by polymerization of the polyketide precursor of 1,8-dihydronaphthalene. Melanin deposits in the appressorium cell wall, which has no contact with the leaf surface, and plays a role of a molecular trap increasing the turgor pressure inside the appressorium and promoting growth of the infectious hypha inside the leaf. The mutants with light-color appressoria are non-infectious. Three unlinked genes *ALB1, RSY1* and *BUF1* are involved in melanin biosynthesis. Their mutations impart a white, pink or dark yellow color, respectively, to mycelium and haustoria. The first of these genes controls the synthesis of polyketide synthetase, the second one – of scitalone dehydratase, and the third one – of the NADP-dependent polyhydroxyl naphthalene reductase.

The protein *Psi1p*, which is homologous to the family of the animal **membrane proteins, tetrasparins**, is expressed in the appressoria of *M. grisea*. These proteins take part in the membrane signaling system responsible for cell adhesion, differentiation and motion. *Psi1p* also controls the same functions, adhesion and differentiation, in haustorial membranes. It is interesting that *psi1* mRNA has been found in all cells, while the protein was found only in haustoria, i.e. its regulation works at the level of translation.

The proteins *cap20* and *cap22* are glycoproteins bound to the appressorium cell wall. One of them is required for development of the haustorium and the other for its functioning. The *PIG1* protein localized in the haustorium membrane is homologous to yeast permease, the enzyme responsible for transport of amino acids through the membrane. The *ACE1* protein with the combined enzymatic properties of polyketidesynthetase/non-ribosomal peptide synthetase was found in appressorial cytoplasm. The product of its enzymatic activity is unknown, however, it may play a role at early stages of the haustorium–host cells interaction.

The morphogenetic processes involved in the formation of appressoria in *Magnaporthe grisea* require the 22 kDa protein metallothionine, which has high affinity to zinc, and peroxysomic proteins, which build the Woronin bodies producing compact plugs in ascomycete septums.

The processes of fungal hypha penetration in plant roots are less explored. Instead of melanized appressoria, small outgrowths, hyphopodia, develop in the mycelium, that channelize the penetration in the root tissue. In *Fusarium oxysporum*, the mitochondrial carrier protein *FOW1* necessary for root colonization has been isolated. Deletions of the gene control lead to reduction of pathogenicity but do not affect the fungus growth *in vitro*.

Mode of pathogen penetration into plant tissues
Thigmotropism and spore transmission

Penetration through wounds is the most passive pathway of penetration. The infectious fungal hypha or bacterial cell is attracted to wounds by metabolites released to the environment. For instance, in the tumorogenetic bacteria *Agrobacterium tumefaciens*, which infects plants only through wounds, *vir-genes* (of *vir*ulence) that are necessary for morphological and metabolic changes related to the plant infection, express under the effect of acetosyringone and other phenols. The synthesis of these phenols is induced in plants by mechanical damage, and they diffuse from wounds to the environment. Phenolic compounds, in particular, flavonoids induce the development of haustoria even in dicotyledonous parasitic plants.

Infection with wound parasites is effective only immediately after wounding, as wound surface stimulates synthesis of wound periderm which first, repairs mechanical openings and, second, prevents the release of chemical attractants. The percentage of successful inoculations of maize with anthracnose (causal *Colletotrichum graminicola*) at the time of wounding was ca. 70%, 1 hour later 22% and 6 hours later less than 5% (Muimba-Konakolonga, Bergstrom, 1990). Therefore, infection of plants with parasites transmitted by phytophage insects is most effective. Bark beetles carrying spores of the pathogen of the Dutch disease of elms *Ophiostoma ulmi* on the surface of their bodies (including the mouth parts) eat flowers of elm varieties and transmit spores to the feeding wounds. Cells of the bacterial fire blight of apple-trees *Erwinia amylovora* are spread on the bodies of bees and other pollinators and are entrained in the wounds made by these sucking insects.

Chemotropism towards stomata

Zoosporous fungi are attracted to stomata by the chemical compounds produced in photosynthesis and reducing the zoospore motility period, and by the substances released from the open stomata and attracting zoospores.

Therefore, zoospores of the hop mildew pathogen *Pseudoperonospora humuli* float in darkness for a long time on a wetted surface of the leaf, with occasional stops, while in the light the duration of motile condition is shortened, and zoospores stop near the stomata. Cells of the phytopathogenic bacteria from genera *Pseudomonas* and *Xanthomonas* are regularly distributed over the leaf surface, but after a while they gather near the stomata under the action of chemical secretion.

Cell wall and cuticle degradation by pathogen-specific enzymatic machinery

At active perforation of the surface, the pathogen has to overcome the cuticle barrier covering most plant organs. The cuticle constitutes the insoluble polymer *cutin* submersed in a hydrophobic complex, wax consisting of non-polar C20 – C32 aliphatic hydrocarbons, fatty acids, alcohols and their esters, which are extracted by organic solvents. The insoluble fraction, cutin, is a C16 – C18 fatty acid polymer (dioxypalmitic, oxyoleic) linked with alcohols by ester bonds. The cutin barrier is destroyed by the enzyme *cutinase* which catalyzes hydrolysis of ester bonds. Cutinase is a glycoprotein capable of being released to the environment. Its sequence has a high degree of homology within various phytopathogenic fungi. Cutinases of fungal species differ in the optimum pH that is likely to be related to the specific features of cutin in different plant organs, and organotrophic specialization of the parasites. For instance, the optimum pH of the leaf-dwelling fungi *Cochliobolus heterostrophus, Botrytis cinerea, Venturia inaequali,* is close to neutral, while in stem-affecting *Fusarium solani f. sp. pisi* (teleomorph – *Nectria haematococca*) and *Rhizoctonia solani* it is in the alkaline area (pH 9.0 – 9.5). *Alternaria brassicola and Colletotrichum lindemutianum,* affecting both stems and leaves, had the optimum pH in subacidic and alkaline areas.

Cutinase is an inducible enzyme expressed in the presence of cutin hydrolysate and repressed by glucose. In the fungus *F. solani,* cutinase is produced constitutively at a small concentration (due to the presence of a non-inducible gene with a weak promoter). In contact with cutin, this enzyme causes destruction of several cutin molecules and production of the monomers that induce expression of the cutinase gene which possesses a strong promoter and produces active cutinase. The following experimental evidence of its pathogenetic role is available: (i) The mutants of *F. solani pisi,* defective in cutinase production, lose their pathogenicity to peas (Table. 5.1).

Table 5.1. Role of cutinase in pathogenesis of *Fusarium solani* f. sp. *pisi* (Kolattikudy PE, Koller W, 1963)

Isolates	Cutinase activity	Percentage of infected hypocotyls	
		Intact	Wounded
T-8	2610000	90	90
T-3	12500	7	83

(ii) The avocado wound parasite fungus *Mycosphaerella* sp., incapable of infecting the intact avocado fruits, began the production of cutinase and infection of intact fruits after transformation by the cutinase gene from *F. solani pisi*.

Penetration through cell wall

Penetration into the plant cell is possible only after overcoming the cell wall barrier, which supports the cell protoplast like a strong frame and protects it from mechanical and osmotic damage. The basic components of the cell wall are carbohydrates. Nine monosaccharides have been found in the plant cell wall: 3 hexoses (D-glucose, D-galactose and D-mannose), 2 pentoses (L-arabinose and D-xylose), 2 uronic acids (D-glucuronic acid and D-galacturonic acid) and 2 desoxyhexoses (L-rhamnose and L-fucose). They produce various polymers with others, and also with peptides (glycoproteins). The cell wall of sycamore (*Acer pseudoplatanus*) contains the following percent of various polymers:

– cellulose: microfibrils of glucose linear polymers of (α-1, 4-glucans), linked with each other by hydrogen bonds over the entire length, 23%;
– rhamnogalacturonan: a linear polymer of methylated polygalacturonic acid (α-1, 4-polygalacturonan), occasionally alternating with rhamnose, 16%;
– galactan: branched ($\alpha 1 \rightarrow 4$, $\beta 1 \rightarrow 6$) polymer of galactose, 9%;
– arabinan: branched ($\alpha 1 \rightarrow 5$, $\alpha 1 \rightarrow 3$) polymer of arabinose, 9%;
– xyloglucan: α-1, 4-glucan with branchings of xylose, galactose and fucose: 19%;
– xylan: α-1, 4-xylan with branches of arabinose and glucuronic acid: 5%;
– glycoprotein: a peptide with a high level of the amino acid oxyproline, linked through oxyprolines with the chains of four arabinose molecules, 19%.

These polymers are linked with each other by covalent or hydrogen bonds, producing a complex network, which constitutes a building material of the primary wall of a growing cell. The secondary more rigid wall governing the cell shape and the middle plate – the intercellular adhesive to make tissue out of separate cells. Thus, cell wall carbohydrates not only determine the cell shape but also the shape of the whole organ, and, in the long run, of the whole plant.

Historically, the cell wall polymers are subdivided according to their extractability by different solvents to *cellulose* (linear fibrils of β-glucan), *hemicellulose* (branched polymers of xylan, xyloglucan, arabinan, etc.) and *pectin* (rhamnogalacturonan). The cellulose girders submerged in hemicellulose matrix make the plant cell wall similar to a reinforced concrete structure. Twigs of the hemicellulose molecules are connected with the cellulose by one end, and with pectin by the other end, ensuring a unified system (Figure 5.1). The middle plate is also made of rhamnogalacturonan (pectin).

Lignin, a three-dimensional polymer of phenolic alcohols which will be discussed later, also deposits in the secondary wall.

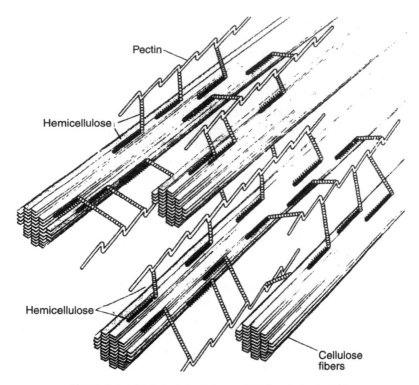

Pectin

Hemicellulose

Hemicellulose

Cellulose fibers

Figure 5.1.   Model of plant stem wall (Albersheim P, 1975).

Role of pectolytic enzymes

Many phytopathogenic fungi and bacteria possess the enzymes destroying cellulose (cellulases), hemicellulose (xylanases) and pectin (pectinases). Wood-destructive fungi possess active cellulases; the most active enzymes in grassy plant parasites are pectinases, while the components of ectotrophic mycorhiza cannot destroy either cellulose or pectin.

Pectolytic enzymes of parasites are the first ones to appear in the infected plant. By degradation of pectin,(a) parasites break links between individual molecules of cellulose, disturb the cell wall composition and provide access to protoplast; (b) by destruction of the middle plate pectin, pectinases allow the parasite to advance inside the plant, and in the case of a strong effect (necrotrophy) they cause death of protoplasts, maceration and disintegration of the tissue; (c) pectinases degradate pectins and produce carbohydrate nutrition.

Fungal and bacterial pectinases have different mechanisms of action upon the pectin molecule and different substrate specificities.

1. *Pectin–methyl esterases* (PME) demethylate polygalacturonide, which results in the production of polygalacturonic acid (Figure 5.2). Free carboxylic groups usually combine with bivalent metals (Ca^{2+}) in a plant to produce salts, pectates. In xylemic vessels affected with vascular infections, calcium

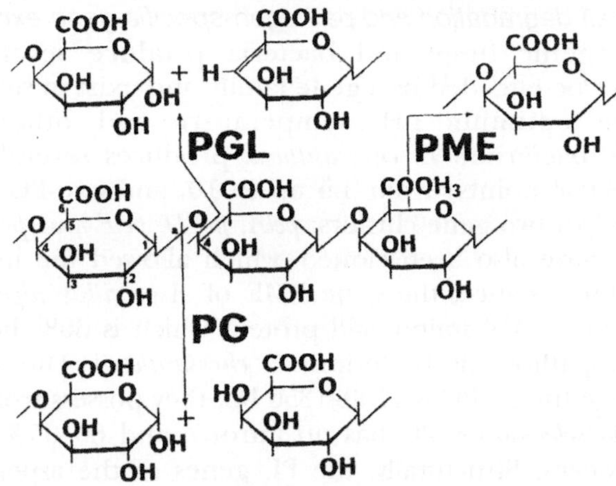

Figure 5.2.   Enzymatic hydrolysis of pectine. PG – polygalacturonase; PL – pectatelyase; PME – pectine methylesterase.

pectates absorb water and produce swelling gel, embolus, which interferes with xylem flow and constitutes a cause of infectious wilt of plants.

2. *Polygalacturonases* (PG) cause decay of polygalacturonide due to the hydrolysis of the glucosidic bonds (Figure 5.2). According to the site of action, enzymes are divided into *endo-PGs* which hydrolyze the pectin polymers of different sizes over the entire molecule and *exo-PGs* which depolymerize pectin from the unreduced ends of the pectic polymer and produce digalaturonates.

3. *Pectatelyase (PL)* catalyzes transeliminative cleavage of pectin into oligogalacturonides. Unlike PG, the resulting products are double-bonded. Its variant, *pectinlyase (PLN)*, possesses substrate specificity for highly methylated pectin.

Further degradation of the pectin molecule is assisted by other enzymes, as follows:

Enzymes (acronym, gene symbol)	Intermediates
Pectatelyase (PL, pel)	Polygalacturonate ↓
Exo-poly-α-galacturonidase (Exo-Pl, peh)	Digalacturonates ↓
Oligogalacturonide lyase (OGL, oge)	4-deoxy-L-threo-5-gexoso-diuloso uronate (DTU) ↓
Ketodeoxyuronate isomerase (KDUI, kdu1)	3-deozy-D-glycero-2,5-hexodiulosonate (DGH) ↓
Ketodeoxyuronate dehydrogenase (KDU, kduD)	2-keto-3-deoxy-D-glucanate (KDG)

Plant cell wall degradation and pathogen-specific gene expression

Most phytopathogenic fungi and bacteria produce several pectolytic enzymes, each can be encoded by a gene family and exist in several isoforms differing in the optimum pH, temperature and other properties. For instance, the bacterium *E. chrysanthemi* produces several PL isoforms with IEPs (isoelectric points) from 4.5 up to 10, and exo-PG with IEP 8.5. Pectate lyase involves two gene clusters: *pelA, pel D, pelE* and *pelB, pelC.* Many fungal pectinases have also been cloned, which allowed the investigation of their structure. For instance, the gene PME of *Aspergillus niger* contains six introns and encodes a 314-amino acid protein which is 30% homologous to PME of the phytopathogenic bacterium *E. chrysanthemi.* The size of all the cloned genes of the fungal PG is 1100–1350 bp, they possess from one to four introns (*Sclerotinia sclerotiorum* PG has no intron), and 60–65% homology in amino acid sequences. Structurally, the PL genes of the aspergillus, *Nectria haematococca* and *Glomerella cingulata,* are extremely diverse in terms of the intron number and arrangement and have low mutual homology.

The high diversity of fungal enzyme sequences is, apparently, a reason of the complex genetic control of phytopathogenic fungal enzymes that are responsible for the degradation of carbohydrate polymers. For the maize parasite *Cochliobolus carbonum,* at least two PGs, encoded by two genes, three glucanases (two genes), five xylanases (four genes) from which *xyl1* is expressed *in vitro* and *in planta, xyl2* – only *in vitro* and *xyl3* – only *in planta,* have been described. Out of five PG genes of *Botrytis cinerea,* one is expressed constitutively, the others are inducible; apparently, their induction occurs under the effect of the oligogalacturonides produced under the action of a non-inducible enzyme. Due to the big variety of enzymes and their encoded genes, elimination of some enzymes due to mutations does not result in a dramatic drop in pathogenicity owing to the compensation effect of the remaining enzymes. In the fungus *Magnoporthe grisea,* the double mutation of the genes *xyl1* and *xyl2* causes activation (or expression) of the "silent" genes that encode additional xylanases.

The variety of hydrolytic enzymes in fungi provides flexible means of attacking the plant cell wall. The parasite can use various types of pectinases depending on the pH, composition and other properties of cell walls in different plant organs and tissues. The substrate specificity of pectinases may determine the age and tissue specialization of pathogens. It has been shown that bean shoots with methylated pectic substances in the cell walls are more susceptible to *Rhizoctonia* than mature plants containing pectin mostly as calcium pectate. Fungal PG actively degrades esterified pectin, but not pectates, hence there is a smaller susceptibility of the mature plants.

Pectinase synthesis is regulated according to the principles of substrate induction and catabolite repression. In *Erwinia,* PL synthesis is induced with plant cell wall preparations, polygalacturonide, purified from cell walls and with products of polygalacturonide degradation (digalacturonide is an especially active inducer). According to the suggested model, low concentrations of PL and exo-PG are produced without induction, by dimer release. The optimum conditions for PL (high pH, availability of bivalent cations) cause accumulation of unsaturated digalacturonides. Under other conditions they are accumulated resulting from the effect of exo-PL. Pectinases are not produced in the presence of more energy-beneficial catabolites (e.g. glucose). An exception from this pattern is the enzyme PNL, which in phytopathogenic bacteria is not induced by pectin, but by DNA-damaging agents and UV light.

In many phytopathogenic organisms, pectinases are exported from the cell due to their need in ensuring microbial transfer through the membrane. Pectinases are eliminated during transporting by cellular proteases. In *E. chrysanthemi*, PL export occurs in two stages: first the enzyme is secreted through the inner membrane to the periplasma, then it is transported through the outer membrane (cell wall) by the gene *out* product.

Some *pel*-genes of *Erwinia* have been transferred to *E.coli*, which resulted not only in its acquiring the ability to macerate potato tuber sections, but also to infect shoots, causing typical symptoms of the bacterial disease "black leg", and to accumulate in the infected plants. Besides, both the disease symptoms and bacteria reproduction were more intensive in the bacteriosis-susceptible potato cultivars, than in the resistant ones. Thus, pectatelyases are the true factors of pathogenicity of the phytopathogenic bacteria of the genus *Erwinia*.

Non-enzymatic mechanisms of plant cell wall degradation by pathogens are less investigated. The protein *expansin* have been found in plants that causes the wall degradation through weakening non-covalent interactions of its components. A similar protein has been also found in the nematode *Globodera rostochiensis*, and its effect on the plant cell wall has also been shown.

Plant–nematode interrelationships

Nematodes can penetrate a plant and feed only after overcoming the plant cell wall barrier. Nematodes produce the enzymes to destroy cellulose (cellulases), hemicellulose (xylanases) and pectine (pectinases). The main and the strongest effect on the plant is rendered by the nematode's hydrolytic and proteolytic enzymes, which are the first cause to the changes in the plant metabolism. Nematode homogenates and exudates were found to contain cellulases, amylase and pectinases. These enzymes destroy plant cell walls, which enables endoparasitic phytonematodes to move both between and inside the cells, as well as to feed. An especially important role in pathogenesis of various diseases is played by pectinases that attack the polymers of the cell walls and middle lamella and thus promote nematode penetration into a plant. Differences in the system of pectolytic enzymes and their activity are supposed to be a cause of essential differences in the mechanism of pathogenesis compared to other pathogens. Pectinase activity in homogenate of the migratory parasite *Pratylenchus pratensis* is seven times higher than that of sedentary *Heterodera trifollii* and *Ditylenchus dipsaci*. The activity of the pectinase enzyme is 28 times higher than in *D. myceliophagus* feeding on fungi.

The biochemical methods of analysis allowed revealing activity of hydrolytic and proteolytic enzymes in homogenates and exudates of parasitic nematodes already for more than 30 years ago, but only the modern molecular genetic methods and technologies, such as the Real-Time PCR and microassay of DNA rearrangements, using the transcripts from various phases of nematodes, allowed receiving sufficiently reliable data regarding the role of certain genes in nematode development and parasitism, as well as characterization of the enzymes secreted by nematodes and revealing their direct role in nematode–plant relationship.

The determinant factors to establish and develop relations between plants and nematodes is the esophageal gland discharge. The genes that encode the gland enzymes inducing parasitic relations between the partners and releasing to the plant tissue through the stylet were called "parasitism genes", and the secreted proteins were called "parasitomes". They are regarded as a component of "secretomes", all proteins secreted by the nematode during parasitizing on the plant (Greenbaum terminology).

The first characterized parasitomes were β-1,4-endoglucanases (cellulases) secreted by two subventral glands of the cyst nematodes *Heterodera glycines* and *Globodera rostochiensis*. It is interesting to note that these nematodes were the first animals in which endogenic cellulases were found. Cellulases of the cyst nematodes were found to be very similar to bacterial cellulases, and they clearly differed from cellulases of other eukaryotes (including the nematode *Caenorhabditis elegans*). This suggested that the cellulase-encoding genes were received by ancestral cyst nematodes from prokaryotic microorganisms by horizontal gene transfer.

Genes encoding various β-1,4-endoglucanases were found in esophageal glands of many sedentary nematodes (Table 5.2). Their primary structure and properties have been investigated in detail. It should be noted that the activity of β-1, 4-endoglucanases was found only in the motile phases of sedentary nematodes (infesting and preinfesting larvae, and males), while in adult females no enzyme activity was displayed. Data are available on the presence of cellulase formation encoding genes also in errant nematodes, such as *Pratylenchus agilis, Paratrichodorus minor, Bursaphelenchus xylophilus, Rotylenchus reniformis* and *Ditylenchus dipsaci*. Today, β-1,4-endoglucanases are regarded as playing the leading part in establishing relations between plants and parasitic nematodes.

Cloning of subventral gland (SVG) proteins also showed the presence of various pectinases. Genes encoding these enzymes were found in the SVGs only at the early phases of development of the infesting larvae of cyst and root-knot nematodes, which confirms their role in penetration and intracellular migration at the early stages of parasitism (Table 5.2).

The gene encoding chorismate mutase, the enzyme directly involved in biosynthesis of the amino acid phenylalanine, was found in the esophageal glands of sedentary nematodes. Phenylalanine is known as a precursor of auxins, the hormones involved in the formation of syncytium, which gave grounds for discussion of a role of this enzyme in host–parasite relationship.

Table 5.2. Products of "parasitism genes" from digestive glands of sedentary phytonematodes (according to different authors)

Compound	Nematode species	Organisms with homologous products	Possible function
1-4 endonucleases (cellulases)	*G. rostochiensis, G. pallida G. tabacum, H. glycines, H. schachtii, Meloidogyne incognita*	Bacteria	Cell wall degradation
Pectate lyase	*M. javanica, M. incognita, G. rostochiensis H. glycines*	Bacteria and fungi	Cell wall degradation
Polygalacturonase	*M. incognita*	Bacteria	Cell wall degradation
Chorismate mutase	*H. glycines M. javanica, M. incognita, G. rostochiensis*	Bacteria	Auxin balance regulation, formation of feeding cells
Thyroxidin peroxidase	*G. rostochiensis*	Animal parasite nematodes	Peroxide destruction, protection against host
Allergen-like proteins	*M. incognita H. glycines*	Same, *C. elegans*	Unknown
Calreticulin	*M. incognita*	Animal parasite nematodes	Unknown
Chitinase	*H. glycines*	*C. elegans*	Unknown

Esophageal glands of the root-knot nematode *M.incognita* were also found to contain the genes that encode chitinases, cellulose-bound proteins, lipoproteins and other proteins. Secreta of the infesting larvae of this parasite as well as other species of the genus *Meloidogyne* contain calreticulin, a calcium-binding protein involved in most regulatory processes in eukaryotic cells, as well as carbohydrates. Thioredoxin peroxidase, a hydroperoxidase-destroying enzyme, was identified in the grains of the SVGs of *G. rostochiensis*. Presence of this enzyme in nematode discharge is supposed to cause inhibition of plant defense response by destruction of the peroxides produced in the plant in response to invasion.

A very interesting phenomenon of molecular mimicry was revealed in investigation of the mechanisms of formation of the feeding sites: the huge cells and syncytium. The dorsal esophageal gland of a parasitizing larva of the soybean cyst nematode *Heterodera glycines* was found to contain gene *Hg-SYV46* encoding a secretory protein, with the function similar to the protein of *Arabidopsis thaliana CLAVA3/ESR*, falling into the family of

CLE-proteins. The *CLE*-proteins are involved in regulation of cell division and differentiation of root and leaf meristem cells of *Arabidopsis*. With the aid of transgenic plants, it was shown that the functional similarity of the parasite gene *Hg-SYV46* and gene *CLE* of the host is determined by the similarity of their conservative terminal sites, which enables the use of plant receptors for parasite protein-binding and induction of a natural plant signal for formation of the feeding cells induced by nematodes in plant roots.

Homogenates and intestine of cyst nematodes *G. pallida* and *H. glycines* contain cysteic (cathepsins), serinic (trypsins) and aspartyl proteinases. These enzymes are involved in the digestion of plant proteins, thus playing a certain role in relation with the host plant.

In addition to proteins, nematode discharge was also found to contain a large amount of other compounds: carbohydrates, lipid components, ammonia, amino acids, amines, 1,2-dicarboxylic acids, aldehydes and organic acids. These products of life activity can cause destruction of host cells, and thus be toxins.

Suppression of plant defenses

The hydrolytic enzymes, which destroy the cell membrane, are not suitable for suppression of plant defenses, since they are high-molecular compounds, and, therefore, they cannot migrate throughout the plant. These enzymes are the close combat weapons. The role of the substances migrating from the infection site over the plant and creating bridgeheads for its further occupation by the parasite is played by *vivotoxins*. This name, suggested by the American plant pathologists Diamond and Waggonner, resulted from the methodology discussion on the criteria of phytotoxins of the microbial origin. In the early studies of fungal toxic metabolites, the plants were exposed to the filtrate of a liquid nutrient medium where the investigated fungus was cultured. If the treatment with the filtrate caused the same symptoms of the disease as the inoculation with the fungus (withering, necroses, etc.), it was believed that phytotoxins, which could be isolated and investigated, had been released to the culture liquid. However, a long growth in an artificial nutrient medium results in the autolysis of individual mycelial cells and release to the medium substances, which normally do not release into the plant. Such methods of investigation have led to the errors in establishing the chemical origin of phytotoxins. Therefore, it has been suggested to accept, as the toxin criteria, first, reproduction of the disease symptoms by the given chemical compound, and, second, its detection not only in a test tube (*in vitro*), but also in the plant infected with the pathogen (*in vivo*).

Vivotoxins are non-specific, they damage not only the parasite host, but also the plant species outside the parasite's food specialization. Therefore, vivotoxins are included in the factors of the non-specific (horizontal) pathosystem.

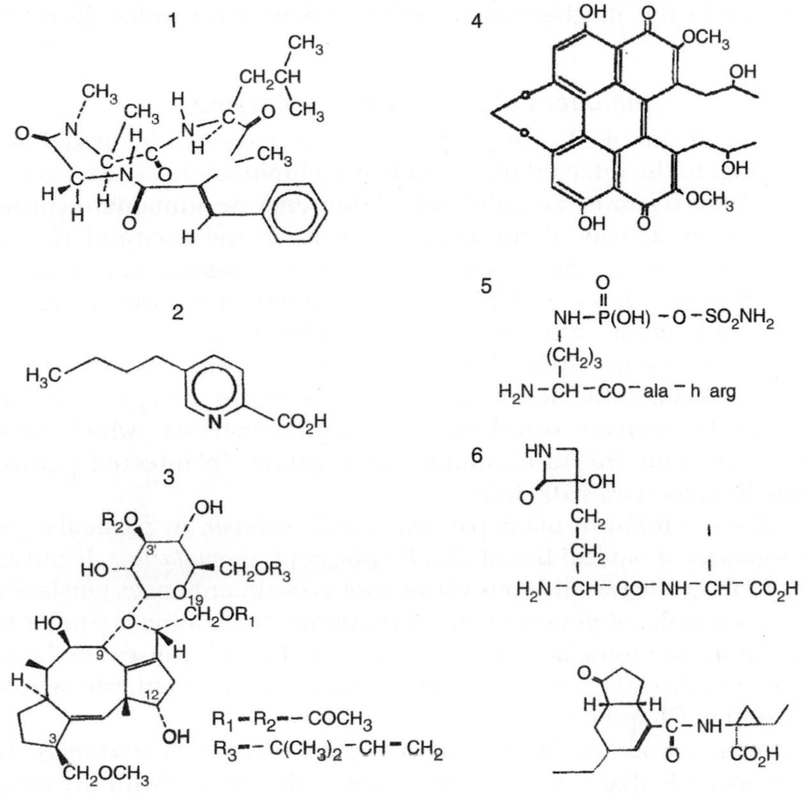

Figure 5.3. Vivotoxins.
(1) Tentotoxin of *Alternaria alternate*; (2) Fusaric acid of *Fusarium oxysporum*; (3) Fusicoccin of *Fusicoccum amygdali*; (4) Cercosporin of *Cercospora* spp.; (5) Faseolotoxin of *P. syringae* pv. *phaseolicola*; (6) Tabtoxin of *P. syringae* pv. *tabaci*; (7) Coronatin of *P.syringae* pv. *coronafaciens*.

Chemically, vivotoxins are divided into several groups (Figure 5.3):

1. Organic acids, e.g. oxalic acid of the white rot causal organism *Sclerotinia sclerotiorum*.
2. Cyclic aromatic compounds (cumarins, nitrogen-bearing alkaloids, etc.). They include alternaric acid (producer: *Alternaria solani*), fusaric acid (producer: *Fusarium oxysporum*), etc.
3. Cyclic peptides consisting of several amino acids closed in a ring: phaseolotoxin (producer: *Pseudomonas savastanoi* pv. phaseolicola), tentoxin *A.tenuis* (syn. *A. alternata*), etc.
4. Glycopeptides, such as the toxins of *Clavibacter michiganense* (causal organism of tomato bacterial cancer) and *Phoma tracheophylla* (causal organism of malsecco citrus).
5. Polysaccharides (toxin of the bacteria *Xanthomonas campestris*).
6. Polypeptides, e.g. ceratoulmin, a 13 kDa toxin of the causal organism of the Dutch disease of elms *O. ulmi*.

According to the mechanism of action, toxins are divided into several groups.

Inhibitors of specific plants enzymes

Tabtoxin, the toxin of *P. syringae* pv. *tabaci*, is a dipeptide linked with a β-lactam ring. In the infected plant, proteases eliminate the active part of the toxin – *tabtoxin-β-lactam*, an inhibitor of the enzyme glutamatesynthetase. The decrease in activity of the enzyme results in the accumulation of its precursor ammonium that causes suppression of phosphorylation, and inhibition of photosynthesis and photobreath. These metabolic changes are manifested as systemic chlorosis and growth inhibition.

Phaseotoxin is a cyclic tripeptide of *P. savastanoi* pv. *phaseolicola*. In the plant, proteases eliminate the active substance *phaseotoxin ornitidine*, a specific inhibitor of the enzyme ornithine carbamoyl transferase, which catalyzes conversion of ornithine to citrulline and arginine. In infected plants, the ornithine level increases 100-fold.

Coronatine is a toxin of many pathovars of *P. syringae*, in particular, of the causal organism of oat leaf fire blight, *P. syringae* pv. *coronafaciens*. It causes yellowing of tobacco leaves, inhibits wheat root growth, enhances amylase activity and hypertrophy of potato tubers. Structurally coronatine is similar to the plant hormone jasmonic acid, overproduction of which causes similar symptoms in plants. Probably, coronatine substitutes jasmonic acid, whose role will be discussed in Chapter 8.

Tentoxin is a toxin of the fungus *Alternaria tenuis* (=*A. alternata*). It is a cyclic tetrapeptide that causes chlorosis due to the linkage with the coupling factor (CF1) and inhibition of photosynthetic phosphorylation.

Membrane active substances

Many toxins, such as fusaric acid (*F. oxysporum*), toxins of the bacteria of the genus *Clavibacter*, toxins of many pathovars of *P. syringae* (coronatine, syringomycines), toxins of the phytopathogenic fungi of the genera *Rhinchosporium*, *Cephalosporium*, *Fusicoccus*, *Cercospora*, *Phoma* and others possess a strong membranomtropic effect. They induce loss of metabolites from the cells and cell necrosis, have effects on transmembrane ion transport, and open stomata, which results in plant wilt.

Generators of active forms of oxygen

Some phytopathogenic fungi of the genera *Cercospora, Cladosporium, Alternaria, Elsinoe* and *Hypocrella* form the cyclic compounds perylenequinones (Figure 5.3 (4)), which need light to manifest their toxicity for the host plants, i.e. they belong to the group of photosensitizer substances. Upon light energy absorption, photosensitizers acquire an energy-active state and the ability, when interacting with molecular oxygen, to generate its active forms, radicals, like superoxide (O_2), hydrogen peroxide (H_2O_2), hydroxyl radical (OH) and non-radical-singlet oxygen ($^1O_2$). All of them are highly toxic and damage plant cells.

In *Cercospora nicotianae,* the gene CRQ1 has been found that encodes a protein, which has the DNA-linking motif, Cys6Zn2, i.e. it is a transcription activator. Mutants on this gene are susceptible to the toxin cercosporin and produce 30–55% less CRQ1 protein than the wild strains. The protein *Crq1* is believed to be necessary both for activation of the genes of cercosporin synthesis, and for resistance to the active forms of oxygen produced by the toxin.

Inhibitors of protein synthesis

Regulation of the synthesis of the low-molecular secondary metabolites possessing vivotoxin properties are exemplified here by the trichothecene toxins of fusarial fungi. These toxins are of importance because their producers cause diseases of wheat, barley and maize and, because toxin-containing vegetal products are toxic for people and livestock (causing loss of appetite, dermatites, anemia, hemorrhagic sepsis, and immunosuppression). These compounds, including several antibiotics of the fungal origin, inhibit protein synthesis in eukaryotes.

Figure 5.4 shows biosynthesis of trichothecene toxins. The key enzyme responsible for the farnesyl pyrophosphate ring closure resulting in trichodiene (trichodiene synthetase) is controlled by the gene *Tri5*. After a series of oxidations, isomerization and the secondary ring closure, two classes of compounds have been produced (A and B), differing in the presence or absence of the keto-group at C-8 atom of the trichothecene skeleton: T-2 toxin of *Fusarium sporotrichoides* and deoxynivalenol (DON) of *F. graminearum.*

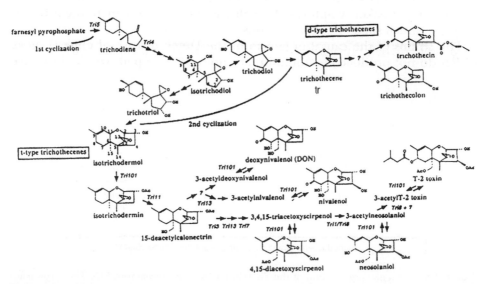

Figure 5.4. Biosynthesis of trichothecene toxins by *Fusarium* fungi. (Kimura M, Anzai H, Yamaguchi I, 2001.)

Molecular studies of the biosynthesis of trichothecene toxins began after the key enzyme trichodiene synthetase had been purified and used to clone the gene *Tri5*. It has been found that the cosmid carrying this gene also complements mutations for other genes that have an impact on the subsequent stages of trichothecene synthesis. This allowed a conclusion regarding the linking of these genes in a common 25 mb cluster. The clusters for both *Fusarium* species (*F. sporotrichoides* and *F. graminearum*) had a high degree of homology, were flanked with the identical genes of tyrosinase and polysaccharide deacetylase, and consisted of at least 11 genes involved in the trichothecene biogenesis (Figure 5.5). They included the biosynthetic genes: *Tri4* (P450 monooxygenase, responsible for trichodiene hydroxylation), *Tri3* (15-0-acetyltransferase), *Tri11* (P450 monooxygenase, involved in C-15 hydroxylation), and the regulatory genes: *Tri6* (transcription activator) and *Tri12* (regulator of trichothecene release from the cell). The gene *Tri7* (4-O-acetyltransferase) of *F. sporotrichoides* is absent from *F. graminearum*, and *Tri13* (type P450 protein) occurs as a pseudogene. These enzymes seem to be unnecessary in the B-type toxin synthesis, therefore, their encoding genes have been lost in the course of evolution.

The role of toxins in pathogenesis is multifaceted, and it has not been completely revealed. It has been established for many parasites that they release toxins to the infected plant and migrate over the plant, outstripping the pathogen migration. Therefore, such systemic symptoms of diseases as wilt and chlorosis may be caused by migration of toxins from the site of infection.

The significance of toxins in pathogenesis has been shown through mutants deficient in the ability of toxin production (Tox). Non-toxicogenic strains were received from phytopathogenic bacteria using transposon mutagenesis (genetic transformation of the cell by a mobile genetic element, transposon, which penetrates to a DNA site and causes mutations of its genes). Such mutants usually cause weaker symptoms of the disease and accumulate in the infected plants in much smaller concentrations.

The strains of the causal organism of the Dutch disease of elms *O. ulmi*, which do not produce ceratoulmin, are not only non-pathogenic, but have

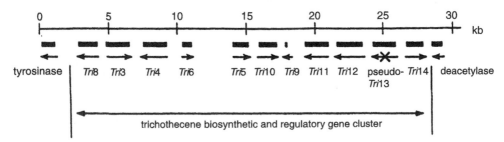

Figure 5.5.   Genomic organization of the *Tri-5*-cluster from. *F. graminearum* F15. The arrowheads indicate the direction of transcription. The symbol X over *Tri-13* indicates that this gene does not code a functional protein. (Kimura M, Anzai H, Yamaguchi I, 2001 (Fig. 4).)

the modified morphological–cultural attributes and grow slowly *in vitro*. The effect of the toxin on the producer morphogenesis stems from the fact that ceratoulmin belongs to the family *of hydrophobins,* the hydrophobic proteins performing morphogenetic functions.

Thus, in many plant pathogens, vivotoxins have narrow specific sites of action (specific enzymes). Their non-specificity is determined by the fact that these sites are available for most plants, and often they are the key sites in the plant metabolism. Some toxins (e.g. ceratoulmin) are bifunctional, and alongside being toxic for the plant, they perform other functions, which may be their primary functions.

Changes in metabolism of infected cells and tissues

Synthesis of the essential plant metabolites, their flow directions, and, hence, the morphogenetic processes are regulated by *phytohormones,* which are divided into several groups of substances similar in the chemical origin and morphogenetic action (Figure 5.6). The most important phytohormones include *auxins* (govern root development and growth, cell division and other processes), *cytokinins* (govern plant top differentiation and growth, delay cell aging, and open stomata), *gibberellins* (cause cell elongation, induce flowering), *abscisic acid* (causes inhibition of cell growth and division, closes stomata, enhances stress resistance, promotes accumulation of reserve products in dormant organs) and *ethylene* (governs fruit ripening, contributes to the autumn leaf abscission, thickens stems and shortens internodes).

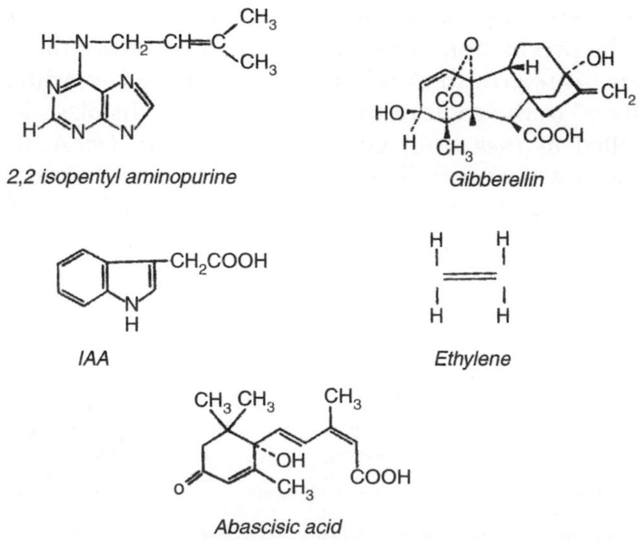

Figure 5.6. Phytohormones.

Many of the events regulated by the phytohormones are characteristic of the sick plants, especially those affected by biotrophs, which change metabolite flows into the direction of the infection, delay cell aging by inducing the "green island" formation, and stimulate plant cell division and elongation, and development of tumors, proliferations, etc. Gibberellin, in particular, was first isolated from the fungus *Gibberella fujikuroi* that causes rice witch (wakanae disease). Rust and powdery mildew fungi secrete cytokinins and auxins into the plant. In cruciferous plants, cytokinins released by the pathogen cause root proliferation (clubroot). The local release of cytokinins by the fungi *Exobasidium* and *Taphrina*, and the bacterium *Corynebacterium fasciens* results in the loss of apical growth and bolting of the affected organs.

Regulation of hormonogenesis has been studied in detail for the gall-forming phytopathogenic bacteria. *Pseudonomas savastanoi* pv. *savastanoi* causes galls in olive and oleander due to the release of an auxin, indolyl-3-acetic acid (IAA), to the affected plants. The IAA synthesis from tryptophan proceeds in two stages catalyzed by the enzymes tryptophan monooxygenase and indoleacetamide hydrolase (Figure 5.7). The genes encoding these enzymes (*iaam* and *iaah*, respectively) in the oleander isolates are in the plasmid *plAA*, and in the olive isolates they can also be in the "chromosome". The oleander isolates can infect both hosts, and the olive isolates only olives. The mutants deprived of the plasmid *plAA* cannot develop tumors, and mutants with superproduction of IAA induce development of larger tumours.

In the causal organism of crown galls *Agrobacterium tumefaciens*, the plasmid T-DNA is integrated in the genome of the infected plant cell. The T-DNA has the genes *tms-1* and *tms-2*, similar to the genes *iaam* and *iaah* P. of *savastanoi* pv. *savastanoi*, the products of which control conversion of tryptophan to IAA, as well as the gene *ipt* controlling synthesis of the enzyme isopentyl transferase necessary for the production of isopentyl-AMP, the key intermediate in the pathway of production of two cytokinins, transribosylzeatin and transzeatin. The production of IAA and cytokinins causes rapid cell division and neoplastic tissue growth. A related species of the bacterium *A. rhizogenes* causes "hairiness" in the roots of the infected plants. Its plasmid localizes the oncogenes called *rol (root locus)* genes. Each of them (*rol* A, B, C, D) transforms the cell to a root cell. The *rolB* and *rolC* are involved in auxin and cytokinin release from the inactive complexes (e.g. *rolC* controls the enzyme β-glucosidase which eliminates cytokinin from an inactive glucoside).

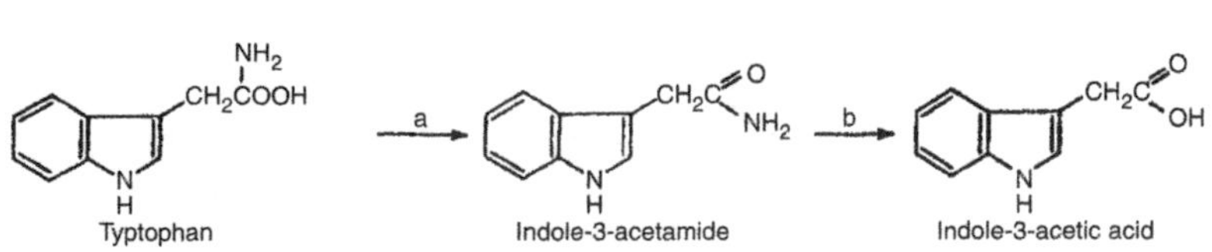

Figure 5.7. IAA biosynthesis in the bacteria *Pseudomonas savastanoi*. (a) Tryptophan monooxygenase; (b) Indoleacetamide hydrolase.

Many vivotoxins also play the role of hormones. For instance, the toxin of the pathogen of peach and almond cancer and wilt *Fusicoccum amygdali* fusicoccin possesses the functions of auxin and cytokinin. The principle effect of fusicoccin consists of the action on the transmembrane transport regulated by hydrogen and potassium ion exchange. This results in an increase in water permeability of the stomatal closing cells, change of their shape and opening of the stomata. Fusicoccin induces tissue elongation and interruption of the dormant period of seeds. Another vivotoxin, helmintosporal, (product of the fungus *Bipolaris sorokiniana*, pathogen of dark-brown wheat spot) causes, similar to gibberellin, rice bolting, necroses of oat and wheat, and stimulates the production of the enzyme α-amylase. Thus, there is no strict functional delineation between vivotoxins and phytohormones: toxic or hormonal action of a compound is determined by its concentration, the plant species, its age and the exposed organ.

Transport in the plant

Fungi and bacteria usually spread in the plant locally, in a short distance from the site of infection, through the intercellular spaces. Xylem fungal and bacterial infections (tracheomycoses and tracheobacterioses) spread throughout the plant by the xylem flow. When the xylem fungi of the genus *Verticillium* spread over the plant vessels, there is an alternation of mycelial and yeast-like growth (mycelial–yeast dimorphism). In the vessels, the fungus reproduces by budding (by arthrospores), which are easily entrained by the xylem flow. Upon arrival to the perforated septums in the vessels, spores germinate there as hyphae that produce pectolytic enzymes dissolving the carbohydrates of the septum and opening further way to the yeast-like spores.

The movement of viruses in the plant poses some particular problems. After reproduction in the primary infected cell, the entire plant is infected due to the movement of the viruses from one cell to another (near transport) and over the phloem (far transport).

The near transport is carried out through plasmodesmas, the membrane strands between the cells containing internal microchannels. Experiments with dextran molecules showed that only very small molecules of ca. 1 kDa, can pass through plasmodesmas. However, the viral nucleic acid encodes a special transport protein (TP) which assists the passage through plasmodesmas. Tobacco mosaic virus (TMV) was the first virus where such a protein was discovered. Its molecular mass is 30 kDa, and it is encoded by a gene located nearer the 3′ end of the viral RNA (see Figure 2.1). This protein has the following properties: (1) It binds with single-stranded molecules of nucleic acids (RNA and DNA) *in vitro* and *in planta* and packs RNA molecules in a long thin nucleoproteid diameter 2–2.5 nm, the center of which is an RNA molecule, and the periphery is made of many TP molecules. (2) It interacts with the cytoskeletal elements: the microtubule material tubulin, actin of microfilaments and also with ER, which enables the nucleoproteid to migrate in the

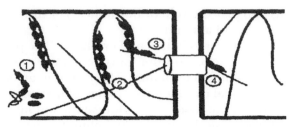

Figure 5.8.   Model for the movement of the P30-RNA complex through the cytoplasm to and through plasmodesmata (Zambrycki PC. 1995 (Figure 8)). Step 1 shows that P30 forms a complex, either in the cytoplasm or on the cytoskeleton, with viral RNA. Step 2 shows that the P30-RNA complex moves long distances through the cytoplasm on microtubules, possible by interacting with a microtubule motor. Step 3 shows that the P30-RNA complex moves short distances to the plasmodesma on plasmodesma associated actin filaments, possible via a myosin motor. Step 4 shows that the P30-RNA complex then may move through plasmodesma to adjacent cell on the same plasmodesma-associated actin filament.

cell by the intracellular tracks. (3) Its receptor in the host plant cell wall is the enzymatic protein pectin–methyl esterase localized in the cell wall. (4) It links with the purified viral RNA *in vitro* and *in planta* and opens the plasmodesmas, which allows the passage of large molecules. In tobacco, where a 30 kDa protein gene was built by genetic engineering methods, the size of the molecules passing through the plasmodesmas increased from 850 Da to 10 kDa and more. (5) The TP negative regulation is provided by its phosphorylation with the plant kinases associated with the cell wall. After the phosphorylation, there is no reverse transport to the already infected cells, and the plasmodesma permeability decreases.

The genes encoding synthesis of transport proteins have been also found in other viruses than TMV. Thus, movement of the virus through plasmodesmas is carried out in the form of nucleoprotein which moves over the cytoskeletal elements: in cytoplasm – over the microtubules, near the plasmodesmas – over the actin filaments permeating the plasmodesmas (Figure 5.8). The velocity of the tobacco etch virus (Poty-virus group) is 1 cell per 2 hours.

The far transport is carried out by phloem flow, therefore, the flow velocity and direction are critical for the virus spread throughout the plant. It was shown long ago that after inoculation of one tomato leaflet with TMV, first the whole leaf is infected, then the stem below the leaf and the roots, then the apex, and, finally, the entire plant. Movement of the virus in the phloem is carried out both as whole particles and as nucleoproteids where the viral RNA is linked with the protein. For instance, in potyviruses, the capsid protein (CP) is involved both in near and far transport, however, mutation analysis showed that these functions are performed by different CP sites.

Selected Literature

Albersheim P. Scientific America 1975; 232:N4.

Daub ME, Ehrenshaft M. The photoactivated cercospora toxin cercisporin: contributions to plant disease and fundamental biology. Ann Rev Phytopathol 2000; 38:461–490.

Davis EL, Hussey RS, Baum TJ, et al. Nematode parasitism genes. Ann Rev Phytopathol 2000; 38:365–396.

Kimara M, Anzai H, Yamaguchi I. Microbial toxins in plant-pathogen interactions: Biosynthesis, resistance mechanisms, and significance. J Gen Appl Microbiol 2001; 47(1):49–160.

Tucker SL, Talbot NJ. Surface attachment and pre-penetration stage development by plant pathogenic fungi. Ann Rev Phytopathol 2001; 39: 385–417.

Wosten Han AB. Hydrophobins: multipurpose proteins. Ann Rev Microbiol 2001; (55):625–646.

Yamada T. The role of auxin in plant-disease development. Ann Rev Phytopathol 1993; (31):253–273.

Zambricki PC. P.C. Plant Cell 1995; 7: 2110, (Figure 8).

Chapter 6

Horizontal pathosystem: resistance factors

Yu. T. Dyakov, V. G. Dzhavakhiya

The Swiss mycologist and plant pathologist E. Geumann (1954) divided the non-specific resistance factors in the horizontal pathosystem into *axeny* and *true resistance.*

Axeny is the resistance that occurs before a pathogen–plant cell contact, and hence, prevents the infection. *True resistance* is manifested after the pathogen contact with the cellular protoplast, and it delays the parasite development and (or) decreases its efficiency. One can understand the difference between axenia and true resistance through the following example: among three cereals – rye, wheat, and maize – the first one is severely affected by ergot, the second one is very seldom affected, and the third one is not affected at all, i.e. the first one is susceptible, and the others are resistant. However, in artificial inoculation of wheat flowers with the parasite conidia, wheat is affected almost as strongly as rye, and maize remains intact. The difference between rye and wheat is that the former is a wind-pollinated plant and its flowers are open for parasite spores, while the latter is a self-pollinator, hence, it has close-up flowering. Thus, the difference between these cereals is not in the cell resistance, but in the chances of the pathogen to reach susceptible cells, i.e. in axenia factors. Unlike these two cereals, maize possesses a true protoplast resistance. As one can see from the above example, most axenia factors are anatomical–morphological, while the nature of true resistance factors is rather physiological–biochemical.

Anatomical–morphological resistance factors

Plant habitus

Potato varieties with spreading haulm are less affected by late blight, than compact-tops varieties, though there is no correlation between the potato plant habitus and its vulnerability at artificial inoculation of leaves in laboratory. The difference stems from the fact that the drops necessary for infection dry up faster on the leaves of tops-spreading varieties, than in varieties with compact-tops plants. Short-stemmed wheat varieties are very vulnerable to the leaf spot fungi, whose spores are dispersed by water splashes (septoriosis, rhynchosporiosis, etc.), as the distance between the leaves (length of internodes) in such varieties is short. Varieties with downward leaves covered, in addition, with dense waxy bloom, are less vulnerable to rust, powdery mildew, and leaf spot, than the varieties with horizontal or ascending leaves, since in the former case the infectious drops roll off the leaves. That is why B.G. Reiter suggests estimation of the leaf drop-retaining capacity as one of the wheat resistance parameters.

Downy leaves

Raspberry varieties with downy leaves are less vulnerable to viral diseases than the varieties with smooth leaves. The reason is difficult feeding conditions for aphids, the virus vectors.

Structure and arrangement of stomata

The tangerine is more resistant to the bacterial cancer pathogen *Xanthomonas citri*, than the grapefruit. The difference stems from the fact that the outer walls of the tangerine stomata have processes preventing penetration of drops of liquid containing bacterial cells, to the substomatal space. The susceptible grapefruit varieties have no such processes. Plants species and varieties with sparse stomata on the leaves are less vulnerable to the fungi, the hyphae of which penetrate through stomata (e.g. diacariotic phases of cereal rust pathogens).

Anatomic features of internal plant tissues

In 1970, E.E. Geshele and L.T. Babayants have shown that the stem rust susceptible varieties of wheat have a developed chlorenchyma, extending as broad bands under the epidermis, while in resistant varieties the chlorenchyma tissue looks like small islands divided into groups of sclerenchyma cells, unavailable for colonization.

Regulation of ontogenesis and damage reparation

Many pathogens, such as the wheat-covered smut, root rots of sprouts, etc. affect plants only at early phases of their development; while other ones – potato late blight pathogens, Verticillium cotton wilt, etc. affect ageing tissues where carbohydrate outflow and decay delays photosynthesis. Therefore, the varieties with fast early phases of ontogenesis are less vulnerable to the former group of diseases, while the varieties with delayed ageing are relatively resistant to the latter group of diseases (M.S. Dunin, 1946). This also makes a basis for agrotechnics of plant protection using regulation of the rate of individual phases of ontogenesis (early seeding of winter cereals and late seeding of summer varieties, to provide simultaneous quickly growing shoots; delay of ageing by nitrogen fertilizers, etc.).

Many plant parasites can penetrate into tissues only through wounds (wound parasites). The rate of suberization of wound surfaces correlates with plant resistance for such parasites. Therefore, buildings for long-term storage of vegetables and potatoes are equipped with air blowers supplying air at a certain temperature and humidity through a layer of stored produce (active aeration), as suberine formation is an energy consuming process.

Resistance to biotroph parasites with haustorial feeding may be governed by the rate of development of the *papilla,* a callous plug closing a hole in the cell wall. A strong silicon-containing papilla provides resistance to the barley varieties with the non-specific resistance gene *ml0* to powdery mildew.

Nutrient value of infected organs and tissues

Though for parasites, plants constitute, first of all, a nutrient substrate, the feed value is not likely to be critical in vulnerability of plant species or varieties. As died-off plants or their parts are occupied by the microorganisms unable to infect the same plants in *live* condition, plant resistance is governed not by the content of the particular nutrients, but by the antibiotic properties of the living tissue. However, in some cases it is the feed value that determines the degree of plant susceptibility to one or other disease. A number of special tests have been suggested to prove the relation between a nutrient component in a plant and its vulnerability: (1) increase in the growth rate in the minimal nutrient medium with added extract from a susceptible plant; (2) demonstration that the extracts from susceptible hosts are more effective at higher dilutions, than the extracts from the resistant ones; (3) isolation and identification of growth stimulators from extracts; (4) demonstration that in susceptible plants, concentration of the stimulating substance is higher than in resistant ones; (5) detection of specificity of the stimulating substance for a given parasite as compared with non-parasites (parasites of other plants); (6) demonstration of the food limit for the given substance in relatively resistant species, varieties, or tissues; (7) identification of the way of action of the stimulating substance. The majority of the tests were carried out to find the relation between the level of the nitrogen-bearing compound *choline* and its ether *betaine* in plants and their vulnerability to *Fusarium graminearum.* A strong damage of cereal flowers is caused by a high level of choline and betaine in anthers.

Growth of the plum parasite *Rhodostricta quercina* requires lipoinositol, the level of which in different plum varieties correlates with the severity of the disease.

Cases of active plant response have been observed: decrease in the synthesis of the products necessary for the parasite, as response to the infection. For instance, the cotton wilt pathogen *Verticillium dahliae* needs the vitamin biotin, and its level grows after infection in susceptible varieties, and falls in the resistant ones. Elimination of the substances necessary for the parasite from the lesion is a plant defense response.

Parasites from the family Pythiaceae (genera *Pythium* and *Phytophthora*) have no enzyme squalene synthetase and, consequently, they do not produce sterols which they need for development of asexual and sexual sporogenesis and pathogenicity. However, they use phytosterols of host plants and transform them to the substances they need. Infection of the potato with late blight activates synthesis of terpenoid metabolism and accumulation of sterols necessary for the parasite in the infection area. Russian scientists L.V. Metlitsky and O.L. Ozheretskovskaya found that synthesis of terpenoids in resistant and susceptible potato varieties proceeds in a similar way up to the stage of C_{15} farnesyl pyrophosphate production, however further, in susceptible varieties, two molecules of farnesyl pyrophosphate condense under the effect of the enzyme squalene synthetase, with the production of C_{30} squalene, from which phytosterols are produced (Figure 6.4), while in the resistant varieties under the effect of another enzyme – cyclase – the farnesyl pyrophosphate molecules close up in two rings and produce bicyclic sesquiterpenes, which are very toxic phytoalexins. Thus, instead of providing the required nutrient, the plant palms off a toxin on the parasite.

Biochemical resistance factors

The chemical factors available to the plant for their own defense against diseases are diverse in chemical composition, conditions of production, and mechanisms of toxicity. Two studies have played an important role for their nomenclature and classification.

In 1928 the zoologist B.P. Tokin discovered toxicity of volatile compounds from onion and garlic homogenates for Protozoa. He called toxic volatile compounds from plants *phytoncides*. In 1941 the German plant pathologist K. Muller discovered that potato tubers, in response to infection with the avirulent race of *Phytophthora infestans,* accumulate in their tissue the toxic substances, which he called *phytoalexins*. Using these terms, L.V. Metlitsky and O.L. Ozheretskovskaya suggested that all antibiotic plant substances (not only volatiles) that are produced constitutively, irrespective of infection, be called phytoncides, and the compounds whose synthesis is induced by infection be called phytoalexins. More complex systems of classification have also been suggested. For instance, G. Ingham isolated, in addition to the constitutional and induced compounds, the semi-constitutional compounds that are present in an uninfected plant, but their synthesis increases after infection, and semi-induced ones, which are produced under various types of stress (infection, mechanical damage, etc.) by disintegration of more complex compounds, and not by synthesis from less complex ones. However, later in 1994 a group of researchers (G. van Etten, G. Mansfield, etc.) suggested that the previous classification should be restored, with two groups of plant antibiotics, which they named *phytoanticipins* and *phytoalexins.* The former are "the low-molecular antimicrobial substances present in a plant before infection or produced after infection from the precursors", and the latter are "the low-molecular antimicrobial substances that are synthesized and accumulated in a plant after the exposure to the parasite". The synthesis of phytoalexins is determined by interaction of the plant gene products and the parasite, i.e. it belongs to a specific pathosystem and will be considered later.

Phytoanticipins are a diverse group of substances which includes the compounds present in intact cells, and the compounds produced only after the cell damage. Besides, alongside the low-molecular compounds, there are also high-molecular compounds, e.g. proteins with pronounced antibiotic properties.

The division of antimicrobial plant substances into phytoanticipins and phytoalexins is provisional, as most biological classifications are provisional. For instance, in the shoots of *Arabidopsis,* synthesis of the antimicrobial protein *thionine* is controlled by two genes, one of which *(Thi2.2)* is constitutional, and the expression of the other *(Thi2.1)* is induced by infection. Some substances are constitutional in certain plant species, i.e. they are phytoanticipins, and in others they are induced by infection (phytoalexins). Since the plant possesses a whole system of defenses, the above division was used only for the educational purpose.

Chemical classification of phytoanticipins

Phenols. Phenols are the cyclic compounds containing the aromatic ring with the phenyl hydroxyl or its substituted radicals (acid, aldehyde, etc.). Many plant phenols contain, in addition to phenyl (C_6), the propane radical (C_3), and they were called *phenylpropanoids*

There are two pathways of the phenol synthesis (Figure 6.1): condensation of acetic acid molecules (*acetate pathway*) and metabolism of phosphorylated sugars (glucose-6-phosphate) through shikimic acid and aromatic amino acids (*shikimate pathway*). Many complex plant phenols containing two phenol rings (flavonoids) are produced by bonding the rings synthesized by the *acetate* and *shikimate* pathways.

Investigation of plant phenols provides a good illustration that the division of antibiotic substances into constitutional and inducible ones is provisional. Phenols are normal plant metabolites since they are contained in pigments, growth regulators, and structural elements of cell walls. At the same time phenols are stress metabolites, whose synthesis grows sharply at injury or infection. Therefore, the key enzymes involved in the synthesis and transformation of phenols, as well as the regulation of enzyme activity, will be discussed in the description of the vertical pathosystem, and in this chapter we will only discuss their impact on the infection process.

Generally, the redox reaction of phenol transformation is described as follows:

$$\frac{1}{2}O_2 + \text{diphenol} \xleftrightarrow[\text{quinone reductase}]{\text{polyphenol oxidase}} \text{quinone} + H_2O$$

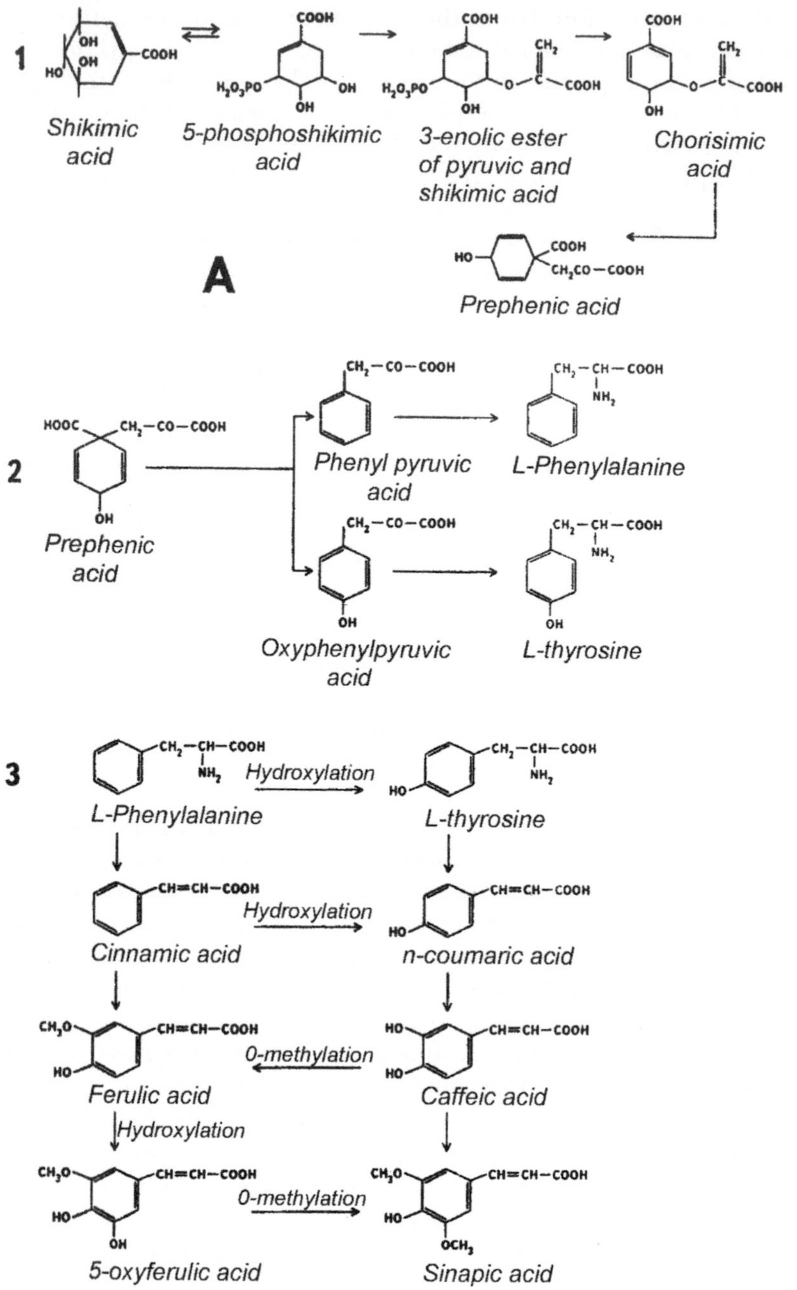

Figure 6.1.   Pathway of phenol biosynthesis in plants: (A) shikimate pathway (stages 1, 2, 3);

(*Continued*)

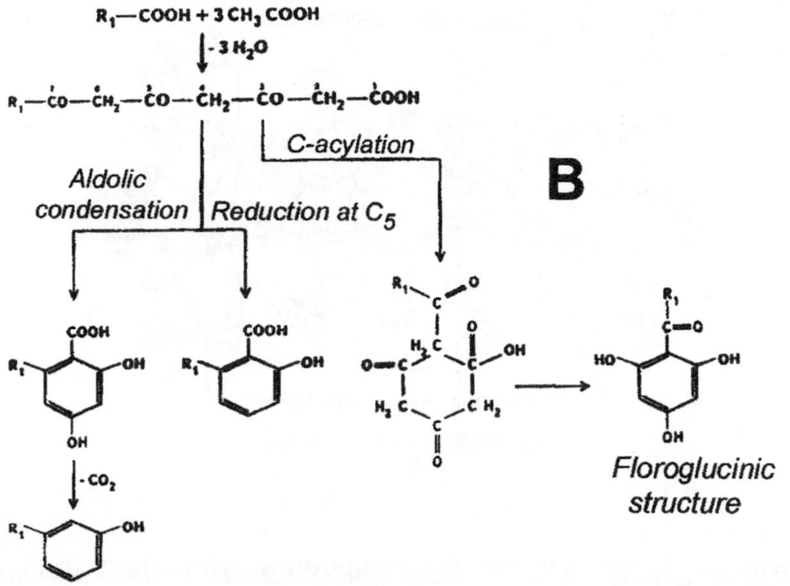

$$R_1\text{---COOH} + 3\,CH_3\,COOH$$

$$\downarrow -3\,H_2O$$

$$R_1\text{---}\overset{\cdot}{C}O\text{---}\overset{\cdot}{C}H_2\text{---}\overset{\cdot}{C}O\text{---}\overset{\cdot}{C}H_2\text{---}\overset{\cdot}{C}O\text{---}\overset{\cdot}{C}H_2\text{---}\overset{\cdot}{C}OOH$$

C-acylation

Aldolic condensation | Reduction at C_5

B

Floroglucinic structure

Figure 6.1. cont'd. (B) acetate pathway.

Hence the basic mechanisms of the phenol antimicrobial action are:

1. Inactivation of proteins due to bonding of the quinones with SH– and NH_2 groups and formation of substitution products.
2. Non-specific oxidation effect of quinones. Since quinones are active oxidizers, they are electron acceptors in many compounds.
3. Effect on the transmembrane transport of substances. Some substituted phenols serve as proton conductors in membranes, and consequently, work as separating agents.
4. Regulation of growth processes. Intermediate compounds of the phenol metabolism may be nontoxic, however, their low concentrations may influence the phytohormone metabolism. Many phenols inhibit IAA-oxydase (indoleacetic acid), thus promoting IAA accumulation. And as it was noted earlier, phytohormones provide a considerable effect on the host–parasite relationship.
5. Creation of chemical barriers. Oxidation products of the phenolic alcohols condensed from the free radical reactions, produce the amorphous 3-dimensional polymer *lignin* (Figure 6.2) deposited in the cell walls (usually in the secondary walls and walls of the dead cells). Lignin provides a diverse defense action. It has covalent bonds with the carbohydrate polymers of the plant cell wall and protects them from the attack of fungal depolymerases; it prevents diffusion of the metabolites of pathogenic organisms to plants cells; it possesses direct toxicity for microorganisms; it can lignify the intraplant fungal hyphae.

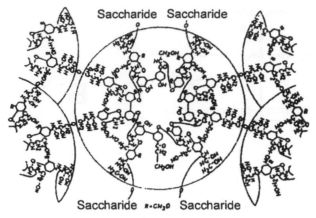

Figure 6.2.   Lignin structure.

Terpenoids. Terpenoids are the compounds synthesized from acetic acid by condensation of its molecules. The first product of condensation – *isoprene* – contains 5 carbon atoms (C_5, Figure 6.3.1). The construction of terpenoids involves two groups of enzymes: polymerases chaining the isoprene molecules, and cyclases transforming the chains to closed rings.

Condensation of two isoprene molecules results in a group of compounds named *monoterpenes* (C_{10}, Figure 6.3.2). These are low-molecular monocyclic volatiles, which frequently confer specific odour to plants.

Condensation of three isoprene molecules results in formation of *sesquiterpenes* (C_{15}). Many bicyclic sesquiterpenes possess high toxicity and are important plant resistance factors (Figure 6.3.3).

Four isoprene molecules produce *diterpenes* (C_{20}), the basic components of woody plant resins. Many diterpenes have a bitter taste (e.g. the bitter factor of the cucumber).

Six isoprene molecules make the basis of a large group *of triterpenes* (C_{30}). They include pentacyclic triterpenoids (five hexatomic cycles) and steroids, the nucleus of which consists of three hexatomic and one pentatomic cycles. Phytosterols (Figure 6.4) are an essential component of plant membranes,

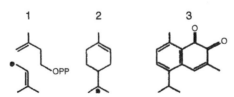

Figure 6.3.   Terpenoids: (1) condensation of two isoprene molecules; (2) monoterpene structure; and (3) sesquiterpene manzonone.

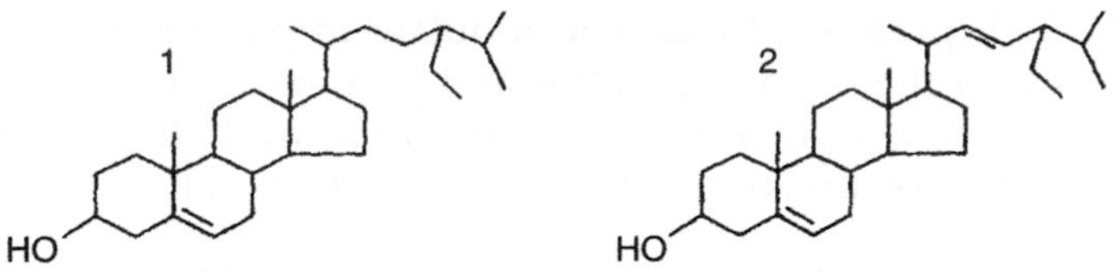

Figure 6.4.   Phytosterols: (1) sitosterol; (2) stigmasterol.

and ergosterol are essential in many fungal membranes. Toxicity of many triterpenoids is caused by their ability to link with membrane sterols, which results in pore formation in the membranes and release of the cellular content to the environment (lysis).

In erythrocyte membranes, triterpenoids link with cholesterol, causing leakage of hemoglobin (hemolysis of erythrocytes). The minimum concentration of triterpenoids that causes staining by the blood plasma hemoglobin (supernatant liquid resulting from precipitation of blood cell elements), is called *hemolytic index* and it constitutes a toxicity indicator.

Finally, the linking of eight isoprene molecules provides a large family of *tetraterpenoids* (C_{40}). Their linear molecules (Figure 6.5) make a large group of the plant pigments *carotenoids*, which colour flowers, fruits, root crops, and autumn leaves of plants.

Proteins and amino acids. The plant resistance factors against pathogens may include the families of low-molecular base proteins, rich in cystein, which include plant *defensins*; proteins responsible for the transmembrane *lipid transport*; proteins inhibiting *protein synthesis* by alien ribosomes at the stage of elongation; and *thionines* accumulating in seeds and sprouts of many plants. Their antimicrobial effect has been shown in many experiments. They have been found in seeds, leaves, flowers, and tubers. Most of these proteins accumulate in cell walls or even on seed surface, and perform the first line of defense functions.

Defensins are low-molecular proteins (45–54 amino acids), rich in cystein. They are produced as protective proteins by mammals, insects, fungi, and plants. Plant defensins can cause morphological changes in susceptible fungal species (morphogenic defensins), inhibit mycelium growth without causing morphological changes (non-morphogenic defensins), and inhibit fungal α-amylase. They link with fungal and bacterial membrane receptors and enhance the loss of K^+ and arrival of Ca^{2+} to cells.

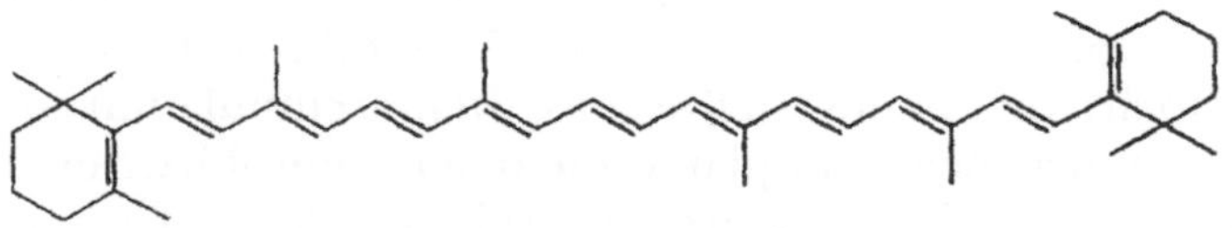

Figure 6.5.   β-carotene.

The sulfur-rich proteins *thionines* accumulate in plant seeds and sprouts. They interact with the membrane phospholipids and cause pore formation in membranes. These proteins have been found in seeds, leaves, flowers, and tubers of different plants; they often accumulate in cell walls providing a protective function.

An unusual amino acid, *L-canavanin*, has been found in the seeds of some legumes, the level of which can be as high as 2–3% of the seeds' dry weight. Since it is an antimetabolite of arginin, it is very toxic and protects seeds from damage by fungi and insects.

Resistance strategy

Plants possess not only chemically diverse defense compounds, but also various strategies of their application. Using military terminology, one can say that plants have the in-depth defense with flexible control, capable of building various defense barriers in the way of the parasite advance, in the necessary place and at the necessary time.

Phytoncides. It was written earlier that B.P. Tokin used the term phytoncides to designate the plant volatiles that kill microorganisms. Later, this term was used as a generic name for all antimicrobial substances present in uninfected plants, which resulted in a great confusion, as they were first macerated for the study of phytoncidity of plant tissues, following B.P. Tokin's approach. The substrates and enzymes, which were in the different compartments of the intact tissue, came into contact at cell destruction, resulting in the production of new chemical substances that were completely absent in the biotroph parasitism, and could not occur at the necrotroph parasitism either, due to the enzyme inactivation by the parasite toxins. Therefore, we will use the term "phytoncides" in B.P. Tokin's original meaning.

A Russian botanist B.M. Kozo-Polyansky called phytoncides "the first line of defense", as they surround plants with a toxic cloud. For instance, 270 organic compounds were found in the extracts of the acacia root volatiles using gas chromatography, and many of them suppressed the growth of mycelium and germination of zoospores of the root parasite *Phytophthora cinnamomi*. Basically, these are the low-molecular compounds: organic acids, aldehydes, monoterpenes, and simple phenols. Antimicrobial leaf exudates develop a protective film on the surface. Release of many antimicrobial substances on the leaf surface occurs from the leaf hairs serving as a depot for phenols and other compounds.

Antimicrobial substances of plant covers. Many plant organs: tubers, bulbs, and trunks of tree species are covered with a layer of dead cells. Cell dying-off, as well as their mechanical damage, results in destruction of intracellular membranes and mixing of the chemical compounds, which are separated by membranes in the live cell. This results in chemical reactions and accumulation of very toxic compounds. In live cells, their accumulation could damage the cells, but in dead cells they play a role of a chemical barrier in the way of the parasite advance. For instance, the yellow colour of onion glumes is determined by presence of the flavone pigment quercetin in vacuoles, which

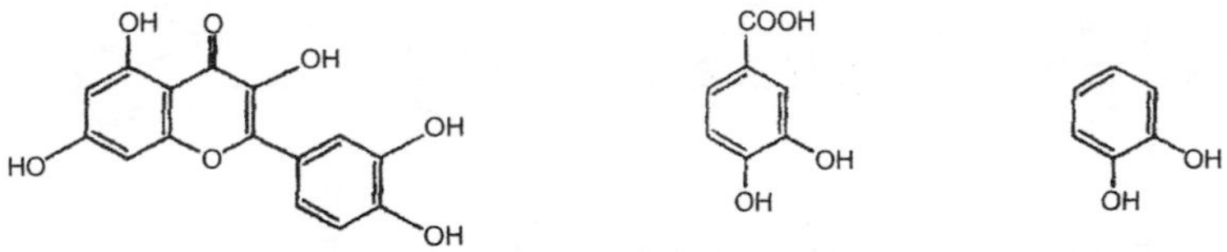

Figure 6.6.   (1) Onion phenol glycoside quercetin and its decomposition products: (2) protocatechuic acid and (3) o-catechol.

constitutes the complex phenol quercetin linked with the monosaccharide rhamnose by a glycoside bond. In dying-off onion glumes, a series of chemical reactions occur due to the vacuole destruction: first, the enzyme glucosidase eliminates the sugar residue, then phenol oxidases split quercetin down to low-molecular water-soluble monophenols: the protocatechuic acid and *o*-catechol (Figure 6.6), very toxic for microorganisms. When they diffuse on the surface of an infectious drop they kill the spores of phytopathogenic fungi. That is why the coloured onion varieties are resistant to the bulb rot fungi. The onion bulbs with damaged covering glumes are affected irrespective of the colour, since simple phenols do not accumulate in live cells.

The dead bark cells of many trees contain diphenol named pinosylvin toxic for microorganisms (Figure 6.7); elm bark cell contains the resistance factor against the Dutch elm disease sesquiterpene manzonone (Figure 6.3.3).

Antimicrobial substances of live cells. Most toxic substances occur in live cells as glycosides, which are first, less toxic than their aglycons, and therefore not hazardous to cell metabolism, and second, they are water-soluble, consequently, they present a form of transport for toxins. They are usually located in vacuoles and separated from the other cell components by tonoplast, which also protects the cell from damage. When the cell membranes are damaged, the vacuole content is released to the cell and glycosides mix with the enzymes glycosidases contained in other membrane vesicles *lysosomes*, or covalently bonded with the cell wall components. The chemical reaction results in the release of very toxic aglycons that kill the cell and the parasites in the cell. This is the general strategy of defense against infection and other stresses. A specific case of such defense is the above discussed, release of aglycons in naturally dying-off covering cells.

Classification of glycosides is based on the chemical structure of aglycons.

Phenol glycosides. The reactions of transformation of the phenol glycoside quercetin in covering glumes of the coloured onion bulbs were described

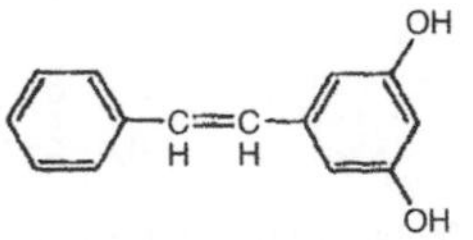

Figure 6.7.   Pinosylvin.

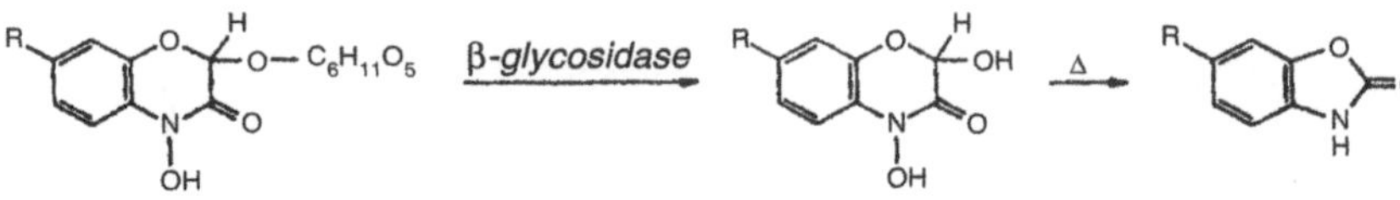

Figure 6.8.   Formation of 2-benzoxazolinon, aglycon benzoxasin.

earlier in the chapter. Different cereal species (rye, wheat, maize) contain the family of the glucosides *benzoxasins*, the aglycon of which in addition to phenol contains a nitrogen cycle (Figure 6.8). In damaged cells, the aglycons benzoxazolinons are released, toxic for rust fungi, pathogens of snow mould, aphids, and other insects. These compounds are likely to be an important factor of non-specific resistance. For instance, the correlation between their content in the leaves of the maize inbred lines and vulnerability of these lines to leaf aphids was –0.72.

Cyanogenic glycosides. Many plants (sorghum, bird's-foot trefoil, millet, flax, etc.) contain cyanogenic glycosides (durrhinin, linamarin, lotaustrolinin), which are produced from the amino acids (tyrosine, etc.) during seed germination. The enzymes responsible for stage-by-stage transformation of the amino acid to a cyanogenic glucoside, are located in the intracellular membranes as a complex, effectively channelizing the carbon flow from *L-tyrosine* to a glycoside. At damage of the membranes, β-glycosidase first eliminates glucose, and then the hydrocyanic acid is eliminated under the action of oxynitrilase (Figure 6.9).

Glycosides of aliphatic compounds. Tulip bulbs accumulate the glycoside *tuliposide*, which after sugar elimination closes into a lactonic cycle, very toxic due to the high reactivity (links at the double bond site with protein SH-groups). The lactone production reaction is an important resistance factor of tulips against *Botrytis cinerea* (Figure 6.10).

Terpenoid glycosides and glycoalkaloids. Triterpenoids and steroids, with sugars, produce glycosides soaping in water, which were named saponins (Latin *sapo* – soap). In most saponins, the oligosaccharide chain is attached to the terpenoid nucleus in the position C-3, though many of them have an

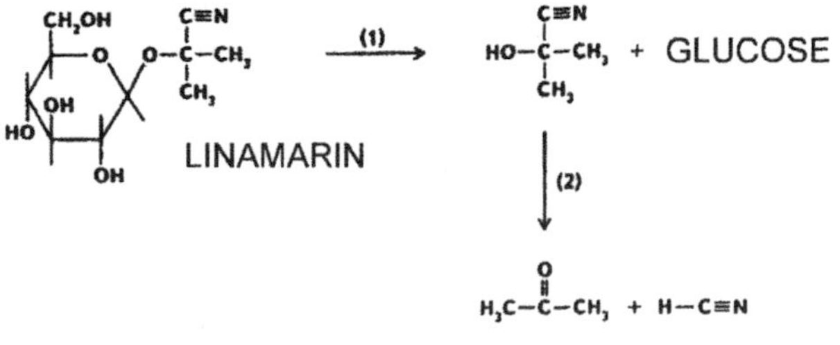

Figure 6.9.   Pathways of linamarin degradation with formation of hydrocyanic acid.
(1) β-glycosidase; (2) oxynitrilase.

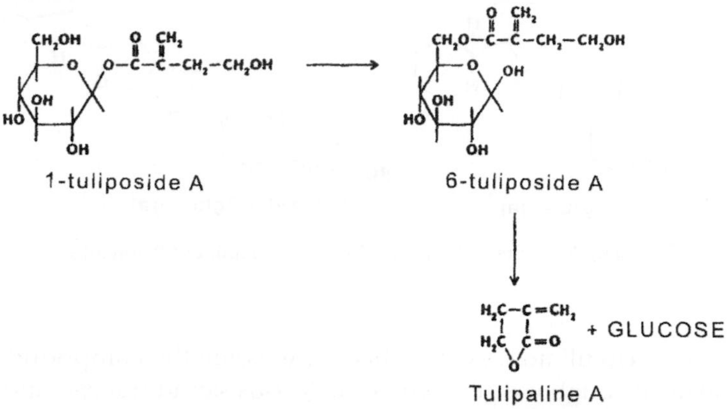

1-tuliposide A   6-tuliposide A

+ GLUCOSE

Tulipaline A

Figure 6.10.   Degradation of the tuliposide molecule by the fungus *B. cinerea.*

additional glucose bond in the positions C-26 or C-28. Saponins are very poisonous as they bind with sterols in the membranes (they have a high hemolytic index). The triterpenoid glycoside *avenacin* (Figure 6.11) accumulating in the oat radices is a resistance factor to the root rot caused by the fungus *Gaeumannomyces graminis.* Steroid glycosides of the Solanaceae contain a nitrogen heterocycle, and are named glycoalkaloids. Synthesis of the potato glycoalkaloids *solanine* and *chaconine* (Figure 6.12) is induced by illumination, therefore, they accumulate in leaves, berries (causing their bitter taste), and greening tubers, which become poisonous, but store well, not affected by rots. Solanine and chaconine inhibit growth of phytopathogenic fungi *in vitro* and are a factor of the potato age resistance (Table. 6.1).

The tomato glycoalkaloid *tomatine* (Figure 6.12) accumulates in the vacuoles of tomato leaves and green fruits. It is very toxic for *Cladosporium fulvum* and other fungi (causes loss of electrolytes through membranes). *Demissine,* the glycoalkaloid of the wild potato *Solanum demissum,* is a repellent for the Colorado beetle.

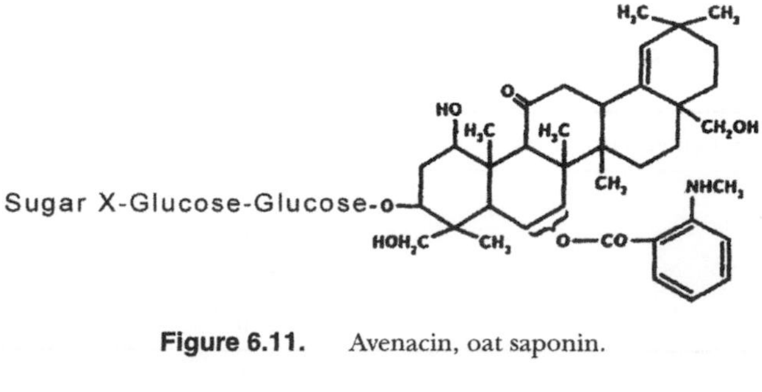

Sugar X-Glucose-Glucose-o—

Figure 6.11.   Avenacin, oat saponin.

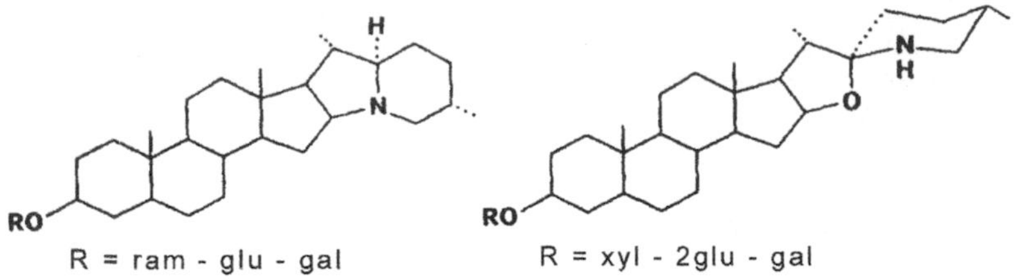

Figure 6.12.   Steroid glycoalkaloids: (1) solanine; (2) tomatine.

Thioglycosides. Up till now, we have been discussing the compounds bonded with the sugar through oxygen. The cruciferous salvadoraceae and cappari- daceae plants accumulate the compounds, the aglycon of which is bonded with the sugar through the sulfur atom. They are called mustard oils, thiogly- cosides, or *glucosinolates* (Figure 6.13). The glucosinolate depot is the vac- uole. At the damage of the tonoplast they contact the specific enzymes, which eliminate the sugar, as shown in Figure 6.13. The resulting product contains the isorodane group R–N=C=S, a very toxic volatile compound affecting the mucous membranes and providing a strong antimicrobial effect. One can experience its lacrymatory effect, e.g. when grating horseradish roots.

Summarizing the aforesaid, it is evident that the system of cell defenses based on the release of toxic substances from glycosides is multi-purpose and very effective. However, many phytopathogenic organisms, in co-evolution with plants, have developed the mechanisms to neutralize this system.

Such mechanisms can be summarized as follows:

1. Gentle effect on the plant cells, without injuring the membranes. It is typical of the biotroph parasites, e.g. the intracellular parasite of the Cruciferae *Plasmodiophora brassicae*, the club root pathogen, stays in live cells with the intact membranes, therefore, its presence does not cause the occurrence of isorhodanides. During the tumour (club root) growth, the integumentary tissues crack, and the wounds are occupied by soil putrefactive microorganisms. And since these microorganisms are necrotrophs, they destroy the infected cells, causing production of toxic aglycons, however, by that time the club root pathogen plasmodia have

Table 6.1.   Role of glycoalkaloids in the potato age resistance to *Alternaria solani* (Sinden SL, et al., 1973)

Period from sprout emergence (days)	Glycoalkaloids (mkg/g)	Lesion size (cm)
30	1570	0.1
70	1020	2.3
90	540	3.1
120	260	3.0

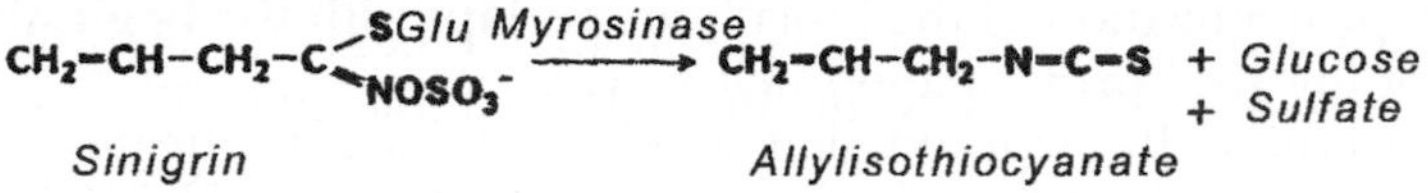

$$CH_2-CH-CH_2-C \begin{smallmatrix} \nearrow SGlu \\ \searrow NOSO_3^- \end{smallmatrix} \xrightarrow{Myrosinase} CH_2-CH-CH_2-N=C=S \quad + \; Glucose$$
$$+ \; Sulfate$$

Sinigrin   Allylisothiocyanate

Figure 6.13.   Formation of isorhodanides from the thioglucoside sinigrin.

been transformed to the spores with thick envelopes and resistance to toxins.

2. Alteration in the site susceptibility to the toxin. In the fungus *Stemphilium loti*, causing leaf spot of the cyanogenic bird's-foot trefoil, an important contribution in breath is made by an alternative, cyanide-unsusceptible breath pathway.

3. Modification of toxic substances to make them less toxic. This mechanism of pathogenicity is very common among plant pathogens, and it will be discussed in Chapter 12.

As a result of co-evolution with the host plants, some chemical immunity factors have turned from repellents (frightening off) into attractants (substances attracting the parasite). For instance, the steroid glycoalkaloids solanine and demissine, similar in the chemical structure, render opposite effects on the Colorado beetle: one attracts, the other repels them. The smell of mustard oils, characteristic of the cruciferous plants, serves as a signal for the white cabbage butterfly looking for an egg-laying place, i.e. a repellent for most insects became an attractant for the white cabbage butterfly; its caterpillars eat even the filtering paper wetted with mustard oil.

Protein inhibitors of viruses

Protein inhibitors of viruses are a special group of vegetal antibiotic substances. A description of these inhibitors and their mechanisms of action deserves a special section in the chapter.

The process of the plant infection with the virus can be broken down into two phases: the stationary phase and the replication phase. The events of the stationary phase include:

- primary mechanical injury of the leaf surface at the site of penetration;
- virus penetration to the cell;
- separation of the virion envelope from the viral nucleic acid (deproteinization).

The replication phase includes:

- synthesis of the viral nucleic acids and proteins;
- assembly of the virion;
- transport of the newly synthesized viral particles in other cells and other parts of the plant.

There is no clear-cut distinction between the concepts, inhibitor of the stationary phase in the life cycle of the virus and inhibitor of the virus replication, because deproteinization of the inoculum overlaps with the beginning of the synthesis of the viral protein and nucleic acids.

It is conventionally believed that a substance is a replication inhibitor if it is able to block the infection through the plant treatment 5–8 hours after the plant inoculation with the virus. We will use this somewhat artificial definition in the discussion of the mechanisms of action of the antiviral proteins, AVP.

The antiviral activity of a particular inhibitor is usually tested by mixing it with the viral suspension, the mix is used for inoculation of the leaves of a supersusceptible host plant, and then the number of the infection spots occurred is counted. The reduction of the number of the infection spots means that the stationary phase in the development of the viral infection has been blocked. However, the number of the infection spots may also be reduced if the virus replication is inhibited to such a degree that there is no development of visible spots, therefore, this test cannot clearly separate the inhibitory effect on the stationary phase of the viral infection from its effect on the replication phase.

The most representative tests enabling judgment about the inhibitory effect on the virus replication are as follows:

– determination of an average size of the infection spots;
– testing the inhibitor effect in protoplasts;
– analysis of the virus accumulation in the systemically infectable tissues;
– plant treatment with the inhibitor several hours after the inoculation with the virus.

All the antiviral proteins known today can be broken down into the *endogenous* (constitutively synthesized) vegetal antiviral proteins (EAVP), and the inhibitor proteins, the synthesis of which is induced in response to the infection (IAVP – *induced antiviral proteins*). Both types of the inhibitor proteins have many similar physical, chemical, and biological characteristics, such as:

– resistance to pH changes and thermal stability;
– ability to be precipitated with high concentration of ammonium sulfate;
– activity at very low concentration;
– broad spectrum of antiviral action;
– influence of the amino groups on the biological activity.

This section discusses only the constitutional (endogenous) antiviral proteins, and properties of the induced antiviral proteins will be analyzed in Chapter 10, "The immune response".

The EAVP are the inhibitor proteins continuously produced by the plant, and not stress-induced in the plant. If EAVPs are extracted from the host plant tissue they can block development of the viral infection in the plants treated with the EAVPs.

Some plants can contain more than one EAVP. For instance, extracts from the leaves of pepper, geranium, and datura contain high-molecular and low-molecular components possessing antiviral activity.

At least four different mechanisms of the EAVP protective action are known:

1. aggregation of the inhibitor with the virion;
2. inhibition of the stationary phase of the viral infection;
3. induction of the systemic resistance to the virus;
4. inhibition of replication by blocking protein synthesis.

Some EAVPs are multifunctional, i.e. possess more than one type of biological activity, which in some cases complicates the study of the mechanisms of their antiviral action.

Consider each of the above listed types of antiviral activity.

Aggregation. It is purely a physical phenomenon depending on the ion force within the infected cell, and concentration of the inhibitor and virions. The EAVPs from *Phytolacca esculenta* and *P. americana*, PAP-I and PAP-II, can aggregate with viral particles to produce insoluble sediment. However, the virion aggregation does not seem to be the only antiviral mechanism of these inhibitors; they may also induce resistance of the host plant.

Inhibition of the stationary phase of the infection process. It was said earlier, that a particular effect of the inhibitor proteins on the stationary phase of the development of a viral infection is a decrease in the number of the infection spots on the leaves treated with these inhibitors. Typical inhibitors of the stationary stage are the EAVPs isolated from *Datura stramonium* and *Dianthus caryophyllus*. The plant treatment with these inhibitors results in reduction of the number of infection spots, though the spot size does not change. The EAVPs from *D. stramonium* and *D. caryophyllus* have no effect on the virus titre in a systemically infected plant.

The evidence found from a series of experiments on the ratio between the logarithm of the number of infection spots on the leaves and the logarithm of the virus concentration in the leaf tissue shows that there are a certain number of sites on the leaf surface, which can potentially develop into an infection spot. In other words, the number of receptors capable of interacting with viral particles is limited, and they are located in particular points of the leaf. Preliminary treatment of the leaf surface with a homologous protein of the viral envelope results in reduction of the number of infection spots upon inoculation, which can be attributed to the competition between the viral envelope protein and the intact viral particles for interaction with the receptors. This interaction is highly specific, as there is no such competition after the treatment with heterologous viral protein.

It is quite possible that the hypothetical receptors may be the application points of the endogenous antiviral inhibitor proteins. Evidently, different receptors possess a different degree of affinity to EAVPs, as treatment of leaves of the same host plant induces a different degree of resistance to different viruses. Similarly, a different degree of resistance is also observed in the treatment of different host plants species with the inhibitor.

Amino groups are likely to play an important role in EAVP linking with the host plant receptors. Presence of these radicals seems to enable the inhibitor to link with a receptor and thus block the virus-specific sites on the leaf surface. A chemical reduction of the amino groups in the inhibitor molecules results in the decrease in its antiviral activity. On the other hand, it has been shown that the amino groups are necessary for successful inoculation of tobacco plants with the tobacco mosaic virus. Thus, there are grounds to believe that the amino groups of the antiviral protein molecules, in particular, are the determinants responsible for competitive linkage with the plant receptors.

Some plants (for instance, *Chenopodium amaraun-ticolor* or *Physalis floridana*) are susceptible to many viruses differing structurally and biologically. Apparently, these plants have different receptors specific for each virus.

Induction of systemic resistance. The process of development of induced resistance in plant systems may be broken down into three consecutive stages:

1. recognition of a biotic or abiotic stress by the plant;
2. generation of a systemic signal, as a stress response at the cell level;
3. inducing, by the signal, a new physiological condition in the tissues that protects the tissues from a further effect of the same stress and, sometimes, from other stresses.

It has been found that some vegetal extracts, when tested in test plants, induce a systemic resistance to viruses, and it has been shown that the resistance inducer is a protein substance. The plant tissue that has acquired resistance as a result of stress effect and generation of the systemic signal initiated by the given inducer, can be used to isolate the proteins capable of inhibiting the development of a viral infection. The protein virus inhibitors are termed by the acronym VIA (*virus inhibitory agents*). The VIAs inhibit a viral infection at plant inoculation with VIA–virus mix. They can also induce systemic resistance at plant treatment before inoculation with the virus. The VIA molecular weight varies from 13.5–15 kDa.

Synthesis of VIA in plant tissues and the respective systemic resistance is completely blocked, if actinomycin D is administered in these tissues immediately after treatment with the inducer. It is well known that the antibiotic actinomycin D is an inhibitor of protein synthesis. Actinomycin D will partially block the VIA induction if it is used within eight hours after treatment of the leaves with the inducing extract. However, if actinomycin D is applied 24 hours after the treatment with the inducer, there is no decrease in VIA concentration or drop of the level of systemic resistance.

These experimental data show that:

1. the level of VIA accumulation in plants correlates with the development of systemic resistance;
2. the protein synthesis related to both processes terminates 24 hours after the induction.

The examples of VIA are the endogenous protein inhibitors from the extracts of *Capsicum frutescens* and *Chenopodium ambrosoides* which induce systemic resistance in upper leaves against the virus, if used for treatment of the lower leaves before the inoculation with the virus.

Ribosome inactivation. Ribosome inhibitory proteins (RIP) are antiviral substances as they inhibit protein synthesis by blocking the final phases of the virus intracellular development. The RIPs specifically inactivate ribosomes, which results in blocking the protein synthesis at the phase of elongation. The primary structure of the RIP isolated from different sources is highly homologous, and this enables in drawing a conclusion that the inhibitory activity of the given peptides correlates to a certain degree with the primary structure of their active sites responsible for ribosome linking.

It was initially assumed that the RIPs are inactive in homologous ribosomes. However, the low ribosome activity of the pokeweed (*Phytolacca americana*) in the extracellular translation experiments led to the discovery that the pokeweed ribosomes are actually inactivated by the pokeweed RIPs (PAP-1) during their release. Other plant ribosomes are also blocked by the effect of their own RIP in similar conditions. However, it should be noted, that the pokeweed ribosomes are still resistant enough against the addition of their own RIPs (PAP-1) and are much more susceptible to the effect of the RIPs isolated from other plants. Yet tritin, a wheat RIP, does not inactivate the wheat ribosomes, and so the isolated wheat ribosomes retain high activity.

The experimental data gained to date allows a conclusion that RIPs have a specific activity towards the ribosomes isolated from different plant species. Apparently, ribosomes have certain structural features which can be recognized or not recognized by different RIPs. The mechanisms of functioning of these proteins, however, are not so simple: data are available that PAP-S (RIP from pokeweed seeds) inhibits carrot cell growth in the liquid culture, yet the same concentration of PAP-1 stimulates rice cell growth.

The time interval of antiviral activity is rather narrow, therefore, antiviral activity of most RIPs is tested by plant inoculation with a mix of these proteins and the viral preparation or a viral RNA preparation. Plant treatment with RIPs some time after the inoculation will not prevent development of the viral infection. For instance, the protective effect cannot be detected, if pokeweed RIP (PAP-1) is applied 30–50 minutes after the inoculation of tobacco protoplasts with tobacco mosaic virus (TMV). Hence, RIPs are active only at very early stages of the virus life cycle. It is well known, that the host ribosomes can bind a viral RNA almost immediately after the RNA virus has lost its envelope. Possibly, the translation complex (viral RNA–ribosome) is already unsusceptible to the RIP action, and while this complex exists, RIP are inactive.

Selected Literature

Bochlmann H, Apel K. Thionins. Ann Rev Plant Physiol Plant Molecular Biol 1991; 42:227–240.

Broekaert WE, Terras FRG, Cammue BPA, Osborn RW. Plant defensins: novel antimicrobial peptides as component of the host defense system. Plant Physiol 1995; 108:1353–1358.

Chappel J. Biochemistry and molecular biology of the isoprenoid biosynthesis pathway in plants. Ann Rev Plant Physiol Plant Molecular Biol 1995; 46:521–548.

Dixon RA. Natural products and plant disease resistance. Nature 2001; 411:843–847.

Herrmann KM, Weawer LM. The shikimate pathway. Ann Rev Plant Physiol Plant Molecular Biol 1999; 50:473–504.

Sinden SL, et al. Phytopathology 1973; 63:303–307.

Wetter RW, MacKay JJ, Sederoff RR. Recent advances in understanding lignin biosynthesis. Ann Rev Plant Physiol Plant Molecular Biol 1998; 49:585–610.

Chapter 7

Vertical pathosystem: avirulence genes and their products

Yu. T. Dyakov, O. L. Ozeretskovskaya

Molecular interpretation of genetic data

From Flor's "gene-for-gene" chart shown as a square net (Table 4.11) it follows that the resistance or *incompatibility* response occurs only if the partners possess two interacting dominant genes while any other combination of these genes will produce the susceptibility response, *compatibility*. A combination of two genes resulting in incompatibility is called *definitive*, and the other combinations *indefinitive*. Such designation emphasizes the decisive role of incompatibility, as in Flor's host–parasite interaction incompatibility which occurs in the presence of dominant alleles of the interacting genes, i.e. the alleles carrying information about functionally significant gene products. It follows from the theory that resistance and avirulence genes possess the "plus" functions, for if the host and parasite genomes contain only one pair of definitive genes (for instance, R3/A3 in the chart below), an incompatibility response will develop despite of the presence of many indefinitive gene pairs:

Plant genes:	R1	r2	R3	r4
Parasite genes:	a1	A2	A3	a4

The "plus" functions of avirulence and resistance genes have been confirmed by numerous experiments. In 1960, Flor reported about receiving the mutants virulent to the flax cultivar Koto after X-ray irradiation of the spores of race 1 of the rust fungus *Melampsora lini* avirulent to the flax cultivars Koto and Leone. After inbreeding of the mutant isolates, 191 strains *M1* were tested for virulence. All of them were found to be virulent to cultivar Koto, like the parent mutant; however, 12 strains also became virulent to cultivar Leone. The results of the observations were interpreted as follows: race 1 is heterozygotous (more correctly, heteroallelic, as rust fungal cells contain two haploid nuclei each, the heteroallelity of which is phenotypically indistinguishable from heterozygosis) in avirulence to cultivar Koto and homozygotous in avirulence to cultivar Leone. Since virulence is recessive, this race is virulent to both cultivars. Genes of avirulence to cultivars Koto and Leone are linked. X-ray irradiation resulted in a deletion of the nucleus carrying the gene of avirulence to cultivar Koto. The deletion covered both linked avirulence genes (to cultivars Koto and Leone). The spores became heterozygous in deletion and virulent to cultivar Koto due to the presence of a recessive gene of virulence to this cultivar in the second (unaffected) nucleus of the dikaryon. The inbreeding produced three types of dikaryotic cells: (1) homozygotes in the gene of virulence to Koto (virulent to Koto and avirulent to Leone); (2) heterozygotes in deletion (also virulent to Koto and avirulent to Leone), and (3) homozygotes in deletion (virulent to both cultivars due to the absence of dominant alleles of avirulence genes). These data suggest that a virulence gene constitutes an inactive allele of the avirulence gene, phenotypically indistinguishable from deletion.

An evidence of the functional importance of resistance gene products were received in specially planned crossing of the wheat lines possessing genes of resistance to stem rust *Puccinia graminis*.

Ditelosomic lines without the chromosomal arm carrying the resistance gene were exposed to the fungal races possessing avirulence alleles definitive to these genes. Thus, the absence of a resistance locus is phenotypically manifested as a "susceptibility" allele. That was the basis for a conclusion that the "susceptibility" allele of euploids can have a nonfunctional DNA sequence (nonsense mutation), or, if it is functionally significant, its gene product does not interact with the product of the definitive allele of the parasite (missense mutation).

Another evidence of the "plus" functions of resistance genes is temperature sensitive (ts) genes, the products of which are functionally active at the optimal (permissive) temperature for this given organism but lose their activity at elevated (restrictive) temperature due to conformational changes of the protein. In cereals, such genes were first observed in year-round growing of plants in greenhouse. It was noted that some cultivars inoculated with the same races of rust fungi are resistant in winter but are affected in summer when the temperature in greenhouses is higher. For instance, the wheat gene of resistance to stem rust *Sr6* controls the resistance response (0) to inoculation by avirulent races at a temperature of 20°C and the susceptibility response (3–4) at 26°C. Among the *ts* resistance genes, the most extensively studied is gene *N* of wild tobacco *Nicotiana glutinosa*. Inoculation of this tobacco leaves with tobacco mosaic virus (TMV) in permissive conditions (25°C) results in formation of small necrotic spots, the number of which is proportional to the concentration of viral particles in the inoculum (hypersensitive response). The leaf cells adjacent to the necrotized cells contain TMV particles unable to propagate over the leaf (they are blocked in the necrosis area). If the inoculated plant is placed in conditions of a restrictive temperature (35°C) the virus will spread from the inoculation areas to the neighboring cells (systemic plant infestation). Upon recovery of permissive conditions, all the areas where the viral particles have spread are necrotized, so the whole plant may die. Since a normal functionally active protein is synthesized in permissive conditions, and defective protein in restrictive conditions, the hypersensitive response (incompatibility) requires an active product of the resistance gene carrying the "plus" functions. A defect in the protein results in the susceptibility response.

It follows from the genetic evidence that interaction of resistance and avirulence gene products results in incompatibility, while the loss or change of the definitive gene by one (or both) partner(s) modifies the response to compatibility. The simplest biochemical interpretation of these data is an assumption that some structures of the host plant, the synthesis of which is controlled by the vertical resistance genes, recognize the structures or metabolites of the parasite, controlled by the complementary avirulence genes, as foreign structures and trigger defense responses. If the host has no recognizing structure (receptor) due to the change or loss of the resistance gene, or if the parasite has no metabolite to be bound by the receptor due to the loss or change of the avirulence gene, the host stops recognizing the parasite as a foreign body and the defense responses are not triggered (or are triggered too late), which leads to compatible relations between the partners (Figure 7.1). During formation of the haustorium, carbohydrates of the parasite fungus cell wall closely contact the plant plasmalemma. H. H. Flor assumed that in the presence of complementarity in the structure of carbohydrate chains of the fungus cell wall and tertiary structure of the plasmalemma protein, the parasite is recognized and a signal to trigger the defense responses is generated. Thus, the dominant alleles of the resistance genes control the synthesis of receptor protein on the cytoplasmic membrane of the host, while the dominant alleles of the virulence genes control the synthesis of the enzymes glucosyltransferases that connect monosaccharides in a fixed order and build carbohydrate chains in the fungus cell walls.

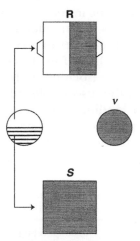

Figure 7.1. A model of the host–parasite interactions based on "gene-for-gene" diallele interactions. Boxes are plant genotypes and circles are parasite genotypes. (R = resistance) the dominant allele and its product, a specific receptor (on the surface of the box); (*s* = susceptibility) the recessive allele (lack of receptor); (A = avirulence) the dominant allele of the parasite and its product, a specific elicitor (arrow); (*v* = virulence) the recessive allele (lack of elicitor). Shaded: compatible responses due to absence of gene product interactions.

The fact that complementary genes of the parasite and the host have a different degree of relation with the interacting products (direct for the resistance gene and indirect – through transferases – for the avirulence gene) should not impose constraints on this hypothesis. As it was noted earlier, resistance and avirulence genes differ also in the chromosomal arrangement: the former frequently form series of alleles while the latter are usually unlinked. Reverse cases are also known: the fungus *M. lini* was found to have a close linking of the two genes providing the definitive response with the flax genes mapped in two unlinked loci.

It would appear that such hypothesis can be easily verified. It is necessary to isolate cell wall preparations from a parasite race virulent and avirulent to some plant cultivar and show the differences in their induction of defense responses. For instance, basidiospores of cowpea rust pathogen *Uromyces vignae* were germinated in starved agar, the shoot exudates were collected and used to treat resistant and susceptible cowpea cultivars (Table 7.1).

In Table 7.1, A is an avirulent race; V is a virulent race; R is a resistant cultivar; S is a susceptible cultivar.

Table 7.1 shows that an incompatible response to the treatment (hypersensitive response) occurs only in the leaves of a cultivar possessing a dominant allele of the resistance gene, in response to the exudate of a race possessing a dominant allele of the avirulence gene. In other cases, according to Flor's rule, the response to exudate does not differ from the response to water treatment. In many experiments, attempts were made to determine specificity of the parasite metabolites not by the number of necrotized cells but by accumulation of plant chemical resistance factors, phytoalexins (FA), in the inoculation area.

The next step after demonstration of specificity is the preparation of fractioning and isolation of the specific inducer (or, according to the commonly used terminology suggested by N. Keen, elicitor), labeling it with a radioactive isotope, and treatment of the leaves of a resistant plant or (rather) protoplasts produced from its cells. After fractionating of plasmalemma proteins, radioactive label enables isolation of the protein binding to the elicitor, and it can be shown that the susceptible cultivar either has no such protein, or, if it does, its elicitor binding constant is considerably low. Next, the receptor protein can be sequenced and its structure can be used to synthesize a complementary DNA strand (cDNA), using it as hybridization test to isolate the resistance gene from the plant genome.

Table 7.1. Average number of mesophyll necrotic cells of the cowpea lines treated with spore exudates of *U. vignae* or water (Chen CY, Heath MC, 1990)

Cowpea cultivar	Race sporeling exudates		
	CPR-1 (A)	CPR-2 (V)	Water
Dixon Cream (R)	296.1	40.0	26.5
California Blackeye (S)	21.0	32.0	23.3

Table 7.2. Highly purified fungal elicitors inducing formation of phytoalexins (FA) in plants (Kogel, Beibman, 1995)

Compound	Pathogen	FA – plant
Heptaglucosyl oligosaccharide	*Phytophthora soyae* (*cw*)	Glyceollin – soybean
β-1,4-glucosamine	*Fusarium solani* f.sp. *pisi/phaseoli* (*cw*)	Pisatin – peas
Glycoprotein	*P. graminis* f.sp. *tritici* (*cw*)	Lignin – cereals
Glycoprotein	*P. megasperma* (*cw*)	Furanocumarins – celery
Peptide	*Cladosporium fulvum* (*if*)	Rishitin – tomato
Protein	*P. cryptogea* (*cf*)	Necroses – tomato
Epoxystearic acid	*P. graminis* (*us*)	Lignin – cereals
Arachidonic acid	*P. infestans* (*cw*)	Rishitin, lubimin – potato
Eicosopentaenic acid	Same	Same

Cw is mycelium cell wall; *if* is intercellular fluid; *cf* is culture filtrate; *us* is urediniospores.

Though by now many elicitors have been discovered, only some of them have been produced as highly purified preparations (Table 7.2). Crude preparations bear many hazards: (1) unpurified extract may contain various elicitors with a synergistic effect; their structure may be similar, which complicates purification of individual compounds; (2) extraction may produce a complex of the elicitor and another substance; (3) the elicitor may be active only in the presence of a certain cofactor which may be lost in purification; (4) the elicitor yield may depend on culturing conditions, crop age, and other factors.

In addition, scientists from different countries have faced other types of difficulties. First, it was not just one parasite metabolite that induced defense response but many metabolites localized in different cell compartments (the membrane, the wall, the periplasm, the cytosol) and belonging to different classes of chemical compounds (carbohydrates, proteins, fatty acids, etc.). Second, elicitor compounds have been found both in avirulent and virulent races, and even in saprotroph fungi (the example in Table 7.1 is a rare exception). Third, these elicitors induce defense responses both in resistant and susceptible plant cultivars, including species and cultivars without genes of resistance to a given parasite. Models were suggested where specific or intracellular non-specific elicitors are produced or released to the environment only *in planta*, as the effect of the host plant metabolites. If so, in studying the elicitor compounds produced by a parasite *in vitro*, the effect of specific elicitors will be masked by non-specific elicitors. However, such hypotheses also resist verification.

Proceeding from the above, elicitors can be classified into several groups. First, there are elicitors isolated from the pathogens or media where they have grown (exogenous biogenic elicitors); second, the elicitors formed in an

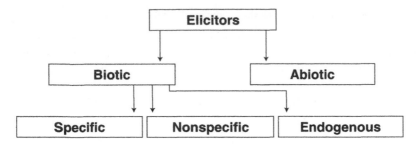

Figure 7.2.   Elicitor classification.

infested plant (endogenous biogenic elicitors); and third, the abiogenic elicitors not involved in pathogenesis. Biogenic or biotic elicitors can, in turn, be
divided into non-specific and specific. Abiogenic elicitors are always non-specific
(Figure 7.2).

The results of the above described studies on nonspecificity of most isolated pathogen elicitors brought many biochemists studying plant immunity
mechanisms to believe that induction of defense responses by the parasite
and their development in a plant are non-specific, and resistance and avirulence genes are indefinitive, while *suppression* (inhibition) of defense
responses by the parasite and responsiveness of the host to the suppressor is
specific. Flor's diallelic pattern of host–parasite interaction (the square grid)
has been turned over (Figure 7.3). In some fungus–host plant combinations

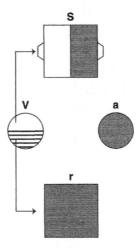

Figure 7.3.   A model of the host–parasite interactions based on diallele interactions opposite to
"gene-for-gene" rule. (S = susceptibility) the dominant allele and its product, a specific receptor;
(r = resistance) the recessive allele (lack of receptor); (V = virulence) the dominant allele and its product, a specific suppressor; (a = avirulence) lack of suppressor. Shaded: incompatible combinations
due to interaction of gene products.

such patterns have been confirmed experimentally, and the specific suppressors have been identified.

Apparently, in different plant-pathogen partnerships both interaction patterns can work: (1) non-specific induction of defense responses and specific suppression by the parasite; (2) specific induction of defense responses and its loss due to mutations of the avirulence gene. Genes controlling formation of specific elicitors are usually called *avirulence* genes (*avr*-genes); genes controlling synthesis of specific suppressors are called *virulence* genes (*vir*-genes). Plant–parasite interaction undergoes the following stages: release of the elicitor → its interaction with cell receptors → transduction of the signal in genes → activation of the immune response genes → accumulation of toxic and barrier compounds (stress metabolites) → suppression by the parasite of various stages of stress metabolism, from elicitor reception to detoxification of antimicrobial substances. We will discuss, in succession, these stages and the compounds involved.

Abiogenic elicitors

Abiogenic elicitors are the substances not involved in pathogenesis processes but inducing defense responses in plants, usually in relatively small concentrations. They include ions of heavy metals, inhibitors of certain stages of metabolism, metal-containing enzymes, and thiol groups as well as some antibiotics, phenols and quinones, UV rays, and a number of fungicides (for instance, Bordeaux liquid, copper oxychloride, zineb, captan, etc.).

According to the ability to induce fytoalexin (FA) formation in potato, ions of heavy metals are arranged in the following order:

$$Hg^{2+} > Cu^{2+} > Fe^{3+} > Ni^{3+} > Co^{2+}$$

Mercury and copper salts possess the greatest induction activity; therefore, probably, defense responses are induced by fungicides containing these metals. However, even the highest induction activity of chemical compounds amounts to only one-third of the activity of FA produced by spores of incompatible fungal races.

One of the possible causes of abiotic induction of defense responses is activation, by abiotic elicitors, of endogenous hydrolases of plant tissues: nucleases, peptidases, glycanases, and other enzymes responsible for degradation of biopolymers that perform the functions of storage, transmission, and use of information. And this, in the long run, induces protective effects in plant tissue.

Nonspecific biogenic elicitors

Nonspecific elicitors induce defense responses irrespective of plant resistance or parasite virulence. They can include polysaccharides, proteins, polypeptides, glycoproteins, lipid-containing compounds, or other type of substances.

Polysaccharides

Polysaccharides from fungal cell walls have been best investigated as elicitors. Fungi are usually subdivided into several groups according to the type of structural polysaccharides in their cell walls. The two best known types of cell wall polysaccharides possessing elicitor functions are β-1,3-glucans and chitins (or chitosan). True fungi contain chitin and glucan in the cell wall. Cell walls of oomycetes contains up to 25% of partially crystallized cellulose (β-1,4-glucan) and the major noncellulose compound β-1,3-glucan.

The elicitor isolated from cell walls and culture fluid of *Phytophthora megasperms* f. sp. *glycinea*, which induces formation of the soybean phytoalexin glyceollin, has been studied in detail. The elicitor is a heptaglucoside where five glucose residues are connected in a linear chain by β-1,6 bonds, and two side residues are connected by β-1,3 bonds. Incomplete acid hydrolysis of fungal cell walls yielded 300 elicitor analogs, with only one of them being active. It induced synthesis of glyceollin at a concentration of 10 nM per cotyledon, while the others displayed no activity even at a concentration of 200 μM. An active heptaglucoside differed from inactive only in the position where two side residues bind to the base consisting of five residues (Figure 7.4). The heptaglucoside did not possess species-related specificity, as it induced not only formation of glyceollin in soya but also rishitin in potato,

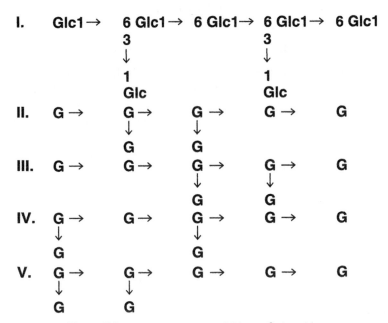

Figure 7.4. Primary structure of 5-hepta-β-glycosides.

phaseollin in beans, and capsidiol in pepper. In addition to phytoalexins, the heptaglucoside also induced synthesis of oxyproline-rich proteins in soybean, formation of ethylene, as well as protected tobacco from viral infection.

Soybean plasma membrane was found to contain centers of specific binding of heptaglucoside. The binding occurs at lower saturations than required for FA formation (10^{-10}–10^{-9} M), and is reversible, which points to noncovalent binding of the elicitor and the receptor. The binding sites were inactivated at treatment with proteolytic enzymes, which confirms their protein or glycoprotein nature. The ability of various heptaglucosides to bind to a receptor correlates with their ability to induce glyceollin synthesis.

N. Keen and collaborators isolated glucomannan from cell walls of *P. megasperma* using β-1,3-glucanase. Glucomannan was found to be 10 times more active than the above-described heptoglucoside.

The culture filtrate and cell walls of *Colletotrichum lindemutianum* were found to contain an elicitor, a 3-bound and 4-bound glucan. It induced accumulation of faseollin and hydroxyfaseollin in beans. Culture filtrates and extracts of various species of *Colletotrichum* also contained polysaccharide elicitors. The most purified fractions contained glucans with the bonds β-1,3 and β-1,4. Most likely, these glucans were not structurally different from the glucans of *P. megasperma*. Later, galactoglucomannan possessing elicitor properties was isolated from *C. lindemutianum*. It contained 10% galactose, 38% glucose, and 45% mannose.

Components of cell walls of *P. infestans* were shown to be elicitors in formation of the phytoalexin rishitin of potato. Accumulation of rishitin was stimulated by a fraction containing β-1,3-β-1,6-bound glucose. However, the ability of the glucan elicitor to stimulate FA accumulation was considerably lower than the activity of the other elicitor of *P. infestans*, arachidonic acid.

The second group of polysaccharide elicitors includes amino sugars contained in the fungal cell walls: β-1,4-bound polymer of acetylglucosamine, chitin, and its deacetylated derivative chitosan. Chitin is an unsoluble polyaminosaccharide made of an unbranched chain of β-1,4-bound residues of N-acetyl-D-glucosamine. In fungal cell walls, chitin is connected by covalent and ion bonds to other polysaccharides, pigments, and proteins, which makes it especially resistant to lytic enzymes.

In higher plants, chitin and chitosan are either absent or occur in minor concentrations, however, the range of enzymes capable of digesting their chains is very broad. For instance, different forms of chitosan are digested by lipase, protease, cellulase, and hemicellulases. Constitutionally, chitinases/chitosanases occur in plants in small quantities, but their level sharply increases under biotic and abiotic stresses.

Chitosan and its hydrolytic products possess antifungal activity. Oligomers of chitin and chitosan were found to be able to bind to DNA, thus impeding formation of RNA. In addition to antifungal activity, chitin and chitosan possess elicitor properties. For instance, chitosan products are efficient elicitors of the pea phytoalexin pisatin. If this elicitor was applied before or even during inoculation, it protects pea tissues from infection by Fusarium.

Japanese scientists showed that N-acetylchitooligosaccharides (especially heptamers and octamers) induced FA formation in suspension cell culture of rice. Sub-nanomolar concentrations of these oligomers also induced fast and short depolarization of membranes. Deacetylated chitooligosaccharides were found to be completely inactive.

The molecular mechanisms of action of acetylated and deacetylated chitin derivatives can be essentially different. In the former case, probably, highly specific binding with the membrane receptors of lectin nature is involved, while fragments of chitosan, apparently, are active due to electrostatic inter-action of the positively charged elicitor molecules with the negatively charged membrane components or DNA molecules.

All the above provide grounds to believe that plants widely use polysac-charides of fungal cell walls for pathogen recognition.

Polypeptides and glycoproteins

The first discovered FA elicitor was monolicollin A, a 8 kDa polypeptide. It was isolated by the Australian plant pathologists Perrin and Cruikhank from the stone fruit rot pathogen *Monilia fructicola* in 1968. Monolicollin induced forma-tion of faseollin in beans in a concentration of 10^{-8} M. This elicitor was found to be specific only for beans, and did not induce formation of pisatin in peas or vieronic acid in horse beans. As beans are not a host for *M. fructicola*, the physiological role of monolicollin A in resistance of this plant was questioned.

An example of peptide elicitors are *elicitins*, a family of hydrophilic pro-teins with a molecular weight ca. 10 kDa formed by all species of the genera *Phytophthora* and *Pythium* studied till now. All elicitins are highly homologous.

Elicitins most distinctly manifest themselves as elicitors in treatment of tobacco. They cause necrosis of plant tissues, formation of phytoalexins, induc-tion of active oxygen forms, as well as systemic acquired resistance.

Isolates of *P. parasitica*, which do not attack tobacco, produce the acidic elicitin *parasiticein*. The isolates pathogenic to tobacco do not produce this peptide, which can point to a negative relation between the production of parasiticein and specific pathogenicity of the organism to tobacco. However, there are no grounds to believe that the genes encoding various peptides in the *Phytophthora* fungi can be alleles of avirulence genes. *P. cryptogea* and *P. capsici* produce two closely related elicitins: cryptogein and capsicein. Each of them consists of 98 amino acid residues. Treatment of tobacco with these peptides provides protection against pathogenic strains of *P. nicotianae* that do not produce elicitins. Cryptogein causes formation of necroses in tobacco in a concentration of 1 µg per plant, while capsicein is 50 times less active. Both peptides have identical sequence of amino acids in the internal area and differ in carboxy and amino terminations. Elicitins of the late blight of potato *P. infestans* are synthesized under the control of the *INF* genes that express in infested plants and in mycelia (but not in spores) growing in arti-ficial media. To understand the functional role of elicitins, two properties are important for their producers: (1) before they reach the membrane of the infested plant, they bind to the regulatory protein of the cell wall;

(2) elicitins can bind to sterols and carry them between artificial membranes, biologically active being only the sterol-loaded molecules. Since pythiaceous fungi (genera *Pythium* and *Phytophthora*) cannot synthesize the sterols necessary for development of spore-bearing (asexual and sexual) and pathogenicity, probably, elicitins are required to transport sterols from an infested plant to the mycelium.

Orthologic proteins NIP1 of *P. parasitica* and PsojNIP of *P. sojae* induce necroses in all tested dicotyledons, including hosts of these fungi. The gene that controls synthesis of PsojNIP protein expresses at a late stage of soybean infection, with the production of a toxin, which provides colonization of the host tissue during the necrotrophic growth phase. Structural analogs of these proteins were found in the fungi (*F. oxysporum*) and bacteria (*Bacillus halodurans, Streptomyces coelicolor, Vibrio* spp), but not in higher plants or animals. Also the 42 kDa enzyme transglutaminase (or, rather, its 13 kDa fragment (Pep-13)), bound to the cell wall of *Phytophthora* fungi, possesses elicitor activity towards dicotyledons.

Fungal enzyme, 22 kDa xylanase (EIX), induces biosynthesis of ethylene, leakage of electrolytes, and hypersensitive response (HR) in tobacco and tomato cultivars. Its receptors are signal glycoproteins bound to the cell surface and regulating the mechanism of endocytosis.

Elicitor activity of culture filtrates of various strains of *F. solani* seems to depend on the presence of a protein component, too, as this activity was partially lost under the effect of the proteolytic enzyme pronase, and was completely blocked in addition of the protein synthesis inhibitor cycloheximide in the growth medium. The high-molecular-weight inducing factor of the culture filtrate of *Botrytis cinerea* that promotes accumulation of the phytoalexin faseollin in tissues of beans has the same nature. At the same time, the treatment that could be expected to denature a larger part of the protein (heating, extreme pH values, organic solvents, detergents) had no impact on the induction activity of these preparations. This questions the importance of native conformation of the protein for expression of elicitor activity.

Flagellins (protein subunits for building bacterial flagella) also possess elicitor activity. Flagellins isolated from various phytopathogenic bacteria have varying central domains and conservative N and C terminal ranges that possess elicitor activity. On the basis of such a structure, a model of interaction of the elicitor and membrane receptor of a plant cell (protein FLS32) was constructed, which includes two stages: the first stage is connection of the N end with the accepting site of the receptor, and the second stage is connection of the C end with the transmitting site.

Many phytopathogen elicitors are *glycoproteins* (GP). For instance, a glycoprotein has been isolated from cell walls of *P. megasperma*, capable of inducing formation of the phytoalexin glyceollin in soybean cotyledons. However, the GP concentration necessary for induction was considerably higher than in the earlier described β-1,3-glucan and glucomannan. The GP activity was completely suppressed by the carbohydrate inhibitor periodate, while the treatment with pronase had no effect on induction activity. Therefore, manifestation of elicitor ability may rather require the carbohydrate and not protein part of the molecule.

Elicitor activity of the culture filtrate of *P. infestans* is connected with a 10 kDa fraction, which probably constitutes a glycoprotein without race-specific effect. β-1,3-glucanase did not suppress but rather strengthened the induction activity of the elicitor, which may be explained by removal of the masking groups covalently bound to the elicitor with β-1,3-bonds.

Nematode amphids localize protein exudates in the area of the head of a female in the intercellular space between the root cells, which gives ground to believe that they are involved in the relations with a host plant. Molecular genetic studies showed that cuticular exudates of the females of the root-knot nematodes *Meloidogyne incognita* are low molecular weight proteins (14 kDa). Proteins with the molecular weight 24, 50, and 46 kDa, as well as lipoproteins containing retinol or fatty acids, were found on the body surface of the larvae and females of two species of potato cyst nematodes. The substances released by nematodes in many cases can bind to plant cell receptors, and further influence, through the plant signal systems, the expression of the genes that determine plant responses.

Lipid-containing elicitors

A special group of elicitors is the lipid-containing metabolites. Lipids play an extremely important role in adaptation of all forms of life. The lipid strategy of biochemical adaptation apparently evolved at the first stages of life on Earth (E.M. Krebbs). Adaptation functions of lipids are intrinsic in organic world, from prokaryotes to the highest eukaryotes, especially important being the lipids of cell membranes. Compensatory changes in the composition of the fatty acids of membrane phosphatides are believed either to eliminate or reduce the distortion in the liquid crystal structure of the membranes.

Inclusion of unsaturated fatty acids in plasmalemma results in dropping viscosity and growing flow, thus increasing lateral motion of proteins, which leads to a greater exposure of the membrane receptors.

A general response of organisms to stress is a change of the degree of unsaturation of the fatty acids of the lipids. The change of the degree of unsaturation of fatty acids maintains the membrane permeability and viscosity irrespective of the environment. Therefore, **desaturases** increasing unsaturation of fatty acids can be regarded as anti-stress enzymes.

Best explored are the lipid-containing elicitors of *P. infestans*. From the mycelium of this fungus, a lipoglycoproteid complex with the activity several times higher than that of cell wall glucans was isolated. The lipid component making up 60% of all the complex is responsible for the induction activity of the complex. Two polyunsaturated fatty acids: 5,8,11,14-eicosatetraenoic acid (arachidonic acid – AA) and 5,8,11,14,17-eicosapentaenoic acid (EPA) were found to be the active substance of the complex . These polyene fatty acids induced defense responses in potato. A very important property of AA and EPA was their ability to induce systemic long-lasting disease resistance in potato.

The AA and EPA are found among all acid lipids of *P. infestans*, as well as in unsaponifiable lipids (ceramide aminoethyl phosphonate and inositol phosphoceramide). Elicitor activity is manifested only in the presence of a free carboxyl group in the lipid. Its substitution with an alcohol group strongly reduced the induction activity. Methyl esters of AA and EPA produced FA accumulation only after a short lag.

A 20-carbon-atoms chain length was found to be an optimum length for the elicitor activity. None of the saturated and unsaturated acids of C-16 and C-18 series were active. Crucially important for biological activity are the position and number of double bonds. The acids with the double bond positions D 5,8,11 possess the greatest induction activity.

The ability to induce phytoalexins depended on the class of lipids where AA was present. For instance, relative activities arranged in the following order: lysophosphatidyl choline: phosphatidyl choline: monoarachidonyl glycerin: diarachidonyl glycerin: triarachidonyl glycerin = 140: 95: 100: 20: 1.

The AA and EPA contained in a fungus are involved in host–parasite interaction. The scientists inoculated potato with suspension of late blight pathogen spores containing radioactive labeled AA. The acid quickly withdrew from the pathogen spores and accumulated in several rows of the cells adjacent to the infection site, but no more than 1 cm from the place of inoculation.

Unlike the carbon-containing elicitors of the cell walls, AA does not interact with the receptor sites in the plasmalemma but deposits in plant cell membranes and replaces linolic and linolenic acids characteristic of the plants. Further transformations of arachidonic acid in plant tissues will be discussed in Section "Systemic Acquired Resistance" in Chapter 8.

Endogenous plant or secondary elicitors

In the middle 1970s the American biochemists Albersheim and Valent discovered a new class of regulatory molecules constituting oligosaccharides of plant and fungal cell walls. These compounds were found to be able to control defense functions of plants as well as processes of plant growth and differentiation. The oligosaccharides possessing regulatory activity received a generic name *oligosaccharines* (OS). The OS from fungal cell walls (β-1,3-glucans, chitin, and chitosan) discussed in the previous section are exogenous in relation to the plant (microbial), while the OS from plant cell walls are called endogenous, constitutional, or secondary. Though oligosaccharines are rich in sugar residues, they can contain noncarbohydrate components: phenols, peptides, and acyl groups.

The discovery of oligosaccharines was considerably promoted by the use of cell wall culture of sycamore (*Acer pseudoplantanum*). The sycamore cells are homogeneous and possess only primary cell wall. It is also important that the cells release to the environment the polysaccharides that are structurally affine to the polysaccharides of the cell wall matrix, which essentially simplifies investigation.

The starting point in discovery of oligosaccharines by P. Albersheim was the fact that polysaccharide fragments of the cell walls of *P. megasperma* can induce phytoalexins in soybean. This fact, as well as a complex structure of polysaccharides of the cell wall matrix initiated a revision of the earlier views that cell walls perform only mechanical functions of maintaining the shape and structure of an organism.

The carbohydrates of plant cell walls consist of the structural part, cellulose, and matrix polysaccharides. It was the matrix structure that turned out to be so unexpectedly complex. For a long time matrix polysaccharides have been divided, according to solubility, into pectic substances and hemicelluloses (Chapter 5). The pectic substances of the matrix carry cellulose microfibrils. Pectines are also among the components of the middle lamella, which provides communication between cells and integrity of the plant tissue.

Pectic substances of the primary cell walls of dicotyledons consist of homogalacturonan, ramnogalacturonan I, and ramnogalacturonan II. Homogalacturonan contains unbranched chains consisting of approximately 25 residues of galacturonic acid connected by α-(1–4)-glucoside bonds. Ramnogalacturonan I and II are made of more than 13 monomers connected by 20 different ways, including α and β bonds.

Hemicelluloses of dicotyledons consist of several groups of polysaccharides: xyloglucans, arabinogalactans, and glucuronoarabinoxylans, named so after the prevailing monosaccharide.

Functionally, hemicelluloses are defined as a class of very branched polysaccharides forming noncovalent transverse bonds with cellulose microfibrils. The dynamic character of xyloglucans transverse bonds is believed to be the main factor to control the cell wall distention, and, hence, plant cell growth. Therefore, xyloglucan is considered the supporting structure of primary plant cell walls.

Oligosaccharines possess unique properties: numerous hydroxyl groups and oxygen atoms, as well as hydrophilic areas characteristic of glucoside residues create conditions for stereochemical groups, providing OS ideal opportunities for precise interaction with the sites on protein molecules (lectin–ligand interactions). For instance, presence or absence of one monomer or change of the position of the side disaccharide, changes regulatory properties of the molecule in a polar way. It is obvious that such a complex structure cannot be needed just for reinforcement of the cell wall.

Formation of oligosaccharines *in vivo* results from degradation of cell wall polysaccharides, and not from synthesis of small monomers. Exposure to stress agents, including phytopathogen attack and wounding, causes intensive decomposition of cell walls and release of oligosaccharines.

Binding of oligosaccharines to the specific receptor sites is supposed to provide plant survival by inducing formation of FA, PR-proteins – proteinase inhibitors, formation of lignin, and other defense responses. Oligosaccharines also regulate plant cell distention stimulated by growth hormones, control processes of morphogenesis and rhizogenesis, cause prompt changes in ion flows and permeability of plasma membranes, and induce protein phosphorylation *in vitro*. Oligosaccharines are released not only in stress conditions but also constitutionally, in the latter case they regulate normal physiological processes. Some scientists hold the opinion that such pleiotropic hormones as auxins and gibberellins are actually enzyme activators that, in turn, release more specific mediators, oligosaccharines, from cell walls.

Many oligosaccharines are active in concentrations of 2 or 3 orders lower (10^{-9} –10^{-8} M) than the phytohormones known till now. Synergism of the heptaglucoside of *P. megasperma* and the endogenous elicitor from soybean cell walls is of interest. The concentration of either one is too small to induce formation of an appreciable amount of the phytoalexin glyceollin, but their combined effect results in a high induction activity.

Discovery of oligosaccharines allowed explanation of the elicitor properties of some carbohydrases. For instance, the cotton phytoalexins were induced by pectatelyase and polygalacturonase from the culture filtrate of the fungus *Verticillium dahliae*. In the culture fluid of *Rhizopus stoloniferum*, the FA elicitor of castor-oil plant casben was discovered and found to be polygalacturonase. *Erwinia carotovora* produces endopolygalacturonase which causes formation of glyceollin in soybean. Inactivation of enzymes is accompanied by the loss of FA-induction ability. Apparently, carbohydrases digest cell wall polysaccharides and form the active oligosaccharines that induce protective effects.

Some scientists (e.g. P. Albersheim and I.A. Tarchevsky) also suppose that oligosaccharines are depots of regulatory molecules.

At present, several major OS groups are known: the earlier-mentioned exogenous (microbial) fungal cell wall oligosaccharines (β-glucans, chitin, chitosan) as well as endogenous plant oligosaccharines, oligogalacturonides and xyloglucans. However, it becomes clear now that the OS group is not limited to only these structures. Cellobiose oligomers as well as oligomers of xylan, arabinoxylan, and probably, some other structures can be included, with good reasons, in the group of oligosaccharines.

Oligogalacturonic (pectic) oligomers

Bioactive oligogalacturonides can be extracted from plant cell walls by autoclaving, and also by enzymatic or chemical hydrolysis. Oligogalacturonides obtained by hydrolysis from cell walls of both dicotyledons and monocotyledons were found to be structurally similar, though the pectin content in the primary cell wall of dicotyledons is ca. 35%, with only 9% in monocotyledons.

Oligogalacturonides induce FA synthesis in various plants: glyceollin in soybean, faseollin in beans, casben in castor-oil plant, etc. The degree of polymerization of the most active elicitors is 12 galacturonic residues for FA induction in soybean, 9 in beans, and 13 in castor-oil plant. The optimum degree of polymerization (9 to 15) necessary for the elicitor molecules, apparently, promotes stable conformation of oligogalacturonides, which provides a fixed spatial position of the charged groups. The enzymes involved in FA biogenesis are also activated. Since pectic OS are relatively undemanding in terms of structure and work in higher concentrations than other oligosaccharines, they are supposed to be able to function without specific receptors, probably, by variation of physiological properties of the membranes.

Pectic OS induce peroxidase isoforms catalyzing the final stages of lignin production and cause formation of the active forms of oxygen. They induce an oxidative burst (see next chapter), the efficiency of which is comparable with the protective production of the active oxygen causing local inflammation in animal cells.

A series of oligomers with the degree of polymerization from 2 to 20 also possessed the ability to induce proteinase inhibitors; recovery of the reducing end of both saturated and unsaturated di- and trigalacturonides completely destroyed the activity. Galacturonides will be able to induce formation of proteinase inhibitors if they have an intact semiacetal ring and a free carboxyl in the sixth atom of carbon (C-6). A change of stereoisomerism of the hydroxyl group in position C-3 or formation of a double bond at C-5, decreased, but did not completely destroy the activity.

The active pectic fragments were obtained from the exposure of polygalacturonic acid to both polygalacturonase and pectatelyase. Probably the cell wall depolymerases produced by different pathogens, form products varying in the degree of degradation and capable of inducing various protective effects. However, most defense responses were induced by oligogalacturonides with the degree of polymerization exceeding nine. Oligogalacturonides exactly of such dimension are known to be able to form intermolecular complexes (egg-box structures) with Ca^{2+} as the binding ion. It is supposed that the effect of oligogalacturonides on plant cell surface correlates with the defense responses they induce. It is even more likely, because there is a similarity between the structural requirements for the oligogalacturonides inducing short-term responses in plasma membrane and the oligogalacturonides producing longer-term growth responses. The latter permits an assumption that the effect of these oligomers on membranes results in the generation of the secondary messengers inducing long-term responses.

The ability of oligogalacturonides to cause prompt short-term changes in K^+ and Ca^{2+} flows, depolarization of the membrane, and alkalinization of the environment confirm that they act at the membrane level.

The oligogalacturonides inducing synthesis of proteinases in tomato leaves simultaneously enhance phosphorylation of plasma membrane proteins. Induction of phosphorylation depends on the structure and dimension of the oligomer. Only oligomers with the degree of polymerization 13 and more were active, as well as those with OH group position like in polygalacturonic acid. The oligomers inducing synthesis of proteinase inhibitors had the same parameters. Thus, similar oligomer specificity was found at phosphorylation and in induction of protection and growth processes.

Hemicellulose oligomers

The main component of hemicellulose is xyloglucan. Its basis is a chain where β-1,4-D-glucose units in position 6 are replaced by α-D-xylose (Figure 7.5). About half α-D-xylose residues in xyloglucan of dicotyledons are bound to β-D-galactose or α-L-fucose-β-1,2-galactose in position C-2. Xyloglucan chains bind adjacent cellulose microfibrils by hydrogen bonds, thus restricting cell distention.

Endo-β-1,4-glucanase digests the β-1,4-bond only of those β-D-glucose residues which are not substituted by the residues of α-D-xylose, α-D-xylose-β-D-galactose, or α-D-xylose-β-D-galactose-α-L-fucose. Usually every fourth residue in the main chain of xyloglucan is not a branching point; therefore, the oligosaccharide products resulting from enzymatic digestion contain a cellotetraose basis.

The best studied effect of xyloglucan nonasaccharide (XG-9) and xyloglucan pentasaccharide (ZG-5) is inhibition of growth induced by herbicide 2,4-D. This is an exclusive property of the xyloglucan oligomers containing fucose and showing activity in a concentration of 10^{-9}–10^{-8} M. The fucose-free xyloglucan oligomers did not possess antiauxin activity. Hence, the trailer fucoside is decisive in growth inhibition activity. The enzyme α-fucosidase, found in plants, splits off the fucose residue from the active fucose-containing

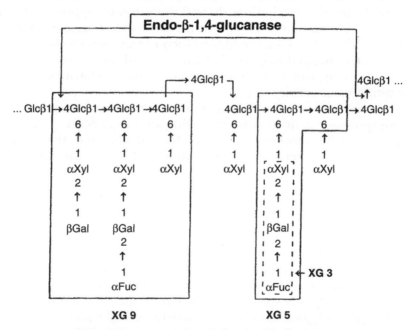

Figure 7.5. Structure of xyloglucan and its fragments.

xyloglucan oligosaccharine, thus making it inactive. The result is that plants are capable of converting biological signals, and the information they carry is not stored longer than necessary.

For a long time, biological activity of xyloglucan oligomers was only interpreted as their effect on the processes of plant growth and morphogenesis. However, evidence has recently appeared regarding immunomodulating role of these fragments. XG-9 and XG-5 were found to possess immunostimulating properties, thus improving resistance of wheat shoots to *F. colmorum* and inducing phytoalexins in soybean cotyledons. It was found that the initial stages of hydrolysis of xyloglucan by cellulase result in the fragments showing elicitor activity in potato tubers. However, as hydrolysis proceeds, xyloglucans digest into oligomers with the degree of polymerization less than 12, which not only possess elicitor activity, but, on the contrary, strengthen the signs of the disease, thus showing immunosuppression properties. The effect was exhibited at extremely low concentrations (1–10 pg/ml).

The data obtained, as well as analysis of pathogenesis processes suggest possible elicitor activity of the oligosaccharines generated at early stages of enzymatic degradation. At a later stage, when depolymerases begin in-depth destruction of cell walls, small molecules are formed that suppress defense responses of plant tissue and facilitate transmission of the parasite.

Two individual xyloglucan oligomers, containing fucosyl residue: trisaccharide (XG-3) and pentasaccharide (XG-5), possessed immunosuppression activity. Both the oligosaccharines showed the greatest suppressor activity in potato at a concentration of 10^{-8} M. At the same time tetrasaccharide (pentasaccharide without terminal residue of fucose) was found to be biologically inert. The data obtained suggests that the terminal fucosyl residue is necessary not only for antiauxin effect of xyloglucan oligosaccharines, but also for immunosuppressor activity.

Trisaccharide was found to be the minimum structural unit determining immunosuppressor activity. It is interesting that replacement of xylose in molecule XG-3 by another structural element changed the suppressor activity to the reverse elicitor activity. Thus, the oligomers containing a part of xyloglucan structure, α-L-fucose-β-D-galactose, can possess either suppressor or elicitor activity, depending on the structure of the third sugar residue.

The fucosyl residue within xyloglucan oligomers plays an extremely important role in relations of potato and late blight pathogen. The oligomers without terminal fucosyl residue become biologically inactive, losing both the elicitor and suppressor properties. Such dependence of biological activity on the structure, combined with extremely low active concentrations, suggests the presence of a highly specific site in potato, the interaction with which results in some biological activity. Since labeled xyloglucan oligosaccharines penetrate the protoplast with difficulty, they are most probably received on the plasma membrane.

Regulation of oligosaccharine formation in a plant

Formation of oligosaccharide fragments possessing suppressor functions is of a special interest, as all endogenous oligosaccharines known till now, either stimulated plant protective functions or had no effect on them. Thus, endogenous oligosaccharines intrinsically possess a broader activity during pathogenesis than it was earlier assumed. This activity consists not so much in stimulation but rather in regulation of immune responses of plant tissue.

The delay in enzymatic degradation of polysaccharides at the stage of formation of elicitor oligomers is especially important. Plants were found to contain proteins specifically inhibiting the activity of cell wall depolymerases. According to the model suggested by Cervone and other scientists, inhibitors of plant cell wall depolymerases can regulate disease resistance through promoting formation of the elicitor-active molecules.

A polygalacturonase-inhibiting protein, PIGI, which is affine to the endopolygalacturonases *C. lindemutianum, F. moniliforme,* and *Aspergillus niger,* was isolated from bean cell wall extracts. It was noted that PIGI usually acts on polygalacturonases of the fungi which do not affect the plant species from which they are isolated, and render smaller inhibition on polygalacturonases of the pathogens of this species. For instance, PIGI preparation partially purified from potato cell walls, inhibited polygalacturonase isolated from the cotton parasite *verticillium dahliae* and the saprotroph *A. alluacianum,* but had no effect on the enzymes of the specific potato parasites *P. infestans* and *Rhizoctonia solani.* Thus, PIGI plays an important role in species immunity and a smaller role in cultivar resistance.

PIGI was found in cell walls of many dicotyledons and monocotyledons, both infested and healthy. This inhibitor is present virtually in all plant tissues and organs, with localization mostly in apoplast. PIGI is a leucine-rich glycoprotein. It is thermostable and resistant to proteases. In the presence of PIGI, the number of elicitor oligomers steadily grows in 24 hours, and only after 48 hours the enzyme hydrolyzes the substrate to smaller, elicitor inactive fragments. In the absence of the inhibitor, molecules of polygalacturonic acid depolymerized to a level of inactive oligomers after 15 minutes of hydrolysis.

Polygalacturonase inhibitors were found in infected fruit, onion, sweet pepper, and cabbage. Roots of sugar beet contain a protein that inhibits pectatelyase of *R. solani* and *Phoma betae.*

PIGI responds to infection, wounding, or treatment with salicylic acid by increase in mRNA, with its level increasing mostly in the cells closely adjacent to the injured area. This evidence indicates that PIGI synthesis is a constituent part of active plant defenses. However, in most cases the PIGI levels in resistant and susceptible cultivars are similar; therefore, this protein cannot be regarded as a specific determinant of resistance or susceptibility.

Apparently, plants possess the mechanisms regulating activity of the depolymerases that degrade cell wall polysaccharides. This degradation promotes accumulation of the elicitor-active molecules and prevents formation of oligomers that are either inactive or possess immumosuppression properties. Pathogens, in turn, possess an ability to secrete proteins which efficiently inhibit host enzymes and use the cell wall polysaccharides as substrate. Thus, not only plants are capable of synthesizing enzyme inhibitor proteins against pathogens, but pathogens also possess a similar means for regulation of plant depoly-merases. Their inhibition prevents enzymatic degradation of plant cell walls and production of elicitor oligomers from these cell walls.

The oligomers produced from degradation of polysaccharides possess clear regulatory functions. The above theses are an illustration to the concept, suggested by I.A. Tarchevsky, that postulate using molecules in the catabolic of stream as the signals acting onto various stages of anabolism. From this point of view, it is unlikely that the evolution could have missed the opportunities of exploiting the information involved in the molecules of the catabolic stream.

Avirulence genes and specific elicitors

Investigation of the structure of numerous non-specific elicitors in the absence of the specific elicitors resulted in the interpretation of the resist-ance induction process as non-specific, with specificity seen in defense response suppression. Contradictions between the data of genetic and bio-chemical tests regarding what is specific, induction or suppression, have been resolved using the molecular genetics methods. The question was addressed using two methodologies.

Receiving mutations of the gene responsible for specificity and studying the plant response to inoculation with a mutant strain. Modified sequences of information molecules or mobile elements of the genome *(transposons)* can be used as probes for identification of the specificity genes.

In most phytopathogenic viruses, the genome contains just few cistrons, therefore, it is available for complete sequencing (determination of the sequence of the nucleic acid bases). Artificially produced mutants or even natural variants of the virus with modified virulence can be used for identifi-cation of *avr*-genes.

In cellular organisms, *avr*-genes are identified with the *transposon* mutage-nesis. Transposons (tn) or insertion sequences are DNA sites with repeating base sequences at the ends. Due to such a structure, they can be built into different sites of the DNA strand, and the gene, where the transposon has entered, can be fully or partially inactivated. Among the cells transformed by the transposon, cells with modified virulence are selected on the basis of the response to inoculation, their DNA is isolated, and the DNA site responsible for virulence is determined using hybridization with the transposon (probe).

Transformation of the virulence gene from one strain to others. This process needs the following: (1) DNA molecule-cutting enzymes (restric-tases) to break the parasite genome into separate fragments; (2) to clone the fragments in an organism like *E. coli* and to set up a gene library; (3) to connect the fragments with a vector where a marker gene has been built-in necessary for selection of the transformed cells (marker gene of resistance to antibiotics or prototrophicity); (4) to treat the recipient strain protoplasts with the vector, and select the transformed cells by the marker gene; (5) according to the change in the plant response to inoculation, to select the clone where the DNA site responsible for specific pathogenicity is built-in (Table 7.3). If the specificity is determined by induction of resistance (specific induction), response of the cultivar *A* to inoculation with the transformed race *b* will change from susceptible to resistant, while the cultivar *b* will retain the resistance response (variant 1). If the specificity is determined by suppression of resistance, then conversely, the response of the cultivar *B* will change from resistance to susceptibility, while the cultivar *A* will remain susceptible (variant 2).

Transformation of the virulence is a complex methodology, as it requires selection of the specific virulence-related fragment among a huge number of the DNA fragments in the gene library. The question can be formulated in the following way: how many transformants have to be tested in a plant to find the required one with certainty? The number of transformants can be evaluated by the following equation:

$$N = \ln (1-P)/\ln (1-f)$$

where P is probability of the gene presence, and f is the part of the genome found in one clone. For example, if the size of the genome of the genus *Erwinia* bacteria is 4.5×10^6 base pairs (bp), and the average size of the cloned fragments is ca. 15 thousand base pairs (tbp), the probability (p) of finding the required gene in one of 1500 transformants will be P = 0.99. Hence, it is necessary to select, according to the marker gene, more than 1500 transformed clones where the vector is built-in, and infect the host plant with each clone individually to isolate a clone with modified virulence. To facilitate this work, it is recommended to: (a) use the restrictases that cleave the DNA molecules into larger fragments; (b) use the constructs capable of carrying large DNA fragments (cosmids and phage λ DNA) as the vectors.

We will next discuss the results of using the above and other methods of analysis of *avr*-genes and specific elicitors in different parasites.

Table 7.3.   Possible changes in the plant response to inoculation with the pathogen after its transformation by the specific pathogenicity gene

Parasite strain	Response to inoculation	
	Cultivar *A*	Cultivar *B*
A	R (I)	S c
b	S c	R (I)
b + fragment *a* variant 1	R	R
variant 2	S	S

Table 7.4.   Types of interaction of TMV strains with *Nicotiana* species

TMV strains	Tobacco species*		
	N. tabacum (n)	*N. sylvestris (N′)*	*N. glutinosa (N)*
U1	S	R	R
U2	S	S	R
Ob	S	S	S

*In brackets: alleles of the tobacco vertical resistance gene.

Viruses

Tobacco mosaic virus (TMV) can be used as an example of the methodology of searching for *avr*-genes in the phytopathogenic viruses. The genus *Nicotiana* includes susceptible (cultural tobacco plant *N. tabacum*) and resistant (*N. sylvestris, N. glutinosa*) cultivars. TMV strains virulent to resistant species were found in the natural virus populations. They were also obtained by artificial mutagenesis and their interaction with the tobacco species complied with the "gene-for-gene" rule (Table 7.4).

The TMV genome contains only 6400 nucleotides and consists of several cistrons (Figure 2.1). Genes adjacent to 5′ end belongs to 126 kDa RNA polymerase and 183 kDa auxiliary protein gene necessary for RNA replication. These are the early proteins emerging in the infected cell first. Closer to 3′ end there are genes of the 30 kDa transport, necessary for migration of the rims, and the 17.5 kDa cepsid protein (CD). The whole genome has been sequenced.

Mutants virulent for *N. sylvestris* (they overcome the effect of gene *N′*) occur naturally, and a large number of them are generated after mutagenic treatment. The genome of many such mutants has been sequenced to find the genes, responsible for the virulence. All the virulent mutants contain changes in the capsid gene, with two hot spots where nucleotide replacements occur most frequently. Replacement of only one nitrogen base (cytosin to uracil and the amino acid serine to phenylalanine, respectively) in position 157 leads to the change of virulence. Consequently, it is the coat protein, not the RNA, that serves as the specific elicitor which recognizes the *N′* resistance gene product, as in the absence of the expression of the capsid protein there is no induction of hypersensitive response.

Mutants virulent to *N. glutinosa* occur very rarely. Their investigation showed that the specific elicitor for the *N* gene product is a 50 kDa helicase domain of the 126 kDa protein that is contained in the TMV RNA polymerase complex. Mutations in the *p*-loop motif that disrupt the ATP-ase and helicase activity of this domain do not reduce the elicitor activity.

In addition to tobacco plants, tomatoes are strongly affected by TMV. Two resistance genes, *Tm1* and *Tm2*, have been described for the resistant tomato cultivars. Mutations that make the virus virulent to the cultivars with gene *Tm1* are mapped in a gene of the 30 kDa transport protein. Thus, all TMV proteins generated in an infected cell are engaged as elicitors.

Capsid proteins (CP) of other viruses also possess elicitor properties. The CPs of strains of the mild pepper mottling virus induce a hypersensitive response in *Capsicum frutescens* that possesses the resistance genes *L²* or *L³*.

It is interesting that the same elicitor (capsid protein of the potato X-virus with threonine in position 121) induces a hypersensitive response in the potato with gene *Rx* and in the species *Gomphrena globosa*, the latter belonging to a different botanic family than potato. Thus, two resistance types earlier regarded essentially different, *varietal* and *specific* (according to N.I. Vavilov's terminology) are induced by the same viral elicitor.

Bacteria

Avr-genes

The first study where cloning of the *avr*-genes was successful was the investigation of the bacteria *Pseudomonas syringae* pv. *glycinea*. The bacterial races and soybean cultivars interact according to the "gene-for-gene" rule. From the race 6 gene library, a fragment was isolated (pPg6L3) which modified the soybean leaf response to inoculation. This fact was discovered during the testing of 600–700 transformed clones. Insertion of this fragment in the other races of the bacteria modified their virulence, which is shown in Table 7.5.

Table 7.5 shows that fragment pPg6L3 does not change the plant response to inoculation from incompatibility to compatibility in any case. Therefore, the gene in this fragment does not induce susceptibility but resistance, i.e. it is the *avr*-gene. The change of response from susceptibility to resistance is observed only in the cultivars resistant against the fragment donor, race 6 (cultivars Acme, Chippewa, Harosoy, Peking), which is indicative of the cultivar specificity of the race 6 elicitor.

Table 7.5. Response of soybean cultivars to inoculation with the natural and transformed strains of *P. syringae* pv. *glycinea* (Keen NT, Stackawicz BJ, 1984)

Bacterial races	Response of soybean cultivars						
	Acme	*Chippewa*	*Flambeau*	*Hardee*	*Harosoy*	*Norchief*	*Peking*
Wild type 1 (wt)	S	R	S	R	R	R	S
wt 1 + pPg6L3	R	R	S	R	R	R	R
wt 4	S	S	S	S	S	S	S
wt 4 + pPg6L3	R	R	S	S	R	S	RS
wt 5	R	R	R	S	S	R	S
wt 5 + pPg6L3	R	R	R	S	R	R	R
wt 6	R	R	S	S	R	S	R

Insertion of transposon Tn5 in fragment pPg6L3, resulting in the gene inactivation, caused the loss of ability for induction of incompatibility. Therefore, the active specificity gene induces resistance, and the loss of the active form of the gene makes the plant susceptible to bacteria. In that way, the hypotheses based on the "gene-for-gene" theory were experimentally confirmed in the first time. The use of a transposon as hybridization probe showed that ca. 1.4 kb DNA site is responsible for avirulence.

The methods designed were used in studying dozens of phytopathogenic bacteria and confirmed the above results. For instance, race 2 of the pepper parasites *Xanthomonas vesicatoria* is avirulent to the cultivars with resistance gene *Bs1*, however, spontaneous mutants frequently occur, with various degree of virulence to these cultivars. They were found to contain ca. 1.2 tbpkb insertion in avr-locus IS476. If this insertion is located in the regulatory site of the gene, the mutant has low virulence, and its introduction in the structural area leads to a high virulence. This is one additional experimental confirmation of Flor's hypothesis that deletion or mutation of the avirulence gene is phenotypically manifested as virulence, and the possible nature of the transient responses to inoculation was shown, between high resistance and complete susceptibility, described for many diseases. Another possible mechanism of acquiring virulence is deletion, during replications, of the repeating sequences contained in the internals of many *avr*-genes.

Gene *Avr-Bs1* from *X. vesicatoria* was transformed in other xantamonads and they began to induce a hypersensitive response in the plant species to which these bacteria were virulent (hosts plants). Thus, the product of the same gene induced hypersensitive response (HR) in the resistant cultivars of the host plant and in the non-host plants of this bacterium. Hence, there is no important difference between the *varietal* and *specific* immunity regarding bacteria and viruses.

Cloning of *avr*-genes showed that their characteristic homologous sequences are available in the genomes of various bacterial species. Families of homologous Avr-genes have been found. The properties of some families are described below.

Family *avrBs2.* It includes gene *avrBs2* of *X. vesicatoria* and its homolog in *X. campestris* pv. *alfalfae*. Protein AvrBs2 is homologous to the enzyme involved in synthesis and hydrolysis of the phosphodiester bonds between carbohydrates or phosphatides.

Family *avrBs3.* It includes avr*Bs3* of *X. vesicatoria;* *AvrB4, Avrb5, AvrB101, AvrB102, AvrBln, avrB7,* and *PthN* of *X. campestris* pv. *malvacearum; PthA of X. citri; AvrXa7, AvrXa10,* and *avrxa5* of *X. oryzae* pv. *oryzae*. Proteins encoded by these genes induce resistance responses, however, their inactivation by mutations leads to the general decrease in pathogenicity. The family of AvrBs3 proteins is necessary for reproduction of the phytopathogenic bacteria in a plant and generation of the signs of diseases. These proteins have the central domain that includes the repeats of 34 amino acids, with their number varying from 13.5 in AvrB6 to 25.5 in AvrXa7. The structure of the C-end site shows that it is functioning as an acidic factor of transcription activation. Mutation analysis showed that the central domain is necessary to manifest specificity of the protein, and C-end, to manifest the elicitor activity.

Family *avrPto* of *P. syringae*. Protein AvrPto disrupts the assembly of the glucan components of the cell wall, thus preventing development of the protective structure, papilla.

Family *avrD*. AvrD gene of *P. syringae* pv. *tomato* is homologous to many genes found in the genome of *P. syringae* pv. *glycinea*, in particular, *avrPg4*. Homology in the protein is 86%, and in the reading frame, 98%. Therefore, homologous genes are present in most bacteria, but due to point mutations their products have ceased to induce a hypersensitive response in their hosts (for instance, avrD does not induce incompatibility in tomato), i.e. they are recessive (inactive) alleles. Gene *avrD*, as well as some other *avr*-genes, is localized on the plasmid. It encodes the 34 kDa protein that does not contain the sites necessary for transmembrane transmission from the cell to the environment (leader peptide and hydrophobic domain). Although it can be transported from the cell by a special secretory system (see below) it is believed to be an intracellular protein. Since gene *avrD* is the first of the five tandemly located reading frames, its product may be an enzyme important for secondary metabolism. The avrD-mediated elicitor activity was found only after its transformation in the genome of *E. coli* which began releasing a metabolite that induced a hypersensitive response in the soybean cultivars with gene *Rpg4*. Crossing of the soybean cultivars with and without gene *Rpg4* produced segregation in F2 into resistant and susceptible progeny in the ratio 3:1, with all resistant segregants (and only they) causing necroses at infiltration of the elicitor from the avrD-transformed *E. coli* cells in the leaf intercellular space. It was found that the elicitor was the low-molecular-weight C-glucosylated lipid syringolide, and the *avrD* gene product was the enzyme that catalyzed synthesis of syringolides (condensation of xylose with β-oxydecanoid acid). Syringolides (Figure 7.6) belong to the family of signal molecules, such as butanolides, inductors of sporulation and formation of the antibiotics actinomycetes, jasmonic acid in plant cells. In the growth medium of the phytopathogenic bacteria containing gene *avrD*, the syringolide content is many times less than in the growth medium of the transformed *E. coli*, but it sharply grows in the infected plants. Thus, two earlier discussed hypotheses: of indirect involvement of virulence gene products in host–parasite relationship (enzymatic "production" of low molecular weight extracellular elicitors) and of the plant metabolite-induced elicitor release only *in planta* have been experimentally supported.

Figure 7.6.   Syringolide, elicitor of *P. syringae* pv. *tomato*.

Almost all investigated *avr*-genes also encode the intracellular proteins. The avrRxv gene product of *X. vesicatoria* is homologous to the virulence factor of the bacterium *Yersinia pseudotuberculosum Yop J.* which causes intestinal infections. This factor, as well as other Avr-proteins, is transported through the bacterial membrane by a special genetic system described below.

Hrp-genes

A group of genes regulating the relations between the phytopathogenic bacteria and plants is called *hrp*-genes (hypersensitive reaction and pathogenicity). The mutants in these genes lose the ability to induce a hypersensitive response in the non-host plants and will be pathogenic for the hosts of a given bacterium. In all investigated bacteria of the genera e.g. *Erwinia*, *Pseudomonas*, *Xanthomonas*, the *hrp*-genes are closely linked and form clusters (groups) of several genes. Expression of a cluster of six *hrp*-genes of *Ralstonia solanacearum* requires activity of one more gene, *hrpS^C^*, whose product is homologous to the two-component prokaryote regulatory protein. The products of many bacterial *hrp*-genes are thermostable hydrophobic glycine-rich proteins, *harpins*. They induce a hypersensitive response in non-hosts. The gene *hrpA1* product of *X. vesicatoria* is an outer membrane protein, and that of gene *hrpB6* is ATP-ase necessary for protein transport from the cell. Some *hrp*-gene products are homologous to secretory proteins of the virulence factors of bacteria from the genera *Yersinia*, *Shigella*, and *Salmonella*, pathogenic for humans, and these factors, in turn, are homologous to some Avr-proteins. Some harpins (HrpW of *E. amylovora* and *P. syringae*) are bifunctional: their N-domain is an HR elicitor, and C-domain possesses pectatelyase activity. Two *hrp*-genes of *P. syringae* pv. *syringae* encode the proteins with a high degree of homology to glucosyl transferase of *E. coli*, the enzyme involved in building carbohydrate polymers of the cell wall. Apparently, harpins transmit the bacterial specific elicitors inside the plant cells through the protein filaments, pili, produced on the surface of the bacterial cells. *Hrp*-genes, like avr, express *in planta*, but not on a rich nutrient medium. Apparently, their inductor is starvation, for they express on a minimum medium, though not so actively as in a plant.

Secretion of the proteins responsible for pathogenic properties. Most avr- and hrp-proteins have no structures for their transmembrane transmission resulting from processing. However, numerous data show that they are not only secreted from a bacterial cell but also pass through the cell wall inside the plant cell and provide manifestation of the signs of disease or resistance response. For example, the presence of the transcription activation factor in the proteins of AvrB3 family is indicative of their interaction directly with the plant cell genome. Several systems of protein secretion have been described for human pathogenic bacteria. One of them, called type III secretion, is involved in transmission of both human and plant pathogenicity factors. More than 20 proteins are involved in the secretory system of this type, most of them localized in the inner membrane, and the functioning of

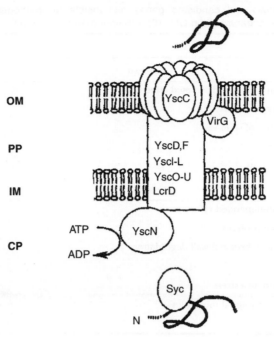

Figure 7.7. Type III secretion system of *Yersinia* (Hueck CJ, 1998 (Figure 1)).
OM – outer membrane; PP – periplasm; IM – inner membrane; CP – cytoplasm; ATP hydrolysis by YscN is indicated. N – amino terminus in secreted protein. Syc – cytoplasmic chaperon, which bind to presecretory protein. Secretion share multimeric outer membrane component YscC and accessory protein VirG.

some of them requires the membrane-associated ATP-ase. Some proteins are shown in Figure 7.7. The primary signal transduced from mRNA to the N-end of the secreted protein results in the protein binding to a low molecular weight chaperon Syc in the cytoplasm. Then the complex is transferred to the accessory protein YscN on the inner membrane of the cell, and through a series of proteins, to the accessory protein of an outer membrane and the multidimensional component YscC that carries the secreted protein from the cell. Finally, the "translocator" proteins facilitate transit of the transported effector proteins through the host eukaryotic cell membrane. Harpins and Avr-proteins function as effectors. Passing through the cell wall, apparently, involves the Hrp-pili found in *P. syringae* pv. *tomato* and *R. solanacearum*. The pili are built from the proteins essential for manifestation of the factors that influence the hypersensitive response and pathogenicity and are secreted as type III (HrpA in *P. syringae* pv. *tomato* and HrpY in *R. solanacearum*).

The number of bacterial pathogenicity factors and their structure
Various bacterial genes and gene products are involved in relationship with plants (Table 7.6).

Table 7.6. Known and candidate genes responsible for pathogenesis in the genome R. *solanacearum* strain GMI 1000 (Salanobath et al., 2002)

Pathogenicity genes	No. of genes
Known	
Type III secretion system and secreted effectors	31
Global regulatory functions	11
Exopolysaccharides of stem wall	18
Hydrolytic enzymes	4
Hormone production	1
Candidate	
Type III secretion-depended effectors	51
Adhesion/surface proteins	93
Hydrolytic enzymes/host cell wall degradation	5
Toxins	13
Resistance to oxidative stress	10
Plant hormones and signal molecules	7
Other	18

Position of the genes that determine the relationship between the pathogenic bacteria and the hosts, in the genome is of interest. Sequencing of bacterial genomes showed the presence of two types of genes. The first type, core sequences, homogeneous in the content of G + C and codons used, and with low mutational capacity, includes mostly the genes with housekeeping functions. The second type contains the genes that provide adaptive properties, including pathogenicity. They vary in the content of G + C and codons, they are highly mutable, and apparently, they evolved in genomes due to horizontal transmission. The areas containing these gene clusters are called *genomic islands,* or in a more narrow sense, *pathogenicity islands.* They are present in the genomes of the bacteria pathogenic both for humans and plants. A simplified model of the pathogenicity island is shown in Figure 7.8.

The direct repeats at the flanks of the pathogenicity islands and availability of mobile items determine frequent genetic rearrangements inside the islands and variation of the virulence factors.

Fungi

A large size of the genome of mycelial fungi and difficulties of their genetic transformation impede molecular genetic manipulations with their avirulence genes. That is why fewer such genes have been cloned in the fungi than in bacteria. The first investigated *avr*-gene was isolated from the genome of the tomato mold pathogen *C. fulvum.* This fungus infects the leaves of susceptible tomato cultivars through the stomata, grows in the

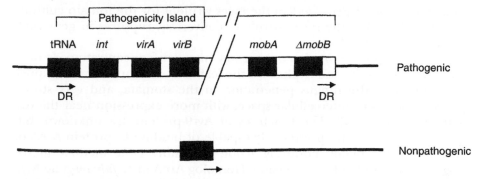

Figure 7.8. Model of a bacterial pathogenicity island (Hacker J, Kaper JB, 2000 (Figure 1)).
The *thin bold line* represents regions of the core genome4 pathogenicity island-specific sequences.
The *box* represent genes. The *arrows* indicate the presence of direct repeats at the ends of the pathogenicity island. DR – direct repeats; *int* – integrase gene; *vir* – virulence-associated gene; *mob* – mobility gene, Δmob – pseudo-mobility gene. *Mob* genes encode integrase, transposases, or other proteins involved in mobility of the prokaryotic genome.

intercellular space without forming the haustorium and without infecting cells, and feeds on cellular exudates. In resistant tomato cultivars, growth of the fungus causes a hypersensitive response, with the relations between the tomato cultivars and fungus races complying with the "gene-for-gene" pattern. The Dutch plant pathologist P. de Wit found a protein in the intercellular fluid of the tomatoes infected with race *A9* of *C. fulvum* that was absent both in the filtrate of the fungus growing on artificial nutrient medium and in uninfected tomato leaves. Infiltration of the intercellular fluid which contained this protein, in the leaves of various tomato cultivars, caused a hypersensitive response only in the cultivars incompatible with race *A9* (those with the resistance gene *Cf9*), i.e. that protein possessed the properties of a race-specific elicitor. It consists of 28 amino acids, six of them being cysteine. After its sequencing, a complementary polynucleotide was synthesized on the basis of the genetic code, which was used as a probe for isolation of the avirulence gene *avr9* from the fungal genome using DNA hybridization. The primary product of gene *avr9* contains 63 amino acids and consists of a signal peptide necessary for transport through the membrane (23 amino acids) and an extracellular protein of 40 amino acids. This protein undergoes the post-translation modifications: first, the fungal protease produces a 32–34 amino acid intermediate; then the plant protease (now outside the fungal hypha) produces a 28 amino acid peptide, the elicitor. It has a barrel-shape structure from three antiparallel lines producing a β-structure and connected by two loops and three disulfide bridges that bind all six cysteines in a cysteic node. In animals, the proteins with cysteic nodes serve as signal molecules. In Avr9-protein, such node, possibly, enhances stability of the elicitor molecule in the leaf apoplast and provides interaction with R-protein.

Transformation of gene *avr9* in the races virulent for the tomato cultivars with gene *Cf9* makes the races avirulent; conversely, ruptures in gene *avr9* sequences restore virulence.

Gene *avr9* expresses *in vitro* only at low nitrogen concentration. It does not express in conidia and hyphae on the leaf surface. Considerable expression occurs after the fungus penetration in the stomata, and very strong expression occurs in intercellular space, with more expression near the vessels than in mesophyll. The functions of Avr9-protein are unknown, but its promoter has 12 sites presumably capable of binding to protein AreA of *A. nidulans* – the major, positively acting regulatory gene mediate global nitrogen repression and derepression. Homolog AreA of *C. fulvum* gene *Nrf1* expresses at nitrogen starvation and its protein NRF1 is *Avr9* activity regulator. Apparently, Avr9-protein is involved in passing nitrogen to mycelium from substrate or induces nitrogen release and redistribution in the plant. An increased expression of avr9 in the vessel area is probably due to the product interference with the transport of nutrients in the plant.

Gene *avr9* is flanked by the direct repeats, where recombination can occur, concurrent with the gene deletion (Figure 7.9). Hence there is a frequent occurrence of the virulent races with this gene lacking in the genome.

The second avirulence gene isolated from *C. fulvum* was *avr4*. The study also began with isolation of the protein elicitor from apoplast fluid of the tomato that possesses the resistance gene *Cf4* and is infiltrated by the spores of an avirulent race. This protein turned out to be larger (86 amino acids), therefore, the sequence of only N-end amino acids was determined, which was sufficient for building a polynucleotide, used as a sample for isolation of gene *avr4*. Its product is *pro*-protein from 135 amino acids with an N-end signal peptide. Like the previous protein, during its maturation it is processed by fungal and plant proteases. However, it has two differences: (1) it has no homologies in the protein databanks and (2) in the virulent races, the encoding gene is not deleted but contains point mutations resulting in replacement of one cysteine in positions 64, 70, or 109 by thyrosine (replacement of codon TGT by TGA). The mature protein has 8 cysteine residues connected by disulfide bonds and is homologous to the chitin-binding protein of the invertebrates. It protects *Trichoderma viride* and *F. solani* from the lytic effect of plant chitinases, which, probably, determines its functional role in the pathogenesis of *C. fulvum*.

The product of the third avirulence gene *Avr2* is an extracellular protein with 58 amino acids. Eight cysteic residues of this protein are also connected in a node by disulfide bonds. This protein is an inhibitor of cysteic proteases of the papain family, which probably, determines its functional role in pathogenesis.

The similar procedure for isolation of proteins from intercellular fluid of the plants inoculated with incompatible races was used by the group of U. Knogge to study the avirulence genes of the barley leaf spot pathogen *Rhynchosporium secalis*. This fungus excretes *in planta* a family of small proteins called *NIP* (necrosis inducing proteins), nonspecifically toxic for monocotyledon and

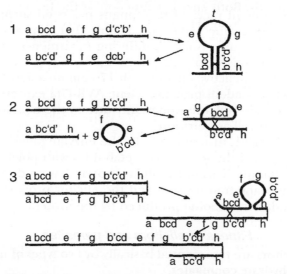

Figure 7.9.   Diagrams of chromosome rearrangements through crossing-over between the repeating DNA sequences.

1 – Intramolecular recombination between inverted repetitions bcd and d′c′b′ in a site between b and c (apostrophes allow to distinguish identical parts of two repetitions) is accompanied by rotational displacement on 180° (inverse) of site efg, made between repetitions. 2 – An intramolecular recombination on direct repetitions bdc and b′d′c′. For their occurrence in a synapsis the duplex of the DNA should be incurvated as a loop. The crossing-over descends on a site between b and c, that leads to cutting of a segment of chromosome between crossing-over sites and coupling the fragments posed outside of a crossing-over site. The cut out sequence (efg together with one copy of repetition bcd) is made in the ring. 3 – A recombination on direct repetitions bcd and b′c′d′ between sister chromatids (an unequal crossing-over). Its result is the deletion of a segment efg and one copy of repetition bcd in one chromosome and their addition (duplication) to another. Multiple repetition of an unequal crossing-over can lead to augmentation of number (amplification) of genes in a chromosome.

dicotyledon plants due to stimulation of H⁺-dependent ATP-ase of the plasmalemma. One of these proteins, *NIP1*, was found to be a race-specific elicitor for the barley cultivars with resistance gene *Rrs-1*. The mutation resulting in replacement of one amino acid in this protein eliminates the incompatible response of the resistant cultivars but reduces pathogenicity even towards the susceptible cultivars. Thus, *NIP1* protein plays a role of a non-specific pathogenicity factor, along with specific avirulence. Studying amino acid sequences of the mutants and artificially synthesized oligopeptides showed that the determinants of toxicity and HR induction are situated at the different ends of the molecule. Apparently, the plant cell contains different receptors of these determinants.

The structure of several *avr*-genes and the proteins they encode have been found for the fungus *Pyricularia oryzae* (teleomorph of *Magnaporthe grizea*). Gene *AVR-Pita* is localized in the telomere region of the chromosome, which determines the gene instability due to frequent rearrangements.

Instability of the virulent races of this fungus has been surprising to plant pathologists for a long time; data are available that from one lesion many monosporous strains can be isolated, differing in virulence in rice cultivars. Pro-elicitor Avr-Pita consists of 223 amino acids, but passing through the membrane produces an active elicitor with 176 amino acids. Functionally, it constitutes a zinc-dependent protease. Gene AVR-Pita expresses at the later stages of pathogenesis, which is apparently due to the necessity to use proteins of the infected cell as nutrition.

The *avr*-protein, Ace1, turned out to be the enzyme polyketide synthetase involved in synthesis of the secondary metabolites with polyketide structure (in particular, melanin). This intracellular protein has no domains responsible for transmembrane transmission and the elicitor properties seem to be due to the secondary metabolite produced by the protein.

Fitness of elicitors and parasites

The specific elicitors are represented basically by two types of molecules: proteins and carbohydrate conjugates.

In fungi, elicitor proteins have the structures to mediate their passing through the membranes. These structures are eliminated by the proteases during maturation of the elicitor. In bacteria, elicitors are transported from the cell by a special secretory mechanism. Outside the host cell protoplast, these proteins interact with the environment (for instance, product *avr9* of *C. fulvum* is involved in nitrogen metabolism in conditions of nitrogen starvation). When they arrive to the host plant cell from the intracellular parasitic structures (haustorium) or use the secretory systems (Hrp-pili of bacteria), they get involved in the pathogenesis by decomposing the host cell polymers (AvrPi-ta of *Magnoporthe grisea*) or by suppressing protective properties (NIP1 of *R. secalis*). The early protein of *P. parasitica* (CBEL) that binds to cellulose and provides adhesion with cellulose-containing substrates, induces defense responses, i.e. serves as a non-specific elicitor for the infected plants. AvrPto of *P. syringae* is responsible for the parasite growth and development of the signs of disease (AvrB3 protein family in xantomonads). Both structural and enzymatic proteins play the role of elicitors in viruses.

The low-molecular-weight elicitors are the cell wall components or signal molecules involved in life cycle regulation. They are also necessary for interaction with the environment.

In the process of coevolution, plants have developed receptor molecules tuned to recognize the parasite compounds on their surface or excreted in the environment. Eukaryotic fungi have quite a wide choice of such compounds, and the future studies will show the diversity of their specific elicitors in terms of the chemical nature and biological functions. Prokaryotes have a limited choice, and a greater part of the investigated elicitors are pathogenesis-related proteins. Viruses have an extremely narrow choice of elicitors: it was shown earlier that all TMV proteins are used as elicitors.

Since all elicitor molecules are not functionally indifferent for parasites, their modifications resulting in drop of recognition by the receptors, or their complete loss affects fitness of the virulent races. This can be epidemiologically important. The contribution of an individual to the genetic pool of the next generation is called *Fitness* (*W*). Fitness of a population for any interval of time equals the ratio of the magnitudes of the population over this time (ln $W = r$). If the population is represented by one clone all its members will have identical fitnesses. However, actual populations consist of the genotypes with different fitness. The difference between the fitness of the best genotype and the average fitness of the population is called *genetic load* (*L*). It can be segregation or mutation load. The former is determined by the advantage of the heterozygotes whose growth produces homozygotes with lower fitness. As most phytopathogens are haploid, and many of them are agamous, the segregation load makes a small contribution to the total fitness.

The mutation load is caused by continuous spontaneous mutation process resulting in the mutants with lower fitness. For instance, two alleles of the same locus occur in a population: normal *A* and mutant *a*. Fitness of the former is $W_A = 1$, and of the latter, $W_a = 1 - S$, where *S* is the selection coefficient (characteristic of the mutation load). If occurrence of the wild phenotype equals *p*, and that of the mutant is *q* ($p + q = 1$), the average fitness of the population will be $1 - Sq$. If *S* is greater than *q*, mutant phenotypes will not accumulate in the population. The total number of mutants (*q*) will be governed by the mutation rate $A \rightarrow a$ and the coefficient of selection against the mutant phenotype (*S*). Hence, $q = m/S$ (1+m). As *m* is a very small value, $1 + m \approx 1$, $q \approx m/S$, and $S = m/q$. Consequently, with the known occurrence of the mutant genotype and mutation rate, *S* can be calculated.

As growth of virulence is accompanied by the loss or structural change of the elicitor, it is accompanied by the drop of fitness in the parasitic or saprophytic phase of the parasite life cycle. In epidemiology, there is even such a concept as "cent of virulence". Table 7.7 shows diallelic relations between the parasite and host plant in the "gene-for-gene" system. If the same relations are represented not qualitatively but quantitatively (as population characteristics), they will look as follows (Table 7.7): *t* is an index of the cultivar resistance which can be calculated by the comparison of the parasite growth in a resistant and susceptible cultivar. According to numerous data, t approaches 1, and 1–*t* approaches 0. *S* is the cost of virulence, i.e. the tax paid by the parasite for the ability to grow in a resistant host genotype. Various methods of its calculation are available. Some *S* values are given below.

P. graminis avenae: 0.126 (against unnecessary virulence genes);

P. striiformis: 0.435 (against virulent race);

Cochliobolus heterostrophus: 0.302 (against race 0 on *T*-cytoplasm);

0.120 (against race *T* on normal cytoplasm);

C. Carbonum: 0.58 (against race 0);   0.17 (against race 3);

P. infestans: 0.622 (against virulence gene of race 1).

Table 7.7.   Parasite fitness in different host genotypes in the "gene-for-gene" system

Parasite fitness	Host plant genotype	
	RR (Rr)	Rr
W_A	1	1–t
W_a	1–S	1–S

It can be seen that the cost of virulence of different genes of the same parasite can differ.

If the sequences recognized by the receptor are in the site of the elicitor molecule which is not functionally necessary, the cost of virulence will not be very high, and the virulent mutants can promptly accumulate and attack the resistant cultivars. If the elicitor molecule variation dramatically affects its primary functions, virulent races may not survive at all. For instance, many natural and artificially produced TMV mutants are known which are virulent to *N. sylvestris* but there are practically no mutants virulent to *N. glutinosa*. Modification of the coat protein of this virus seems to affect its fitness much less than modification of the enzyme molecule necessary for RNA replication. There are many intermediate conditions between these extremes. For instance, in the races of *C. fulvum* virulent to the tomato cultivars with gene *Cf9*, gene *avr9* is deleted; hence, its product is not essential, at least in infected tomato plants. However, as gene *Cf9* has been inserted in many tomato cultivars that retain resistance for a long time (virulent races rarely occur), it could be assumed that the lack of this gene affects the fungus fitness (the gene may be necessary for interseasonal conservation of the fungus outside of the plant in nitrogen-lean environment). Besides, the normal supply of nitrogen is necessary not only for the growth processes and normal morphogenesis but also for manifestation of the fungal pathogenic properties. For instance, it was shown that preliminary cultivation of *C. truncatum* on the nutrient media with carbon to nitrogen ratio varying from 80:1 to 10:1 influenced the length and width of the conidia, the number of their nuclei, the rate of germination and formation of appressoria on the leaves of *Sesbania exaltata*, and pathogenicity of the fungus.

It was established by epidemiologists that the cost of virulence can be very high if one parasite genotype combines several particular virulence genes (recessive mutations of avirulence genes). Table 7.8 shows that each of the two virulence genes of *P. graminis tritici* fairly frequently occurs in the Canadian stem rust populations, but their co-occurrence in genotypes is extremely rare. The search of such virulence gene combinations is interesting in terms of mechanisms of plant–microbe interactions and is extremely useful for development of the cultivars with durable resistance.

Another situation was described for gene *avrB31* of *X. vesicatoria*. This gene product in race 2 induces resistance of the pepper cultivars with gene *Bs1*, and virulence to these cultivars is caused by insertion of *IS476* in *avr-locus*. It was

Table 7.8.   Effect of some combinations of virulence genes of *P. graminis tritici* on parasitic fitness (Van der Plank, 1982)

Virulence genes	Years					
	1970	1971	1972	1973	1974	1975
P6	9.3	31.2	16.8	11.3	11.1	14.2
p9d	86.2	68.8	82.8	88.7	92.3	93.6
p6 + p9d (design)	8.0	21.2	13.9	10.0	10.2	13.3
p6 + p9d (actual)	0	0	0	0	0	0.9

shown that the strains with insertion have a better fitness both in parasitizing in susceptible plants (without gene *Bs1*) and in survival in soil. Thus, the cost of virulence can sometimes have a negative effect.

Selected Literature

Ahro R, Dreier J, Eichenlaub R. Genetics of phytopathogenic bacteria. Progress in Botany 1997; 58:410–427.

Boller T. Chemoreception of microbial signals in plant cells. Ann Rev Plant Physiol Plant Mol Biol 1995; 46:189–214.

Bonas U. Hrp-genes of phytopathogenic bacteria. In: Dangl J, eds. Top Microbiol Immunol. Springel Press, 1994; 192:79–98.

Chen CY, Heath MC. Physiol Molec Plant Pathol 1990; 37:169–177.

Cornelis GR, VanGijsegem F. Assembly and function of type III secretory system. Ann Rev Microbiol 2000; 54:735–774.

DeWit PJGM. Molecular characterization of gene-for-gene systems in plant-fungus interactions and the application of avirulence genes in control of plant pathogens. Ann Rev Phytopathol 1992; 30:391–418.

Gabriel DW, Roelfs BG. Working models of specific recognition in plant-microbe interactions. Ann Rev Phytopathol 1990; 28:365–391.

Hacker J, Kaper JB. Ann Rev Microbiol 2000; 54:643.

Hahn MG. Microbial elicitors and their receptors in plants. Ann Rev Phytopathol 1996; 34:387-412.

Hueck CJ. Type III protein secretion systems in bacterial pathogens of animals and plants. Microbiol Mol Biol Rev 1998; 62:379–433.

Keen MT, Dawson WO. Pathogen avirulence genes and elicitors of plant defense. In: Boller T, Meins F eds. Genes Involved in Plant Defense. Springer Verlag, 1992:85–114.

Keen NT, Stackawicz BJ. In: Struct, Funct Biosynth Plant Cell Walls. Proc Annu Symp Bot Dep Plant Pathol. Univ of California. Riverside. 1984; 344–358.

Leach JE, Vera Cruz CM, Dai J, et al. Pathogen fitness penalty as a predictor of durability of disease resistance genes. Ann Rev Phytopathol 2001; 39:187–224.

Leach JE, White FF. Bacterial avirulence genes. Ann Rev Phytopathol 1996; 34:153–179.

Lindgreen PB. The role of hrp-genes during plant-bacterial interactions. Ann Rev Phytopathol 1997; 35:129–152.

Ush DS. Calcium regulation in plant cells and its role in signalling. Ann Rev Plant Physiol Plant Mol Biol 1995; 46:95–122.

Chapter 8

Vertical pathosystem: resistance genes and their products. Signal transduction

Yu. T. Dyakov, O. L. Ozeretskovskaya

Resistance genes and susceptibility genes

For growth in a plant, the parasite needs to: (1) overcome protective covers and other living cell defenses to receive "an access to the pantry" and (2) have complete nutrition for growth in a plant and generation of adequate progeny. In this connection, plant genomes contain the genes the products of which promote parasite growth and feeding (we will call them susceptibility genes), and the genes the products of which interfere with these processes (resistance genes).

Susceptibility genes

The products of susceptibility genes can be involved in the following processes:

1. To control synthesis of the receptors specifically binding to the parasite toxins, which results in destruction of the cell immune properties (see details in Chapter 11).
2. To repress defense response processes in the infected cells. Such mechanism exists in the barley *Mlo* gene product – a membrane protein with seven transmembrane domains. MLO protein binds to the Ca^{2+} sensor calmodulin and prevents accumulation of the reactive oxygen species (ROS) and hypersensitive response in the infected cell. Various *mlo* mutants possess race-nonspecific resistance to the mildew pathogen *Blumeria (Erysiphe) graminis f.* sp. *hordei*. The resistance is manifested as deposit of polysaccharide (callose) barrier in haustorium penetration areas and as death of the infected cells. Mutant plants show growth delay, early leaf senescence, grain yield drop, as well as systemic necrosis at lower temperatures – which shows an important role of *Mlo* gene for general fitness of barley. The primitive varieties (field races) of barley in its native Ethiopia contain 0.2–0.6% of mildew resistant plants, and in the areas of annual infestation their occurrence exceeds 20%. The nonspecific resistance

gene from the Ethiopian barleys, called *mlo-11*, unlike the genes modified using artificial mutagenesis, provides partial but not complete resistance and has less adverse effect on viability. It has been transferred to many modern resistant barley cultivars. The *MLO*-site in *mlo-11* mutants consists of a number of tandem repeats of the truncated *MLO* sequence flanked by the upstream sequence. The repeats are separated by the GT dinucleotide. The progeny of self-pollination of the plants homozygous in *mlo-11* contain completely susceptible plants, where the structure of *MLO* gene has recovered due to repeat excision (meiotic instability).

3. To enrich the parasite feeding conditions. Gene *Arabidopsis PMR6* controls pectate lyase synthesis. *pmr6* mutants are resistant to the mildew fungi *Erysiphe orontii* and *E. cichracearum* but not to bacterial or oomycete infections. Cell walls of the mutant plants are rich in pectin that inhibits the haustorial functions, in particular, development of feeding channels through the plasma membrane. Thus, the resistance is recessive and is not connected with induction of defense responses in the infected cells. Therefore, gene *PMR6* can be viewed as the dominant susceptibility gene.

Resistance genes

Resistance genes can be divided into two functional groups:

1. Their products are the enzymes degrading the parasite attack factors. Such genes are not numerous and some examples will be discussed in Chapter 12.
2. Their products are the receptor molecules recognizing the parasite metabolites controlled by *avr* genes (elicitors). The activated Ribosomal proteins (R-proteins) transmit excitation to the genome using various signalling systems, which results in the expression of immune response genes and hypersensitive reaction (HR). That is, these are the genes functioning in the "gene-for-gene" system. They are found virtually in all plants and they will be discussed later in this chapter.

Investigation of resistance genes by the methods of classic genetics

Investigation of resistance genes using the methods of classic genetics has brought the following conclusions:

1. Resistance genes are sited on chromosomes in a nonrandom way. Analysis of the numerous publications on mapping of the genes of plant resistance to phytopathogens carried out by the Argentine geneticist E. Favre has shown that their distribution considerably differs from the Poisson distribution. The most common three types of resistance gene location are: single diallele (one locus with two alleles controlling resistance and susceptibility, with resistance usually being dominant); single multiallele (one locus with many co-dominant alleles controlling resistance to the different races of a pathogen), and

a linked siting, when many di- and multiallele loci are linked and form a block of phenotypically similar genes that determine resistance to one or several diseases.

Some examples:
Flax was found to contain about 30 co-dominant genes of resistance to rust races (pathogen: *Melampsora lini*) located in five loci: 13 allelic genes were mapped in locus *L*; seven closely linked genes, in a complex locus *M*; five, in locus *P*; three, in locus *N*; and one in locus *K*.

Genes *H, M, T* and *R*, controlling wheat resistance to smut, are located in the same linkage group and recombine with 15–37% frequency.

The short arm of the barley V chromosome contains a five-locus isophenic block that determines resistance to mildew; in one of loci – *Mla* – more than 30 co-dominant alleles have been mapped.

Clover resistance to races 1, 3, 7, 8, and 10 of the mildew pathogen is controlled by allelic or closely linked genes.

Corn resistance to the races of the common rust pathogen *Puccinia sorgi* is controlled by five loci. Two of them – *Rp1* and *Rp5* – are linked, and 14 allelic and closely linked genes have been mapped in locus *Rp1*.

The genes of tomato resistance to *Cladosporium fulvum* form several linkage groups. Genes *Cf1, Cf4*, and *Cf9* in the short arm of the first chromosome, as well as *Cf2* and *Cf5* in the short arm of chromosome VI are closely linked. The latter are likely to be allelic, as they do not recombine in the hybrid progeny.

The gene of rice resistance to the bacteria *Xanthomonas oryzae Xa21* belongs to a multigene family comprising at least eight members that are mapped in the same locus in chromosome XI.

Development of clusters of linked genes that are functionally and, probably, structurally similar, is caused by intragenic or intergenic exchanges of the DNA sites with direct (a b c d – a 'b' c 'd') or inverted (a b c d – d 'c' b 'a') repeating sequences (*ectopic recombination*), which results, in particular, in unequal crossing-over (see Figure 7.9).

2. Resistance genes can be multifunctional and provide resistance to a number of pathogens. For instance, rye genes of resistance to stem, brown and yellow rust pathogens are located in a very short arm of chromosome IR. Probably, it is the same gene.

3. Resistance genes have a high mutation rate. For instance, γ-irradiation was shown to result in 2.3% of mildew-resistant barley mutants (0.7% in control), 2.7% of rust-resistant flax mutants, and 0.1% of rice blast disease resistant mutants.

4. Apparently, resistance gene products perform at least two functions: receptors recognizing the pathogen elicitors and mediators transmitting signals from an excited receptor to the genome. Sometimes these functions can be separated using mutagenesis. An example is a series of corn and *Arabidopsis thaliana* mutants, with development of necroses and intracellular processes in uninfected tissues similar to

those at incompatible infections. Mutations in the recognition site can reduce reception specificity, and mutations in the activation site can induce constitutional activation of the immune response. For instance, in the above described corn locus *Rp1*, controlling resistance to rust, many allelic or closely linked genes have been mapped. Some alleles are instable and produce susceptible mutants due to nonuniform crossing-over. Some individual mutations in locus *Rp1* cause spontaneous necroses; mutant plants lose race specificity and respond to inoculation with any pathogen race by a hypersensitive response. The barley locus *mlo* provides nonspecific resistance to all races of the mildew pathogen *Blumeria graminis,* and at a low temperature it causes formation of necrotic spots without inoculation. Tomatoes, homozygous in recessive gene *ne,* form necrotic spots without inoculation, but only in the presence of the dominant gene of resistance to *C. fulvum Cf2* in the genome.

Molecular genetic studies

The molecular studies of Flor's resistance genes are a much greater challenge than studying parasite avirulence genes. Products of the resistance genes – hypothetic receptors – are unknown and cannot serve as molecular probes, like a protein, a specific elicitor of *C. fulvum*. The huge size of the higher plants' genome prohibits isolation of the resistance gene by such methods as total checkout of the clones in the gene library. As it is unknown whether the *R*-gene expression will enhance at inoculation, these genes cannot be studied using the methods that involve isolation of the phase-specific *m*RNA fractions. That is why it took scientists more than forty years to begin cloning *R*-genes and studying the structure of their products, and the first reports on this subject in mid-1990s stirred a great interest.

To isolate the resistance genes from different plants, the following methods and their combinations were used:

1. Use of the plant species with a small size of genome, such as flax and, especially, *Arabidopsis* (the size of *Arabidopsis* genome is 1×10^8 bp, 7×10^8 bp in flax, and 1.7×10^{10} bp in wheat).
2. Marking of the resistance genes with overlapping segments of DNA cut by restriction enzymes (restrictases) (RFLP-loci) in distributed progeny F2.
3. Marking of the resistance loci using transposon mutagenesis.
4. Use of a radioactive-labeled specific elicitor as a probe to isolate the receptor, and reconstruction of the encoding gene by the structure of its product.

The isolated products of the investigated *R*-genes (R-proteins) incorporate several structures (Figure 8.1):

Large C-terminal site. It contains a large number of repeating sequences with a high content of the amino acid leucine (*l*eucine *r*ich *r*epeats, LRR). Each repeat consists of 23–24 amino acids, and the number of such repeats varies from 14 ones to 37 in various R-proteins. This structure, repeated in

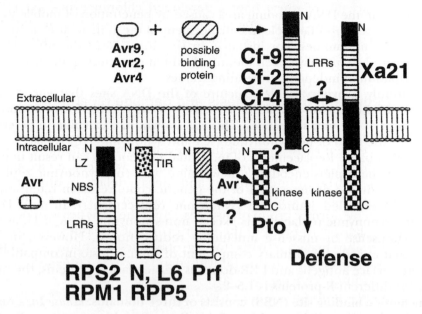

Figure 8.1. The structures of *R*-gene products and a model coupling the recognition of microbial Avr-dependent ligand and activation of plant defense. *Pto* can directly bind to AvrPto. The other R-proteins probably bind the corresponding *Avr* gene products, either directly, or in association with a binding protein. Both Pto and Xa21 have a protein kinase domain. RPM1, RPS2, N, L6, and Cf proteins also activated defense through protein kinases. Prf is required to Pto-mediated resistance. Abbreviation: LZ, putative leucine zipper region; TIR, region with homology to the cytoplasmic domain of the Drosophila Toll and human interleukin-1 receptors; LRR, leucine rich repeat motif; N, amino terminus; C, carboxyl terminus (Hammond-Kozak KE, Jones JDG, 1997(Figure4)).

many eukaryote proteins, implements protein interactions, i.e. serves as a receptor binding to a ligand, a pathogen protein elicitor. The LRR structure of the porcine inhibitor of ribonuclease PRI is a part of β-chain with a consensus sequence xxLxx (L is leucine, x is any amino acid), where the leucine residue is included in the hydrophobic core, and the other amino acids make a hydrophilic surface binding to the ligand. The LRR structure is very promising for genetic reorganizations due to point mutations, deletions, or inversions caused, in particular, by the above described events of ectopic recombination, which results in variation of the receptor properties of the molecule (see Figure 7.9). It has been experimentally shown that the mutations that modify the response to inoculation with the virulent pathogen races are mapped in this structure.

The LRR regions of the tomato proteins Cf9, Cf4, Cf2, and Cf5 have the sites that attach carbohydrates through glycosyl bonds, hence the R-proteins inducing the tomato resistance to *C. fulvum* are glycoproteins, and their carbohydrate residues can also provide binding to the ligands. Thus, the post-translational modifications also enhance variability of the receptor molecules.

Ruptures in the DNA encoding area caused by penetration of mobile items can also play an important role in enhancement of the genetic diversity of the resistance gene families. For instance, in rice locus *Xa21* the transposons *Truncator* and *Retrofit* have been identified, which can cause gene disruption and formation of independent reading frames.

Apparently, a hypervariable structure of the DNA sites that encode the domains of the leucine-containing repeats, is most suitable for evolutionary generation of the new sequences. This was shown, in particular, in determination of the ratio of nonsynonymic (K_a) and synonymic (K_s) replacements of the DNA bases. Replacements of individual nucleotides can result in generation of the triplets encoding a new amino acid (nonsynonymic replacements), or, due to degeneration of the genetic code, result in retention of the earlier encoded amino acid (synonymic replacements). In most DNA sites, the synonymic replacements exceed non-synonymic (K_a:K_s < 1), as the latter (nonsense or missense mutations) reduce fitness. However, in the genes that encode the primary component of human histo-incompatibility, parasite surface antigens and LRR-domains of the plant R-proteins, the ratio K_a:K_s for different R-proteins is 1.5–2.

Nucleotide binding site (NBS) consists of three domains: *kinase 1a*, a phosphate binding loop (P-loop); *kinase 2*, binding a metal ion necessary for the phosphorus transmission reaction; and *kinase 3a* where arginine provides interaction with the ATP purine base. NBS is a signal site that binds ATP and GTP, therefore, it can activate kinases or signalling G-proteins. NBS is homologous to genes *Ced* of the nematode *Caenorhabditis elegans* and human genes *Apaf-1*, which regulate the activity of caspase (a proteolytic enzyme). Caspase is one of the factors of apoptosis (programmed cell death) that will be described in more detail in Chapter 9.

Leucine zipper region (LZ). It promotes formation of helical structures that provide dimerization or specific interaction with the other proteins. The R-proteins residing in an intact cell as monomers in inoculation can form, with involvement of LZ region, homo-oligomers or hetero-oligomers interacting with the elicitor, or vice versa, inoculation stirs dissociation of the earlier available oligomers. Heterodimerization of R-protein with other proteins is also possible.

Region of homology with the cytoplasmic domain of the drosophila Toll–protein and mammalian interleukin I receptor, called TIR (Toll/Interleukin I Resistance). The Toll receptor protein controls dorsoventral polarization of the drosophila embryos, and it is also involved in nuclear localization of *Dif* factor (*dorsal related immunity factor*) that activates protective properties of the adipose body. In particular, the Toll protein regulates synthesis of the fruit fly antifungal peptide drosomycin.

Interleukin 1 is one of the essential *cytokines* (intercellular signal transmitters) that activates mammalian immune cells due to interaction with cell surface receptors. The binding of both receptor proteins is accompanied by the activation of serine/threonine protein kinases. The plant R-protein TIR region has similar functions.

Table 8.1. Plant R-protein grouping based on their structure

Protein regions	Group 1	Group 2	Group 3	Group 4	Group 5
LRR	+		+	+	+
NBS				+	+
LZ				+	
TIR					+
PK		+	+		
Examples of R-proteins	*Cf2, Cf4, Cf5, Cf9* (tomato – *C. fulvum*)	*PTO* (tomato – *P. syringae* pv. *tomato*)	*Xa21* (rice – *X. oryzae* pv. *oryzae*)	*RPS2* (*Arabidopsis* – *P. syringae* pv. *tomato*); *RPM1* (*Arabidopsis* – *P. savastanoi* var. *maculicola*); *I2* (tomato – *F. oxysporum* f.sp. *lycopersici*) *Mi-1* (tomato – gall nematode)	*N (N. glutinosa* – TMV); *L6, M* (flax – *M. lini*) *RPPS* (*Arabidopsis* – *Peronospora parasitica*)

Serine/threonine protein kinase (PK) found in some R-proteins possesses distinct signalling properties. It is a factor of transcription activation and, simultaneously, an activator of other signalling pathways.

Depending on the combination of the above listed regions in the molecule, R-proteins can be divided into several groups (Table 8.1).

Group 1 R-proteins possess extracellular LRR glycoprotein anchored in the membrane with the C-end of the molecule. It follows from the above that these proteins function as elicitor-binding receptors (membrane localization of LRR), but they cannot perform the signal transduction function. These proteins are probably associated with the membrane protein kinases, or their short cytoplasmic N-end domain can itself interact with protein kinases.

The gene *Pto* product, on the contrary, cannot perform the receptor functions, as this product is an intracellular serine/threonine protein kinase, i.e. the postulated double-domain structure of R-proteins in tomato seems to be divided between two factors of resistance to different parasites. However, transformation of the yeast genome by the tomato *Pto* gene and pseudomonad *avrPto* gene showed that their products can directly interact after autophosphorylation of Pto protein. The interaction occurs through the AvrPto protein region called GINE motif which is localized in a flexible loop, suited for interaction with other proteins.

To study the effect of Pto superproduction, the encoding gene was transformed into the tomato genome. This manipulation resulted in necrotized palisade mesophyll cells. Simultaneously, accumulation of salicylic acid and

PR-proteins, callose deposition, and cell wall lignification was observed, i.e. metabolic effects, characteristic of hypersensitive response. The plants became resistant to the virulent races of *Pseudomonas syringae* pv. *tomato, Xanthomonas vesicatoria* and *C. fulvum*, i.e. they have acquired nonspecific resistance. Thus, overproduction of intracellular protein kinase results in the transformation of specific to nonspecific resistance.

Gene *Pto* is linked with gene *Prf*, whose product has regions LRR, NBS and LZ (as Group 4 proteins), as well as a long region with two direct repeats, not homologous to the known proteins. Proteins Pto and Prf appear to work in an infected cell in a coordinated way.

Rice R-protein Xa21 has all the prerequisites to comply with the above model (Figure 8.1). The receptor LRR region is put forward, and the cell inside contains a region that possesses a serine/threonine protein kinase structure and is capable of intracellular signal transduction.

R-proteins from Groups 4 and 5 possess receptor regions and signal transduction sites, but the intracellular localization of these proteins does not fit the earlier suggested hypotheses on the receptor membrane position. For the viruses whose proteins are synthesized inside the host cells the intercellular localization of R-proteins (in particular, gene N product) is natural. Bacterial elicitors can penetrate the plant cell using the excretory system type III (Figure 7.7). The interaction mechanisms of most fungal elicitors with intracellular receptors are not clear. In rare cases, direct interaction of a fungal elicitor with R-protein was reported. For instance, the *Avr Pi-ta* product (neutral zinc-metal protease) *Magnoporthe grisea* interacts with the LRR domain of the rice protein Pi-ta in a yeast two-hybrid system, and the mutant allele protein that failed to provide the plant resistance to infection did not interact with the elicitor. As Pi-ta protein is localized inside the cell, the corresponding parasite Avr protein will be transported to the cells from the interstitial space that accommodates the producer. In most of the other investigated systems no specific interaction has been found. For instance, the elicitor Avr proteins of *C. fulvum* bind to the cells and protoplasts of both resistant and susceptible tomato lines. Consequently, the tomato possesses a nonspecific receptor binding the pathogen extracellular proteins without inducing a hypersensitive response. Some believe that cell signal activation occurs after the interaction of the elicitor and heterodimer that consists of R-protein and a nonspecific receptor, or, more likely, of R-protein and protein kinase (there are several types of kinases connected to the plasma membrane).

Also syringolides synthesized under the control of *avrD* gene of *Pseudomonas syringae* pv. *tomato* induce a hypersensitive response only in the soybean carrying *Prg4* gene, but they bind to the 34 kD protein which occurs in both resistant and susceptible plants. Many (but not all) barley mildew resistance genes, contained in the locus *Mla*, need two unlinked genes, *Rar1* and *Rar2*, to elicit specific resistance.

The assumption that plant–parasite specific interactions are more complex than the mere gene product interaction postulated by the "gene-for-gene" rule has been supported by the experiments on altering the virulence of the *Puccinia graminis tritici* strains treated with X-rays, in near-isogenic wheat lines. The result was the mutants that changed the response to inoculation from 0–3 to 4 simultaneously in the lines carrying different Sr-genes (2, 3, 5, 7, 8, or even 9 different genes). Consequently, in addition to the genes of the "gene-for-gene" system, wheat carries the genes whose products are necessary for expression of various avr-genes of the parasite and Sr-genes of the plant.

To explain these facts, Van der Biezen and Jones suggested the "guard" model. This model predicts that elicitors normally function as virulence factors (see Chapter 6). Pathogenicity targets affected by the elicitor could be proteins involved in defense or in host metabolism. R-protein binds to this complex and turns the cell metabolism towards defense responses. It was shown that overproduction of Pto-protein improves nonspecific plant resistance to various pathogens. Binding of Pto to AvrPto (shown in the yeast-two-hybrid system) degrades the defense potential of the infected cell, i.e. AvrPto is an intrinsic factor of bacterial pathogenicity. However, the product of Pto-linked *Prf* gene (LRR-NBS-LZ) recognizes the Pto-AvrPto complex and activates the pathways leading to defense responses.

Another example, Avr1 protein of the bacterium *Pseudomonas savastonoi* pv. *maculicola* does not bind to the R-protein *RPM1* of *A. thaliana* but binds to protein RIN4 which is involved in the normal plant life (it is required for shoot growth, meristem functioning, and flowering). After binding to the elicitor, this pathogenicity target (PT) protein is phosphorylated and acquires the ability of tri-partite interaction (with R-protein that guards PT protein against damage by pathogens).

Figure 8.2 shows the interaction pattern of the Avr products and R-genes based on the guard hypothesis.

The above material permits a number of important conclusions:

1. The hypothetical structure of R-proteins suggested from studying the phenomenology of plant–parasite relationships (Albersheim model, etc.) was found to be a particular case differing from many investigated R-proteins.

2. Most investigated R-proteins are madeup of a combination of a number of similar blocks; their structure does not reveal the taxonomic relations either between plants or between their parasites. Homologous R-proteins were reported in the plants that belong to different orders; they provide specific resistance to viruses, bacteria, fungi, nematodes, and insects. This shows that, first, they emerged a long time ago, before the divergence of plants into the modern taxons, and, second, they initially performed not protective but other intracellular functions. For instance, the extracellular LRR region of the rice protein Xa-21 was shown to be similar to the products of the *erecta* and *clavata* genes of Arabidopsis that govern the shape and size of the flowers. The Pto protein of tomato is homologous to the cytoplasmic

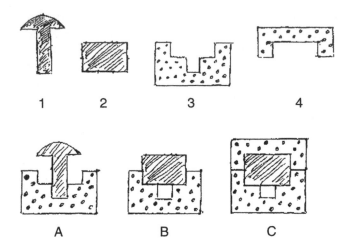

Figure 8.2. Scheme of the guard hypothesis on plant immunity.
1 – Parasite nonspecific elicitor. 2 – Parasite effector (suppressor). 3 – Receptor of nonspecific elicitor (guard protein). 4 – R-protein. A – Reception of nonspecific elicitors induced the cascade of signal transduction and expression of genes of immune response (incompatibility). B – Pathogens effector suppress of immune response by mask the recognition sites to elicitors or to first kinase of signal transduction (compatibility). C – R-protein is recognized by the pathogen effector or its conjunction with receptor as a specific elicitor and recovered the signal transduction cascade (incompatibility).

domain of the gametophyte incompatibility gene product of cabbage SRK (incidentally, gametophyte self-incompatibility, as well as resistance to pathogens, is phenotypically manifested as a necrotic response to certain gene products). Expression of another gene from the S-gene family of self-incompatibility of cabbage, SFR2, which also encodes synthesis of the receptor protein kinase, sharply increases at wounding and infiltration of phytopathogenic or saprotroph bacteria, i.e. in addition to preventing self-pollination, it also performs the functions of defense against infection. However, though most resistance genes (and the proteins they encode) have similar structures (made of identical blocks), this does not yet prove their common origin. For example, investigation of the structure of two genes *Rpg1 Glycine max* and *RPM1* of *A. thaliana* which have a similar structure (NBS-LRR) and recognize the same protein *Pseudomonas syringae*, showed homology only in the conservative sites, and the absence of orthology. The structure of these genes is evidence rather in favor of the convergent evolution than common origin.

3. The data has been confirmed that the same gene can control resistance to different pathogens. For instance, the tomato gene *Mi* controls resistance to the nematode and the aphid, and the gene *PRR8/HRT* alleles of *A. thaliana* provide resistance to the oomycete *Peronospora parasitica* and to the virus.

4. At the molecular level, the data on the cluster arrangement of the resistance genes has been confirmed. Each investigated gene (*Cf, M, N, Pto, Xa-21*, etc.) constitutes a complex locus, which encodes several structurally similar or identical proteins. Such structure has developed due to the above described (Figure 7.9) events of ectopic recombination (unequal crossing-over, etc.), caused by the direct or inverted repeats inside the genes or at their ends. For instance, the loci *Cf4* and *Cf9* of tomato consist of two tandemly repeated genes; however, in addition to the actively functioning sequences (*Cf4, Cf9*) their homologues – called *Hcr* (*h*omologues *C*ladosporium *r*esistance) – also occur which do not induce a hypersensitive response. Experimenting in transposon mutagenesis showed that inactivation of some *Hcr*-genes causes loss of resistance to individual strains of *C. fulvum*, which, consequently, produce *Hcr*-product-specific proteins in the plant. Recombination in the locus of corn resistance to rust *Rp1* is accompanied by development of the new resistant phenotypes. The high degree of homology between the flax loci *H* and *L* which control resistance to rust is evidence of their evolution from the same precursor. However, since these loci are located in different chromosomes, their development can be induced by intergenic recombinations.

The gene family formation can be also caused by duplications with subsequent development of structural and even functional differences. Apparently, this was the way of development of the *Pto*-gene family in tomato, including the *FEN*-genes which are responsible for susceptibility to the insecticide phenthionin, and not resistant to bacteria. Such a structure can promptly respond to emergence of new races and species of parasites, similar to multiple genes of immunoglobulins in mammals.

The total number of the loci controlling R-proteins is very large. In a small genome of *Arabidopsis* more than 150 R-loci distributed over all the chromosomes have been found. About 60% of them encode such structures as TIR-NB-LRR, and 40% encode LZ-NB-LRR. Other plants' genomes possess many more R-loci.

Signal transduction

Signalling systems

Cells of a multicellular organism can maintain their homeostasis due to the property of selective recognition of extracellular substances. These can be elicitors or hormones, cell metabolism mediators, proteins and peptides, low-molecular-weight metabolites, as well as environmental physical factors (light, high and low temperature, etc). The process of recognition is carried out using the so-called "signalling systems", which determine the cell response to various chemical and physical exposures. Pathogenic microorganisms or their elicitors were found to induce a cascade of defense responses in the plant cell long before the full manifestation of resistance or susceptibility. Moreover, signalling systems determine the early course of plant responses resulting either in compatible or incompatible interaction with infection. The purpose of these systems is transduction and multiplication of the signal generated by the pathogen or its elicitor.

The signalling systems operating within a cell begin functioning starting from the contact of the pathogen or elicitor with the receptor, which is usually located on the cytoplasmic membrane, and finalized by the plant cell defense response. Most nonspecific elicitors are connected with the external receptor site localized on the plasmalemma, which causes autophosphorylation and change in the receptor conformation. The phosphoric acid residue is next transferred to the inner area of the receptor and activates the receptor-associated enzyme.

The receptor, regardless of the nature of contacting effector, has a common arrangement: an area outside the cell, an intramembrane area, and an area submersed in cytoplasm. The outer and inner receptor areas are variable, and its middle part is constant. The outer N-end of the receptor is elicitor-specific, while the inner C-end is specific to the enzyme associated with the receptor. The latter determines the signalling system to interact with.

As shown in the previous chapter, many R-proteins, unlike the reception of nonspecific elicitors, are not connected with the membrane, therefore, the specific elicitors can contact them only using some mechanisms, for example, hrp-proteins of bacteria. The elicitor reception appears to need preliminary dimerization or heteromerization of the R-protein LRR region.

There are dozens of various protein kinases and phosphoprotein phosphatases in the signal transmission pathway that regulate the degree of protein phosphorylation, and thus regulate their activity. Selective phosphorylation of free side OH-groups of serine, threonine or tyrosine residues in proteins is a common event in organisms. Phosphorylation is carried out by the protein kinases using the ATP as phosphate donors. Depending on the type of protein, phosphorylation either enhances or lowers its functional activity. Activity of protein kinases, in turn, is regulated using multipurpose intracellular messengers. Protein phosphorylation is reversible, as cytosol contains numerous phosphoprotein phosphatases that digest phosphate from phosphorylated enzyme (protein), which reverts its activity to the initial level.

A special role in the signalling systems is played by the regulation, with phosphorylation-dephosphorylation, of the nuclear factors of transcription regulation that interact with the gene promoter sites and either enable or inhibit their expression. The transcription regulation factors are thought to have several phosphorylation sites in amino acid residues where orthophosphate can be added due to the effect of the "own" protein kinases, which determines the nature of activation of the transcription factors. A necessary condition for regulation of the transcription factor activation is their polymerization and linking with DNA in oligomers. In R-proteins, LZ site serves this purpose. Transfer from oligomers to monomers depends on the dephos-phorylation processes.

The major signalling systems known till date include:

- Cycloadenylate;
- MAP-kinase;
- Ca^{2+}-phosphoinositol;
- Lipooxygenase;
- NADPH-oxidase (superoxide synthetase);
- NO-synthetase.

In many signalling systems, a complex of G-proteins work as the interme-diate link between the cytoplasmic part of the receptor and the first activated enzyme.

G-proteins

Many of the secondary messengers known today work by means of the key molecules of a special class called G-proteins, which direct signalling from the receptors into the cell. These proteins involved in signal transduction are regulated by guanyl nucleotides, hence their name, G-proteins. Various receptors transmit commands of the elicitors, hormones and other effectors by way of stimulation of a G-protein. Since early 1980s, more than 100 recep-tors signalling through G-proteins have been found. A number of effectors dependent on these proteins have been identified.

The G-proteins involved in transmembrane signalling are located on the internal surface of the cytoplasmic membrane. Their molecules consist of three subunits – the polypeptide chains called (from the larger towards the smaller one) α, β, and γ. Among all the G-proteins isolated till date, subunit α is specific, while β and γ may differ.

Subunits α, β and γ at rest produce a complex where a subunit is bound to guanisidine diphosphate (GDP). After an elicitor or another primary effec-tor is added to the receptor, the receptor changes its conformation and binds to the G-protein. The binding results in GDP release by the subunit. The vacancy is immediately taken by guanisidine triphosphate (GTP) whose con-centration in the cell is very high. The binding to GTP causes activation of subunit α.

The GTP-bound subunit α separates from subunits β and γ and moves by diffusion on the internal surface of the cytoplasmic membrane till it binds to an effector, for example, adenylate cyclase. After a few seconds, subunit α hydrolyses GTP to GDP, inactivates, separates from the effector, and again binds to the loose subunits β and γ.

Thus, G-proteins serve as switches or timers which determine when and how long the signalling pathways are enabled or disabled. Signalling is enabled when the GTP-bound subunit α connects to the effector, and is disabled after GTP has been hydrolyzed to GDP. Thus, the duration of the "enabled" or "disabled" condition is determined by the rate of GTP hydrolysis.

G-proteins also possess an ability to enhance signals. For instance, one sub-unit α, binding to a adenylate cyclase molecule, can stimulate synthesis of many molecules of cyclic adenosine monophosphate before the "enabled" condition expires and GTP turns to GDP.

Thus, the cell membrane works as a switchboard for numerous diverse signals coming from outside. It assesses their relative intensity, adds and transfers the signals to the secondary messengers whose properties deter-mine the cell response to the changes in the environment.

Cycloadenylate signalling system

Many chemical mediators work by activation of adenylate cyclase, which leads to the formation of cyclic adenosine monophosphate (cAMP) from adenosine triphosphate (ATP) (Figure 8.3). The two ATP reactive terminal phosphates are easily hydrolyzed, and the energy released in hydrolysis of one or both high-energy phosphate groups can be used for energy-consuming biochemical processes.

cAMP regulates extracellular processes in many investigated organisms. An external stimulant may bring a ten-fold increase in the cAMP concentra-tion within a few seconds. Adenylate cyclase used to synthesize cAMP is con-nected to the cytoplasmic membrane. Phosphodiesterase digests cAMP to adenosine-5-monophosphate. The adenylate cyclase signalling system is enabled as a result of interaction between the elicitor and the receptor and activation of adenylate cyclase. A higher cAMP content leads to activation of cAMP-dependent protein kinases, and, consequently, to phosphorylation of various proteins, including the factors of transcription regulation and expression of the defense response genes. Functioning of the cycloadenylate signalling system was found to be related with induction of resistance in many plants.

Plants, as against animals, showed a low concentration of cAMP, which stirred doubts of its functional role. However, the use of phosphodiesterase

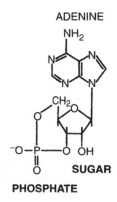

Figure 8.3. Cyclic adenosine monophosphate (cAMP).

inhibitors that transform cAMP to a noncyclic form showed that a low concentration of cAMP is caused by its intensive transformation into an inactive form.

MAP-kinase signalling system

Experiments with the inhibitors of protein kinases and phosphatases showed that protein phosphorylation and dephosphorylation is the most common cell response to various stresses, such as environment alkalization, generation of ROS, pathogens and specific and nonspecific elicitors, and hypersensitive response. The central role in these processes is played by mitogen activating protein kinases of serine-threonine type (MAPK). They were called so due to activation of mitosis in various organisms, from yeast to mammals. The MAP-kinase pathway of signal transduction is a cascade of three signalling modules. The last stage in the cascade, MAPK, is activated following the double phosphorylation of the tripeptide motif localized in the T-loop of the kinase catalytic domain. Phosphorylation is provided by another stage of the cascade, MAPK-kinase (MAPKK), which, in turn, is phosphorylated by MAPKK-kinase (MAPKKK). The phosphorylated MAPK triggers the transcription activation factor. The cell contains several such modules responsible for different functions. Their association using the scaffold proteins brings a very prompt and efficient cell response to the signals received.

Protein phosphorylation is also important in plant defense response signalling. Mutations in *Pto* gene (serine-threonine kinase of tomato) and *Xa21* gene (LRR-kinase of rice) lead to the loss of race-specific resistance of these plants, and mutation in gene *FLS2* (LRR-kinase of Arabidopsis) makes the plant insensitive to the nonspecific bacterial elicitor flagellin. The MAP-kinase activity is usually accompanied by activation of intracellular biochemical processes involving accumulation of salicylic acid, a universal low molecular weight messenger. On the other hand, mutations in FDR1 genes (MAPKK) of MAPK-4 make Arabidopsis resistant against virulent pathogens, which indicates a role of the MAP-cascade in the negative control of the defense responses. In both these cases, resistance was also accompanied by accumulation of salicylic acid. The MAP-kinase that induces synthesis of salicylic acid was named SIPK (*salicylic acid-induced protein kinase*). SIPK plays a role both in defense response gene activation and in HR induction.

Ca-phosphatase signalling system

In eukaryotes, the major phospholipids, such as phosphatidyl choline, phosphatidyl ethanolamine and phosphatidyl inositol can be hydrolyzed in 4 areas by the phospholipases possessing A_1, A_2, C & D activity (Figure 8.4).

For the subject in question, most interesting is phospholipase C which is specific for phosphoinositols. It was found for a number of cases that phospholipase C is activated at the earliest HR stages, while the other phospholipases are activated only several hours later. Phosphatidyl inositol, exposed to kinase, transforms into inositol diphosphate (Figure 8.4).

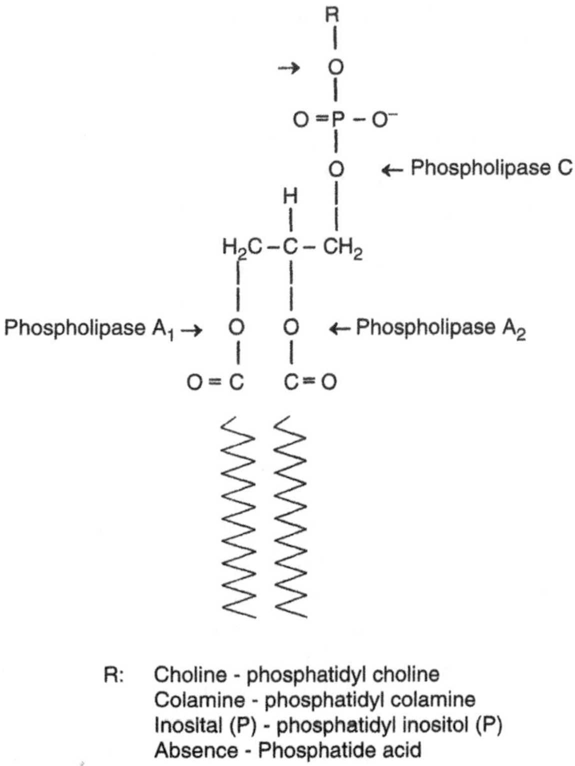

Figure 8.4.   Structure of the main phospholipids in plant cells and their hydrolysis by phospholipases.

The latter under influence of phospholipase C decomposes into solvable inositol triphosphate (IP_3) and membrane-bound diacyl glycerin (DAG). DAG activates protein kinase C and next, through a cascade of phosphory-lated proteins, leads to the expression of protective genes.

IP_3, in turn, opens the calcium channels of the cytoplasmic chains and tonoplast, where Ca^{2+} concentration is three orders higher than in cytoplasm. As a result, Ca^{2+} is released in cytosol where it activates various Ca^{2+}-dependent enzymes, including Ca^{2+}-dependent protein kinase. The channels are only open for a short time, and the calcium that enters the cytosol is either promptly removed or bound by the intracellular molecules. The calcium ion, similar to cAMP, is an important intracellular regulator and serves as a sec-ondary mediator for some extracellular signalling systems.

A second way to regulate Ca^{2+} is the Ca^{2+}-binding protein *calmodulin* which was found in all plant and animal cells investigated till date. This ubiquitous intracellular receptor of calcium ions is involved in most regulatory processes in eukaryotic cells. Plants were found to have many calmodulin isoforms, some of them activating and the other inhibiting Ca^{2+} calmodulin-dependent response. Concentration of the calmodulin isoforms can activate or inhibit the enzymes of the different signalling systems. Different calmodulin isoforms may possess different intensity of expression, which is determined by the signalling system enabled. For instance, the difference in the degree of activation of adenylatecyclase and phosphodiesterase depends on Ca^{2+}-calmodulin concentration. Calcium ions and calmodulin molecules influence the degree of phosphorylation of different proteins. Ca^{2+}-calmodulin can activate both protein kinases and phosphoprotein phosphatases. Ca^{2+}-dependent protein kinases can phosphorylate and, thus, activate the protein transcript regulation factors and defense response gene expression.

Lipooxygenase signalling system

A characteristic feature of plant cells (as against animal cell) is a high content of unsaturated C-18 fatty acids (linolic and linolenic), which are oxidized by lipooxygenase (LOG). In literature two names, *octadecanoids* and *oxylipines* are used to describe all oxygenated products of the lipooxygenase pathway of plant fatty acid oxidation. As not all products of lipooxygenase oxidation in plants are C_{18} compounds, lipooxygenase metabolism researchers more often use the term *oxylipines.*

Unsaturated fatty acids in membrane phosphotides are released by phospholipase A_2 (Figure 8.5), which is activated by many elicitors, phytohormones, as well as biogenic and abiogenic stress factors. The central role is played by the linolenic acid and, especially, linolic acid, which are LOG substrates that attach molecular oxygen in the positions near the double bonds. The result is formation of hyperperoxyl derivatives. The lipooxygenase signalling system – a major plant cell signalling system – was named after LOG. It was found that not only free but also unsaturated fatty acids within triglycerides can be LOG substrates, which speeds up the lipooxygenase signalling pathway.

LOG activation is induced by various effectors, with the most active stress response shown by 13-LOG attaching oxygen in position of carbon atom 13 (Figure 8.6). An important role in lipooxygenase metabolism is also played by 9-LOG and plant hydroperoxidelyases that catalyze conversion of 9-hydroperoxilinolenic acid to C_9 carbon compounds, *nonadienals.*

Physiologically active compounds generated using hydroperoxidelyase include the traumatin acid and traumatin. They can induce cell division and callus formation in the area of mechanical damage. Under the effect of hydroperoxidelyase, C6 aldehydes – *hexynals* – are formed from 13-hydroperoxilinolenic acid as soon as 15 s after mechanical damage of a leaf.

Allenoxide synthase (earlier called hydroperoxydehydrase) is one of the most important enzymes of fatty acid hydroperoxides. This enzyme degrades hydroperoxides, with formation of allene oxides. The latter undergoes hydrolysis and ring closure, with formation of 12-oxophytodienic acid. The recovery of the double bond in α and β oxidation cycles leads to the conversion of this compound to *jasmonic acid* (JA) and its *methyl ester* (Me-JA).

Biotic and abiotic stress factors cause strong activation of the lipooxygenase pathway, with an increase in the content of both free fatty acids and

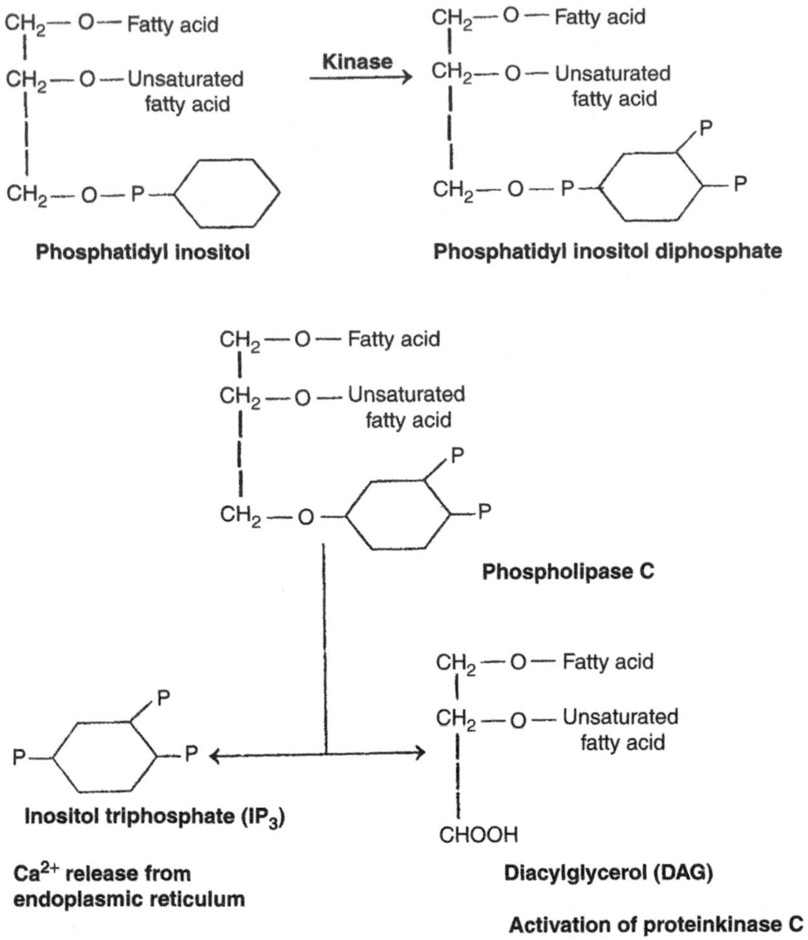

CH$_2$— O — Fatty acid

CH$_2$— O —Unsaturated
            fatty acid

CH$_2$— O — P —⬡
Phosphatidyl inositol

$\xrightarrow{\textbf{Kinase}}$

CH$_2$— O — Fatty acid

CH$_2$— O — Unsaturated
            fatty acid

CH$_2$— O — P —⬡— P
                     P
Phosphatidyl inositol diphosphate

CH$_2$ — O — Fatty acid

CH$_2$ — O — Unsaturated
            fatty acid

CH$_2$ — O —⬡— P
                   P
Phospholipase C

P—⬡— P
      P
Inositol triphosphate (IP$_3$)

**Ca^{2+} release from
endoplasmic reticulum**

CH$_2$ — O — Fatty acid

CH$_2$ — O — Unsaturated
            fatty acid

CHOOH
Diacylglycerol (DAG)

Activation of proteinkinase C

Figure 8.5.   Phosphatidyl inosital pathway of signal transduction.

their derivatives. The cause of increase in oxylipids is activation of enzymes by various stress factors, especially those catalyzing the initial stages of lipooxygenase metabolism: phospholipases and LOG. Various stress factors and related signal molecules can cause LOG gene expression (especially 13-LOG). These stress factors include many pathogens and elicitors, systemine, chitosan, oligogalacturonides, salicylic acid (SA), jasmonates, abscisic acid, mechanical damage, aqueous stress, high temperature, and ultraviolet light.

Many peroxy and oxyderivatives of linolic and linolenic acid possess high fungicidal activity. A primary chemical protection of plant wound surface is provided by hexynals which are active antimicrobial agents. Nonodienals also possess bactericidal and fungicidal properties.

A special role in plant resistance is played by JA. Interaction of the JA carboxyl group with amino acids results in synthesis of various conjugates.

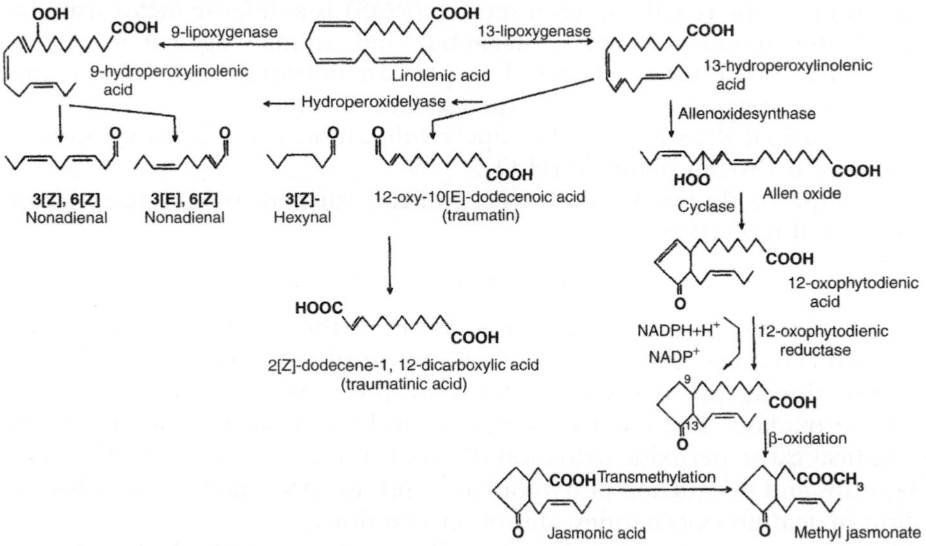

Figure 8.6. Formation of oxylipids from linolenic acid in plants.

Jasmonates are often viewed as stress phytohormones. Me-JA is a highly active signalling compound, in many cases mediating the activity of some stress factor. Me-JA pairs modulate enzyme activity in lipooxygenase pathway, activate 13-hydroperoxidelyase but not 9-hydroperoxidelyase, which leads to an increase in production of bactericidal and fungicidal hexynals. In other words, Me-JA not only can activate the lipooxygenase system, but also alter the direction of its functioning.

The free unsaturated fatty acids released by phospholipases A_2 can independently activate protein kinases or, using LOG, attach oxygen and turn into hydroperoxy derivatives. In the course of lyase, hydroperoxygenase and allene oxidecyclase reactions, these derivatives form the compounds toxic for microorganisms (hexynals and nonadienals) and nonvolatile compounds (oxy- and epoxy derivatives), as well as cyclic phytodienic and JA that can induce, with protein kinases involved, defense response gene expression.

Today, enough information is available to view the lipooxygenase pathway of membranous lipid conversion as an independent signalling system.

NADPH-oxidase (superoxide synthase) signalling system

Among the environmental factors affecting an organism, the so-called *oxidative burst* has a special effect. An elevated concentration of the ROS is recorded in human and animal cells exposed to such important factors as radiation, pesticides, many drugs, industrial waste, ischemia, carcinogenesis, cataract, diabetes, and, finally, upon bacterial ingress inside the immune cell phagosoma. Oxidative burst is the most pronounced component of the infection process.

Generation of the reactive oxygen species (ROS) is a defense factor used by cells of the immune system of warm-blooded animals against infection. Free radical processes are believed to play an important role among the mechanisms of aging and *apoptosis* (programmed cell death).

The principal ROS include the superoxide anion (O_2^-), hydroxyl radical (OH), and hydrogen peroxide (H_2O_2).

The superoxide anion is a free radical formed by single-electron reduction of molecular oxygen:

$$O_2 + e^- \rightarrow O_2^-$$

The superoxide is generated in interaction of the oxygen molecules with the NADPH oxidase system as well as with reduced flavines, quinones, thiols, and also during the reaction catalyzed by the flavine-containing enzyme xanthine oxidase. The superoxide radical and other ROSs generated from the radical cause peroxide oxidation of unsaturated fatty acids, oxidation of SH-groups and destruction of tryptophan residues, DNA damage, depolymerization of acid polysaccharides, and other reactions.

The strongest oxidant that destroys cell biopolymers is the hydroxyl radical. Its oxidative activity is higher than in superoxide anion, however, it is inferior in stability (the life of a hydroxyl radical is only 10^{-9} s, whereas for superoxide anion it is 10^{-3} s). A hydroxyl radical is produced in interaction of superoxide with hydrogen peroxide:

$$O_2^- + H_2O_2 + H^+ \rightarrow O_2 + H_2O + OH$$

The transfer of the second electron to the oxygen molecule leads to formation of the third ROS ingredient, hydrogen peroxide:

$$O_2 + 2e^- \rightarrow O_2^{2-}$$

$$O_2 + 2H^+ + 2e^- \rightarrow H_2O_2$$

However, in most cases H_2O_2 is produced in interaction of the superoxide radicals among themselves (dismutation reaction):

$$2H^+ + 2O^- \rightarrow H_2O_2 + O_2$$

Hydrogen peroxide causes oxidation of SH-groups in proteins – the peroxide oxidation of unsaturated fatty acids – however, these reactions are relatively slow, and the main hazard of hydrogen peroxide to the cells is its ability to generate hydroxyl radicals. ROS are capable of interconvertibility, therefore, they are often viewed as a complex.

The discovery of the ROS anti-infection function in animals motivated search for analogies also among plants. In a normally functioning plant cell there is a balance between oxygen activation and inactivation, therefore, the amount of its reactive species is at a safe level. However, structural and functional failures in plant tissues usually lead to oxygen activation. The normal balance between ROS formation and degradation may be broken at very diverse plant pathological conditions.

As oxygen activation is one of the earliest plant cell responses, ROS may be hypothesized to play an important role in pathogen growth suppression. For instance, hypersensitive response yields phenol release from vacuoles and their enzymatic oxidation. Since this process is accompanied by generation of the ROS in toxic concentrations, this oxygen may be the cause of death of the host cells and the pathogen they bear.

Generation of superoxide anions in a resistant host was found to be localized around the penetrated hyphae. Hence, a susceptible and non-resistant plant responds to inoculation by promotion of its oxygen activation where necessary and when necessary, and this activation is systemic.

In earlier days, ROS were only viewed as a source of highly toxic though short-lived compounds which inhibit the parasite growth, and today another, more essential, ROS function has been investigated: their involvement in the superoxide synthase signalling system. Oxidation of NADPH contained in cytoplasmic membrane, by molecular oxygen (Figure 8.7) produces the radical superoxide anion that turns into hydrogen peroxide through the reaction catalyzed by superoxide dismutase. Superoxide anion and hydrogen peroxide are the secondary messengers in the superoxide synthase system. Hydrogen peroxide causes activation of the transcript regulation factors, and, consequently, expression of the immune response genes. In *Arabidopsis*, tobacco, and some other plant species, the genes have been found that are homologous to the human gene *gp91 phox* which controls synthesis of the membrane protein of neutrophil NADPH oxidase. In potato this function is performed by two genes: StrbohA and Strboh2. Expression of StrbohA

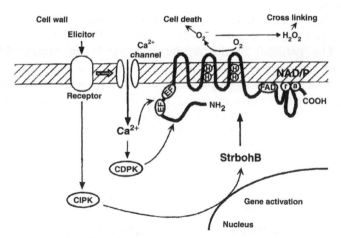

Figure 8.7.   Structure of membrane binding NADP-oxidase and regulation of its activity. (Is modified from Joshioka H. et al. Molec. Plant–Microbe Interactions. 2001. V.14. N6.)

CDPK – calcium-dependent protein kinase, CIPK – calcium-independent protein kinase. StrbohB gene expression is induced by *Phytophthora* nonspecific elicitor. StrbohB-product – protein with six transmembrane domains and the putative position of the two hemes, the FAD and NAD/P (r-ribose, a – adenine) binding sites and the EF hands.

is necessary for the early phase of the oxidative burst (immediately after the elicitor penetration) – H_2O_2 accumulation, and expression of Strboh2 is required for generation of a massive oxidative burst 6–9 h after the treatment.

An essential role in this signalling system is played by salicylic acid (SA), with its concentration multiply growing not only in infection sites but also in the tissues far from the infection site. As SA binds catalase that decomposes hydrogen peroxide, the amount of hydrogen peroxide grows even more. Intensity of apoptosis (cell death) of *Arabidopsis*, caused by the toxin of the fungus *Fusarium moniliforme* correlates with the SA level. No necroses from fumonisin have been observed in mutants with a low SA level. Data are also available on existence of the protein kinases directly activated by salicylates, which can explain the gene expression of the defense response against both SA and hydrogen peroxide.

NO-synthase signalling system

In animal cells, nitrogen oxide or nitroxide (NO) forms from L-arginine using NO synthase. The substrates of this reaction are arginine, NADPH, and molecular oxygen, and its products are NO, citrulline, and NADPH:

$$\text{L-arginine} + NADPH + O_2 \rightarrow \text{citrulline} + NO + NADPH^+$$

In plants, in addition to this route, nitrogen oxide is produced from nitrite reduction by the enzyme nitrate reductase in the presence of NADPH.

The radical NO is a major neuromediator, intermediator of the immune system, and regulator of the human cardiovascular system. In an animal body NO works as a transmitter of nerve impulses in the peripheral and central nervous system.

It was already mentioned that one of the most prompt plant defense responses to pathogen invasion is the oxidative burst which leads to ROS formation. The product of the oxidative burst is the superoxide radical which by itself cannot cause cell HR death as its life is too short. The superoxide next turns into hydrogen peroxide which is toxic only in relatively high concentrations, and, consequently, can hardly cause a hypersensitive response, either. Hence, it is clear that plant cells should possess, together with ROS, some concurrent factor intensifying their effect.

In mammalian macrophages ROS works together with nitrogen oxide that kills bacterial cells. It has been shown that in plants, alongside ROS, also works NO which is involved in growth processes and differentiation, opening of stomata, and causes transcription gene activation in response to wounding, abiotic stresses, or HR.

ROS and NO interaction is synergistic, as the amount of nitrogen oxide in a supersensitive response grows tenfold. The inhibitors of NO synthesis reduce the HR by the *Arabidopsis* cells inoculated with *Pseudomonas syringae*, and thus promote transmission of disease. Conversely, treatment of soybean cell suspension with Na-nitroprusside, which does not cause HR but is a NO donor, leads to the hypersensitive response.

NO synthase is activated only in resistant but not in susceptible TMV-inoculated tobacco leaves. The same leaves were also found to contain a large amount of NO. Injection of animal NO synthase, its substrate and cofactors in tobacco leaves lead to transcription of the protective gene *PR-1*.

Nitrogen oxide synergistically interacts with the endogenous factor dependent on salicylic acid (SA). SA accumulation in infected tissue inactivates catalase and thus strengthens the concentration of hydrogen peroxide which enhances the oxidative burst, and the latter, in turn, increases SA synthesis. Probably that is why SA is called "signalling amplifier".

NO reacts with oxygen superoxide and forms peroxynitryl (ONOO), an extremely toxic compound that inhibits pathogens. Next, hydroxyl radical forms from peroxynitryl, and it increases toxicity even more.

The increase in NO concentration leads to production of guanylate cyclase that catalyzes conversion of GTP into cGMP. cGMP was found to activate protein kinase either directly or through a longer signalling pathway including the cyclic adenosine diphosphate ribose (cADFR). The latter opens the Ca-channels of intracellular calcium depots, which leads to an increase in calcium concentration in cytosol, activation of Ca-dependent protein kinases, phosphorylation of the transcription regulation protein factor, and initiation of synthesis of protective proteins.

Signalling genetics

Signalling transcription of a plant defense response involves more than 1500 transcript factors. Mutation analysis allowed identification of some genes regulating signal transduction in a cell. These genes were called RDR (Required for Disease Resistance). In tomato, 4 unlinked genes have been identified: *Rcr1*, *Rcr2*, *Rcr3*, and *Rcr5* (Required for *C. fulvum* resistance). Rcr3 controls synthesis of cysteic protease which is secreted in the apoplast space. HR induction in tomato with resistance gene Cf2 requires proteins Rcr3 (cysteic protease) and avr2 (cysteic protease inhibitor). Therefore, Cf2 is a guard protein that recognizes interaction of avr2 and Rcr3 and induces HR. Rcr3 has no effect on resistance controlled by the other Cf-genes; mutations in the other genes cause partial reduction of resistance. The RDR functions in barley are possessed by genes Ror1 and Ror2 (necessary to elicit resistance by gene *mlo*) and Rar1 (encodes a small, highly conserved Zn^{2+} binding protein), and Rar2 (involved in generating oxidative burst in defense responses caused by gene *Mla*).

In *Arabidopsis*, gene *Eds1* (*e*nhanced *d*isease *s*usceptibility) is necessary to elicit the protective functions of the genes with TIR domain, its product, as well as a product of another gene, *PAD4* (*p*hytoalexin *d*eficient) – triacyl glycerol lipase – is involved in the SA signalling pathway (with salicylic acid). Functioning of *R*-genes with LZ-domain requires product of gene *Ndr1* (*N*o *d*isease *r*esistance) encodes a small, highly basic, putative integral membrane protein, as well as gene *PBS2*. The effect of the former occurs at the earliest stages of signal transduction and is caused by coupling of the processes of ROS generation and SA production. At the same time, the products of such

Arabidopsis genes as *PBS3, NPR1, SID1, SID2,* and *EDS5* are necessary for SA signalling irrespective of R-protein structure. Unlike gene *PAD4* that encodes the regulatory component of SA accumulation, Sid-proteins are directly involved in SA synthesis. Of interest are the functions of protein Npr1 (*non-expressor of PR-genes*) playing a dual role: it is involved both in the SA signalling pathway and in SA-independent pathway, which needs JA and ethylene signalling (more detail below). Overproduction of *Npr1* makes a plant resistant even to virulent pathogen races. This gene product has a bipartite nuclear localization sequence and a sequence typical of the membrane proteins ankirines that provide protein–protein interaction. In particular, *Npr1* interacts with the transcript factor family TAG. Nuclear localization of protein Npr was found using conjugation with the green fluorescent protein (GFP). An important role in SA-related signalling is played by two amino transferases encoded by genes *ALD1* (*AGD2-*like *d*efense response protein) and *AGD2* (*a*berrant *g*rowth and *d*eath). *ALD1* mutations lead to reduction in SA accumulation and make *Arabidopsis* susceptible to *Pseudomonas syringae*, while *AGD2* mutations, on the contrary, enhance SA accumulation and resistance. Both proteins have high homology regions but show amino transferase activity in opposite directions, which appears to explain the opposite effect of these proteins on the processes in infected plants. Anyway, at the early stages of induction of SA synthesis, amino acid imbalance is required.

Thus, different signalling regions of R-proteins induce different initial stages of transduction (with similar final stages), as shown in Figure 8.8.

More essential differences in signalling pathways resulted from the effect of various pathogenic factors. Despite common components, such as ROS,

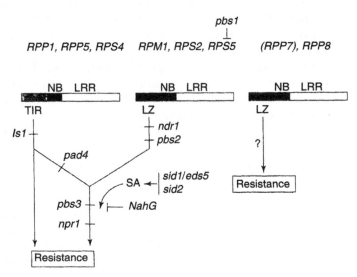

Figure 8.8.   Dissection of *R* gene-mediated signalling pathways in *Arabidopsis*. (Feys BJ, Parker JE, 2000 (Figure 3).)

Upper line – R-genes, next line – R-proteins, below – the genes, and their products that take part in signal transduction.

defense responses to inoculation with race-specific biotrophic parasites, on the one hand, and to nonspecific necrotrophs, insects, as well as mechanical damage, on the other hand, run through different signalling pathways. It was shown earlier that signalling caused by R-protein activation in *Arabidopsis* and other plants by race-specific elicitors is connected with SA synthesis. However, the relation between the type of parasitism and induction of a particular pathway (through SA or JA) is not always traceable. Besides, some pathogens induce both pathways.

In infection of *Arabidopsis* with the necrotroph parasite *Alternaria brassicola*, signalling is connected with the synthesis of JA and ethylene. Microarray analysis showed that the same transcript factor (ERF1) regulates expression of many genes required for synthesis of ethylene and JA. At the same time, the protein inhibits synthesis of both ethylene and JA. The mutant in gene *cev1* of cellulososynthetase produces ethylene and Me-JA constitutively, i.e. inhibition of cell wall synthesis induces expression of these phytohormones. Activation of defense responses needs expression of genes *PDF1.2 (PIN)* and *COI1 (JAR)*, dependent on the presence of JA, while mutations in genes *NPR* and *NahG* have no effect on the course of the defense response. Gene *COI1* of *Arabidopsis* encodes the protein that contains a region rich in leucine repeats, and an F-box which is selectively bound to repressor proteins and activates the genes necessary for jasmonate synthesis. Induction of the genes involved in jasmonate synthesis also requires the mitogen-active protein kinase 4 (MPK4).

SA and JA signalling pathways interfere with each other. It was shown in earlier studies that SA or its acetylated derivative aspirin inhibits JA expression in tomato caused by wounding. SA increases the level of the key JA biosynthesis enzyme allenoxyde synthase by promoting formation of the JA precursor oxyphytodienoic acid (OPDA) but inhibits the final step of JA biosynthesis (Figure 8.9). On the other hand, JA-dependent expression of gene *PDF1.2* suppresses accumulation of salicylic acid, which, in turn, represses *PDF1.2*. This repression did not show in the genotype carrying mutations in genes *cpr6* (constitutive expressor of *PR* genes) and *NPR*. Thus, during activation of SA-signalling either *NPR1* or signalling component downstream of *NPR1* can cause a down-regulation of the wound response pathway. The key factor regulating the course of defense responses in SA or JA pathways is the transcript factor *WRKY70* whose expression is activated by SA and repressed by JA. As this factor is an activator of SA-induced genes and repressor of JA-induced genes, it works downstream the synthesis of SA and JA signalling molecules (Figure 8.10). Data is available on interference between signal transduction pathways induced by SA and by ethylene. SA inhibits activity of 1-amine cyclopropane-1-carboxylic acid, a key enzyme in the ethylene synthesis.

Interactions of the signalling systems induced by nonspecific and specific elicitors

It was shown in the previous chapter that parasites possess a large number of molecules capable of causing immune responses in infected plants (elicitors). Some types of elicitors (nonspecific) are intrinsic in many pathogen species

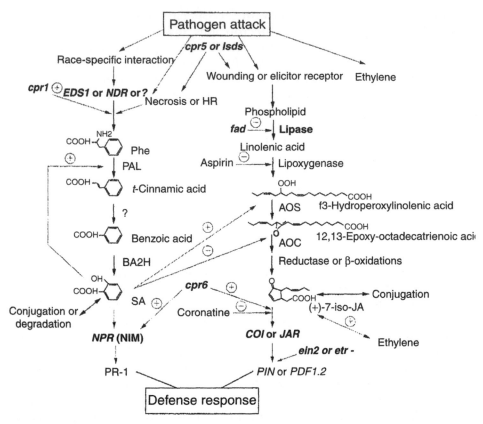

Figure 8.9.   Interactions between the defense signalling cascade in *Arabidopsis* (Maleck K, Dietrich RA, 1999 (Figure 2)). Elements in black bolds are mutants. *NPR1* is allelic to *NIM*.
SA – salicylic acid, PAL – phenylalanine ammonia lyase, BA2H – benzoic acid 2-hydroxylase,
AOS – allene oxide synthase, HR – hypersensitive response.

and races and cause defense responses in many plant species and cultivars, while other elicitors (specific) function only in certain combinations described by Flor's "gene-for-gene" rules. Elucidation of the relative role of the nonspecific and specific defense systems is important for understanding such basic phenomena as specific and nonspecific resistance, induction, and suppression. Consider a model of functioning of specific and nonspecific resistance, suggested by L. Navarro et al. (2004).

Pathogenic organisms possess a number of molecules recognized by their hosts as nonspecific elicitors (for example, components of cell walls). These molecules: 1) are highly conservative; 2) are present in various organisms; and 3) play an important role in life support. In animals and plants, receptors for such nonspecific elicitors are constituted by Toll-like molecules. In particular, the *Arabidopsis* receptor to the 22-amino acid motif of bacterial flagellin is a

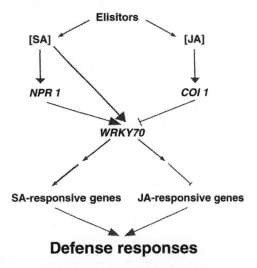

Defense responses

Figure 8.10.   Role of WRKY70 in the regulations of SA and JA signal transduction pathways (Li J, Brader G, Palva ET, 2004). High *WRKY70* levels activate expression of SAR-related genes while repressing JA-responsive gene expression. Low *WRKY70* levels favor JA-responses over SAR.

protein complex including a LRR-receptor of kinase FLS2. The relation with flagellin activates the mitogen-activated protein (MAP) and calcium-dependent (CD) kinases (Figure 8.11). In the figure this transduction pathway is designated by the letter (A). Phosphorylation results in destruction of the relation of the transcript factors (TFs) and kinases with the negative regulator (Neg. reg), as well as in *FLARE* (*Fla*gellin *R*apidly *E*licited) gene activation and synthesis of a set of proteins: 1) ubiquitins responsible for degradation of the transcript factor negative regulator (pathway F); 2) signalling proteins that control intensity of the immune response (pathway E); 3) various R-proteins whose expression in infected plants is higher than in uninfected (for example, expression of the TMV N resistance gene in the transgenic tobacco leaves inoculated with the virus increases 38-fold, and in the upper noninoculated leaves even as much as 165-fold) (pathway G); 4) various antibiotic products (PR-proteins, etc.) (pathway H). Phytopathogenic bacteria, through type III secretion system, bring to the cells the effector molecules (suppressors) that inhibit a signalling pathway involving reception of nonspecific elicitors (pathway B); however, specific R-proteins of a plant can recognize some pathogen effectors (suppressors) as avr-proteins, including the same signal transduction pathways and immune response expression (pathway C).

Conclusion

Figure 8.12 summarizes the above discussed plant cell signalling systems. It was shown earlier that the individual signalling systems can function both

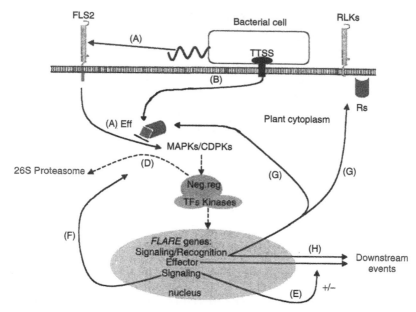

Figure 8.11. Model of interactions of bacterial and plant metabolites (Navarro L et al., 2004). Dashed arrows indicate hypothetical processes. Explanations in the text.

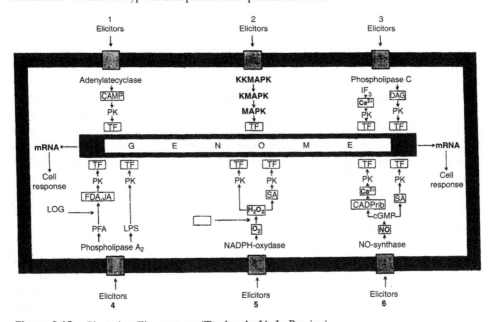

Figure 8.12. Plant signalling systems (Tarchevsky IA, In Russian).

cAMP – cyclic adenosine monophosphate; PK – protein kinase; KKMAPK – kinase of kinase of MAP kinase; KMAPK – kinase of MAP kinase; MAPK – mitogen-activated protein kinase; IP$_3$ – inositol triphosphate; DAG – diacyl glycerine; PFA – hydroperoxy forms of polyenic fatty acids; LOG – lipoxygenase; PDA – 12-oxyphytodienic acid; JA – jasmonic acid; LPS – lysophosphatides; SA – salicylic acid; SOD – superoxide dismutase; NO – nitrogen oxide; cGMP – cyclic guanine monophosphate; cADPrib – cyclic adenosine diphosphate ribose.

independently or interacting with each other. Signalling systems can be activated, or, on the contrary, inactivated by the secondary messengers of the other signalling systems. Integration is possible for most signalling systems, except the NO system, which not only fails to promote but, vice versa, suppresses the other systems. Besides, the longer is the host–parasite interaction, the greater is the number of the signalling systems. This can lead to simultaneous functioning of a large number of the cell signalling systems. According to I. A. Tarchevsky, this is exactly how the phenomenon of a nonspecific integral plant immune response to inoculation with various pathogens or treatment with elicitors can be explained.

Not always the transcript regulation factor can be activated only by one type of signalling system – several signalling systems need to be involved for this purpose. Possibly, each of these systems determines phosphorylation of its own amino acid residue of the transcript regulation factor, which totally results in expression of several protection genes. Thus, plant cells possess a coordinated signalling network where some signalling systems can activate, or, reversely, inactivate the other systems.

One of the features of the signalling systems is not only signal transfer to the genetic system but also its considerable amplification. A good illustration is the cycloadenylate system, with all its components integrated in such a way that a previous link of the signalling system does not turn into the next one but activates it by multiple amplification of the signal. Thus, interaction of one signalling molecule with a receptor can lead to a million of molecules governing the cell response.

All the signalling systems known till date can be classified into evolutionarily earlier and later ones. For instance, the cycloadenylate and Ca-phosphoinositol signalling systems are evolutionarily older than the others as they do not use molecular oxygen for their functioning, while the lipooxygenase, superoxide synthase and NO synthase signalling systems need oxygen.

It is very likely that efforts of the researchers studying plant signal transduction will lead to discovery of both the new signalling systems and their interactions.

Selected Literature

Asai T, Tena G, Plotnikova J, et al. MAP kinase signalling cascade in Arabidopsis innate immunity. Nature 2002; 415:977–983.

Backer B, Zambryski P, Staskawicz B, et al. Signaling in plant–microbe interactions. Science 1997; 276:726–733.

Bogdanove AJ. Protein–protein interactions in pathogen recognition by plants. Plant Molec Biol 2002; 50:981–989.

Dangl JL, Jones JDG. Plant pathogens and integrated defence responses to infestion. Nature 2001; 411:826–833.

Dixon RA, Harrison MJ, Lamb CJ. Early events in the activation of plant defense responses. Ann Rev Phytopathol 1994; 32:479–501.

Feys BJ, Parker JE. Interplay of signalling pathways in plant disease resistance. Trends Genet 2000; 16:449–455.

Fluhr R. Sentinels of disease. Plant resistance genes. Plant Physiol 2001; 127:1367–1374.

Greelman RA, Muller JE. Biosynthesis and action of jasmonates in plants. Ann Rev Plant Physiol Plant Mol Biol 1997; 48:355–382.

Hammond-Kozak KE, Jones JDG. Plant disease resistance genes. Ann Rev Plant Physiol Plant Molec Biol 1997; 48:575–607.

Hardie DS. Plant protein serine/threonine kinases: classification and function. Ann Rev Plant Physiol Plant Mol Biol 1999; 50:97–132.

Hubert SH, Webb GA, Smith SM, et al. Resistance gene complexes: evolution and utilization. Ann Rev Phytopathol 2001; 39:285–312.

Hutcheson SW. Current concepts of active defense in plants. Ann Rev Phytopathol 1998; 36:59–90.

Keen NT. The molecular biology of disease resistance. Plant Mol Biol 1992; 19:109–122.

Lamb CH. Plant disease resistance genes in signal preception and transduction. Cell 1994; 76:419–422.

Lamb CH, Dixon RA. The oxidative burst in plant disease resistance. Ann Rev Plant Physiol Plant Mol Biol 1997; 48:251–275.

Li J, Brader G, Palva ET. Plant Cell 2004; 16: 319–331.

Maleck K, Dietrich RA. Defense on multiple fronts: how do plants cope with diverse enemies? Trends Plant Sci 1999; 4:215–219.

Michelmore RW, Mayers BC. Cluster of resistance genes in plants evolve a divergent selection and birth-and-death process. Genome Res 1998; 8:1113–1130.

Moffat AS. Mapping the sequence of disease resistance. Science 1994; 265:1804–1805.

Navarro L, et al. Plant Physiology 2004; 135:1113–1128.

Pieterse CMJ, Ton J, VanLoon LC. Cross-talk between plant defense signalling pathways: boost or burden? AgBiotechNet 2001; 3:1–6.

Ryan CA, Farmer EE. Oligosaccharide signals in plants: a current assessment. Ann Rev Plant Physiol Plant Mol Biol 1991; 42:651–674.

Takker FLW, Joosten HAJ. Plant resistance genes: their structure, function and evolution. European J Plant Pathol 2000; 196:699–713.

Tarchevsky IA, Plant Cell Signaling Systems. Moscow. Nauka 2002; p. 292 (In Russian).

Thaler JS, Owen B, Higgins VJ. The role of the jasmonate response in plant susceptibility to diverse pathogens with a range of lifestyles. Plant Physiol 2004; 135:530–538.

Tian D, Traw MB, Chen JQ, Kreitman M, et al. Fitness costs of R-gene-mediated resistance in Arabidopsis thaliana. Nature 2003; 423:73–77.

Tichtinsky G, Vanoosthuyse V, Cock JM, et al. Making inroads into plant receptor kinase signalling pathway. Trends Plant Sci 2003; 8:231–237.

Wenzel G. Function of genetic materials responsible for disease resistance in plants. Progress in Botany 1998; 59:80–123.

Chapter 9

Hypersensitivity

S. F. Bagirova

Hypersensitivity is a natural defence for plants in response to a variety of pathogens (e.g. viruses, bacteria, fungi, nematodes), which is characterized by rapid cell death accompanied by an accumulation of toxic compounds within the dead cell. It may be said that under such circumstances, the plant sacrifices parts of itself for the sake of the whole.

The hypersensitive reaction is a feature of resistance, among various types of incompatible interactions between plant and pathogen. A complex signal cascade in the cells of a resistant plant, initially coming from an avirulent pathogen leads to cell death at the site of infection, so that those pathogens, which have penetrated would perish within the cells. This is effectively suicide. Hypersensitive response (HR) exhibits cell-wall modification, accumulation of PR-proteins (chitinase, gluconase), phytoalexins as well as phenolic compounds. This local response induces systemic alterations known as systemic acquired resistance (SAR). As a result the plant acquires resistance to a pathogen, which would previously have caused a disease.

How many guises does cell death have? What are the precise mechanisms of the hypersensitive response? Are there any differences between cell suicide and cell death as a consequence of direct destruction by a pathogen? On the one hand, cell death is attributed to resistance, incompatible combination between plant and pathogen, and is characteristic to the hypersensitive response. On the other hand, it has been regarded as a symptom of disease (compatible interaction). In both cases plant cell death has commonly been referred to as necrosis. The rate of cell death cannot apparently be a major sign of the hypersensitive response. Pathogens can cause steady death in cases of necrotrophy but, alternatively, can stimulate the plant at the primary stage of disease before causing cell death in the case of biotrophy. Consequently, the question arises whether there are remarkable differences between these two types of interactions (compatible/incompatible) or they are two sides of the same coin, no more than the same process at different speeds. To answer these questions we should describe in detail the issue of plant cell death in different interactions between host and pathogen.

First, let us examine the history of research on cell death. Cell death has been known since the discovery of the cell itself. Dead cells of cork tree were described by Hook, who was the first to discover the cell in 1665. In the nineteenth century scientists described cell death as a process. A century later two types of cell death were found in the animal kingdom: apoptosis,

Figure 9.1. *Monstera deliciosa.* Characteristic leaf blades evolve in ontogenesis as a result of apoptosis.

defined as programmed cell death, and necrosis, or pathological death accompanied by inflammation. This discovery had a great impact on modern views on cell death and made a more clear understanding of concrete mechanisms of cell death possible. Currently, programmed cell death is known in a range of species from distant taxonomic groups: animals, plants, fungi, and other microorganisms including prokaryotes. Apoptosis originated in prokaryotes, presumably, as a defence against virus invasion or against damage by reactive oxygen species (ROS), i.e. as mechanism for the elimination of cells with excessive production of toxic oxygen species. As it evolved, apoptosis' functions become wider. Cell death is of remarkable significance for embryogenesis, morphogenesis, differentiation, response to unfavourable climatic conditions as well as in immune response. Examples of plant apoptosis are well-known – development of xylem and phloem, leaf differentiation (e.g. in *Monstera* (Figure 9.1)), germination of pollen tube (death of pistil cells in the path of the tube growth), falling leaves, and ripening fruits. Programmed cell death (PCD) in a healthy plant involved in the maintaining of homeostasis. Alternations in the PCD process can lead to damage of homeostasis. Recent studies revealed many similarities between apoptosis in animals and hypersensitive response in plants. The following aspects of both PCD and necrosis: morphological, genetic, biochemical, and signalling with comparative analysis of cell death in animals and plants, will be considered below.

Morphological alterations in cells during apoptosis and necrosis

Typical signs of apoptosis in animals are the condensation and destruction of the nucleus, DNA, which splits into oligonucleosomic fragments and can produce a DNA ladder on electrophoresis and the formation of apoptotic bodies (membrane-wrapped structures containing DNA fragments). The plasma membrane stays intact up to the moment of formation of apoptotic bodies. The phospholipid phosphatidylserine, which is normally hidden within the plasma membrane, is exposed on the surface. The cell which is undergoing apoptosis becomes wrinkled, sharply decreases in volume, and the protoplast shrinks (Figure 9.2). Its chromatin becomes less condensed. Concentration of heterochromatin increases and it moves to the primary nucleus borders. Nuclear DNA divides into fragments of about 50 kb. Following this, the fragments are cut into oligonucleosomic fragments of 180 bp with the help of Ca^{2+}-dependent endonucleases. Apoptotic bodies containing DNA fragments migrate from the cell centre to its periphery. It has been discovered that all these features of apoptosis are displayed during hypersensitive response in plants.

DNA restriction into long fragments has been observed during hypersensitive response caused by infection with viruses, bacteria, and fungi. In infected plant cells, free 3'-OH of nuclear fragments can be identified, along with activation of Ca^{2+}-dependent endonucleases, and fragments of 50 kbp and oligonucleosomic fragments, which can produce a typical DNA ladder when injected to an agarose gel electrophoresis. In the case of tomato and *Alternaria alternata* f.sp. *lycopersici*, it was shown that DNA fragmentation resulting in oligosomic fragments occur only in incompatible combinations of plant–pathogen (at infection with an avirulent fungal strain, but not with a virulent one). Moreover, remnant apoptotic bodies migrate to the cell's periphery. Toxin from this tomato pathogen (AAL-toxin) caused apoptosis even in animal cells. Studies on bean and rust system showed that 3'-OH oligonucleosomic fragments are identified only in cells containing fungal haustoria. This indicates a

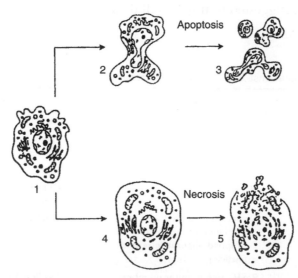

Figure 9.2. Morphological changes in cells during apoptosis and necrosis (from Darzynkiewicz. The Cancer Research Institute, NY Medical College, Elrnsford, NY 10523, darzynk@nme.edu).

high-specificity of the process. In animal kingdom apoptotic vesicles are utilized by neighbouring cells or are bound by receptors on specialized cells such as macrophages or dendritic cells, which then engulf the cell fragments. Phagocytic cells secrete cytokines (such as TGF-b or IL-10) that inhibit inflammation. In plants, phagocytosis is prevented by cell wall fortification. Instead, during hypersensitive response, plasmodesmata are broken and around the site of infection periderma is formed (as a consequence, healthy cells take on meristematic activity). A damaged cell wall is either completely destroyed by means of hydrolytic enzymes acted very locally or fortified with silica, cellulose, lignin, and proteins. Fortification of the wall isolates an invaded pathogen within the cell and makes the invasion of new pathogen almost impossible.

An essential role in programmed cell death, both in animals and plants, is played by mitochondria. Inducers of apoptosis (see discussion below) can sharply decrease membrane potential in mitochondria. This drop (depolarization) results from an increase in permeability in the inner membrane due to formation of huge pores. Their diameter allows molecules with molecular weight of less than 1.5 kDa to cross the membrane. Mitochondrial matrix swells. This results in mitochondrial outer membrane breaking. As a consequence, proteins emerge from the space between inner and outer membrane and this is considered to be a crucial stage in the progression of apoptosis. It is suggested that proteins emerge from the space between inner and outer membrane also as a result of hyperpolarization (sharp increase in membrane potential of inner membrane) or formation of huge channel in the outer membrane. Chloroplasts in plant cells may as well play a role in direct regulation of PCD in plants at the hypersensitive response by generating ROS and as a coordination centre. Other membrane structures such as central vacuole membrane (tonoplast) are suggested to trigger cell death in plant cells.

Table 9.1. The characteristics of apoptosis and necrosis

Feature	Apoptosis	Necrosis
Cell volume	Decrease	Increase
Integrity of plasma membrane	Plasma membrane stays intact up to the moment of formation of apoptotic vesicles	Damaged plasma membrane tears
Cell contents	Split into apoptotic vesicles	Appear into intercellular space
State of the nucleus	Nucleus fragmented with formation of membrane-wrapped oligonucleosomic particles	Nucleus destroyed
Organelles	Pores open in outer membrane to liberate apoptosis inducers	Swollen and destroyed

What happens to the cell at necrosis? (1) Cell volume considerably increases; (2) the permeability of plasma membrane and other membranes alters; (3) cytoplasm bubbles; (4) tonoplast breaks down; (5) organelles swell and are destroyed; (6) the cell disintegrates; (7) remnants of organelles degrade; and (8) the cell contents leak out, leading to inflammation of surrounding tissues.

Typical features of apoptosis and necrosis are compared in Table 9.1.

Genes involved in plant cell death – paranoid mutants

Genes regulating necrosis do not exist, as cell death at necrosis is reluctant, sudden, not planned. None of the specific genetic programmes are activated at necrosis. In contrast, apoptosis results from a certain genetic program. Numerous regulating genes in apoptosis have been found with positive and negative regulating roles. These genes can directly activate apoptosis via inducers, block via repressors, alternate plant metabolism, affect signalling molecules and their receptors or influence R-genes themselves. Areas of homology between sequences of apoptotic regulation genes in animals and plants are rare. This indicates that either plants developed their own genetic mechanisms for regulation of programmed cell death or plant/animal genes have diverged too significantly to be identified as homologous.

As discussed above, apoptosis is not only a defensive response, but also a normal process in healthy multicellular organisms. Studies on paranoid mutants, or mutants, which mimic resistance, allow us to distinguish apoptosis genes connected to the defence response. Paranoid plants have been defined as mutants which in the absence of pathogens sporadically form lesions, which are either similar to disease symptoms or resemble the hypersensitive response. The plants appear as if they are constantly under pathogen attack simulating disease or resistance to disease. Some of the mutations in paranoid mutants are assigned to the loci which control specific resistance, i.e. the hypersensitive response. Such mutants have histochemical or molecular features typical to resistance, characterized by local and systemic resistance to pathogens; even those which normally interact compatibly (i.e. cause the disease). Mutated genes may be connected to salicylic acid production, other signalling molecules, or their receptors, and mutations are conditional. Mutant phenotypes can be effected by applying hormones, light, or temperature. Mutants may result from mutagenesis, transgene expression, or can arise spontaneously. Major mechanisms of spontaneous mutations are gene conversions, non-homologous crossing-over, and transposon mutagenesis.

There are two classes of mutations.

Class I: Initiation mutations Mutants of this type randomly form local unexpanded lesions around the sight of infection (targeted cell). These mutations are most frequently mapped into R-loci. They have been referred to as "R-genes gone bad". Initiation mutations are best studied in *Zea mays*. More than twenty different mutations of class I have been investigated in *Z. mays* plants. For example, *rp1* mutations mapped in *Rp1* locus of resistance to *Puccinia sorghi*. Deletion of this gene triggers the hypersensitive response in the absence of the pathogens by activating a specific receptor. This locus is very unstable: high rates of mutation, such as gene conversion or non-homologous crossing-over, are very likely. *Zea mays Rp1* paranoid mutants demonstrate resistance to different rust species.

Class II: Propagation mutations Mutants of this type give wide and extensive lesions and large areas of dead tissue. These mutations have been attributed to negative regulators of PCD in ontogenesis or pathogenesis. Below are some examples of this class of mutations.

The lls (lethal leaf spot) mutation in *Z. mays* is induced by the endogenous transposon-mutator (Mu). The gene was cloned using the transposon mutagenesis approach. The phenotype of each mutant depends on light intensity and development of mature chloroplasts. In these mutants ultrastructure of mitochondria seems to stay unchanged, but chloroplasts are found to be swollen (Figure 9.3), followed by an increase of heterochromatin level and transportation of it to the borders of the nucleus. This indicates that chloroplasts have an important role in PCD of lls-mutants. This data shows that chloroplasts may operate as a coordination centre with an alternative function in PCD mechanisms.

In *Arabidopsis*-lsd (lesions simulating disease), mutation induces local and systemic resistance to virulent pathogens in wild type plants. It is intriguing that transgene lsd-mutants are not able to accumulate salicylic acid and exhibit a low expression level of PR-proteins, but show resistance to *Peronospora parasitica*. This suggests that two independent pathways of signal

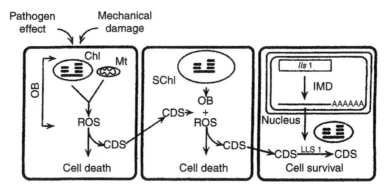

Figure 9.3.   The effect of chloroplast morphological alteration on cell fate in lls 1 mutants of maize (from Brent Buckner. Division of Science, Truman State University, Kirksville MO 63501, USA bbuckner@truman.edu).

Chl – chloroplast; Mt – mitochondrion; SChl – swollen chloroplast; OB – oxidative burst; ROS – reactive oxygen species; CDS – cell death signals; IMD – induction of mechanical destruction.

transduction exist: the first leads to cell death, the other to activation of PR-genes and synthesis of salicylic acid (Figure 9.4).

The Rt (root necrosis) mutation in soya bean illustrates an example of hypersensitivity, which occurs in roots. Mutations of propagation occurring in a single cell lead to the death of a vast plant area. In reverse mutation of such mutations autonomic sectors of living tissue can be identified. This can be explained by so called "biosocial events". The cell monitor gathers signals from a large group of cells. If a signal from one single cell is absent, altered or damaged, then the group of cells as a whole will undergo suicide.

Studies on paranoid mutants are one of the major approaches used to investigate signal transduction. Plant-resistance genes appear to be the centre of signal interplay in plant immune response (Figure 9.5).

It has been discovered that an early event (within the first ten minutes) of the hypersensitive response is the direct or indirect interaction of an elicitor with an R-gene product (Avr-dependent signals) and the alteration of the ion levels the cell contain (Ca^{2+}, H^+/K^+ exchange). This leads to an increase in pH, and anion channels (in particular chloride) opening. The role of numerous signal molecules, such as G-proteins, kinases, phosphatases, phospholipases, and the products of lipid oxidation (see discussion below) have been revealed using such studies. Some signals generate ROS (see Chapter 7). Experimental data on signals shows such a typical sequence: ion channels – ROS – genes

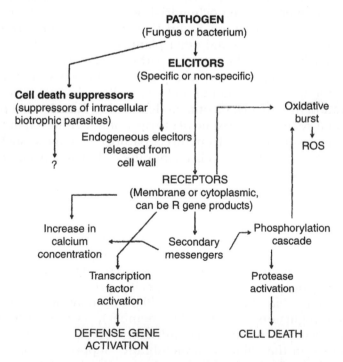

Figure 9.4. Diagram of signal transduction at hypersensitive response.

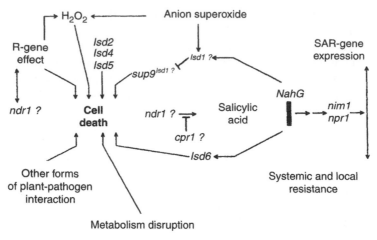

Figure 9.5.   Tentative death circuitry in disease resistance in *Arabidopsis* (from Jeffery Dangle: Department of Biology, University of North Carolina, Chapel Hill, North Carolina 27599 USA).

transcription, followed by the synthesis of salicylic acid, phytoalexins, and PR-proteins. A similar, but independent signal pathway – the first step of which is recognition of an avirulence factor or molecular complex by an R-gene product, leads (via a shift in Ca^{2+}, H_2O_2 concentration, and protease activity (see below)) to opening of mitochondrial pores, matrix swelling, and other events of PCD triggering. It was found that the early stages are multifunctional (general for apoptosis and other plant responses) and occur at transduction of signals by several interactive pathways (Figure 9.6).

During necrosis similar events take place (alterations in permeability of plasma membranes and membranes of organelles, changes in ion content, ROS generation by NADPH, nicotinamide adenine dinucleotide phosphate-oxidative plant system, cytoplasmic pH turns acidic). However, in the case of necrosis these events resulted from direct pathogenic effects (such as toxins or mechanical destruction of cell membranes), but not from active plant defence (see also Conclusion).

Plant NADPH-oxidative system is homologous to that one in macrophages or neutrophils in animals: it is inhibited by the same specific inhibitors, and protein components of the system, interact with the similar antigens, and have similar molecular weight. Membrane proteins of the NADPH-oxidative system in plants and animals contain Ca^{2+}-binding domains. This indicates that oxidative systems in plants and animals are regulated by Ca^{2+}. A number of other homologous zones between specific proteins related to PCD in plants and animals have been identified. Plant proteins, which are similar to antiapoptotic regulators of BCl-2 (Bax member), have been found. These proteins are localized in the mitochondrial membrane, nuclear membrane, and in plants – in the chloroplast membrane. Suppression of apoptosis in response to viral infection in *Nicotiana* transgenic plants ranges depending

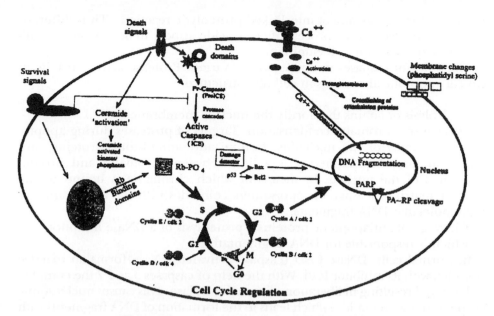

Figure 9.6. Cell cycle regulation (Gilchrist, 1998).

on the expression level of Bax-gene. This data indicates that plants and animals share components of PCD system.

Biochemistry of apoptosis

Activation of cysteine (and some other) proteases is the most universal feature of PCD. The major components of PCD, the caspases (from cysteine aspase)is a family of evolutionarily conserved cysteine proteases, which specifically cleave proteins after aspartic acid residues. Currently 10 caspases have been identified. On the basis of structural homology, the family of caspases are divided into subfamilies 1 (caspases 1,4,5), 2(caspase 2), and 3(caspases 3, 6–10).

The spectrum of enzymes involved in apoptosis of plants is different from animals. For example, in plants elevation of serine, metalloproteases, and aspartic proteases activities can be identified. Notably, plant aspartic proteases are in many ways similar to animal aspartic proteases, involved in PCD of HeLa human cells. Moreover, other proteases appear to play a role in apoptosis, first are the calpains, or, Ca^{2+}-dependent cysteine proteases, and proteases of ubiquitin–proteasome pathway. These proteases are the necessary components of a cascade of proteolytic enzymes. Thus, they are inhibitors of calpain blocking apoptosis. The ubiquitin–proteasome mechanism of protein degradation is also activated during apoptosis.

Different proteases interact with each other: they may step by step activate each other, participate in consequent transmission of apoptotic signal, or can be a component of the same united proteases complex – proteasome.

Proteases form a cascade of interlinked proteolytic reactions. Thus different proteases involved in the process of cell suicide appear to be a strong demolishing weapon during cell elimination.

The roles of caspases in apoptosis are varied. Protease activity results in typical apoptotic alterations in cell ultrastructure.

1. Hydrolysis of lamins that fortify the nuclear membrane.

This causes chromatin condensation. Targets of proteases during apoptosis are proteins of the nucleolus, histone and non-histone proteins, and topoisomerase. Topoisomerase is a link between chromatin and protein structures of the nucleus, by means of which chromatin is bound to the nuclear matrix. Cleavage of topoisomers coincides with the appearance of high-molecular DNA fragments.

2. Cleavage of antiapoptotic proteins – proteolysis of a DNase inhibitor, which is responsible for DNA fragmentation.

In normal cells, DNase CAD (caspase activated DNase)forms an inactive complex with its inhibitor Icad. With the help of caspases 3 and 7 the complex is destroyed resulting in liberation of an active DNase. This causes nucleosomic chromatin fragmentation, which leads to the formation of DNA fragments with a molecular mass which is similar to the molecular mass of DNA in nucleosomic particles – 180–200 nucleotide pairs. These fragments develop into a typical DNA ladder during the gel electrophoresis. Apoptosis is also possible without DNA fragmentation. The nuclear protein Acinus (apoptotic chromatin condensation inducer in the nucleus) with the help of caspase 3 and an unidentified protease is cleaved into fragments. One of these fragments in the presence of additional non-nuclear factors, causes apoptotic chromatin condensation and nucleus fragmentation without DNA fragmentation (karyorexis). In addition to direct activation of nucleases and proteases, by means of local proteolysis, there is elimination of free space between nucleases and DNA in chromatin by removal of proteins, which protect DNA.

3. Inhibition of DNA repair: the inactivation and distortion of the regulation of proteins involved in DNA repair, in mRNA splicing, and DNA replication.

The target of caspases is poly-(ADP-ribose)-polymerase (PARP), which is involved in DNA repair by catalyzing protein poly ADP-ribosylation linked to DNA. The donor of ADP-ribose is NAD^+. The activity of PARP-polymerase increases by over 500 times after binding to fragments of damaged DNA. The PARP is involved in the repair of damaged DNA, regulation of endonucleases activity, and the maintenance of chromatin structure by means of ADP-ribosylation. Apoptotic cell death is accompanied by cleavage of PARP by caspases. In case of massive DNA fragmentation, activated PARP markedly decreases the contents of intracellular NAD^+. This leads to inhibition of glycolysis and mitochondrial oxidation and causes necrotic cell death.

4. Destruction of cytoskeletal proteins.

Degradation of structural and functional proteins, and proteins of mitotic apparatus have been found to occur during apoptosis.

5. Involvement in gene expression.

This function is related to the proteolysis of repressors with the formation of proteins regulating transcription (modification of transcription factors). Targets of proteases, for example, can be histones, which play a significant role as a gene repressor.

6. Among the various functions of proteases is the transmission of the apoptotic signal.

Signals may be transmembrane and receptor-coupled. Transmembrane proteins are involved in both direct inductions of apoptosis along with receptors or via activation of protein kinase, which plays an essential role in transduction of the transmembrane signal from a wide range of receptors.

Localization of proteases in different cell compartments makes transmembrane signal transduction more efficient. Some proteases, the so-called transmembrane proteases, bind to the membrane. They are integrated into the plasma membrane, tonoplast, membrane of nucleus, or organelles. Others are localized in the nuclear matrix, cytoplasm, or organelles. Asparagine plant proteases are presumably localized in vacuoles. Serine proteases are suggested to exist in the cytoplasm and in the nucleus. It is well-known that proteases in the nucleus are tightly linked to chromatin, including histones. Translocation of the proteases can result in their activation, for instance, the transport of a Ca^{2+}-sensitive protease and protein kinase from the cytoplasm to the membrane which is triggered by Ca^{2+} accumulation within the cell (influx). At the same time, autocatalysis of inactive protease forms takes place.

The activity of some proteases might be caused by an increase of concentration of Ca^{2+} in the cell. This is typical for cells that have undergone apoptosis (see section above). Reactive oxygen species may also trigger protease activation. Activation of DNA-sensitive serine proteases that are specific to serine can be generated by fragments of single-stranded DNA molecules.

A lot of different pathways serve to prevent cells from spontaneous and unnecessary death, and ensure the correct algorithm of events prior to apoptosis occurring. Several mechanisms (described below) have been discovered which help cells to avoid unnecessary self-destruction by means of proteases.

First, proteases are synthesized in an inactive form. Processing of inactive proteases may be maintained via autolysis or proteolysis with the help of other proteases. Caspases, for example, are synthesized in cells in the form of procaspases, which are inactive monomers with molecular weight 30–50 kDa. The active form is tetramer. This consists of two subunits $(p10–p20)_2$ (Figure 9.7). The activity of procaspases is only 1–2% of caspase activity.

Proteolytic auto- or trans-cleavage, and the subsequent assembly of the two subunits will lead to the formation of active caspase. The regulatory N-domain (prodomain) disconnects from the procaspase molecule and the rest of the molecule disintegrates to form small (about 10 kDa) and large (about 20 kDa) subunits. Two heterodimers form a tetramer with two independent catalytical centres. Initially, concentration of active endogenous

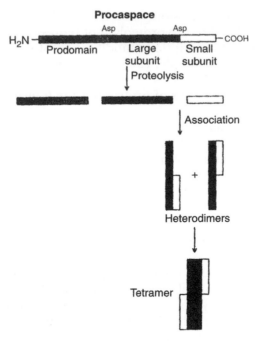

Figure 9.7.   Formation of active caspase from procaspase.

caspases is very low. Nevertheless, by means of autocatalysis the activity can increase rapidly.

Second, proteases join with inhibitors forming inactive complexes, such as the latent complexes found in cysteine Ca^{2+}-sensitive protease and in some other proteases. Inactive [inhibitor-proteases] complexes dissociate due to different apoptosis triggers. For instance, reversible association of a Ca^{2+}-sensitive protease and endogenous inhibitors is regulated by Ca^{2+}. Cysteine protease is covalently attached to the inhibitor via a disulfide link. Liberation and activation of caspases can result from thiol-disulfide exchange and is associated with glucose metabolism and the redox state of the cell.

Third, proteases can be a component of a specific receptor-coupled complex. That is, [receptor+ligand+adapter+procaspase] form a specific complex in which caspase activation takes place. Such a complex is called apoptosome, or apoptotic chaperone. The most intriguing fact is the conservative regions of the homology (including the NB-domain) between the adapter in animals and R-proteins in plants, such as tomato, *Arabidopsis*, and *Nicotiana*. Moreover, the proteins are structurally similar (Figure 9.8).

R-proteins have been suggested to play a role as adaptors in apoptosome. This means that the plant apoptosome presumably dissociates, thereby initiating the programmed cell death, once the receptor recognizes Avr-protein.

R-proteins, therefore, appear to be responsible for initiating the cell death at the site of infection: recognition of signals and switching on the cell death machine, the first and the most important steps towards rapid cell death.

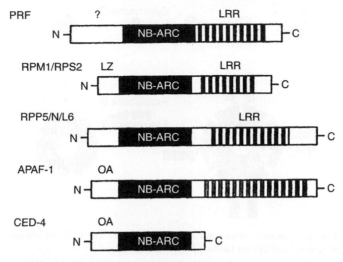

Figure 9.8.   Domain structures of some NBS-LRR plant resistance proteins and pro-apoptic adaptor proteins (van der Biezen EA, Jones JDG, 1998).
PRF – R-gene of tomato; RPM1, RPS2, and RPP5 – R-genes of *Arabidopsis*; N – R-gene of tobacco; L6 – R-gene of flax; APAF1 – human gene; CED4 – gene of *Caenorabtilis elegans*. OA–the caspase recruitment domain.

There are several pathways for programmed cell death. The signal transduction can take the following pathway: inducer – receptor – adapter – caspases of 1st stage – regulators – caspases of 2nd stage. The receptor interacts with the ligand.

Is the cell death a reversible process? At the stage of the activation of the 1st stage caspases, the cell can still be rescued. Regulators were found which can either block or, just the opposite, enhance demolition activity of 1st stage caspases. After the 2nd stage caspases are activated by 1st stage caspases via the proteolysis of procaspase, the process becomes irreversible. Those caspases are then capable of activating themselves (autocatalysis and auto processing) and activating factors of DNA fragmentation into nucleosome fragments.

Now we turn our attention again to the role of mitochondria in apoptosis. Typical apoptotic alterations in mitochondrial ultrastructure can be induced by oxidative stress and increased Ca^{2+} concentration. While the cell undergoes apoptosis, apoptogenic factors are liberated from the intra-membrane space of mitochondrion. These include:

1. Procaspases,
2. Cytochrome C,
3. AIF (apoptosis inducing factor) – a 57 kDa flavoprotein.

Having supplemented with an isolated nucleus, this flavoprotein can cause chromatin condensation, and, when added to isolated mitochondria, it results

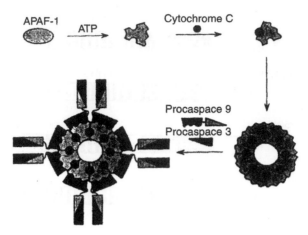

Figure 9.9.   Caspase activation involving APAF-1 and cytochrome C (from Michael Hengartner. Cold Spring Laboratory, Cold Spring Harbor, NY 11724, USA).

in the liberation of cytochrome C and caspase 9. Liberated cytochrome C in combination with cytoplasmatic factor APAF-1 (apoptosis protease activating factor-1) form a complex with procaspase (Figure 9.9). The APAF-1 serves as an armature where autocatalytic processing of caspase 9 takes place (multimer armature APAF-1–cytochrome C complex resembles a propeller). Inhibitors for cytochrome C liberation, which block apoptosis, such as Bcl protein, were found.

Signalling

How do cells decide whether it is time to die or survive? They react to a network of multiple and diverse internal and external signals. Cells are equipped with a complex and precise mechanism for the tuning, adjustment, balance, and correction of this response. Protein kinases and phosphatases, G-proteins, cell-adhesion molecules, active biolipids, and nitric oxide have been recognized as important components of this mechanism in both plants and animals (see also Chapter 8).

Overall, apoptosis is a highly regulated process. It can be activated by intrinsic (the leakage of cytochrome C and perturbation of mitochondria) or extrinsic way (direct induction of cell receptors containing the death domain). The activity of caspase is modulated by a number of adapters and inhibitors. As a rule, inhibitors of apoptosis are less effective in intact cells than in in vitro tests.

Expression of genes involved in apoptosis, altered cell adhesion, and changed cytoskeleton organization are among those signs of apoptosis which are controlled by the phosphorylation/dephosphorylation of cell proteins, with the combination of two types of enzymes, protein kinases and phosphatases. Direct interaction of a protein kinase receptor with an extracellular first messenger or intercellular second messenger and autophosphorylation are involved in the mechanism of protein kinases activation. It has also been shown that the switch between apoptosis and cell survival can be regulated by the balance of different protein kinase isoforms which result from alternative splicing and are characterized by altered specificity to transcription factors. Inhibitors of phosphatase suppress cell death. Inhibitors of protein kinases may induce or accelerate cell death.

A modification of proteins during apoptosis is their nitration or nitrosylation. After the discovery of intracellular synthesis of unstable, highly reactive radical nitric oxide (NO), a lot of research was devoted to the involvement of NO in the signal transduction of apoptosis during the hypersensitive response. It appears that NO is one of the key molecules involved in apoptosis signalling: it can trigger cell death, activate G-proteins, mobilize Ca^{2+}, and regulate SAR. It can also inhibit caspases by nitrosylation of the active cysteine site. Lipid modifications of proteins (e.g. G-proteins), such as myristoylation is also essential for the interaction of proteins with membranes and other components of the signalling pathway.

Alteration in cell–cell and cell–matrix adhesion is an important step of apoptosis. Cell adhesion receptors and their ligands function as signalling molecules during PCD. It is essential that some of the receptors are connected to cytoskeleton inside the cell and closely linked to phosphatases and protein kinases. Some adhesion receptors are involved in caspase activation. Apoptosis can also be activated by DNA-methylation inhibitors.

Another important component of signal transduction during the hypersensitive response is G-proteins (the heterodimeric guanine nucleotide-binding regulatory proteins) and G-protein-coupled receptors. G-proteins mediate signalling from a number of seven-transmembrane spanning cell surface receptors to diverse intracellular effectors such as K^+ and Ca^{2+} channels, adenylcyclases, phospholipases, cGMP, receptor kinases. G-proteins are composed of α-subunit and assembled β and γ-subunits (by dimer). The G-protein signalling system involves cycles of association/dissociation of the subunits. The α-subunit is active when GTP (guanosine triphosphate) is bound and inactive when GDP (guanosine diphosphate) is bound. The function of the α-subunit can be determined by the covalent attachment of various lipids.

During apoptosis remodelling of the cell membrane takes place. Activated lipases are involved in the generation of bioactive lipids which play an important role in PCD and may serve as mediators (intra- and extracellular) in apoptosis signalling.

Many pathogens can manipulate processes involved in apoptosis within host cells. For example, viral inhibitors of apoptosis can represent pseudo substrates of caspase, prevent the binding of caspase to death receptors, mimic Bcl-2, prevent mitochondrial damage, antagonize death receptors or block caspase activation.

Even though caspase plays a central role in both initiation and execution of apoptosis, the inhibition of total caspase activity does not prevent cell death by alternative pathways.

Conclusion

We have described two completely different pictures of cell death: apoptosis and necrosis. These pictures are mosaic and seem to consist of numerous different elements. However, if we examine more closely these elements, we discover that apoptosis and necrosis share many common molecular and histochemical markers. The generation of ROS (generally referred to as a death elixir because they are universal), cytoplasmic acidification, alterations in the permeability of plasma membrane and organelle membranes, and ion contents are important components of the death process in both apoptosis and necrosis. There are common inducers for apoptosis and necrosis. Intracellular ATP and NAD^+ concentration may turn apoptosis into necrosis. Hormones also regulate apoptosis (gibberellin stimulates and abscisic acid inhibits). Hydrogen peroxide in low concentration is an apoptosis inducer, while in high concentration leads to necrosis. Apoptosis, in contrast to necrosis, is inhibited by protein synthesis inhibitors and is an energy-dependent process, as it demands constant replenishment of ATP in cells. That is why if there is a shortage of ATP in cells or excessive production of ROS, cells of a resistant plant, which normally would have undergone apoptosis, perish in a necrotic manner.

Deficiency in energy and other resources force the cell not to undergo the process of apoptosis (in spite of R-gene activation and the initiation of other PCD events), but (with other independent signalling pathways taking place) follow a different algorithm of signals and events. Thus, the cell death mechanisms are similar – both necrosis and apoptosis can be considered as two extremes of the same curve. This may illustrate a principle of biological economy. Cell suicide appears to adopt mainly the same mechanisms that are involved in cell death under pathogen attack.

The above scheme, which implies two cell death types (apoptosis and necrosis), is obviously not universal. Plants, presumably, evolved several pathways to interact with pathogens and different defence mechanisms. Furthermore, cells under pathogen attack would have perished in an apoptotic manner, and hence trigger, in turn, the death of the surrounding cells by other mechanisms. The type of response would depend on external factors (such as the dose applied) and internal resources of the cell itself.

Selected Literature

Chichkova NV, Sang Hyon Kim, Titova ES, et al. A plant caspase-like protease activated during the hypersensitive response. Plant Cell 2004; 16:157–171.

Dangle JL, Dietrich JL, Richberg MN. Death don't have no mercy: cell death programs in plant-microbe interactions. Plant Cell 1998; 6:1793–1807.

Eckardt N. Ins and outs of programmed cell death and toxin action. Plant Cell 2005; 17:2849–2851.

Gilchrist DG. Programmed cell death in plant disease: the purpose and promise of cellular suicide. Ann Rev Phytopathol 1998; 36:393–414.

Greenberg JT. Programmed cell death in plant-pathogen interactions. Ann Rev Plant Physiol Plant Molecular Biol 1997; 48:525–545.

Hara-Nishimura I, Hatsugai N, Nakaune S, Kuroyanagi M, Nishimura M. Vacuolar processing enzyme: an executor of plant cell death. Curr Opin Plant Biol. 2005; 8(4): 404–408.

Hatsugai N, Karoyanagi M, Yamada K, Meshi T, Tsuda S, Kondo M, Nishmura M, Hara-Nishimura I. A plant vacuolar protease, VPE, mediates virus-induced hypersensitive cell death. Science 2004; 305:855–858.

Heath MC. Apoptosis, programmed cell death and hypersensitive response. Eur J Plant Pathol 1998;104:117–124.

Hoeberichts FA, ten Have A,Woltering EJ. A tomato metacaspase gene is upregulated during programmed cell death in *Botrytis cinerea* infected leaves. Planta 2003; 217: 517–522.

Jones AM. Programmed cell death in development and defense. Plant Physiology 2001; 125:94–97.

Levine A, Pennell RI, Alvarez ME, Palmer R, Lamb C. Calcium mediated apoptosis in a plant hypersensitive disease resistance response. Current Biol 1996; 6:427–437.

Liang H, Yao N, Song JT, Luo S, Lu H, Greenberg JT. Ceramides modulate programmed cell death in plants. Genes Dev 2003; 17:2636–2641.

Lorrian S, Lin B, Auriac MC, Kroj T, Saindrenan P, Nicole M, Balague C, Roby D. Vascular associated death1, a novel GRAM domain-containing protein, is a regulator of cell death and defense responses in vascular tissues. Plant Cell 2004; 16(8):2217–2232.

Mittler R. Lam E. Identification, characterization and purification of a tobacco endonuclease activity induced upon hypersensitive response cell death. Plant Cell 1995; 7:1951–1962.

Morita-Yamamuro C, Tsutsui T, Sato M, Yoshioka H, Tamaoki M, Ogawa D, et al. The Arabidopsis gene CAD1 controls programmed cell death in the plant immune system and encodes a protein containing a MACPF domain. Plant and Cell Physiology 2005; 46(6):902–912.

Richberg MH, Aviv DH and Dangle JL. Dead cells do tell tales. Curr. Opin. Plant Biol. 1998; 1:480–485.

Rojo E, Martin R, Carter C, Zouhar J, Pan S, Plotnikova J, Jin H, Paneque M, Sanchez-Serrano JJ, Baker B, et al. VPE gamma exhibits a caspase-like activity that contributes to defense against pathogens. Current Biol 2004; 14:1897–1906.

Salvesen GS, Dixit VM. Caspase activation: the induced-proximity model. Proc Natl Acad Sci USA 1999; 96:10964–10967.

Sanmartin M, Jaroszewski N, Raikhel N, Rojo E. Caspases. regulating death since the origin of life. Plant Physiology 2005; 137:841–847.

Shiran Y, Kachroo P, Shah J, Klessig DF. A gain-of-function mutation in an Arabidopsis Toll Interleukin1 receptor-nucleotide binding site-leucine-rich repeat type R gene triggers defense responses and results in enhanced disease resistance.Plant Cell 2002; 14(12):3149-62.

Solomon M, Belenghi B, Delledonne M, Menachem E and Levine A. The involvement of cysteine proteases and protease inhibitor genes in the regulation of programmed cell death in plants. Plant Cell 1999; 11:431–443.

Strasser A, Liam O'Connor, Dixit VM. Apoptosis signalling. Ann. Rev Biochemistry 2000; 69:217–245.

Van der Biezen EA, Jones JDG. The NB-ARC domain: a novel signaling motif shared by plant resistance gene products and regulators of cell death in animals. Current Biol 1998; 8:226–267.

Yao N, Imai S, Tada Y, Nakayashiki H, Tosa Y, Park P, Mayama S. Apoptotic cell death is a common response to pathogen attack in oats. Mol Plant Microbe Interact 2002; 15(10):1000–1007.

Chapter 10

Immune response

V. G. Dzhavakhiya, O. L. Ozeretskovskaya, S. V. Zinovyeva

Interaction of the secondary messengers with the genome of an infected cell results in a change of the genome activity: expression of some genes will be suppressed, and of the others, strongly activated. This leads to accumulation of new products (stress metabolites) in the cell, and many of them are toxic for parasites. These products will be discussed in this chapter.

Phytoalexins

The honor of discovering phytoalexins (FA) belongs to the German plant pathologists Muller and Borger. According to their definition, FAs are antibiotics of vegetal origin synthesized in a plant de novo in response to microbial infection and involved in the mechanisms of plant disease resistance. The American plant pathologist Joseph Kuc provides another definition of phytoalexins: he believes that FAs are the low molecular weight antimicrobial substances accumulated in plants in response to infection or stress. The rate and amount of their accumulation in a plant are determined by their release (or release of their close precursors) from the conjugates or by synthesis de novo, as well as their degradation by plant enzymes and microorganisms.

For a long time FAs were regarded as the key mechanism of plant resistance, however, today, other defense ingredients, in addition to FAs, have been found in plants, such as PR-proteins, proteinase inhibitors, reactive oxygen species, oxyproline-rich glycoproteins, mechanical barriers in the infection pathways, and other immune responses. Certainly, this does not mean that FAs have lost their importance, however, their protective action should be viewed against a background of a multicomponent response of the plant cell.

K. Muller, in his classic studies, that made the basis for prediction of phytoalexins, applied suspension of spores of an incompatible race of the late blight pathogen on the surface of a potato tuber section in the form of a cross. A day later the whole section surface was infected with a compatible race of the same pathogen. Several days later the fungal mycelium grew on the entire section surface, except for the areas previously infected with the incompatible race where hypersensitive response (HR) had developed. The picture looked like a black cross of necrotized tissue on the white background of the fungal mycelium. The protected tissues inhibited not only the growth of the compatible Phytophthora race but also growth of the causal agent of fusarial wilt and some other pathogens.

In the late 1950s, Muller developed a classic method of FA production and assessment. Infectious drops containing suspension of pea pathogen (or non-pathogen) spores were put in horse bean seed recesses instead of the removed seeds. After a certain time, a fungitoxic substance was released from the plant tissues to the infectious drops (diffusates), the substance not being produced if the seed chambers contained water. On the basis of these experiments, phytoalexins were defined as the compounds produced from interaction of two metabolic systems: the host and the parasite, inhibiting the parasite's growth. An advantage of this methodological approach is the use of the undamaged tissue, i.e. the metabolites produced in mechanical damage of the cells do not get in the drop, and there is no cuticle preventing metabolite interchange between the germinating spores of the parasite and the plant cells.

The method suggested by Muller soon brought appreciable results. In 1960–1962, the Australians Cruikshank and Perrin isolated a phytoalexin from infectious tissues of peas, which was called pisatin, and identified its structure. Next, other FAs were characterized: phaseollin from beans, capsidiol from pepper (suspension of the inductor fungus spores were placed in the halves of the cut fruits), etc.

The list of elicitors capable of inducing FA formation is very long. For instance, more than 200 compounds, microorganisms, and physiological stresses can cause accumulation of pisatin in peas, phaseollin and kievitone in beans, and glyceollin in soybean. These compounds may temporarily interfere in the plant metabolism, chemically manifested as generation of an alarm signal. Such signals can initiate a cascade of events resulting in FA synthesis and accumulation.

To date, about 350 FAs have been characterized in the plants of 30 families. They are mostly legumes (about 130 FAs), owing to the use of Muller's method for their isolation. Most phytoalexins have been isolated from dicotyledons, though some FAs from monocotyledons have also been characterized: such as FAs of rice, maize, sorghum, barley, oat, wheat, onion, and lilies. Phytoalexins were isolated from plant leaves, stems, roots, and fruits, though they are not always synthesized in all infected organs. Phytoalexins are lipophilic compounds, localized around the infection site. Because of the diversity of their structure (Figures 10.1 and 10.2) it is difficult to judge about the mechanisms of their antimicrobial activity.

Phenylpropanoid FAs are confined to the plants of the Leguminosae, Solanaceae, Convulvolaceae, and Umbelliferae families. Phytoalexins possess some structural specificity related to the taxonomic position of their producers. For instance, isoflavonoid FAs mostly occur in legumes, but they are absent in the Solanaceae; on the contrary, sesquiterpenoid FAs are typical of the Solanaceae but do not occur in legumes. As to the structural specificity of phytoalexins in the plants of the same family, here is an example: the

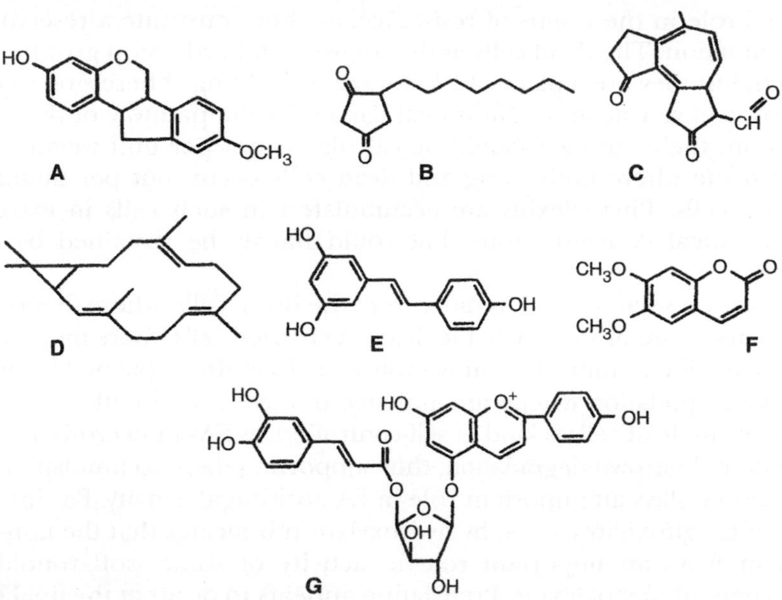

Figure 10.1. Phytoalexins, in which molecules are only carbon, hydrogen, and oxygen (Kuc J, 1995). A– medicarpin (red clover, white clover, alfalfa, cowpea); B – tsibulin 1D (onion); C – letucenin A (lettuce); D – casben (castore bean); E – resveratrol (arachis, grape); F – scoparon (citrus); G – caffeic acid ester arbinizyl-5-0-apogenidine (sorghum).

norsesquiterpenoid alcohol rishitin was found in potato and tobacco plant but not in pepper, whereas the sesquiterpenoid phytoalexin capsidiol is present in tobacco and pepper but not in potato.

FA formation is closely connected with HR, therefore, if HR is coordinated to control the course of disease, FA formation can also be controlled. The dead (necrotized) cells resulting from the hypersensitive response play a special

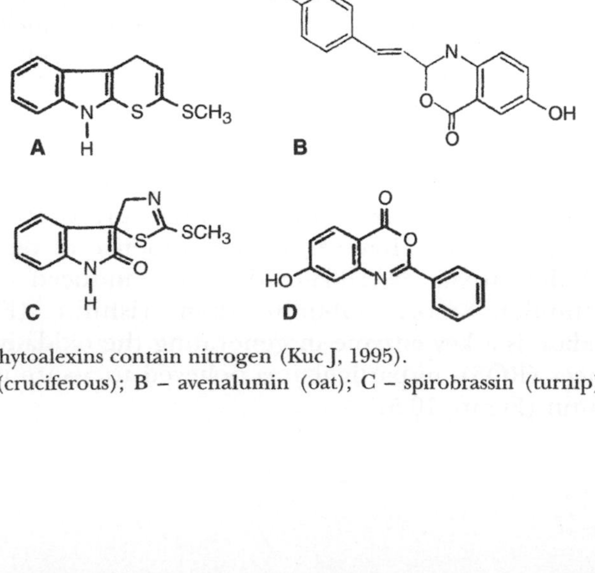

Figure 10.2. Phytoalexins contain nitrogen (Kuc J, 1995). A – cyclobrassin (cruciferous); B – avenalumin (oat); C – spirobrassin (turnip); D – diantalexin (carnation).

functional role in the events of resistance, as they constitute a reservoir for FA accumulation. The dead cells in themselves can hardly stop growth of the pathogen, but they are a place of FA storage or holding. Therefore, to judge the FA protective role as an antifungal barrier in the pathway of the spread of infection, their number should be calculated not per unit weight of the infected tissue where both living and dead cells occur, but per number of necrotized cells. Phytoalexins are accumulated in such cells in extremely high, antifungal concentrations that could hardly be sustained by living plant cells.

It is unknown what makes FAs move from the living cells, where they or their close precursors are produced, to the dead, necrotized cells where they accumulate. Probably, FA accumulation in necrotized cells is also explained by the fact that the FA degradation systems are inhibited in these cells. In other words, the FA presence in dead cells is kind of self-controlled, as FAs in necrosis inactivate the system of their own degradation, thus supporting their accumulation.

Prenylation plays an important role in FA antifungal activity. For instance, the loss of fungitoxicity caused by hydroxylation indicates that the non-polar side chain plays an important role in activity of some isoflavonoid and sesquiterpenoid phytoalexins. Prenylation appears to occur at the final stages of FA biosynthesis.

The principal pathways of biosynthesis responsible for generation of most FA precursors include the shikimate, acetate-malonate, and acetate-mevalonate pathways (Figure 10.3). FA biogenesis is frequently connected with the enzyme activation at the key stage of biosynthesis. This enzyme can control the early stages of biogenesis but can also be involved in the final stages. An example is biosynthesis of sesquiterpenoid phytoalexins of potato. 3-hydroxy-3-methylglutaryl coenzyme A reductase (HMG-CoA reductase) is a regulatory enzyme of the early stages of the acetate-mevalonate pathway (Figure 10.4). Activity of this enzyme grows both in mechanically wounded and infected tubers. However, the isoenzyme produced in response to tuber cutting differs from the respective enzyme produced in response to infection. In the latter case, sesquiterpencyclase activity also grows (sesquiterpencyclase, SC) that catalyzes the conversion of franesyl pyrophosphate to a cyclic precursor of terpenoid phytoalexins. Simultaneously, squalene synthetase (SS) activity decreases, which leads to formation of triterpenoid sterols and steroid glycoalkaloids. Both enzymes function at the branching stage of the acetate-mevalonate pathway (Figure 10.4). The enzymes regulating the final stages of synthesis and conversion of phytoalexins are also controlled by various systems of signal transduction. For instance, potato tubers treated with the elicitor from the *Phytophthora infestans* cell wall preparation accumulate more rishitin than lubimin. However, in the presence of diphenyl iodine, an inhibitor of the enzyme NADPH-oxidase, the induced tubers, on the contrary, accumulate more lubimin than rishitin (Figure 10.5). As NADPH-oxidase is a key enzyme in generating the oxidative burst, reactive oxygen spices (ROS), in particular, is believed to assure metabolism of lubimin to rishitin (Figure 10.5).

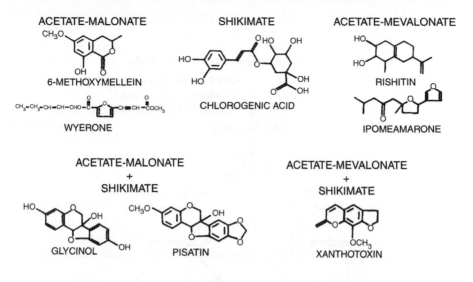

Figure 10.3. Biosynthetic pathways of some phytoalexins (Kuc J, 1995).

Some FA precursors can be produced through a combination of one to three biogenetic pathways, which provides additional opportunities for regulation of their biosynthesis. For example, synthesis of the phaseollin precursor requires involvement of several pathways of biosynthesis (shikimate, malonate). In most cases regulation is carried out at the transcription level, therefore, the inhibitors of transcription (Actinomycin D) and translation (Destroyicidine) suppress the FA formation, and accordingly, suppress resistance to pathogen growth in the infected tissues, which can be seen from the data in Table 10.1.

However, some elements of post-translational control can also take place in FA synthesis. J. Kuc believes that FA formation is not always carried out as their de novo synthesis, and that their accumulation sometimes occurs as a result of conjugate hydrolysis. For instance, soybean contains a constitutionally large quantity of the isoflavonoid conjugant of daidceine, which is a close precursor of glyceollin. At an incompatible combination of soybean with the late blight pathogen, the conjugant of daidceine is rapidly hydrolyzed, which leads to accumulation of glyceollin, and the fungus growth terminates. In a compatible combination, hydrolysis slows down; therefore, a low level of glyceollin is noted after the front edge of infection has passed. Such data were also received for the phytoalexins medicarpin

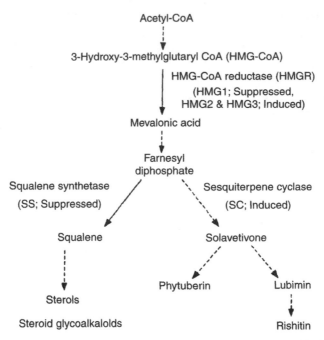

Figure 10.4.   A regulation isoprenoid biosynthesis in potato tuber. Wound-induced sterol and steroid glycoalkaloid synthesis are suppressed in favor of sesquiterpenoid phytoalexin synthesis during expression of the HR (Yoshioka et al., 2001).

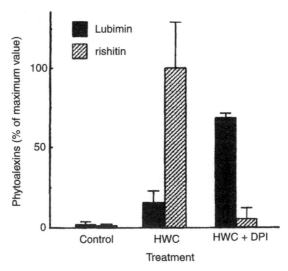

Figure 10.5.   Induction of phytoalexin production in potato tubers by Phytophthora unspecific elicitor (HWC) and NADP-oxidase inhibitor diphenyliodine (DPI) (Yoshioka H, 2001).

Table 10.1. Effect of antibiotics on hyphae growth of *Phytophthora megasperma* and accumulation of the FA glyceollin in soybean tissues (Yoshikawa M, Yoshikawa K, Masago H, 1979)

Soybean cultivar/treatment	Length of interstitial hyphae (microns)	Glyceollin (mkg/ml of tissue raw weight)
Harosoy 63 (resistant)		
Control (water)	36	1252
Actinomycin D (100 mkg/ml)	>1500	74
Blasicidine (0.5 mkg/ml)	>1500	101
Harosoy (susceptible)		
Control (water)	>1500	138

and maackiain in chick peas. The rate of the precursor hydrolysis depended on the elicitor concentration: at low elicitor content there was almost no hydrolysis of the glucoside, while high elicitor doses lead to glyceollin formation. Thus, pterocarpane phytoalexins cannot only be synthesized de novo, but can also be produced from conjugates as a result of their hydrolysis.

An indicator of chemical compound toxicity for fungi is ED50 (concentration of the chemical inhibiting colony growth by 50%). The ED50 of phytoalexins for different fungi varies from 10^{-3} to 10^{-5} M, i.e. FAs are rather weak antifungal agents. There is no evidence of their transport in plants. However, the FA accumulation rate and their localization in infection site maintain their concentration in the plant at a level toxic for the fungi. The outcome of the plant–parasite relationship depends on three factors: amount of the phytoalexin produced, rate of its production, and susceptibility of the parasite to the phytoalexin. The degree of FA involvement in the plant resistance events is found on the basis of many tests: antimicrobial activity measured in experiments in vitro, rate of FA synthesis, localization in the pathogen growth sites, presence of the compounds that influence their antimicrobial activity, as well as FA detoxification by the fungus and plant enzymes.

Phytoalexins play a role in the species-specific and cultivar immunity (host and non-host resistance). The FA role in species-specific immunity is clearly shown in Table 10.2.

It can be seen that synthesis of pisatin in response to pea inoculation with the cereal parasite *Erysiphe (Blumeria) graminis* begins much earlier than in response to infection with the "own" pathogen *E. pisi*; pisatin concentration can be 25 times greater, and it is 13 times more toxic. That is, pisatin concentration in the

Table 10.2. Formation of the FA pisatin in pea leaves after their inoculation with two species of mildew fungi (Oku H, Oushi S, Shizaishi T, Baba T, 1975)

Fungal species	Beginning of pisatin formation (hours after inoculation)	Maximum concentration (mkg/ml)	ED50 (mkg/ml)
Erysiphe graminis	15	>2000	40
E. pisi	48	78	530

peas inoculated with *E. graminis* 50-fold exceeds the dose lethal for the pathogen, whereas in the peas inoculated with *E. pisi* it is 7 times less than the lethal dose.

The relationship between FA and the cultivar immunity is shown in Table 10.3.

It can be seen that upon inoculation of all investigated cultivars with an avirulent race, the potato FAs rishitin and lubimin were accumulated in the concentrations greater by an order of magnitude than upon inoculation with the virulent race.

In some cases the substances present in the infection area can influence the fungus susceptibility to phytoalexins. For example, inclusion of polar lipids from peas in the growth medium of some fungi decreased their susceptibility to pisatin, maackiain, and phaseollin. Commercial phosphatidyl choline also decreased susceptibility of some fungi to pisatin. Such sterols as phytosterol, stigmasterol, and cholesterol reduced susceptibility of *Phytophthora infestans* to rishitin and lubimin. Thus, efficiency of phytoalexins as protective agents depends on the presence in the environment (including in the infected plant) of the compounds that influence antifungal activity.

Table 10.3.   Concentration of the FAs rishitin and lubimin (g/ml) accumulated in response to potato tuber inoculation with compatible (0) and incompatible (1.2.3.4) races of *P. infestans* (Ozeretskovskaya OL, 1973)

Potato cultivar/resistance genes	Race	Rishitin	Lubimin
Uralsky/R3R4	0	49	41
	1.2.3.4	2	4
Krasnoufimsky/R2R3	0	47	66
	1.2.3.4	2	3
Epoka/R3R4	0	63	51
	1.2.3.4	3	3
Stolovy 19/R4	0	44	35
	1.2.3.4	4	5
Veselovsky/R2	0	23	33
	1.2.3.4	4	2
Lava/R3R4	0	52	45
	1.2.3.4	10	5
Vulcan/R1R3	0	42	50
	1.2.3.4	7	6
Kameraz/R1	0	23	40
	1.2.3.4	9	7
Olev/R1R4	0	25	37
	1.2.3.4	5	3

Phytoalexin stereoisomers can also noticeably differ in antifungal activity. The pterocarpane FAs possess both (+) and (−) stereoisomerism. Most legumes accumulate the (−) forms, except peas which produce (+)pisatin, and sophora (*Sophora japonica*) which produces both (+) and (−) maackiain. The antifungal activity of maackiain and pisatin towards some fungi is stereo specific. Some pathogens isolated from alfalfa and red clover are more inhibited by the (+) maackiain than by the (−) form of this phytoalexin. The differences in susceptibility probably depend on the pathogen's inability to degrade the (+) isomer. These data are promising for development of the plants producing the same FA with different steric configurations.

Though most plants produce FAs in a classic way, in some species (cucumbers, sweet melons, watermelons) no FA formation has been found, though there are all grounds to assume that they are available.

One of the possible uses of phytoalexins for disease control is transformation of β-glucanases and chitinase genes into a plant that will not only destroy the pathogen covers but also produce the oligomers possessing elicitor activity and inducing FA formation. Another opportunity is transformation into a plant of the genes necessary for synthesis of unrelated FAs. An example can be FA from carrots, which are resistant to many horse bean pathogens. Transformation of the carrot FA synthesis genes into horse beans protected the latter from many horse bean pathogens. A complex aspect of this method is that FA transformation from one plant into another often requires insertion of not one gene but a group of genes responsible for FA synthesis. These approaches will be discussed in more detail in the final chapter.

PR-proteins

PR-proteins (pathogenesis-related proteins) were first discovered by virologists in the leaves of a tobacco plant cultivar that produced hypersensitive response (HR) to inoculation with TMV. Various PR-proteins were later found in many other plant species, but the tobacco proteins have been investigated most extensively. In this connection, this section will primarily discuss the tobacco PR-proteins and the nomenclature accepted for them.

The term "pathogenesis-related proteins" means a group of proteins induced in a plant in response to fungal, bacterial, viral, and viroid diseases, as well as to some chemicals. The first researchers came to a conclusion that all PR-proteins possess certain common properties: they are soluble only in acidic conditions (pH = 3), they are present in extracellular fluid, and they are highly resistant against degradation by proteolytic enzymes. It would seem logical, as extracellular fluid around the necrotic area has low pH and high proteolytic activity, therefore, proteins possessing the necessary properties should be well adapted to these conditions. However, it was found later that some PR-proteins (for example, of tomato and potato) were not acidic but basic, besides, they were destroyable by proteolytic enzymes.

Next, another set of PR-proteins was found in tobacco plants which were not localized in intercellular fluid but in vacuoles, and they constitutionally expressed in the roots. Basic proteins are induced in the leaves upon treatment with ethylene or upon wounding, whereas acidic proteins do not express in these conditions.

All tobacco PR-proteins are divided into several groups. Each acidic extracellular protein is designated with a digit for the group and also with various letters: "a", "b", "c", etc. The numbering is based on relative mobility at electrophoresis in a non-denaturing gel system. Different proteins designated with the same number (for example, PR-1a, PR-1b, PR-1c), are serologically related, have a close molecular weight, and possess partial similarity in amino acid sequence.

The PR-proteins encoding genes express through various signaling pathways. PR-1 and PR-5 are SA-dependent. Gene PR-4 is induced both by SA, and JA, and ethylene; gene PR-12 (Arabidopsis defensin gene) is induced by ethylene and JA, but do not respond to SA. Induction of the tobacco gene PR-2 by ethylene is based on the presence of GCC-boxes in the promoter part of the gene that bind to ethylene.

PR-1 type proteins

Among the proteins accumulating in tobacco leaves in response to TMV, acidic forms PR-1a, PR-1b, and PR-1c have been found which have similar molecular weight and amino acid sequence, as well as serology affinity. In addition to tobacco plant, these proteins were also found in infected cowpea, potato, maize, barley, and tomato. This broad occurrence of type PR-1 proteins both in monocotyledons and dicotyledons confirms their important role in plant responses to stress.

Another subgroup comprises basic PR-like proteins. In the amino acid sequence they are 65% similar to the above mentioned acidic proteins of tobacco. Similar to the acidic PR-proteins, the basic proteins contain a hydrophobic N-terminal site consisting of 30 amino acids, which, probably, functions as a signaling peptide necessary for penetration through the membrane. The tobacco genome can contain more than 6 genes corresponding to the basic PR-proteins. Group PR-1 proteins of tomato leaves inhibit germination of zoospores and pathogenicity of *Phytophthora infestans*.

Accumulation of PR-proteins upon inoculation with the necrotizing pathogens and acquisition of resistance to the subsequent inoculation by the plants suggest that group 1 PR-proteins are involved in systemic acquired resistance.

PR-2 type proteins: β-1,3-glucanases

β-1,3-Glucanases are able to destroy cell walls of some fungi as well as split off the fragments possessing immunoregulating properties (endogenous elicitors and suppressors). Initially, three acidic extracellular proteins were found and designated, respectively, PR-2a, PR-2b, and PR-2c. Later, two other additional acidic proteins were identified, which turned out to be β-glucanases, like the first ones. In addition to acidic glucanases, a basic glucanase has been identified, which, together with glucanase PR-2c possesses a 50–250 times greater activity of laminarin degradation (β-1,3-glucan), reserve carbohydrate of oomycetes) than all the other glucanases. The basic glucanase accumulates in vacuoles and expresses constitutionally in the roots, lower leaves, and flowers of uninfected tobacco plants up to a high concentration, but in the upper leaves it is found only upon inoculation.

On the basis of the amino acid sequencing, the glucanases have been classified into three classes. Class I includes the ca. 33 kDa basic proteins localized in vacuoles. They are generated as protoproteins subjected to enzymatic proteolysis. Classes II and III include the ca. 36 kDa acidic proteins. Micromolar doses of PR-2 proteins (ca. 50 mkg/ml) show antifungal activity against a broad range of the plant and animal pathogens.

In the tobacco cDNA library, three sites of the sequences corresponding to the basic glucanases and five sites encoding acidic glucanases have been identified. β-1,3-Glucanases are taken to play a role in the processes leading to the hypersensitive response in the plant tissue.

PR-3 type proteins: chitinases

Endochitinases are divided into five groups, or classes. Class I molecules (ca. 32 kDa) contain N-terminal cysteine-rich domain (ca. 40 amino acids), chitin-binding hevein-like domain, and highly conserved central portion. Class II (27–28 kDa) lack the N-terminal cysteine-rich domain. Class III (28–30 kDa) do not share amino acid sequence homology with any other class. Class IV proteins are similar to class I but significantly shorter due to four deletions. Class V (41–43 kDa) includes the proteins similar to bacterial chitinases.

In plants, chitinase and β-glucanase express upon pathogenic infection, wounding, and treatment with elicitors, their expression usually correlating with resistance. Therefore, group II and III PR-proteins are believed to be involved in induced plant resistance. A combination of chitinases and β-glucanases is especially efficient in inhibiting the pathogen growth.

PR-4 type proteins

PR-4 proteins are divided into two subclasses. Subclass I comprises 13–14.5 kDa chitin-binding proteins. They are similar to the chitin-binding lectin hevein. Subclass II has no chitin-binding domain. Proteins of both subclasses possess a broad range of antifungal activity. Serologically similar proteins accumulate in the apoplast fluid of the tomatoes inoculated with *Cladosporium fulvum*.

PR-5 type proteins: taumatin-like proteins

This group comprises two proteins of tobacco plant that are also absent in uninfected plants. Their molecular weight without the hydrophobic N-terminal site – which possibly serves as a signaling peptide – is approximately 24 kDa. Investigation of the amino acid sequences of group PR-5 proteins showed their similarity to the sweet-tasting proteins taumatins from the African bush

Taumatococcus daniellii (therefore, they are called TL-proteins), and also to the α-amylase-trypsin inhibitor of maize. The tobacco genome contains two genes of group PR-5 proteins.

The taumatin-like proteins are structurally very similar to osmotin which accumulates in tobacco cell culture – the cells adapted to osmotic stress. It has also been found in uninfected roots of tobacco plants. However, neither osmotin nor PR-5 protein has the sweet-taste characteristic of taumatin.

The similarity of PR-5 proteins to the bifunctional α-amylase proteinase inhibitor of maize (60% identity in amino acid sequences) substantiates the assumption regarding the functions of these proteins in induced resistance against a broad range of pathogens and pests, which probably consists in inhibition of the digestive enzymes.

The use of PR-protein isolation methods in tobacco plant has led to discovery of the numerous proteins in other plant species; these proteins can also be provisionally classified among the pathogenesis-related proteins. They include virus replication inhibitors, products of the genes encoding potato wound responses, proteinase inhibitors, thionines, cell wall coat proteins, aromatic biosynthesis enzymes, peroxidases, superoxide dismutases, etc. It is difficult to say today whether all these proteins could be placed among the PR-proteins only by one feature, induction in response to stresses.

The functions of some of the above described proteins are known, thus, the specific role played by a particular protein in pathogenicity could be understood. For instance, the infection-induced peroxidases are involved in biosynthesis of lignin and suberin, and prevent the spread of the pathogens in the plant and the damage they cause. Similarly, the role of the proteinase inhibitors preventing the pathogen feeding is clear. Glucanases and chitinases are able to destroy cell walls of fungi and probably, bacteria.

However, the functions of many PR-protein groups are still far from being found. Some proteins specifically induced by the pathogens in one organ can constitutionally express in tissues of other plants or even in the other organs of the same plant, at any time or at a certain phase of development. One would hardly disagree with the opinion that investigation of PR-proteins' functions in plants is one of the most complex, unclear, and sometimes also contradictory divisions of phytopathology.

Some additional information regarding PR-proteins will be provided in the section "Systemic acquired resistance of plants".

Antiviral proteins

The section on plant resistance factors of Chapter 4 has been considering the inhibitors of the protein origin, which are plant constitutional metabolites: endogenous antiviral proteins (EAVP), with their basic characteristics and mechanisms of action.

However, infected plants contain antiviral proteins whose synthesis is induced in response to infection and which are absent in healthy plant tissues before the infection. They are called induced antiviral proteins (IAVP). As against EAVP, which influence the early phases of a virus life cycle and which have many common characteristics with induced inhibitor proteins, IAVP inhibit the late stages of viral infection. This is evident from the fact that plant treatment with IAVP considerably reduces the virus titer in the plant. Besides, IAVP remain efficient even if they are used many hours after the inoculation with the virus.

As IAVP in plants are present in the tissues possessing systemic resistance, many investigators view them as possible candidates to the role of signal molecules in development of systemic acquired resistance of plants to viruses. However a low level of expression of these protein-encoding genes considerably impedes the protein isolation from plant tissues in the form of purified agents for further investigation. Nevertheless, several IAVP have been studied in sufficient detail to date.

Inhibitor of viral replication

The inhibitor of viral replication (IVR) from tobacco plants consists of two biologically active proteins, 26 and 57 kDa, which are, respectively, a monomer and a dimer. IVR can be isolated from TMV-inoculated protoplasts and tissue extracts of supersensitive tobacco cultivars. This highly active IVR accumulates in tissues in very low concentrations, 10 ng/g of tissue. Such very low IVR content in plants appears to be the reason why IVR transcripts are not found in the respective tissue extracts.

IVR is non-specific in its antiviral activity because it inhibits replication of several different viruses, except its inductor TMV. IVR can suppress virus replication in protoplasts, if it is used not later than 18 hours after the inoculation, and in leaf disks, not later than 5 hours after the inoculation. In uninoculated distal leaf tissue, IVR is found as soon as 30–36 hours after inoculation with the virus, though in the upper leaves it can be found only 7–14 days after the inoculation.

IVR can be isolated both from inoculated and non-inoculated tissues of an attacked plant, and its formation seems to be one of the host plant defense responses aimed to localize the virus infection.

IVR is a typical inhibitor of viral replication because:

(1) it can decrease concentration of viral particles in the tissues of systemically infected host plants;
(2) it is efficient if applied for five or more hours after inoculation with the virus and even may be efficient 18 hours after the inoculation.

Inhibitors from "green islands"

Light and dark green spots, a classic sign of a mosaic virus, develop on the leaves infected by some viruses. The "green island" cells usually contain less virus particles compared with the light green leaf areas and look cytologically or physiologically uninfected. The "green islands" show systemic resistance to subsequent reinfection by the same or closely related systemic viruses.

The reasons of the virus-free nature of the "green islands" have not yet been discovered. Possibly, IAVP that occur in systemically infected tissues are involved in maintaining a low concentration of the virus in the green islands. "Green islands" were found to contain induced antiviral proteins responsible for suppression of viral infection – they are usually called inhibitors from green islands (IGI). They have many similar features with other IAVP. For instance, IGI isolated from the "green islands" of the cucumber infected with the cucumber mosaic virus is very similar to the tobacco IVR. Like the tobacco IVR, it has two active fractions of close molecular weight, it affects the virus replication, and reacts with antiserum to IVR.

IGI isolated from "green islands" of different plants are serologically different.

It is interesting that tobacco IGI can inhibit TMV replication in protoplasts and leaf disks treated with this inhibitor after their inoculation with the virus, but the "green island" tissues from which IGI was isolated remain susceptible to the necrotizing TMV strain.

Antiviral factors

The term "antiviral factors" (AVF) designates antiviral proteins active due to the presence of the supersensitivity N-gene in tobacco plants. One of such AVF is a 22 kDa phosphorylated glycoprotein that occurs in a plant as a precursor, pre-AVF. Its processing is induced by viral infection, probably, in response to emergence of a double-stranded RNA (dsRNA), a replicative form of the viral RNA, in the cells.

Like tobacco IVR, AVF are induced only by viral infection, they inhibit virus replication, and can play a role in development of systemic acquired resistance. The AVF are found after systemic infection, even at very low degree of inoculation. Apparently, the level of AVF accumulation can determine whether the viral infection will be localized or become systemic.

The AVF, similar to IVR, are powerful inhibitors and express in very small quantities, which impedes isolation and further analysis of these proteins. Just as in the case of IVR, an mRNA, corresponding to AVF-transcripts, cannot be found in the plants infected with a virus.

It is interesting that AVF have many common characteristics with interferons, including temperature and acid stability, glycosylation, broad antiviral range, and induction of oligoadenylate synthesis. In particular, AVF from tobacco plants induce oligoadenylate synthase, and oligonucleotides can activate AVF. Human interferons were shown in some cases to be able to inhibit replication of plant viruses. In addition, it was found that human interferon-induced oligoadenylates, as well as synthetic 2'-5'-oligoadenylates, inhibit TMV replication in tobacco plants.

Proteinase inhibitors

Proteinase inhibitors are a special group of plant proteins brought together due to a common ability to form stoichiometric complexes with proteinases, which leads to competitive inhibition of catalytic activity. Proteinase inhibitors occur in plants of various taxonomic groups. They are abundant in seeds and other accumulating organs of plants where they can make up to as much as 5–10% of water-soluble proteins.

Almost all proteinase inhibitors from plants are carbohydrate-free peptides or small proteins. Proteinase inhibitors are believed to be resistant against activity of proteinases themselves, though there are some exceptions from this rule.

Proteinase inhibitors of cereals and Solanaceae have been investigated most extensively. Among the Solanaceae family, the greatest diversity of the inhibitor forms has been described in potato. Tubers of this plant were found to contain the inhibitors that suppress the activity of serine, cysteine, and aspartyl proteinases and carbopeptidases. Inhibitors of serine proteinases are usually divided, on the basis of common properties, into the following groups of related proteins: family of potato inhibitors I, family of potato inhibitors II, and proteinase inhibitors of the Kuniz family (the first crystalline inhibitor of trypsin isolated by M. Kuniz from soybean in 1945–1946).

The greatest number of studies investigate serine proteinases of the family of potato inhibitors I and II, where a typical feature is the oligomeric structure (potato inhibitor I consists of 4–5 subunits, inhibitor II is a heterodimer).

An overwhelming majority of protein inhibitors of serine proteinases work by the so-called substrate-like mechanism, where the inhibitor acts as a highly specific enzyme substrate exposed to slow limited proteolysis.

The primary structure of inhibitor I from potato is similar to the inhibitors from some other plants. Therefore, they are all included in the same family of the potato inhibitor chemotrypsin I, though not all of them are isolated from potato. Inhibitors of this family are common in the plants of various taxonomic groups.

Serine proteinase inhibitors are concentrated in the central vacuole. Transport of inhibitors from cytoplasm to the central vacuole is believed to occur together with their transformation in active forms as a result of limited proteolysis. For instance, inhibitors I and II result from post-translational limited proteolysis provided by the respective proteinases.

Potato tubers also contain low molecular weight proteinase inhibitors. Some of them may be products of limited proteolysis of high molecular weight inhibitors whereas others can be a direct result of translation.

Many proteinase inhibitors from plants are characterized by "double-headedness", i.e. each molecule of such inhibitor can simultaneously bind two molecules of the same or different proteinases in two reactive centers.

There are a large number of proteinase inhibitors in plants, which provides additional opportunities for their regulation.

(1) Proteinase inhibitors can differ in substrate specificity.
(2) Inhibitors occur in various isoforms.
(3) Oligomers of proteinase inhibitors can dissociate and come together in various combinations, the inhibitor properties depending on the type of oligomers and their combination.

Usually, three key functions of proteinase inhibitors are analyzed:

(1) Proteinase inhibitors as plant reserve proteins;
(2) Proteinase inhibitors as protein or enzyme regulators of the plant;
(3) Proteinase inhibitors as a component of plant defenses.

Obviously, most interesting to us is the protective function of proteinase inhibitors, therefore, the other functions will be discussed briefly.

(1) It was noted earlier that the greatest amount of proteinase inhibitors is contained in the plant accumulating organs: seeds and tubers, the inhibitors being synthesized at the same phase of development as the reserve proteins. Upon grain germination the inhibitor activity drops and dies away. In potato tubers germination, the content of inhibitors continues to decrease till the plant begins generating new tubers where the inhibitor accumulates again.
(2) Possibly, the main function of proteolytic enzyme inhibitors is their involvement in regulation of plant own proteinases. In most cases, plant endogenous proteinases are affected by cysteine proteinase inhibitors, and not by trypsin and chemotrypsin inhibitors that affect proteinases of microorganisms. Indications are available that the cysteine proteinase contained in germinating seeds does not only degrade their protein zein but also destroys the inhibitor of endogenous serine proteinase. Protein bodies of buckwheat seeds simultaneously contain all elements of proteolysis, i.e. the inhibitor and substrate. Proteinase inhibitors can protect plants from uncontrollable proteolysis. Since they are contained in cytoplasm, they are able to protect plant cell proteins in case of damage of intracellular structures and release of proteinases.
(3) Proteinase inhibitors can perform an important protective role in the plants attacked by microorganisms. For instance, many protein inhibitors of trypsin and chemotrypsin can suppress serine proteinases of microorganisms. For example, potato tubers contain inhibitor I capable of suppressing chemotrypsin proteinase, and inhibitor II (double-headed) suppresses both trypsin and chemotrypsin. Both inhibitors actively suppress subtilisin and proteinases of the genus *Aspergillus* fungi.

The effect of inhibitors I and II from potato on the proteinases secreted in culture fluid of *Fusarium solani* and *F. sambucinum* was shown. Suppression of proteolytic activity can impair the pathogen ability to absorb plant proteins, and accordingly, to suppress their growth and development.

Trypsin and chemotrypsin inhibitors directly suppress growth in the culture of some phytopathogenic fungi.

In some cases, interrelation between plant disease resistance and their proteinase inhibitor content has been established. Such correlation was found for wheat resistance to stinking smut, and for lupine and soybean to fusarial wilt.

In germination of legume seeds, proteinase inhibitors rapidly diffuse in surrounding solution and, apparently, contribute to seed defense against infection at early stages of germination.

Potato tubers contain two key reserve proteins. They are patatin totaling ca. 40% of all water-soluble proteins of tubers, and proteinase inhibitor II (ca. 10%). Patatin possesses lipidacyl hydrolase activity which may contribute to protection of potatoes from phytopathogens. Upon potato tuber wounding, patatin formation in leaves and stems terminates, and systemic induction of inhibitor II begins.

From barley seeds a "two-headed" inhibitor was isolated that contained independently working centers of binding subtilisin and α-amylase. Later, similar bifunctional inhibitors were isolated from seeds of wheat, rye, and triticale. The amino acid sequence of two inhibitors of endogenic α-amylase and subtilisin from wheat and barley has been established. The molecule of a bifunctional inhibitor contains two methionine residues. To assess the position of the residue contained in the reactive center, the inhibitor was subjected to limited proteolysis by staphylococcal proteinase, which resulted in complete disappearance of the activity of the α-amylase inhibitor, while the activity towards subtilisin was retained.

The bifunctional inhibitors are supposed to be able to play an important role both in prevention of grain germination and in plant protection against phytopathogens. In dormant barley grain the bifunctional inhibitor is localized in endosperm. During plant germination the inhibitor is transported from endosperm to the germ. Apparently, the bifunctional inhibitor of α-amylase and subtilisin, on the one hand, inhibits grain germination under unfavorable conditions, and on the other hand, suppresses proteinase activity of the pathogenic microorganisms.

A considerable contribution in the study of the role of proteinase inhibitors was made by the research of the American biochemist C. Ryan. It was found that simple mechanical wounding of potatoes or tomatoes causes fast accumulation of proteinase inhibitors over the entire plant. Accumulation of proteinase inhibitors was observed in tomatoes after inoculation with *Phytophthora infestans*. Synthesis of proteinase inhibitors occurred after tobacco plant treatment with cell wall preparation of *P. parasitica*. The plant response was due to formation of a specific wound hormone in the damaged place, designated as PIIF (proteinase inhibitor-inducing factor). PIIF was identified as a soluble pectic polysaccharide enriched by residues of uronic acid, with the molecular weight ca. 5 kDa. 4-α-D-galacturonosyl-D-galacturonic acid was found to be a minimum fragment necessary for PIIF activity. The inhibitor activity works in the presence of an intact hemiacetal ring and a free carboxyl group at C-6.

PIIF is produced as a result of enzymatic degradation of the cell wall, and it belongs to the "endogenous elicitors" discussed earlier. Under the PIIF effect, the content of inhibitors I and II in the leaves of Solanaceae could be as much as 2% of the total newly synthesized proteins. The resulting inhibitors were localized in the vacuoles.

The ability to form PIIF due to damage was found in 10 of 23 investigated plant species. A greater accumulation of PIIF was observed in alfalfa, then in tobacco, tomato, potato, strawberries, cucumber, pumpkin, clover, horse beans, and grapes. The similarity in PIIF induction found in the species of different families is indicative of the uniform principles of plant defenses.

Formation of proteinase inhibitors could be caused, besides PIIF, by citrus pectin, oligouronides with the degree of polymerization 2–20, as well as soluble derivatives of chitin and chitosan. In some cases proteinase inhibitors can be induced by the effect of abscisic acid.

Ryan suggested a hypothetical pattern of transmission of the induction signal of proteinase inhibitor biosynthesis in response to wounding by insects or infection by phytopathogens (Figure 10.6). The first pathway, due to mechanical wounding caused by insects, is induced by the polypeptide systemin which is rapidly transported over the entire plant, interacting with the specific receptors. A second pathway is due to infection by pathogens, and it assumes induction by oligouronides that are released from plant cell walls and interact with other specific receptors. It is needless to say that the suggested pattern of two pathways of proteinase inhibitor induction is hypothetical. However, it is supported by

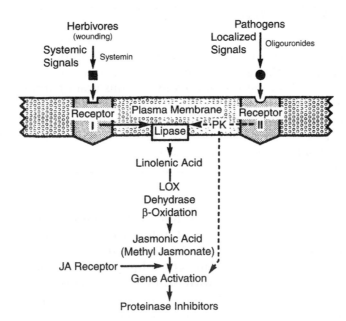

Figure 10.6.   Signal transduction of expression of genes of proteinase inhibitors in plant by insect feeding (I) and by infection of pathogenic microorganisms (II) (Ryan C A, 1992).

the fact that the specific lipoxygenase inhibitor suppresses synthesis of the proteinase inhibitors generated in response to wounding but has no effect on accumulation of inhibitors under the activity of oligouronides.

The role of proteinase inhibitors in pathogenesis processes should be viewed in a very broad range, as they can regulate activity of the enzymes involved in defenses: both inhibit unnecessary proteinases and activate the previously bound proteinases, or, on the contrary, bind the proteinases which could decompose the newly synthesized proteins.

Phenylpropanoids and lignin

Phenylpropanoid compounds constitute a broad range of structural cyclic substances possessing numerous physiological functions. They are formed as a result of deamination of the amino acid phenylalanine by the enzyme phenyl alanine-ammonia lyase (PAL).

The structure of simple phenylpropanoids is C6–C3, the carbon carcass of phenylalanine (Figure 10.7). Hydroxylation, methylation, and dehydrogenation of cinnamic acid results in formation of its phenolic derivatives (*p*-cumaric, caffeic, ferulic, and sinapic acid), and also simple cumarins (umbelliferone, psoralen). Free acids rarely accumulate in plant cells up to high levels: they are usually conjugated with sugars (for example, conjugate of salicylic acid with glucose), cell wall carbohydrates (ferulic acid esters), or with organic acids (esters of sinapic and chlorogenic acid). Salicylic, benzoic, and *p*-hydroxybenzoic acids are not true phenylpropanoids as they have no tricarbonic chain which is typical of cinnamic and *p*-cumaric acid.

Many stress-inducible phenylpropanoids possess a C15 flavonoid carcass synthesized through chalcone synthetase that condenses *p*-cumaryl-CoA and three molecules of malonyl-CoA. Thus, chalcone synthetase connects two phenol rings, one of which is synthesized through shikimic acid, and the other through acetic acid (these two pathways of phenol synthesis were described in Chapter 5). In most plant families, the initial product of chalcone synthetase is tetraoxychalcone which further turns into other classes of flavonoids, such as flavones, flavanols, antocyanes, and 3-deoxantocyanidines. In the plants of the leguminous family, which alongside chalcone synthetase also possess the enzyme chalcone reductase, trioxychalcone can be formed. In a number of species, including pine, grapes, and peanut, condensation of *p*-coumaroyl-CoA or CoA cinnamic acid with three malonyl-CoA molecules that is catalyzed by the enzyme stilbene synthetase can result in formation of stilbenes. In legumes, isoflavone synthetase modifies the isoflavonoid carbon carcass, which results in accumulation of a number of simple isoflavonoids: cumestanes, isoflavanes, and pterocarpanes. Phenylpropanoids can be structurally different due to hydroxylation, glycosidation, alkylation, prenylation, sulfatation, and methylation.

With the sophistication of the methods of enzyme purification as well as use of radioactive labeling, the stages of phenylpropanoid transformation

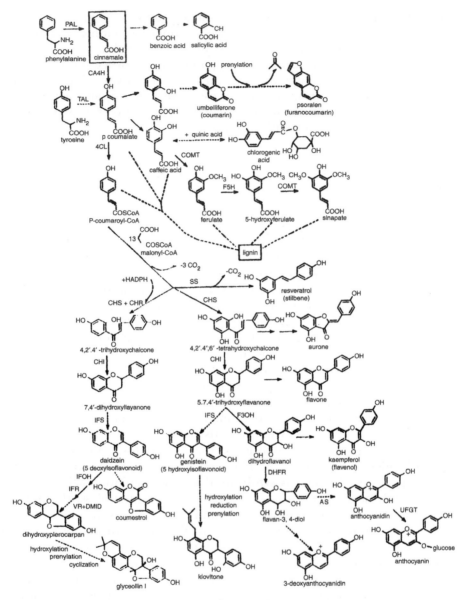

Figure 10.7.   Biosynthesis of stress-induced phenylpropanoids (Dixon R A, Paiva N L, 1995 (Figure 1)). Plain lines indicate the reactions that are catalyzed by single enzymes (many of these genes are cloned). Dashed lines indicate the pathways that are regulated by some enzymes that are not investigated enough or that are differ between plant species.

Enzymes: CA4H – cinnamic acid 4-hydroxylase; CHI – chalcone isomerase; CHR – chalcone reductase; CHS – chalcone synthase; 4CL – 4-coumarin: CoA ligase; COMT – caffeic acid-O-methyltransferase; DHFR – dihydroflavanol reductase; DMID – 7,2-dihydroxy-4-metoxyisoflavanol dehydrase; F3OH – flavono-3-hydroxylase; F5H – flavonic acid-5-hyroxylase; IFR – isoflavon reductase; IFS – isoflavon synthase; PAL – phenylalanine ammonium lyase; SS – stilbene synthase; TAL – tirosine ammonium lyase; UFGT – UDP-glucoso-flavonol-3-O-glycvosyltransferase; VR – vestitin reductase.

have been revealed. Figure 10.7 summarizes various types of phenyl-propanoids induced in a plant by various biotic and abiotic stresses, and the enzymes involved in their transformation.

Many stress-inducible phenylpropanoids are categorized as phytoalexins. They include pterocarpanes (glyceollins), isoflavanes, prenylated isoflavonoids (kievitone), stilbenes, psoralenes, cumarins, 3-deoxyanthocyanidines, flavanols (quercitin, campherol), and aurones. The level of these compounds grows around the infection site up to the concentrations that are toxic for the pathogens in the experiments in vitro. The amount of salicylic acid in tobacco, cucumbers, and Arabidopsis grows in response to infection, exposure to UV radiation, and increased concentration of ozone in the air. It was shown earlier and will be shown later that salicylic acid, in addition to its inherent antimicrobial activity, also constitutes an element of the signaling system that leads to acquiring local and systemic resistance.

Improvement of the modern biochemical equipment enabled discovery of the new enzymes of phenylpropanoid biogenesis. So sulfotransferases were discovered, each of them catalyzing sulfatation of a specific flavonoid hydroxyl. It was also found that plant tissues contain highly specific isoflavonoid prenylisotransferases.

Though the relationship of phenylpropanoids and stressful factors was established a long time ago, it remained unclear for a long time in which cells they are synthesized. It was only known that phenylpropanoid FAs accumulate near the infection site. It was found that UV-beams induce accumulation of antocyanes and other flavonoids in the upper epidermal cells of plants. However, it was not clear whether these compounds are synthesized in the epidermal cells or transported into these cells from the adjacent cells. The hybridization and immunolocation methods enabled the discovery that flavonoids, chalcone synthetase proteins, as well as transcripts of the enzymes phenylalanine ammonia lyase (PAL), chalcone synthetase, and 4-coumarin-CoA ligase are accumulated around the site of infection in the parsley leaves inoculated with a species of *Phytophthora megasperma* nonpathogenic for parsley.

The subcellular sites of phenylpropanoid biosynthesis are still a subject of discussion. Many phenylpropanoids are accumulated in vacuoles, usually in the form of glucosides or conjugates. The latter can then be involved in transport through tonoplast.

Some major enzymes of the phenylpropanoid pathway of biosynthesis are described as cytoplasmic (soluble), however, there is serious evidence that they are related to the membrane structures. Some soluble enzymes are believed to form "metabolones", or metabolic clusters on the membrane surface. The intermediates of the phenylpropanoid metabolism rapidly move from one enzyme to another, are transported inside the membranes, and arrive in a vacuole. This model covers simple phenylpropanoids, whereas prenyltransferases involved in the synthesis of prenylated pterocarpanes and furanocumarins are connected with the plastid membranes.

In many plant species, especially in legumes, such key enzymes as PAL and chalcone synthetase are encoded by a gene family. So far, it is not clear whether

such encoding reflects different forms of enzymes in the metabolic compartments, or stressful situations require so many phenylpropanoids that they can
be provided only by a gene family. PAL is a tetramer enzyme, therefore, expression of many encoding genes (at least four), in combination with possible post-
translational modification, can result in formation of heterotetramer forms
that will respond differently to different stimulants. Chalcone synthetase also
possesses the isoenzyme forms.

As plant defenses are multicomponent by nature, it is not easy to identify
which compound and in what concentration can protect the plant.
Unfortunately, those plants which are investigated as the models for the FA
response, such as soybean, alfalfa, green beans, and parsley, are not ideal objects
for classical and mutation mapping, whereas Arabidopsis, convenient for
molecular and genetic studies, does not form flavonoids and furanocumarins
characteristic for the former species.

The first direct demonstration of potential significance of phytoalexins in
plant protection was introduction of the stilbene synthetase gene from
grapes in a tobacco plant. A foreign gene product was found to be capable
of switching the chalcone synthetase substrate to synthesis of stilbene FAs,
resveratrols, therefore, the plant acquired resistance to the gray rot pathogen
Botrytis cinerea. However, according to some data, a number of individual FAs
cannot play a significant role in the defenses. For instance, investigation of
an Arabidopsis mutant deficient in the indole FA camalechin showed that
this phytoalexin does not constitute a resistance factor against an avirulent
strain of *Pseudomonas syringae*, however, it delays growth of a virulent strain of
the same bacterium.

Analysis of transgenic tobacco plants with a lower level of phenol compounds showed a much higher growth rate of *Phytophthora parasitica* in these
plants compared with the control. In transgenic plants, conversion of tryptophan to tryptomine leads to reduction of the phenylalanine pool, drop in the
level of chlorogenic acid and bound phenols, and consequently to higher
infestation with *Phytophthora*.

Lignin

Lignin is a complex of the polymers formed from a mix of simple phenylpropanoids (see Chapter.6). It is produced in secondary thickening of cell
walls, therefore the intercellular space is impregnated with this phenol polymer. Lignification constitutes enzymatic dehydrogenation and subsequent
radical condensation of the cumaryl, coniferyl, and sinapyl alcohols. The
ratio of these building blocks differs in different plant species. Lignin of
dicotyledon plants contains two main monomer types of the subunits G and
S, termed guaicyl and syringyl. The former is derived from the coniferyl alcohol, and the latter from the sinapyl alcohol. The ratio of G and S subunits is
different in different lignins.

Lignin is covalently bound with cellulose and hemicelluloses as well as with cell
wall glycoproteins, and possibly, with the oxyproline-rich proteins. Incompletely
formed lignin probably plays a role in fungal hyphae growth limitation.

Local lignin impregnations sometimes occur in protoplasm at the early stages of fungal infection with incompatible parasites. It is interesting that the composition of lignin induced by infection frequently differs from lignin of healthy tissues. Rapid accumulation of the lignin precursors is frequently accompanied by a local increase in de novo PAL, peroxidase, and hydrogen peroxide involved in the synthesis and condensation of the lignin components. Treatment of plant tissues with PAL inhibitors suppresses lignin formation and promotes penetration of infection in plant tissues.

An important role in lignin formation is played by peroxidase, the activity of which, similar to PAL activity, grows upon mechanical damage of tissues and infection. Peroxidase: (1) carries out polymerization of cinnamic alcohol to lignin; (2) produces strong cross-links between cellulose, pectin, cell wall glycoproteins, and lignin; (3) is involved in wound healing – deposition of a water-tight barrier from polymerized aliphatic and aromatic compounds (tissue subernization); (4) is involved in catabolism (auxin decomposition). The Arabidopsis genome library was found to contain more than 70 genes that encode peroxidase and proteins structurally close to peroxidase. In the plant cell, peroxidase is localized in various compartments and exists in several isoforms:

(1) Cationic forms (pH 8.1–11). They catalyze formation of hydrogen peroxide from NADH and water. Localization in the central vacuole. They possess indolyl acetic acid-oxidase activity and regulate the level of auxins in a cell.
(2) Weak anionic forms (pH 4.5–6.5). Localization in cell walls. They are moderately active lignificators. They carry out wound subernization.
(3) Anionic forms (pH 3.5 – 4.0). Localization in cell walls. They carry out lignification and cross-linking of cell wall polysaccharides. Their activity is regulated by the level of auxins.

The increase in local lignification is one of the mechanisms limiting the spread of viruses over the plants tissues. It was found that the hypersensitive response of the TMV-affected tobacco leaves involves, in addition to PAL and peroxidase, 4-hydroxylase of cinnamic acid, CoA ligases, and *o*-methyltransferases, their content rapidly growing around the necrosis area.

Oxyproline-rich glycoproteins

The proteins bound to the primary plant cell wall are usually glycoproteins. They are either covalently bound to the cell wall or stay in the cell wall in the dissolved state. Some of them are enzymes, others perform structural functions.

Most cell wall glycoproteins are rich in oxyproline. The oxyproline-rich glycoproteins (ORG) are linear molecules rich in basic amino acids. A high level of these amino acids transform the cell wall into a polycationic barrier binding negatively charged particles or cells that are similar to bacterial cells.

All oxyproline-rich glycoproteins possess common structural features.

(1) High content of oxyproline, β-oxyamine acid serine, threonine, and either glycine or alanine.
(2) Content of arabinose and galactose in equal proportion.
(3) In addition to arabinose and galactose, other sugars and amino acids can be present.

Plant cells secrete at least three classes of glycoproteins that contain a large amount of oxyproline: extensins, arabinogalactane proteins, and lectins.

Extensin is secreted as a soluble protein, weakly bound to the cell walls through isodityrosine bridges; eventually, an insoluble cell component is produced. Two thirds of the soluble form of extensin is carbohydrates, and one third, proteins. The molecular weight of carrot extensin is 86 kDa, and approximately 36 kDa is released in deglycosylation. Extensin cross-linking with the cell wall components seems to be built with involvement of a specific peroxidase.

Extensin has an unusual amino acid composition: 41 mol.% of oxyproline, 12 mol.% of serine, 10 mol.% of lysine, 8 mol.% of tyrosine, and 0–11 mol.% of histidine. The lack of acidic amino acids and the large content of lysine and histidine determine the basic isoelectric point within 10–12. Cell wall extensin can be converted into a soluble glycoprotein using prolyl hydroxylase and oxyprolyl arabinosyl transferase. Both enzymes are localized in the Golgi apparatus. Tri- and tetrarabinoside side chains are mostly connected with oxyproline residues.

As opposed to extensin, arabinogalactane proteins are soluble oxyproline-contained glycoproteins localized both in cytoplasm and extracellularly. The amino acids of arabinogalactane proteins are mostly oxyproline, serine, alanine, and glycine. Arabinose and arabinogalactane chains with high degree of polymerization, as well as the branched carbohydrate core of α-D-galactopyranosyl residue are connected with oxyproline. Arabinoside and glycuron acids are additionally attached to the core. Hydroxylation of proline and transport of arabinoside to oxyproline seem to proceed in the same way as in extensin.

And, finally, the third ORG class is the lectins of the plants of the Solanaceae family that are localized both intracellularly and extracellularly. These proteins contain 50–60% of oxyproline, serine, glycine, and cysteine that are the dominant amino acids. The arabinose residues are bound to oxyproline, and galactosyl residues to serine. ORGs from potato agglutinate avirulent but not virulent strains of the bacterial wilt pathogen *Ralstonia solanacearum*. They were named "agglutinins" to distinguish them from structurally similar lectins that possess only low hemagglutinating activity. Agglutinins are soluble compounds of cell walls that contain up to 61% of carbohydrates. The carbohydrate part consists of arabinose and a small amount of galactose, glucose, and glucosamine. The molecular weight is 91 kDa, and it decreases at deglycosilation to 56 kDa. So far, it is difficult to judge whether agglutinins play any role in plant–parasite relationship.

There are some facts to show a protective role of ORGs in plant resistance.

(1) There is a clear correlation between ORG accumulation in cell walls of the melon and the melon's resistance to *Gloeosporium lagenarium*. Analysis of several plant–pathogen systems showed that cell–wall enrichment with ORG occurs in a number of plant species upon inoculation with fungi, bacteria, and viruses.
(2) The level of oxyproline (ORG marker) in cell walls grows much faster upon inoculation of resistant than susceptible cultivars.
(3) Artificial increase, or, on the contrary, suppression of ORG level leads to respective induction or inhibition of resistance.
(4) Treatment of plant tissues with microbial endogenous elicitors induced ORG formation in the tissues.

The mechanisms of ORG biosynthesis are not yet completely clear. It is assumed that ethylene is involved in this process, as the elicitors induce its formation. Besides, ethylene in itself can induce ORG formation. In cell walls of the melon, the ORG content grows by several times in response to inoculation with *Gloeosporium lagenarium*, which is accompanied by a sharp rise in ethylene production in the plant. The elicitors from this fungus or the melon cell walls caused a similar response. Treatment of the melon plantlets with ethylene before the inoculation increased both the ORG content and resistance, whereas inhibition of ORGs promoted their colonization by the pathogen.

Specific inhibition of ethylene biosynthesis in inoculated or elicitor-treated plant tissues suppressed oxyproline deposition in the walls. The ethylene precursor aminocyclopropanecarboxylic acid induced ORGs in the tissues of the untreated plants, and also restored ORG formation in the tissues treated with the inhibitor.

The elicitor from *Phytophthora megasperma* induced synthesis of ethylene and ORGs in soybean hypocotyls. An increased content of oxyproline also occurred at incompatible but not compatible interaction of the cucumber with *Cladosporium cucumerinum*. On the contrary, oxyproline concentration in the wheat cell walls augmented upon inoculation with a compatible line of *Erysiphe graminis*.

Accumulation of ORGs and their mRNA in infected tissues and cell culture occurred in the beans affected by *Colletotrichum lindemutianum*, accompanied by temporary activation of prolyl hydroxylase and arabinosyl transferase. Arabinosyl transferase is connected with the Golgi apparatus, whereas prolyl hydroxylase is concentrated in endoplasmic reticulum.

An increase in ORG level in the cell walls of infected tissues strongly changes their properties. As polymers, ORGs strengthen the cell surface, and as polycations, they change the surface charge.

ORG genes of carrot, tomato, and beans have been cloned and sequenced. In all three cases, the genes possessed considerable homology to each other and contained a characteristic repeating sequence that encodes serine–four proline pentapeptide. ORG formation is encoded by a family of genes that are regulated in a different way in wounded or infected plants.

The amount and rate of mRNA accumulation for ORGs in the bean hypocotyls inoculated with an incompatible race of *C. lindemuthianum* were found to be considerably higher than in a compatible interaction. It was also shown that mRNA for ORGs was also accumulated in uninfected tissue apart from the inoculation site, which was indicative of a systemic character of their formation.

Thus, the plant cell wall contains glycoproteins, many of them possessing fermentation activity. Some of them, especially peroxidases and glycosylhydrolases, can play an important role in plant defense responses against phytopathogens. For instance, the specific peroxidase is involved in lignin biosynthesis and formation of isodityrosine bridges of extensin, as well as the bridges where cell walls are bound by ferulic acid. Glycosylhydrolases, as a common component of plant cell walls, may be responsible for the release of oligosaccharines from the cell walls of plants or fungi.

Modification of plant gene expression at nematode invasion

The effect of plant nematodes on the host is made up of mechanical damage, chemical action, and use of the feeding resources. The molecular response of a plant includes the reaction to all factors of stressful action of the nematodes. Such response in many respects is similar to plant responses to other phytopathogens, but also has its own characteristics. All phytopathogenic nematodes are obligate biotrophic parasites feeding on the contents of cytoplasm of living plant cells. A unique feature is formation of the special structures, feeding cells (syncytium and huge cells), in plant tissues upon invasion by sedentary nematodes. A complex of morphological and physiological changes in the areas of localization of sedentary nematodes, leading to the development of the feeding structures, results from the changes in the host gene expression. Though the molecular signal for development and building of the huge cells has not been elucidated, there are grounds to believe that their formation is induced by the nematodes themselves. The changes of the gene expression in the feeding areas are, probably, confined to several cells. Identification of such genes is extremely difficult. A great success in understanding of where, when, and with what intensity genes of one or another enzyme express, was achieved using the transformed plants. Experiments with the transformed plants, into the genome of which the genetic construct with a reporter gene had been inserted, for instance, β-glucuronidase (*GUS*) gene under control of a promoter (gene expression regulator), showed where in a plant particular compounds express. The promoter of the plant gene triggers the reporter gene according to the expression characteristics of its own gene, and the β-glucuronidase gene is called a reporter gene as it helps to find out where expression of the gene studied occurs under the promoter. The reporter role of the *GUS* gene shows conversion of a colorless substrate into a bright dark blue product. Plant treatment with this substrate results in appearance of the dark blue color where the *GUS* gene expresses. Thus the *GUS* gene and other reporter genes allow finding out where, when, under what effect, and with what intensity the gene, from which the promoter was taken, expresses in a plant. It was shown using this approach that many genes in the feeding cells modify their expression in response to nematode invasion. In such a way expression of the genes encoding the proteins involved in cell cycle regulation, rearrangement of cell walls, metabolism, osmoregulation, and hormonal response to formation of the huge cells has been studied. The tobacco plant was taken as an example in invasion by the root-knot nematode *Meloidogyne incognita* to show that the root-specific gene *TobRB7*, which encodes the water flow, during formation of the huge cells expresses in the meristem roots and in the area of the vascular cylinder. Transcription activity of the cellular markers, such as cyclin-dependent kinase *CLC2a* and myotic cyclin *CYC1At*, at early stages of formation of the huge cells was observed. It is known that formation of the huge cells and syncytium is accompanied by the structural change of the cell walls. Therefore, it is not surprising that plant invasion enhances expression of the genes encoding the cell wall degradation enzymes, endonucleases, and polygalacturonases. In tobacco plant roots, five genes have been identified that enhance expression in the roots invaded by the root-knot and cyst nematodes. Expression of these genes was observed in the huge cells, syncytium, root ends, and lateral roots. It is supposed that the plant growth and development regulators are involved in the development of the huge cells and root-knots. Model experiments with *Arabidopsis thaliana* as the host-plant of two nematodes *Heterodera schachtii* and *M. incognita*, using the reporter-gene approach, showed that nematode invasion can change the expression of the genes associated with the division of the root end meristem, which can serve as confirmation of their involvement in formation of the huge cells and syncytium. Biochemical analysis of the auxins and cytokinins showed an increase in their number in the invaded plants, especially in the root-knots. Analysis of the mutant and reporter-gene constructs showed that the genes responsible for the auxins and ethylene in the plant can work as positive regulators of susceptibility to cyst nematodes.

It was shown in the model system "Arabidopsis – root-knot nematode *M. incognita*", using the *GUS* reporter gene, that enhancement of gene expression in the huge cells induced by the nematodes is the same as in the healthy plants in various types of cells during their development. These studies support the hypothesis that the "normal" biochemical functions can be involved in the development of the pathogen and morphological and physiological changes occurring during the formation of the parasite feeding sites. This was confirmed by the study of the molecular characteristic of the plant gene *PRE* encoding D-ribulose 5-phosphate 3-epimerase, the key enzyme of the pentose phosphate pathway. This metabolism pathway plays a leading role in the actively growing cells at production of NADPH, a necessary compound for numerous biochemical processes (for instance, in the synthesis of fatty acids and isoprenoid compounds, such as sterols), also involved in the intermediate stages of carbohydrate metabolism in the synthesis of nucleotides and cell wall polymers. Analysis of the homozygous mutants revealed that the gene activity mostly shows at the early phases of the huge cells formation. Finally, it is seen from the expression of the plant gene *PRE* in the proliferating root end cells and the small number of these cells involved in induction of the lateral roots at invasion by nematodes that the genetic control of these processes is carried out by the same way.

The change in the plant gene expression observed after invasion by sedentary nematodes can be connected not only with formation of the feeding cells but also with the stress or defenses. For instance, the change in the gene expression of the potato leaves after invasion of the roots with the cyst nematode *Globodera rostochiensis* included induction of PR-proteins. Larvae of the root-knot nematode, which damaged the cells in a much smaller degree than cyst nematodes, also induced expression of these defense genes.

The roots of tomatoes invaded by root-knot nematodes were found to contain the genes homologous to some known defense genes (including the genes of peroxidase, chitinase, lipoxygenase, and proteinase inhibitors). The catalase-encoding genes are also induced locally (in the roots) and systemically (in the leaves) after invasion by the root-knot (*M. incognita*) and cyst (*Globodera pallida*) nematodes. It is known that catalase is involved in plant protection against the oxidative stress caused by parasitic nematodes. The plants invaded by sedentary nematodes were found to contain the genes encoding extensions, which are hydroxyproline-rich proteins of the plant cell walls.

Expression of individual genes in the huge cells or developing root-knots of plants decreases after nematode invasion. For instance, the gene promoter of vegetative phenyl alanine – lyase 1, highly active in the non-invaded cells, "becomes silent" several days after the plant invasion.

Genetics of plant resistance to nematodes

Three categories of plant genes are involved in response to nematode invasion: genes inducing plant defenses; stress-related genes; and genes involved in nematode feeding, including formations of syncytium. Genetic analysis of the potato plants with hypersensitive response allowed characterization of 14 genes involved in defenses, stress, and susceptibility of a plant cell in response to nematode invasion.

The character of nematode–plant relationship in some cases is determined by direct or indirect interaction of a single resistance gene of the host-plant with the avirulence gene of the parasite.

Products of the plant resistance genes were called R-proteins (from the English word "resistance"). The greatest accomplishment of molecular genetics in the recent years was the cloning and structure functional analysis of R-genes.

It is well-known that products of plant resistance genes to bacteria and fungi (R- proteins) have similar structures. One of such structures is the C-terminal site with a large number of repeating sequences with high content of the amino acid leucine, designated as LRR (leucine-rich repeat region), which carries out protein interactions, i.e. serves as a receptor connecting with the pathogen elicitors. These elicitors are recognized by the host cell, and the signal is transduced to the nucleus, which leads to activation of the genes that govern the plant defense response, similar to local cell death. The LRR proteins are often a part of the signal transduction cascade. Pathogen resistance genes also possess the leucine zipper motif (LZ) and a site of binding with nucleotides. Figure 27 shows a diagram of nematode interaction with plant cells.

The genetic basis of plant–nematode relationship has been investigated only in few parasitic systems (potato – *G. rostochiensis, G. pallida;* tomato – *M. incognita;* soybean – *H. glycines,* and some others) (Table 10.4). Investigation of the structure and functions of the genes responsible for resistance to nematodes began a relatively short time ago, and the first success was obtained in studying the tomato gene responsible for resistance to the root-knot nematode *M. incognita.* Cloning showed that this gene of plant–nematode resistance belongs to the class of nucleotide - binding site – leucine - rich repeat (NBS-LRR) - encoding genes which also includes resistance genes to viruses, bacteria, and fungi. Some nematode resistance genes are similar in sequences and structure to the earlier cloned phytopathogen resistance genes. For instance, the tomato resistance gene *Mi* is close to the fungus resistance genes *Cf2* and *Cf5,* and

Table 10.4. Known plant resistance genes to phytoparasitic nematodes

Species	Locus	Nematode	Localization
L. peruvianum	*Mi*	*M. incognita*	Chromosome 6
		M. javanica	
		M. arenaria	
L. peruvianum	*Mi3*	*M. incognita*	Chromosome 12
		M. javanica	
L. pimpinellifolium	*Hero*	*G. rostochiensis*	Chromosome 4
Solanum tuberosum spp. *andigena*	*H1*	*G. rostochiensis,* pathotypes *Ro1* and *Ro4*	Chromosome 5
S. spegazzini	*Gro1*	*G. rostochiensis,* pathotypes *Ro1* and *Ro4*	Chromosome 7
S. spegazzini	*Gpa*	*G. pallida* pathotypes *Pa2* and *P3*	Chromosome 5
S. vernei	*GroV1*	*G. rostochiensis,* pathotype *Ro1*	Chromosome 5
S. bulbocastanum	R_{Mc1}	*M. chitwoodi*	Chromosome 11
B. procumbens	*Hs1pro1*	*H. schachtii*	Chromosome 9
Glycine max	*Rhg4*	*H. glycines,* race 3	
Tryticum aestivum	*Cre*	*H. avenae*	Chromosome 28
T. tauschii	*Cre3*	*H. avenae*	Chromosome 2D

even to gene *Meu1* responsible for the resistance to aphids. However, the potato gene of resistance to *G. rostochiensis* has more similarities with gene *Rx* responsible for the virus resistance than with other nematode resistance genes.

Interesting data on the structure of nematode-resistant genes was obtained from cloning and molecular analysis of gene *Hs1^{pro-1}* from *Beta procumbens*, a wild ancestor of the sugar beet *Beta vulgaris*, responsible for resistance to *Heterodera schachtii*. The sugar beet resistance gene *Hs1^{Pat-1}* encodes a 282-amino acid protein that possesses receptor regions and signal transduction sites, and, as against other nematode-resistant genes, has a large N-terminal site and membrane localization of the receptor (LRR region) where it can link with the nematode elicitors. Other nematode-resistant genes (genes of resistance to the tomato, potato, or wheat nematodes) have a C-terminal site, and the receptor protein can be localized in the cytoplasm where the nematode elicitors are delivered through the stylet, and the receptor binds with the elicitors inside the cell.

At the genetic level, the relationship in the host–parasite system appears as gene-for-gene, i.e. a parasite virulence gene corresponds to a resistance gene. Recognition of an avirulent pathogen by the resistance gene carrier plant leads to activation of a defense response, which is frequently manifested in the form of the hypersensitive response showing as local cell necrosis in the nematode localization area. Such response is well-known in tomatoes possessing gene *Mi*. So far there are no data on isolation of avirulence genes from parasitic nematodes, but there is some indirect evidence of the presence of such a gene. Though the root-knot nematodes *M. incognita*, *M. arenaria*, and *M. javanica* are parthenogenetic species, they show considerable intraspecific and host-specific variability in parasitizing on particular plants. In natural conditions, some populations of root-knot nematodes show virulence, i.e. ability for reproduction in the plants carrying a resistance gene (gene *Mi* from tomatoes, *Me3* from pepper, or *Rk* from peas). To study the genetic differences between the avirulent and virulent root-knot nematodes, natural populations of these pathogens from various regions were analyzed using RAPD or AFL markers. Though these investigations did not reveal a relationship between the DNA polymorphism and virulence of the nematodes, they have shown that the genetic determinant of virulence can vary in various virulent populations. Today, the best explored are the relations of the root-knot nematode *M. incognita* and the resistant tomato with gene *Mi*. Due to parthenogenesis, genetics of the virulence of the root-knot nematode cannot be tested using hybridological analysis. However, it has been confirmed in experimental conditions that virulence is a hereditable genetic trait. Candidates for a role of an avirulence gene of the root-knot nematodes to gene *Mi* have been recently found. Comparison of the soluble proteins of the virulent and avirulent females of these nematodes in 2-D polyacrylamide gel revealed the presence of an extra protein in the avirulent females. Cloning and molecular analysis of this protein showed that it is one of the proteins released by the amphids. The gene encoding this protein can be a candidate for an avirulent gene of the root-knot nematode.

There is also indirect evidence of the presence of an avirulence gene in the potato nematode *G. rostochiensis* that corresponds to resistance gene *H1*. Genetic analysis of the closely related strains of the soybean cyst nematode enabled identification of the dominant and recessive determinants of the parasite in various soybean lines.

Hypersensitive response

This plant defense response develops in response to attack of a broad range of pathogens (viruses, bacteria, fungi, or nematodes) at incompatible combination of partners. A complex signal sequence in the cells of a resistant plant, which primarily arrives from the parasite, leads to a rapid local death of the plant cells in the areas of the nematode penetration and is accompanied by accumulation of toxic products in the dead cells. The pathogen dies together with the invaded cells. Such form of cell death is like suicide at the cell level. Scientists have recently been showing a lot of interest to the mechanisms of hypersensitive response in connection with apoptosis, genetically programmed cell death.

Apoptosis occurs in the cells surrounding the site of penetration of the sedentary nematodes highly specialized to parasitism (root-knot and cyst nematodes). In apoptosis, the nematode-invaded cells accumulate the pathogen-inducible proteins (PR-proteins), phytoalexins, as well as a number of substances of secondary plant metabolism, also with an increase in activity of some enzymes: peroxidase (PO), polyphenoloxidase (PPO), phenylalanine ammonia lyase (PAL), lipoxygenase (LOG), and some other enzymes. The rise in PO activity in resistant cultivars can be related to the emergence of the new isoforms of this enzyme. Some isoforms can suppress the hydrolytic enzymes of the nematodes, oxidize their toxins to the level of neutral substances enhancing formation of mechanical barriers, and also activate other defense processes. PAL is involved in regulation of biosynthesis of the phenols that control plant–nematode relationship.

Biochemical factors of plant resistance to parasitic nematodes
Phytoanticipins and phytoalexins

In studying plant–nematode relationship, a special emphasis is made on phenols: their elevated content is believed to be responsible for plant resistance both against errant and sedentary nematodes. Among phenols, researchers have been recently focused on salicylic acid (SA), which is explained by its positive role in formation of induced immunity to pathogenic bacteria, fungi and viruses, as well as nematodes. It was found that SA can accumulate in the nematode localization area and systemically grow by several times in the unaffected plant organs. The SA possesses an ability to move in the tissues (mostly in the phloem) and also serve as an elicitor triggering the cell signal systems. Exogenic SA was found to induce gene expression and the formation of a whole series of pathogen-inducible proteins, as well as formation of phytoalexins. The SA content in the roots and leaves of the tomato plants susceptible and resistant against root-knot nematode *M. incognita* differed ten-fold and increased upon invasion, which gives ground to believe that SA is involved in tomato resistance against nematodes.

An important role in plant resistance against phytonematodes is played by isoprenoid compounds. Plant isoprenoids were found to possess a number of properties to assure their existence in stressful conditions caused by parasitic nematodes. They can act on the pathogen as toxins (rishitin, gossypol, odoratsin, tomatin, chaconin), possess antifeedant properties (azadirachtin, saponins, sesquiterpenoid lactones), influence the nematode growth and molting processes (azadirachtin, phytoecdisons, sesquiterpenoid lactones), larvae hatching from eggs (glycinoeclepin A), inhibit the effect of cholinesterase (glycoside from asparagus), or block recognition of the host plant (citral, menthol). Furostanol glycosides isolated from the cells of *Dioscorea deltoidea* possess adaptogenic activity; they act on the intracellular membranes to normalize the photosynthesis processes disturbed by the invasion; and they enhance general metabolism intensity of the infected plants.

However, the absolute number of anticipins in plant tissues does not always correlate with the degree of resistance. Most likely, phytohelminthes possessing wide adaptation capabilities easily adapt to the constitutional antibiotics of the host. Probably, during coevolution these parasites either have acquired the ability to detoxicate such host inhibitors or have developed tolerance to them. Adaptation of the parasites to the induced antibiotics, phytoalexins (FA), is much more difficult.

The first data regarding phytoalexins was obtained in studying plant interaction with phytopathogenic fungi. In 1971 it was found that FA can also be produced at plant invasion by nematodes. To date, phytoalexins produced in response to invasion by phytohelminthes have been found and characterized in the representatives of only four families: Leguminosae, Malvaceae, Solanaceae, and Musaceae.

The chemical structure of phytoalexins is determined by the plant genotype. Usually, several phytoalexins of the same biogenic origin are produced in the invaded plant. For instance, the family Solanaceae is characterized by sesquiterpenoid phytoalexins, with two of them, rishitin and lubimin, found in tomato and potato plants invaded by the root-knot (*M. incognita*) and stem nematodes (*Ditylenchus destructor* and *D. dipsaci*), respectively. Tissues of a cotton plant invaded by *M. incognita* were found to contain five compounds, sesquiterpenoids and triterpenoids. Four isoflavanoids were found in tissues of the Lima beans upon its invasion by *Pratylenchus scribneri*, and two of them were identified as the complex coumestans: coumestrol and psorallidin. The presence of several FA allows a better understanding of their protective function: it is more difficult for the parasite to get adapted to several antibiotics than to one. In addition, various inhibitors simultaneously present in the invaded tissues can render a synergistic effect on the nematode. A report was published recently on finding a new compound, phenyl phenalenon, in a resistant banana hybrid invaded by

Radopholus similis and other pathogens. Phenyl phenalenon is viewed as a new class of phytoalexins.

According to the published data, FA formation in plant tissues in response to bacterial or fungal invasion occurs 8–24 hours after the inoculation. As to nematodes, formation of phytoalexins in plant tissues is determined by their biology. Invasion of plants by errant nematodes can lead to FA formation in the first few hours after the invasion, and their localization is strictly confined to the sites of the nematode penetration. Upon plant invasion by sedentary nematodes, formation of phytoalexins in plants begins at the moment of nematode localization and beginning of feeding (after 72 hours).

The available data on FA toxicity towards nematodes, FA localization, and dynamics of their accumulation show that they can serve as a defense in plant immunity. This is confirmed by a direct correlation between the ability of a plant species (cultivar) to produce phytoalexins in response to invasion and its resistance.

PR-proteins

The information regarding a protective effect of PR-proteins in interaction of plants and parasitic nematodes is extremely limited and contradictory. A number of publications point to the absence of PR-protein induction in plants in response to nematode invasion. Other publications note the PR-protein induction in plants in response to nematode invasion, however, the relation of these proteins with resistance or susceptibility has not been revealed.

At the same time, data are available not only to confirm the induction but also possible involvement of PR-proteins in plant protection against parasitic nematodes. Investigation of chitinase in two soybean cultivars, varying in resistance to *M. incognita*, showed that high activity and early induction of chitinases can determine soybean resistance to root-knot nematodes. The roots of the resistant tomatoes were found to contain a considerable increase in only one enzyme, namely, β-1,3-glucanase, whereas there was no chitinase activity. This seems to be due to the metabolism of the investigated plants which belong to different families. The revealed fact of a considerably higher increase in activity of the PR-proteins in the tissues of resistant (or immunized) plants in comparison with the susceptible plants in response to invasion can confirm their involvement in regulation of the relations in the system "plant–parasitic nematodes". The protective role of chitinase can be determined by its effect on the chitin component of the nematode egg shells or cell walls of the nematodes, where chitin is also one of the basic components.

Proteinase inhibitors

A protective action of the plant proteinase inhibitors was first shown in investigation of *G. pallida* in potato with a gene of trypsin inhibitor. Numerous reports are available on inhibition of nematode growth in plants, with the genes of proteinase inhibitors.

Food factors

In analysis of the factors of plant resistance to phytopathogens, stress is usually made on the inhibitors, while the factors of resistance related to low nutrient value of the host tissues as nematode feeding medium are much less investigated. Meanwhile, phytohelminthes are biotrophic organisms completely dependent on the composition of the host plants. A number of observations show the relation of plant resistance and the presence in their tissues of the essential amino acids, the lack of which makes reproduction and development of the parasites impossible. Of special interest is studying those substances, which are vital to phytohelminthes but are not synthesized by them. A special role among such substances is played by sterols. Nematodes are unable to synthesize these compounds by themselves and, consequently, use the sterols of the host plants. An example of the system "tomatoes – root-knot nematode *M. incognita*" showed that invasion of the tomato roots changes the qualitative and quantitative composition of sterols in the resistant and susceptible cultivars in the opposite directions. In the resistant cultivars, a decrease in both the total sterol fraction and all revealed components was observed, while in the susceptible cultivars usually a reverse event was observed. Especially noticeable is the change in the content of stigmasterol, most preferable to the nematode. It drops by almost 2 times in the tissues of the resistant cultivar, whereas in the susceptible cultivars it increases by 2–4 times. Taking into account that the dominant sterol in the eggs and females of the root-knot nematodes is stigmasterol, it could be asserted with high likelihood that the qualitative and quantitative composition of sterols determines life activity of the nematode.

Systemic acquired resistance

The chemical processes and compounds described above are connected with the local plant resistance, and they work directly in the place of infection. However, more than 100 years ago the researchers knew that the plant which had survived an infection, becomes resistant against a subsequent infection. In 1933 the American plant pathologist Chester, who summarized about 200 publications, described a phenomenon which he termed physiological acquired immunity. At that time scientists believed that plant immunity is similar to the immune response of the mammals. In those days, at least three different processes were included in the category of acquired immunity: cross protection against viruses (resistance to a highly pathogenic strain of a virus in the plants inoculated with a low pathogenic strain of a related virus); microbial antagonism, or biocontrol; and the acquired resistance. The concept of systemic acquired resistance (SAR) as well as the term localized acquired resistance (LAR) was first introduced by the German scientist Ross in 1961 on the basis of investigation of the tobacco plants inoculated with TMV. Systemic acquired resistance developed in the infection-free tissues of an inoculated plant, whereas localized acquired resistance developed directly in the inoculation site. Thus, systemic acquired resistance is a secondary immune response to infection that is generated due to the presence of localized resistance.

Since the mid-1970s, a great success in SAR induction was achieved by the school headed by the American plant pathologist J. Kuc. Their object of research was the plants of the *Cucurbitaceae* family. To induce resistance, they used the non-pathogens of these plants, nonpathogenic strains of the pathogens, pathogens themselves, and, finally, products of the parasite and host metabolism. The researchers established that cucumbers could be systemically protected against the anthracnose pathogen through inoculation of their first true leaf with the fungus. The same result was received in melons and watermelons that could be systemically protected against fungal, bacterial, and viral diseases through a previous local inoculation of the first true leaves. Such experiments were carried out not only in greenhouses but also in the field. Mechanical wounding of the leaves as well as their damage with dry ice and some chemicals did not induce protection. The non-specific resistance induced either by the anthracnose pathogen or the tobacco necrosis virus protected plants at least against 13 pathogen species, including obligate and facultative parasitic fungi, viruses, and bacteria causing local and systemic infections. SAR correlated with systemic accumulation of intercellular acidic peroxidase and chitinase, proteinase inhibitors, and lignin, as well as with deposition of callose, i.e. the immune response factors.

The systemic resistance showed 48–72 hours after inoculation of the inductor leaf, then reached its maximum, this period coinciding with appearance of the signs on the inductor leaf. Systemic resistance was found to be transported over a plant from bottom upwards. For instance, a melon or watermelon scion on a cucumber stock, containing an inductor leaf, induced resistance in the scion. Thermal treatment of the inductor leaf petiole phloem blocked the signaling transport of resistance. It is extremely interesting that resistance is transferred to the regenerant plants through tissue culture.

However, our understanding of the biochemical mechanisms involved in SAR evolved slowly before the discovery of PR-proteins correlating with SAR propagation. Their discovery helped to find the SAR markers that would precede the biochemical responses. For this purpose, cDNA were isolated which are produced in infection-free tissues of the inoculated plants. Thus, the tobacco plant-TMV system was used to find mRNAs of at least 9 gene families that are continuously induced in uninfected leaves of infected plants. They were named the SAR gene family. SAR gene expression begins about 6 hours after the inoculation and corresponds to the period when resistance can be determined phenotypically. Moreover, abiotic agents that induce resistance, such as salicylic acid or 2,6-dichlorisonicotinic acid, cause expression of the same range of SAR genes and approximately with the same intensity. Thus, expression of SAR genes correlates with the state of resistance.

Some SAR genes seem to play an active role in resistance. For instance, a number of cDNA-encoded proteins were found to either possess direct antimicrobial activity or possess enzymatic properties leading to production of antimicrobial proteins. They include the classes of SAR genes encoding β-1,3-glucanases and chitinases.

Another class of the SAR genes encodes the group of cysteine-rich proteins called taumatins. Taumatin-like proteins are known as antifungal products, the activity of which is determined by their ability to destroy membrane integrity, hence the name of this class of SAR genes, permatins.

One of the SAR gene groups is PR-1 proteins, whose function is not yet clear, however, they are known to be common in angiosperms. PR-1 proteins from tobacco plant and tomato suppress growth of *Phytophthora infestans* in vitro.

Further confirmation of SAR gene involvement in resistance is derived from experiments in transgenic plants. Tobacco plant and cabbage, carrying the chitinase gene from beans, were protected against *Rhizoctonia solani*. A high level of PR-1 expression in transgenic tobacco plants lead to a drop in *Perenospora tabacina* and *Phytophthora nicotiana* infestation. Thus, some SAR genes encode the proteins possessing antimicrobial activity, and expression of these genes in plants provides increase in SAR.

Rapid accumulation of the data on SAR gene expression began after introduction in the studies of the model plant Arabidopsis, where the SAR genes belong to the same family as in tobacco plants. In other cases, for example, in cucumber, the SAR genes differ from the tobacco genes. Different taxonomic groups of plants can possibly have their own network of SAR genes, developing in response to evolutionary pressure of specific pathogens.

The basic question which interests all researchers today is how resistance is induced in tissues of the plants some distance apart from the infection site. Plants are supposed to have some mobile molecules that can activate the resistance mechanism in the cells far apart from the infection site. Existence of such mobile signals was predicted by Ross. A number of criteria, have been suggested, to be met by the compounds that could play the role of intercellular systemic signal molecules. The latter should be synthesized in the plant, move over it, quantitatively grow after damage of the plant tissue, and induce defense mechanisms. A number of experiments showed that the SAR induction signal is transmitted over the phloem. Impaired phloem transport in ringed plants, affects SAR induction also.

As it was discussed in the previous chapter, there are at least two parallel pathways of intercellular signal transmission. The first type is connected to the molecules that signal plant tissue wounding, irrespective of whether it was mechanical wounding or it was caused by necrotrophic pathogens or pests. The signal molecule of this type is jasmonic acid (JA). The second type of signal molecules is specific only for pathogenic infection: bacterial, fungal, or viral. The signal molecule of the second type is salicylic acid (SA). Experiments with Arabidopsis mutants show that these two pathways are relatively independent. In Arabidopsis, a signal from jasmonic acid is transmitted to the control protein Coip that induces expression of jasmonate-dependent genes (proteinase inhibitor gene PIN, etc.) Mutation of Coip, which blocks the PIN gene susceptibility to jasmonic acid, suppresses resistance to the necrotrophic parasites *Botrytis cinerea* and *Alternaria brassicola*, but does not influence the resistance to the biotroph *Peronospora parasitica*. On the other hand, mutation of the gene NPR1, the protein of which is an active transmitter of a signal from a SA to SAR genes (PR1, etc.), considerably reduces the basic resistance to *P. parasitica* but has no effect on resistance to *B. cinerea* and *A. brassicola*. Moreover, there is interference between these two signaling pathways. For example, the necrotrophic bacterium *Erwinia carotovora* induces synthesis of some PR-proteins (not SAR) in tobacco plants but inhibits expression of SAR genes, and SAR induction inhibits a response to inoculation with Erwinia. SA renders regulatory effect on JA biosynthesis: it enhances accumulation of an essential biosynthesis enzyme allenoxide synthetase, but inhibits the final steps of the biosynthesis (the enzyme allenoxide cyclase), thus suppressing a response to wounding.

In addition to the salicylic acid and some of its derivatives, jasmonic acid and its methyl ester (Me-JA), the candidates for the role of systemic signals include systemin and, ethylene and, possibly, also some oligosaccharines and abscisic acid (Figure 10.8).

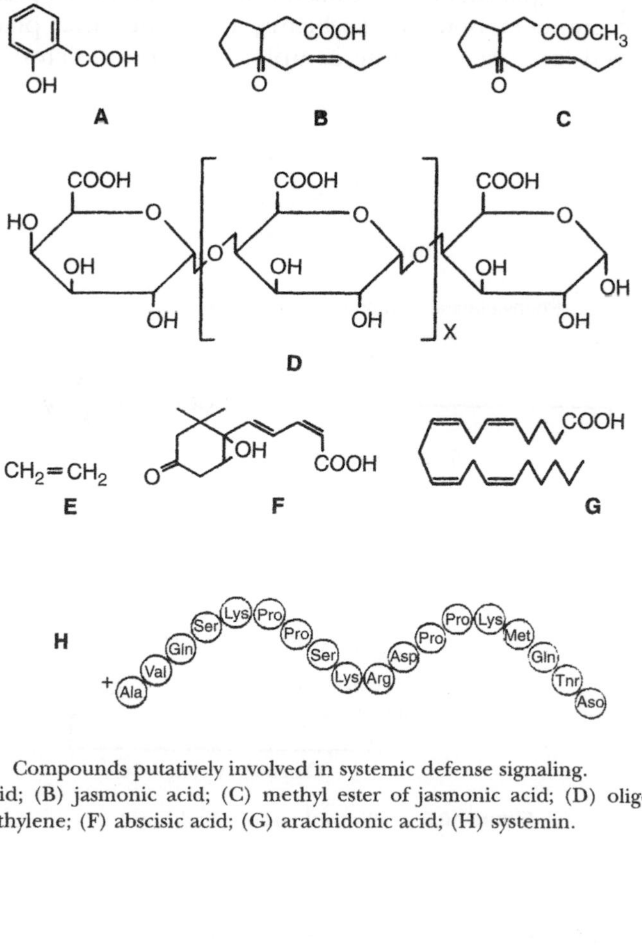

Figure 10.8. Compounds putatively involved in systemic defense signaling. (A) salicylic acid; (B) jasmonic acid; (C) methyl ester of jasmonic acid; (D) oligogalacturonides (X=2–0); (E) ethylene; (F) abscisic acid; (G) arachidonic acid; (H) systemin.

Salicylic acid (SA)

Higher plants are supposed to have two possible pathways of SA biosynthesis (Figure 10.9). In both cases SA is synthesized from transcinnamic acid either via *o*-coumaric acid or via benzoic acid. In the first pathway, transcinnamic acid first turns in oxycinnamic acid, then to SA. Such pathway forms a link between pathogenic induction of phenylpropanoid biosynthesis and SAR signaling induction.

In the other pathway, transcinnamic acid is oxidized to benzoic acid, then hydroxylated in the ortho-position to SA. The final stage of synthesis in the second pathway is conversion of benzoic acid into SA using acid benzoyl 2-hydroxylase, with possible involvement of cytochrome E-450. It was found that activity of acid benzoyl 2-hydroxylase in infected tissues grows approximately 10-fold, and this growth is blocked by protein synthesis inhibitors.

Salicylic acid has been long known in medicine as a therapeutic agent, but it was only recently revealed that it is also an active ingredient for the plants that synthesize SA. Among all tested derivatives of benzoic acid, only SA, its acetyl derivative (aspirin), and 2,6-dihydrobenzoic acid were able to induce resistance to the tobacco necrosis virus and PR-proteins in tobacco plants. Resistance to various pathogens is induced not only in tobacco plants, but also in tomatoes, cucumbers, potatoes, soybeans, and some other plants. The SA role in plant resistance seems quite complex and contradictory, it depends

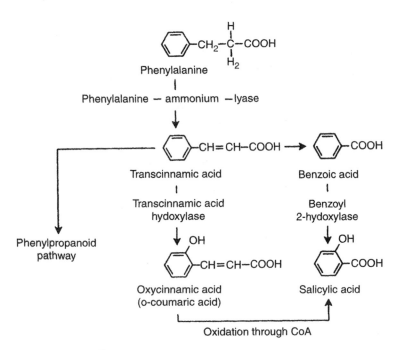

Figure 10.9.   Biosynthesis of salicylic acid.

on many circumstances (SA concentration, interval between treatment and infestation, etc.), and sometimes turns to its opposite: enhancement of susceptibility to disease. SA regulates deposition of the callose polysaccharides in the plasmodesma, thus preventing the short-range transport of viruses through plasmodesmas (Chapter 5); i.e. callose deposition, together with hypersensitive response, delineates an infected from healthy symplast.

SA meets the requirements to the systemic signal molecules well: it easily propagates in the phloem vessels as its physical properties are nearly ideal for remote transport in sieve-like tubes; its amount systemically grows by tens of times under influence of pathogens; and it can induce some defense mechanisms in plants.

Salicylic acid is present in plants not only in the free form but also in the form of glycosides, though it is transported in the phloem only in the free form. It is assumed that glycoside-SA is an inactive accumulating form that can promptly release SA in the area of developing infection. It is extremely interesting that SA accumulates in plant tissues only in response to infection, but its amount does not grow in mechanically wounded tissues.

The most convincing evidence of SA involvement as SAR signal were received in the experiments with transgenic tobacco plants: gene nahG from the bacterium *Pseudomonas putida* that encodes synthesis of salicylate hydroxylase was built in the plant genome. Salicylate hydroxylase catalyzes conversion of salicylic acid to catechol, which is not a SAR inductor. The plants that express gene nahG do not accumulate SA in response to pathogenic infection and are unable to induce SAR in response to viral, bacterial, or fungal infection. These experiments directly demonstrate SA involvement in the SAR signaling system, but they do not reveal, whether SA is a remote, long systemic signal propagating in the phloem, or some other signals induce SA synthesis at a distance from the infected leaf. A number of experiments show that the latter assumption is more likely.

In a number of experiments, concentration of SA and acid peroxidase (as a marker of SAR gene in cucumbers) was measured in various cucumber tissues after inoculation with *Pseudomonas syringae*. It was found that the removal of the inductor leaf 4–8 hours after the inoculation (before SA accumulation) did not prevent systemic induction of salicylic acid and SAR gene expression. These experiments show the presence of the systemic signals different from SA. Similar data were also received in the grafting experiments. SAR signal remained in the stock carrying salicylate hydroxylase gene (nahG), which had been engrafted on a scion inoculated with local infection.

In 1991, Cheng and Klessing isolated from tobacco tissues a protein to which the SA is bound. That protein bound only SA and its analogs capable of inducing resistance and expression of PR-genes. The SA-binding protein turned to be a catalase, the activity of which was blocked after binding to salicylic acid. This leads to an assumption that the blocking of catalase activity results in accumulation of hydrogen peroxide, which either itself or through other reactive oxygen species activates expression of protective genes, acting as an intracellular messenger.

Probably, catalase inhibition blocks the main channel of hydrogen peroxide consumption, thus causing its accumulation. This does not only result in the pathogen intoxication, but also in HR, strengthening of cell walls due to cross-linking of proteins and carbohydrates produced with involvement of peroxidase, formation of proteinase inhibitors, etc.

It is assumed that the reactive oxygen species generated in a cell are not sufficient to cause a hypersensitive response. It was shown earlier (p. 238) that the reactive oxygen species act synergistically with nitrogen oxide. SA is actively involved in the system of transduction induced by the reactive oxygen species and nitrogen oxide, which gave ground to call SA "a systemic enhancer" of this cascade of reactions.

Further studies showed that SA can bind not only to catalase but also to some other Fe-containing enzymes: ascorbate peroxidase, aconitase, and lipoxygenase.

One more SA-binding protein was found in tobacco leaves. It reversibly bound to SA and had affinity to SA 150 times greater than affinity to catalase.

Interesting data were received in investigation of benzothiadiazole, a new synthetic elicitor of plant defense responses, which is a functional analog of salicylic acid. Like SA, this compound is able to inhibit catalase and ascorbate peroxidase, and induce expression of the protective genes (genes of acid PR-1, PR-2, and PR-3) even to a greater degree than SA.

Since SA involvement in the systemic signal induction of PR-proteins and SAR was established, the optimistic mood has been replaced by a sense of mild disappointment.

(1) First of all, it was found that SA is instrumental in SAR induction only in some plants, and sometimes a short period of resistance is replaced by susceptibility to disease.

(2) The experiments in transgenic plants clearly showed that SA is not the only SAR signal molecule, however, its presence is necessary both for signal transduction and resistance induction.

(3) New hopes for understanding the SA role arose when catalase was discovered to be the SA receptor, and it was found that hydrogen peroxide accumulating as a result of blocking catalase induces cell defense responses. However, SA was found to bind not only catalase but also some other Fe-containing enzymes. In addition to SA, 2,6-dichlorisonicotinic acid, benzothiadiazole, and succinic acid are also able to bind catalase, and consequently induce SAR and PR-proteins.

Salicylic acid possesses an ability to bind to catalase only in some plant species or their organs.

All the above said, does not at all deny SA involvement in SAR and induction of protective responses, just questions its all-round action. The SAR character can possibly differ in different plant species or in the same plant in conditions of different stresses. An active SA role in signal transduction from an external ligand up to expression of defense responses is still obvious, though one should not forget that other pathways of transduction of such signals can also work, in parallel to salicylic acid.

Jasmonic acid and its methyl ester

After salicylic acid, jasmonic acid (JA) and its methyl ester (Me-JA) are the most realistic candidates to be involved in SAR. However, it has not been found so far whether they are systemic signal molecules or intracellular messengers, or possibly, combine both functions. This section discusses involvement of jasmonates in SAR events. Jasmonate biogenesis is shown as a diagram in the Section "Signal transduction".

Intensive research in the recent years has considerably added to our understanding of the jasmonate physiological role and mechanisms of action. The data received to date shows that jasmonates can mediate the course of the chemical reactions related to plant stress resistance. Very little is known about their role in induction of plant resistance to phytopathogens.

JA and Me-JA content increases after mechanical wounding of plant tissues as well as after treatment of various plant species with elicitors of suspension cell culture. Exogenic jasmonate leads to synthesis of a whole set of the so-called "jasmonate-inducible proteins" that possess species-specificity. They include proteinase inhibitors I and II, trypsin inhibitor, thionine, napine, cruceferine, some reserve proteins, phenylalanine–ammonia mylase, chalcone synthase, and lipoxygenase. Jasmonate precursors (for example, 12-oxophytodienoic acid) show even greater activity in induction of proteins than jasmonates.

For many plants, induction of specific polypeptides and proteins mediated by jasmonic acid has been shown, but functional activity was found for only some of them. To date, two functions of JA-inducible proteins have been revealed: these are the reserve proteins as well as the proteins involved in protection of plant tissues from disease and stresses (proteinase inhibitors and thionines).

Thionines are functional analogs of proteinase inhibitors. These are low molecular weight polypeptides with ca.45 amino acid residues, rich in SH-groups. Thionines are localized in plant cell walls. They are toxic for fungi and bacteria, and seem to be involved in plant defense responses to phytopathogens, therefore, some information on thionines is provided in Chapter 4 in the description of phytoanticipins.

Some researchers assume that jasmonates are an integral part of the system of transduction of the signal that regulates induction of protective genes in plants. Jasmonates are believed to be able to act at the stage between the elicitor–receptor complex and expression of the genes encoding formation of the enzymes and proteins necessary for defense responses.

Two possible modes of jasmonate involvement in induction of protective genes could be assumed:

– JA and Me-JA can be elicitors by themselves, inducing formation of protective compounds in plants;
– jasmonates can enhance activity of the elicitors, without inducing or weakly inducing resistance responses.

Exogenous Me-JA in a concentration of 10^{-6}–10^{-5} M, even in the absence of elicitors, induces synthesis of the secondary metabolites in suspension cell culture of various plants. Jasmonate and its precursor 12-oxophytodienoic acid induced various FAs in cell cultures of more than 30 plant species. At the same time, in other plant species the exogenously added jasmonate did not induce synthesis of secondary protective metabolites. For instance, Me-JA induced only traces of cumarin in suspension cell culture of parsley. However, preincubation of suspension culture with Me-JA enhanced its sensitivity to low concentrations of biogenic elicitors, leading to intensification of cumarin secretion, phenol deposition in cell walls, and production of activated oxygen. Though Me-JA by itself did not induce formation of rishitin in potato, its presence enhanced rishitin-producing activity of the low concentrations of arachidonic acid and intensified necrosis formation in the potato tubers. Synergism of the activity of Me-JA and arachidonic seems to be possible, or Me-JA enhances the signal transduction from the elicitor to expression of protective genes.

A JA derivative, 12-(5-β-D-glucopyranosyl)-jasmonate was recently found in potato leaves. This compound is an endogenous factor of potato tuber formation. The aglycone part of the glycoside was named tuberonic acid (Figure 10.10).

The level of endogenous jasmonate rapidly grows in the wounded tissues of plants. As JA can be transported in the phloem, jasmonates are assumed to be the intercellular messengers providing information transmission to uninjured cells. Induction of a high level of proteinase inhibitors requires less than 1 nl of Me-JA. Treatment of tomato leaves with JA precursors (linolenic and hydroxyperoxylinolenic acid) induced proteinase inhibitors. The structurally close compounds, which nevertheless, are not JA precursors, did not induce accumulation of proteinase inhibitors. These results, together with JA presence in the phloem, confirm that lipid derivatives can function as systemic signals in plants.

Me-JA is involved in transduction of a wound signal through atmosphere, as it is volatile at ambient temperature. Uninjured tomato plants that were growing in a chamber containing a volatile fraction of Me-JA, accumulated similar amount of proteinase inhibitors upon wounding. Gaseous Me-JA, constitutionally released by sagebrush plants, induced accumulation of proteinase inhibitors in the tomato leaves in the same chamber. Me-JA also rendered systemic effect on interaction of the potato and the late blight pathogen. For instance, if potato tuber disks were incubated in the chamber with the atmosphere saturated with Me-JA, the same defense responses were induced in the disks as in local application of jasmonates (Ilyinskaya et al., 1997).

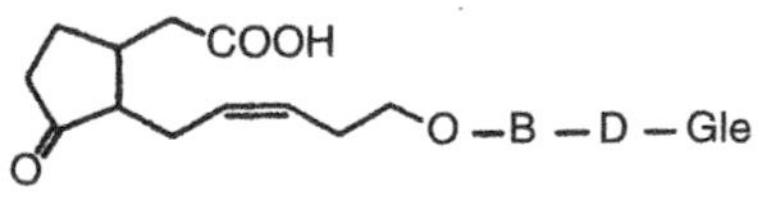

Figure 10.10.   Glycoside of tuberonic acid.

Me-JA volatility, together with its biological activity, permits to assume that Me-JA, similar to ethylene, can perform allelopathic interaction between plants or their organs.

Together with proteinase inhibitors, jasmonates are able to activate a number of enzymes: phenylalanine–ammonia lyase, chalcone synthetase, oxymethylglutaryl-CoA-reductase, etc. Jasmonates are active lipoxygenase elicitors in plant tissues. For instance, prompt lipoxygenase expression was observed in soybean upon wounding or treatment with jasmonates. Treatment of potato tuber disks with Me-JA resulted in increase in both lipoxygenase per se and its mPNA. Moreover, Me-JA lead to an increase in lipoxygenase transcripts even in anaerobic conditions, where the mechanical wounding does not induce activity of this enzyme.

One of the presumable functions of lipoxygenases in plants is their involve-ment in the defense response of plant tissues to a pathogen attack. Lipoxygenase inhibitors suppress the potato tubers resistance to late blight, thus promoting spread of the disease. It is interesting that Me-JA completely eliminates the negative effect of the lipoxygenase inhibitor, and the infestation of the potato tuber tissues with late blight is even less than in control.

Systemin

Systemin is the only peptide possessing hormone-like activity in plants. It was discovered in Rayn's laboratory in 1992, and its structure was identified in the following year. Before the discovery of systemin, the only known plant mole-cules involved in cell communication were relatively low molecular weight, non-protein compounds.

The polypeptide systemin consists of 18 amino acid residues. It was first iso-lated from tomato leaves as a high molecular weight precursor containing 200 amino acids and named prosystemin.

Systemin synthesis is induced either by mechanical wounding or injury by insects. So far systemin has been found only in above-ground organs (but not in the roots) of tomatoes and potatoes. Respectively, wounding of any above-ground part of these plants arouses induction of systemin. Exogenous sys-temin in an extremely low concentration (40 fmoles) induces expression in tomato of the serine proteinase inhibitors, identical to those produced in response to wounding or damage by mandibulate insects. The inhibitors are assumed to be inducible, protective compounds preventing easy digestion of proteins by insects.

Labeled systemin was used to find that this polypeptide spreads from the point of its application all over the leaf within 30 minutes. Three hours later it was already found in the phloem exudates of the cut-off tomato leaves, which shows that systemin can systemically move in the phloem. Today, however, the ability of systemin phloem transport is questioned.

The wounding of one tomato leaf induces transcription of the systemin mRNA over the entire plant, though it does not rule out that accumulation of systemin in different parts of plants may result from some other wound signals. Systemin is believed to bind to the membrane receptor, and thus

activate lipase, which releases linolenic acid. The latter is a source of jasmonates that are the compounds to induce systemin. It is not impossible that
there may be other signals also inducing systemic accumulation of systemin.

It is still unknown whether systemin is induced by the hypersensitive
response or its accumulation is only due to wounding. The relationship
between systemin, salicylic acid, jasmonic acid, and ethylene is not clear,
either. It is difficult to judge whether an increase in the level of systemin in the
entire plant results from its transport from the wounded tissue area, or
whether some unknown signal induces systemin accumulation at a distance,
which leads to systemic activation of proteinase inhibitor genes. Nevertheless,
it cannot be ruled out that the prompt induction of systemin upon wounding,
its ability to move in the phloem, and induce protective proteins makes
systemin a possible candidate for SAR transduction.

Polyamines

Plant polyamines (compounds containing two or more amino groups in
the molecule) constitute the low molecular weight regulators of physiological processes, involved in control of cell division and distention, meristem
activity, morphogenesis, flowering, and aging. Polyamines are assumed
to be the components of the stress-induced signal transduction chain bound
both to ethylene and the messenger molecule. In animal cells, polyamines
generate Ca^{2+}, release from reserve pools to cytosol, and are involved
in signal transduction through the receptors conjugated with G-proteins.
They can transport in the phloem of the plant in stress conditions
(Kuznetsov et al., 2002), which also makes them candidates for SAR
transduction.

Oligogalacturonides

Oligogalacturonides are sometimes viewed as signal molecules that regulate the
systemic response of plant tissues. These pectic fragments of plant cell walls
repeat the effect of wounding, through systemic induction of proteinase
inhibitors in tomatoes. Proteinase inhibitors can be induced by the molecules
of such small size as galacturonic acid dimers. At the same time, the oligogalacturonides with the degree of polymerization more than six were found unable
for transportation in the tissues where they were introduced. However, as small
oligomers of galacturonic acid are mobile in plants, oligogalacturonides should
not be completely eliminated from the range of possible systemic signals.
Admittedly, such small mobile fragments of pectic substances have not been
found in plant leaves, but they may be present in ripening fruits and falling
leaves.

It would be interesting to find out whether wounding induces local induction
of polygalacturonases that release the oligogalacturonides which induce systemic signals. Local formation of oligogalacturonides is quite possible,
as many fungal and bacterial pathogens possess a number of pectolytic enzymes.
Therefore, one should not ignore possible triggering of systemic responses by
the pectic fragments in certain cases. Possibly, oligosaccharide molecules of

different chemical nature and different degree of polymerization can possess different ability to transport in the tissues and induce defense responses.

Ethylene

Ethylene is a gaseous plant hormone that regulates many physiological processes. The interest for ethylene involvement in SAR is determined by the fact that plant inoculation with pathogenic flora is frequently accompanied by ethylene release. The ethylene production was found to be especially intensive several hours prior to appearance of the hypersensitive response and was proportional to the HR intensity. The elicitors from plant and fungal cell walls were found to enhance the ethylene release both from plant tissues and from their suspension cultures.

Ethylene induces several PR-proteins. For instance, treatment of tobacco plant with etaphone (ethylene-releasing substance) induced the transcripts of basic and some acidic isoforms of PR-proteins of groups 1, 2, 3, and 5. In a number of plants, ethylene induced basic chitinase and β-1,3-glucanase.

Ethylene activates phenylalanine-ammonia lyase and peroxidase, thus contributing to strengthening plant cell walls (during lignification).

The tissues treated with ethylene accumulate oxyproline-rich proteins that are involved in transversal binding of cell walls. However, all ethylene-induced structural changes are confined to the area adjacent to infection or wounding. So far there are doubts whether the dose of ethylene produced in HR or wounding is sufficient for induction of systemic resistance.

Nevertheless, there is evidence to confirm the role of ethylene as a SAR signal molecule:

- ethylene rapidly diffuses from the infection area;
- ethylene synthesis becomes more intensive upon a pathogen attack;
- ethylene induces some PR-proteins that play a protective role.

Thus, function of ethylene in SAR still remains unclear.

Abscisic acid

Abscisic acid can play a role of systemic signal in induction of proteinase inhibitors in wounded plant tissues. Tomatoes and potatoes, deficient in abscisic acid, do not possess this ability. Treatment of one leaf with abscisic acid resulted in its accumulation in another untreated leaf, which is indicative of the phloem or xylem mobility of this substance. Local damage of a leaf gives rise to accumulation of abscisic acid and induction of proteinase inhibitors. It is possible, however, that other signals can be translocated through the wound, which leads to accumulation of abscisic acid. This is confirmed by Ryan's data who observed that abscisic acid does not induce accumulation of proteinase inhibitors in tomato plants. Thus, the role of abscisic acid in induction of plant defenses against pathogens is still questionable.

Arachidonic acid and eicosanoids

Arachidonic acid (AA), a component of the membrane and reserve lipids, is an essential fatty acid for humans and animals. AA and to a lesser degree, eicosapentaenic acid (EAA) are the precursors of biologically active eicosanoids. Oxidation of AA through cyclooxygenase results in formation of the multipurpose animal hormones: prostaglandines, tromboxanes, and prostacyclines, whereas lipoxygenase leads to formation of leucotrienes as well as a number of active hydroperoxy and hydroxy derivatives.

Eicosanoids are involved in many physiological effects through indirect influence on the functions of enzymes, hormones, and mediators in humans and animals. Prostaglandines regulate the system of cyclic nucleotides that govern dozens of metabolic processes. AA is part of diacylglycerine that plays one of the basic roles in the phosphoinositol system, transmitting signaling information from the cell surface. Though most eicosanoids provide intercellular contacts, some of them can act as secondary messengers.

AA and EAA have not been found in healthy tissues of angiosperms. Higher plants are believed to have lost the C-20 fatty acids during evolution, using C-18 acids, linolic and linolenic, as an equivalent. After losing AA and EAA, the plants, nevertheless, have maintained the ability for oxidative transformation of these acids.

One may believe that the absence of AA and EAA – far from weakening the plants, but rather gave them certain advantages in the struggle against the pathogenic microflora containing C-20 acids. The plants have possibly acquired the ability to use AA and EAA oxidation products as elements of the systemic signal recognition to notify the plants about the imminent threat.

It has been shown that systemic late blight resistance in potato can be achieved by treating plants with arachidonic acid in small concentrations (10^{-8}–10^{-7} M). These concentrations of AA did not cause plant cell necrotization and FA formation. However, all tissues of the tubers, in which the surface was treated with AA, acquired systemic late blight resistance. The state of resistance gradually propagated from the treated surface to the center of the tuber and continued for several months. Then the resistance gradually regressed in the reverse order. The inducing activity of the elicitor produced the strongest effect on the sporulating capability of the pathogen: the sporulation process has been completely suppressed, while the mycelium growth was only half suppressed.

The tuber treatment with low concentrations of AA did not only protect the tubers from the late blight pathogen but also from a number of other fungal and bacterial diseases. The treated tissues also reacted more promptly and intensively to wounding.

The potato tubers induced with AA enhanced the activity of peroxidase, polyphenoloxidase, lipoxygenase, as well as the content of phenolics. However, out of these four parameters only peroxidase and lipoxygenase activity correlated with the transmission of the systemic inducing effect in potato tissues.

In the systemically induced tissues, the volume of agranular reticulum, the number of mitochondria, and the number of leucoplasts with differentiated stroma increased. Apparently, some parenchymatic cells, earlier carrying out the functions of reserving starch, acquired an ability for active biosynthetic processes, which determined the enhanced resistance of the tubers to infection and wounding.

Thus, several successive stages of development of durable systemic resistance of potato tubers were established:

(1) The time between the elicitor application and formation of a systemic signal;
(2) The time of the signal transmission over the potato tissues, duration of the signal being directly proportional to the tissue distance from the place of the elicitor application;
(3) The time of rearrangement of the metabolism and structure of the plant tissue that lasts from the reception of a systemic signal by the plant cell till the onset of resistance;
(4) The period of systemic induced resistance, which can last from several months to a whole year, depending on the specific circumstances;
(5) The time of gradual withdrawal from the state of immunization, when the tissue structures and all biochemical characteristics revert to normal.

The above data suggests generation of certain intercellular signals in the induced tubers that propagate over all tissues and successively immunize these tissues. Estimation of the propagation rate of such signals seems possible. With the knowledge of the average size of tubers and their cells, the distance of each experimental layer from the treated surface as well as the time for this layer to acquire late blight resistance, it is possible to estimate that the rate of propagation of a systemic signal in the tuber tissues averages 0.2–0.25 mm/hour, or 2–3 cells per hour.

Treatment of potato leaves with arachidonic acid brings only local effect, no systemic protection has been observed.

It was written earlier that arachidonic acid, unlike the elicitors of another nature, does not bind to the receptors but is deposited in plant cell membranes, replacing the linolic and linolenic acid characteristic of the plant. It was discovered that 90% of exogenously added AA is found in the potato lipids two hours later, mostly in the potato phosphotides, whereas 2–5% of AA is oxidized immediately after the treatment.

As arachidonic acid is not characteristic of plants in general and of the potato, in particular, its incorporation in phosphatidylinosite diphosphate and its product diacylglycerine can change their regulatory properties and thus initiate a resistance mechanism. On the other hand, potato phospholipases should gradually release AA from phosphotides, recovering the tuber homeostasis. The arachidonic acid released in portions will undergo oxidative transformations under the activity of lipoxygenase. Potato lipoxygenase was found to be able to convert exogenic AA into biologically active products, close to the eicosanoids generated by animals.

The question whether AA and EAA oxidation products can apply for a role of signal molecules in plants is still open. However, some prostaglandines in a concentration of 10^{-10}–10^{-7} M were capable of inducing late blight resistance in potato. In addition, 5-oxyeicosatetraenic acid, 15-oxyeicosatetraenic acid, and 5- oxyeicosapentaenic acid, as well as leucotriene LTC4 possessed protective systemic properties in the same concentrations. The treatment of potato tissues with lipoxygenase inhibitors blocked the AA-induced potato defenses.

The results obtained show the involvement of eicosanoids in induction of systemic resistance in the potato. A special stress should be made on the extremely low concentrations of the biological activity of these substances.

In addition to potato, AA induces systemic resistance in tomato, sugar beet, cucumber, maize, and some other plants. The period of induction during which the systemic resistance signals are produced may be possibly governed by the duration of AA deposition in the lipids of the treated plants.

RNA molecules

In the large green algae, *Acetabularia*, the nucleus, situated in the basal part, controls the morphogenesis, running in the apex, using the long-lived mRNA molecules. Therefore, the hypothesis regarding the RNA molecules as material carriers of SAR seems so exciting. However, the large size of the RNA molecules prevents their passing through the plasmodesms and migration in the plant. In a previous chapter (Chapter 5) the mechanism of cell-to-cell and phloem transport of plant viruses using special transport proteins has been described. In the nucleus-free cells of the sieve-like tubes of the pumpkin, protein CmPP 16 has been found that is responsible for mRNA transport to a long distance in the phloem. It was discovered using the serum specific to the 35 kDa transport protein of the red clover necrotic mosaic virus.

Another candidate for the role of a SAR determinant can be the small RNA molecules involved in the post-transcriptional gene silence (PTGS). The PTGS phenomenon occurs in the plants and fungi transformed by a foreign or endogenous DNA, and is manifested as decrease in the synthesis of the RNA molecules whose sequences are homologous to the introduced nucleic acid. This leads to the suppression of some viral proteins and, probably, provides a cross protection effect against the viruses of the plants previously infected with the RNA of the affine viruses (see Chapter 2). The PTGS mechanism and its use in the gene engineering constructs will be discussed in more detail in the following chapters. In the laboratory of the American virologist D. Baulcombe, a short molecule of the antisense RNA containing only 25 nucleotides has been found that is involved in the PTGS effect. The molecule is long enough to carry a specific sequence and short enough to move through the plasmodesms and be a SAR agent. Short (mi)RNA and (si)RNA have been found in the phloem of many plants. Proteins that bind small RNA (CmPSR1) and carry them to the cells were also found in the phloem.

Non-chemical signals

When plant physiologists and plant pathologists study plant stress signals, they mostly have in mind the effect of particular substances. It is no wonder, that the chemical compounds can be isolated and characterized. At the same time it is well-known that durable remote signaling information in animals transmitted through muscle fibers and nerve cells as well as in release of hormones is accompanied by electric signals.

Nerve cells are known to be susceptible to chemical or physical stimulants at the one end, and at the other end, release a chemical signal, a neuromediator acting on the other cells. Stimulation of one end of a nerve cell causes electrical excitation (action potential), which rapidly propagates to the other end, where the neuromediator is released.

The flowability of excitable membranes necessary for generating the action potential is based on asymmetric accumulation of ions on one side of the membrane. This difference of potentials can be generated due to the movement of such ions as sodium, potassium, and calcium.

A durable remote electric signal can also play an important role in systemic excitation of plant tissues. Induction of the action potential in response to wounding of the plant tissue has been found. For instance, crushing the tissue of a tomato leaf or its short burn leads to generation of the action potential that propagates over the plant.

Wounding and localized infection cause the withdrawal of K^+ and Cl^- from the cell and input of Ca^{2+}. These physiological changes are necessary for generating the action potential. Many systemic changes induced by wounding can be explained by generation of the action potential in the wounding area. Therefore, it cannot be ruled out that the action potential, rather than any of the compounds moving from the wainding area, acts as a translocation signal responsible for systemic induction of the proteinase inhibitors in the wounded tomato tissues.

The action potential correlates well with systemic accumulation of the proteinase inhibitors induced by the oligogalacturonides that depolarize cell membranes. Systemic induction of proteinase inhibitors was blocked, if the plant tissues had been treated with the inhibitors of membrane ion transport. It is interesting that SA was found to be one of such most effective inhibitors of ion transport.

Convincing evidence of involvement of the action potential as a systemic signal is provided in the experiments of Wildon and colleagues. They cooled a leaf petiole, thus blocking removal of the metabolites from the petiole. Though such treatment prevented the outflow, it did not affect the systemic generation of the action potential. A conclusion was drawn from these experiments that the systemic signal induced by wounding is rather an electric than chemical signal by nature. At the same time, there is an opinion that the action potential is not active enough for systemic accumulation of proteinase inhibitors and SAR. Thus, involvement of the action potential in SAR so far remains unclear.

Apparently, we will see the discovery of many new components of the systemic SAR signals in the near future. But already now we can assert that the research in this area will promote the development of new approaches of plant protection against disease and stresses. Maintaining a high immune status in plants is especially important today due to the unfavorable environmental situation which leads to immune deficiency not only in humans and animals but also in plants.

Selected Literature

Alexander D, Lawton K, Uknes S, et al. Defense-related gene induction in plants. Genetic Engineering 1994; 16:195–212.

Delaney TP, Uknes S, Vernooij B, et al. A central role of salicylic acid in plant disease resistance. Science 1994; 266:1247–1249.

Dixon RA, Paiva NL. Stress-induced phenylpropanoid metabolism. Plant Cell 1995; 7:1085–1097.

Eckardt NA. Small RNA on the move. Plant Cell 2004; 16:1951–1954.

Enyedi AJ, Yalpani N, Silverman P, et al. Signal molecules in systemic plant resistance to pathogens and pests. Cell 1992; 70:879–886.

Hammerschmidt R. Phytoalexins: what we have learned after 60 years? Ann Rev. Phytopathol 1999; 37:280–306.

Kuc J. Phytoalexins, stress metabolism and disease resistance in plants. Ann Rev Phytopathol 1995; 33:275–279.

Linthorst JM. Pathogenesis-related proteins of plants. Critical Rev in Plant Sci 1991; 10:123–150.

Oku H, Oushi S, Shizaishi T, Baba T, Phytopathology 1975; 65: 1263–1267.

Ozeretskovskaya OL. Potato antibiotic compounds add their role in resistance to *Phytophthora infestans*. Ph.D. Thesis M.1973 (In Russian).

Raskin J. Role of salicylic acid in plants. Ann Rev Plant Physiol Plant Mol Biol 1992; 43:439–463.

Ryals J, Uknes S, Ward E. Systemic acquired resistance. Plant Physiology 1994; 104:1109–1112.

Ryan CA, Plant Molec Biol 1992; 19(1):123.

Ryan CA, Pearce G. Systemin: a polypeptide signal for plant defense genes. Ann Rev Cell and Developmental Biol 1998; 14:1–17

Williamson VM, Hussey RS. Nematode pathogenesis and resistance in plants. The Plant Cell 1996; 8:1735–1745.

Williamson VM, Gleason CA. Plant-nematode interactions. Current Opinion in Plant Biol 2003; 6:327–333.

Yoshioka H. Molec Plant-microbe interactions 2001; 14:6.

Yoshikawa M, Yoshikawa K, Masago H. Physiol Plant Pathol 1979; 14:157–169.

Chapter 11

RNA silencing as a general defence mechanism against pathogens

P. Susi

Introduction

RNA silencing is one of the most recent pathogen defence responses discovered and involves a coordinated series of subcellular events that ultimately lead to the post-transcriptional termination of gene expression. RNA silencing process begins when a gene that is homologous to an endogenous "target" gene is introduced into a host cell. This can occur as a result of virus infection or following gene transfer during genetic transformation. The most important feature of RNA silencing shared between the kingdoms is the formation of small interfering RNAs (siRNAs) of 21–23 bp in length. They originate from the target gene messenger RNA (mRNA) by cleavage activity of a double stranded RNA (dsRNA) size-reduction enzyme, Dicer. Homology-based recognition of "foreign" RNA initiates a series of events that results in disruption of the target gene.

Currently, RNA silencing is one of the "hot spots" of modern research subjects, and the use of techniques related to it is growing exponentially in gene analyses. The RNA silencing process is also called co-suppression or post-transcriptional gene silencing in the context of plant- and plant virus-related silencing events, quelling in fungi and RNA interference in vertebrates and invertebrates (Table 11.1). The underlying mechanisms are broadly based on similar principles in which comparable signalling molecules are involved. There are, however, significant differences in the mechanisms of RNA silencing in different organisms as influenced, for example, by the nature of the target gene. It is, therefore, an oversimplification to suggest that observations made in one species are necessarily of direct equivalence in a related species. Because of the conserved nature of the core processes involved in gene silencing, the collective term "RNA silencing" is preferred. This chapter focuses on the general mechanism of RNA silencing in plants, plant viral suppressor proteins of RNA silencing and applications in plant breeding.

Coexistence and discovery of RNA silencing

The story of RNA silencing got started in early 1990s from attempts to overexpress gene constructs encoding key enzymes in the anthocyanin

Table 11.1.   Nomenclature of "RNA silencing" vs. organisms

Species	Organism/context
Post-transcriptional gene silencing (PTGS)	Plant
Co-suppression	Plant
Quelling	Fungus
RNA interference	Mammal, man
Virus-induced gene silencing (VIGS)	Plant virus–plant

biosynthesis pathway, in transgenic petunia. Contrary to expectation, the pigmentation in the flowers of transformed plants was not enhanced. Instead, the flowers were de-pigmented, and significantly, endogenous gene mRNA transcript levels were greatly reduced. Because both the transgene and the endogenous gene were suppressed, the observed phenomenon was termed "co-suppression" (Table 11.1).

A few years after the discovery of co-suppression, a similar negative corre-lation was described between the infectivity of *Tobacco etch virus* (TEV) and the expression of the untranslatable coat protein of TEV in transgenic *Nicotiana*; in brief, the less the TEV coat protein was expressed, the more resistant the transformed plants were to virus infection. Given that the silen-cing phenomenon operates at the RNA level (i.e. post-transcriptionally), the silencing effect was named post-transcriptional gene silencing (PTGS). At the time when plant scientists made their first observations concerning "co-suppression", researchers working with the fungus *Neurospora crassa* described an analogous phenomenon, quelling, in which overexpressed transgenes induced gene silencing at the post-transcriptional level. However, the significance of these observations went unnoticed for several years until the discovery that dsRNA was even more effective in silencing gene expres-sion than single-stranded antisense-RNA. The silencing phenomenon in *C. elegans* was termed RNA interference. It is instructive to note that although antisense-RNA silencing and RNA silencing are mechanistically dissimilar, they share common features in terms of the response that they raise. Thus, a general RNA silencing mechanism appears to account both for antisense-RNA-mediated silencing and RNA silencing.

Key characteristics and triggers of RNA silencing

Early events in triggering RNA silencing

In order to initiate target mRNA destruction i.e. RNA silencing, it is first necessary for an introduced gene and the homologous endogenous target gene to specifically interact. Plausibly, the recognition/initiation phase of RNA silencing is promoted by the presence of homologous RNA transcripts alone (perhaps above a certain threshold level), or alternatively, by the expression of aberrant and/or incorrectly processed RNA transcripts that operate through a dsRNA intermediate, inducing the formation of siRNAs. Both small sense- and small antisense-RNA molecules have been detected during early stages of RNA silencing in various organisms. There is, therefore, a strong likelihood that a dsRNA intermediate is first formed prior to the formation of siRNAs (see below). Consistent with this view is the observation that dsRNA is a more potent activator of RNA silencing than single-stranded RNA or antisense-RNA. In addition, transgenic plants that possess inverted copies of a transgene (raising the possibility of read-through antisense-RNA production) are more likely to be silenced than plants that contain directly repeated copies of the transgene.

A completely homologous interaction between the introduced gene and the target gene is not an absolute requirement for the successful initiation of RNA silencing in plants. The homologous segment required to initiate RNA silencing can be as short as 23 nucleotides (i.e. less than 10% of the size of the target gene) but there appears to be strict species-specific requirements both in terms of the length of sequence and the degree of homology shared, to initiate RNA silencing. Generally, more than 75% homology is needed to induce RNA silencing against a specific gene.

Virus infection, cross-protection and RNA silencing

Virus infection is an efficient trigger of RNA silencing particularly if viruses share sequence homology with nuclear genes. In plant pathology, "weaker" virus strains have been used to protect plants against "stronger" viruses with which they share a degree of sequence homology. This "cross-protection" phenomenon is related to RNA silencing and explains a number of reported cases in which transformed plants that contain a virus-derived gene(s) show resistance to infection from a related virus.

Small interfering RNAs (siRNAs) and micro-RNAs

A unifying feature of RNA silencing across kingdoms is the production of small (21–26 nucleotide) RNAs that act as specificity determinants that downregulate gene expression. An amplification step is required for production of a significant amount of "secondary" siRNAs specific to the target gene and also to areas extending outside the target gene (i.e. transitive RNA silencing; see below). Coupled with systemic signalling, this explains why the induction of RNA silencing, even in single cells, can result in a huge systemic response.

In contrast to the production of siRNAs in *Drosophila* embryos and mammalian cells, two distinct classes of siRNAs, short (21–23 bp) and long (24–26 bp), are produced in the fungus *Mucor circinelloides*, and also in wheat germ lysates during transgene silencing; short and long siRNAs apparently serve different functions. In fungi, both short and long siRNA species have been shown to accumulate differentially during vegetative growth, but their role in silencing signalling is unclear. In plants, short siRNAs are involved in local RNA silencing, whereas the long siRNAs appear to have an explicit role in systemic signalling and also in RNA-directed DNA methylation.

In trypanosomes and in plants the formation of long siRNAs corresponds to the *de novo* synthesis of siRNAs that are homologous to genome-encoded retrotransposons. This observation suggests that RNA silencing may act to restrict the activity of these potentially deleterious (genome restructuring) mobile genetic elements. The distinct classes of siRNA molecules appear to be produced by different Dicer orthologs.

Currently, more than 300 micro-RNAs (miRNAs) have been identified. Unlike siRNAs, micro-RNAs are processed by Dicer-mediated cleavage of stem-loop precursor RNAs that are transcribed from genomic loci distinct from other recognized genes, and are predicted to form dsRNA hairpin structures. The miRNA can reside on either the 5′ or 3′ side of the double-stranded stem. Plant miRNAs differ from animal miRNAs, in that they are far more complementary to cellular mRNAs; it has been proposed that plant miRNAs mediate target mRNA cleavage *via* an RISC-mediated RNA silencing mechanism. It was originally thought that siRNAs and miRNAs have different and distinct physiological roles. However, it appears that siRNAs and miRNAs may be functionally interchangeable, i.e. some siRNAs may act like miRNAs to repress translation, and some miRNAs may act like siRNAs to silence transcription of the target gene. Both pathways also require Dicer-like enzymes to produce the active RNA component(s) to repress/silence gene expression. Furthermore, some viral suppressors of RNA silencing can differentially regulate the accumulation of siRNAs and miRNAs generated by these pathways. The degree of homology shared with target RNA may be the key factor that reduces the activity of the RISC, leading to the inhibition of target mRNA expression.

The Dicer enzyme, the RNA-induced silencing complex and RNA dependent RNA polymerase

The Dicer enzyme, originally precipitated from the RNA-induced silencing complex (RISC) in *Drosophila* and also its family members, are large, ATP-dependent (adenosine triphosphate) RNase III-type enzymes that distinctively contain one or two dsRNA-binding domains, putative RNA helicase, RNase III and PAZ domains. The formation of siRNAs from dsRNA during the RNA silencing process is indicative of the precision cutting activity of an RNase III-type (nuclease) enzyme; RNase III-type enzymes cut dsRNA into 21–26 nucleotide long siRNAs molecules characterized by a double-stranded body with 5′-phosphate and 3′-hydroxyl ends and a 2-base 3′ overhang on each strand of the duplex. Processing of dsRNA into siRNA duplexes starts from the ends of both blunt-ended dsRNAs or dsRNAs with short (1–5 nucleotide) 3′ overhangs, and proceeds in 21–26 nucleotide steps.

ATP also is required in multiple steps of the RNA silencing cascade, including the size-reduction of initiator dsRNA into siRNAs, the unwinding of dsRNA to activate the RISC and the maintenance of the 5′ phosphate group of siRNA, a process that is essential for cells to be able to check the authenticity of siRNAs with regards to their targeted-destruction siRNA activity. Proteins homologous to *Drosophila*-derived Dicer have been isolated from fungi, plants (e.g. CARPEL FACTORY from *Arabidopsis thaliana*) and animals.

In wheat germ extracts, exogenous single-stranded RNA is copied into complementary RNA by an RNA-dependent RNA polymerase, and is then preferentially cleaved into long siRNAs suggesting that extracts of the RNA-dependent RNA polymerase are physically linked to a specific Dicer ortholog. RNA silencing of endogenous mRNA requires the putative RNA-dependent RNA polymerase, SGS2/SDE1. In contrast, transgenes that generate dsRNA and viruses that replicate through a dsRNA intermediate bypass this requirement. RNA-dependent RNA polymerase may have an important role in transitive RNA silencing (see below).

Transitive RNA silencing

A phenomenon known as transitivity increases the initial pool of siRNAs by producing new siRNAs corresponding to sequences located outside the primary targeted regions of a transcript. The direction in which transitive RNA silencing proceeds depends on the organism in which it occurs. In plants, transitive RNA silencing is dependent on RNA-dependent RNA polymerase activity and is bi-directional, i.e. it can travel both in $5' \to 3'$ and $3' \to 5'$ directions. Transitive RNA silencing does not occur in miRNA-directed cleavage of mRNA targets even though miRNAs act like siRNAs in wheat germ extracts.

Local and systemic silencing signal(s)

This is known to occur in plants, fungi and *Caenorhabditis elegans*. Systemic signalling of RNA silencing typically involves new (i.e. previously unaffected) cells that recognize the silencing signal(s) and then become a secondary "beacon" of the silencing event. As a result, the maintenance of RNA silencing becomes independent of the original initiator event. The signalling pathway or genes involved are largely unknown. Evidence obtained in plants indicates that the local silencing signal(s) that moves between cells, possibly *via* plasmodesmata, can extend outside the area targeted for silencing, and that a systemic silencing signal(s) moves within the vascular system, mimicking the movement of viruses or virus particles. This is suggestive of a mode of signalling that uses RNA and/or RNA-based molecules. Because siRNAs can induce RNA silencing when introduced into living tissues, siRNAs are considered prime silencing signal candidates. It is also noted that treatment of plants with non-toxic levels of cadmium, which affects the vascular transport of viruses, prevents systemic but not the cell-to-cell spreading of silencing, suggesting that the systemic silencing signal(s) is transported *via* the phloem.

It has been suggested that 25 nucleotide siRNAs are indispensable in long-range or systemic signalling in plants. However, the initiator and the site of action of such molecules are currently unknown. One possibility is that 25 nucleotide siRNAs that are synthesized in infiltrated tissues, act as a phloem-specific silencing signal(s) inducing the production of 21 nucleotide siRNAs in or near the vascularity in young leaves. Cell-to-cell movement of silencing would then proceed independent of 25 nucleotide siRNAs. This view is supported by the observation that some viral suppressors, that suppress systemic signalling (see below) but that do not affect the cell-to-cell movement of the silencing signal(s) at the edge of suppressor infiltrated areas on leaves, prevent the production of 25 nucleotide siRNAs (possibly by binding to them), whereas leaves infiltrated with suppressors that have no effect on systemic silencing, produce 25 nucleotide siRNAs.

There are also reports that indicate that siRNAs may not be the systemic signals, and it is therefore likely that there are many mobile systemic silencing signals that are precursors or intermediates of siRNAs (such as aberrant RNAs and dsRNA), and that they mediate different responses that trigger RNA silencing.

Why RNA silencing mechanism exists?

The protective effect of RNA silencing in reducing virus infectivity supports the view that PTGS has evolved as a mechanism to defend plants against virus infection and also to moderate the possible deleterious, genome-restructuring (insertional) activity of virus-like mobile genetic elements e.g. retrotransposons. The former suggestion is consistent with the fact that several viruses have evolved a direct countermeasure against RNA silencing by producing proteins that suppress RNA silencing. Many suppressors of RNA silencing are indispensable for the ability of viruses to move within the plant.

Suppression of RNA silencing by plant virus-encoded proteins

Many viruses appear to be able to negate the effect of RNA silencing by producing RNA silencing suppressor proteins. Currently more than 20 RNA silencing suppressor proteins have been described. Although most RNA silencing suppressor proteins share little sequence homology, many of them possess the general feature that they positively influence the *in planta* movement of viral genomes or virus particles. Most RNA silencing suppressor proteins have been identified by their ability to prevent local or systemic *GFP*-induced (green fluorescent protein) silencing of *GFP* expressing transgenic *Nicotiana benthamiana*. However, not all RNA silencing suppressor proteins are capable of suppressing virus-induced RNA silencing. This most likely reflects the fact that RNA-dependent RNA polymerase is not required by plant viruses to mediate RNA silencing of host genes because viruses trigger silencing by virtue of their dsRNA replication intermediate. This in turn appears to reflect the underlying competency of plants to engage in systemic signalling. If the suppressor does not prevent systemic signalling, then the silencing signal(s) that spread ahead of the virus may prevent expression of the viral suppressor, thereby inhibiting suppression.

On the basis of suppressor protein activity, virus-derived RNA silencing suppressor proteins can be assigned to four groups. Members of group I are characterized by the fact that they affect the maintenance of RNA silencing. Group I suppressor proteins possibly function before the formation of siRNAs, which suggests that they prevent the perception of the silencing signal(s). Another feature of group I suppressors, including the HC-Pro protein from *potato virus Y* (PVY) and *tobacco etch virus*, the AC2 protein from *African cassava mosaic virus* and the P1 protein from *rice yellow mottle virus* is that they can revert silencing in previously silenced tissue in a "reversal assay". Although PVY–HC-Pro protein eliminates the formation of siRNAs, it has no influence on DNA methylation suggesting that epistatic interactions are irrelevant to the primary mode of action of PVY–HC-Pro protein.

Group II suppressor proteins prevent the translocation of the silencing signal(s) or inhibit the action of the silencing signal(s), and characteristically, cannot revert the RNA silencing effect once it has been established. A notable group II suppressor protein member is protein 2b from *cucumber mosaic virus* (CMV). The CMV–2b was the first suppressor of RNA silencing shown to be translocated to host cell nuclei, a process that is essential for its suppressor activity. Given that CMV–2b prevents the translocation of the silencing signal(s) and that the translocation of CMV–2b to the cell nucleus is essential for its suppressor activity, then the signal(s) involved in silencing most likely navigates *via* nuclear pores. In additional grafting assays, the silencing signal becomes non-functional after passing through the CMV–2b-expressing inter-graft, suggesting that CMV–2b inhibits the activity of the silencing signal(s).

Group III suppressor proteins prevent the formation of the silencing signal(s) but cannot revert the silencing event. Currently, protein 25 (P25) from *potato virus X* (PVX) is the only known member in this group.

Group IV suppressor proteins exert their effect early, during the initiation phase of RNA silencing. The single class member of this group, *turnip crinkle virus* CP, suppresses local and systemic RNA silencing, possibly by interfering with the function of the Dicer-like RNase in plants.

Suppressor activity has been related to the seed or pollen transmission of some plant viruses. In the case of seed- or pollen-mediated virus transmission, it is essential for the survival of the virus population that the host plant does not accumulate a lethally high virus titre. Virus infected plants that suitably moderate virus titres to a non-lethal level are said to "recover" from infection.

RNA silencing in plant breeding

The analysis of the plant phenotype and the corresponding gene/protein is usually difficult, and requires complex molecular – biological and biochemical techniques. The generation of transgenic plants and especially the transformation of recalcitrant plants may be the limiting step in the gene analysis. Therefore, the advent and use of RNA silencing mechanism may be advantageous to analyze the genes for which no function is known. Herbaceous species like *N. benthamiana* and *N. sylvestris* have been used in laboratory experiments but there are only few examples of use of RNA silencing in crop plants, and it is not guaranteed that all genes can be effectively silenced.

In plant breeding, attempts are being made to remove the undesired features by crossing to native species or other traits or by screening mutants that have the desired characteristics. This is, however, expensive and often even impossible to achieve since the phenotype may be due to multiple alleles or genes that are essential to the characteristics to a certain extent. Selective suppression of the activity of the genes that are involved in complex meta-bolic reactions (e.g. those involved in sugar and lipid metabolism) in crop plants have been achieved using RNA silencing. For example, the fatty acid composition of canola (*Brassica napus*), soybean (*Glycine max*), cotton (*Gossypium indicum*) and sunflower (*Helianthus annuus*) has been altered by selectively silencing specific desaturase enzymes. Use of tissue-specific promoters in silencing constructs would allow tissue-specific silencing that would not affect the metabolism in other parts of the plant. Considering the systemic spread of silencing signals in plants, this may be impossible to achieve.

High expression level of a transgene (mRNA) is often a desired feature of a transgenic plant. Plant viral suppressor proteins can be used in co-transformation/expression assays to suppress RNA silencing, and also when expressed stably in plants, they may keep transgene active over generations. It has not been extensively tested whether suppressor proteins can overcome silencing caused by multiple copies of a transgene, but if that would be the case it would allow high level expression of a transgene in a single plant. This would be especially useful in production of biomolecules used in biomedical industry. Thus, high amounts can be produced at reasonable cost and possible toxic side-products can be avoided.

Enhancement of heterologous gene expression by RNA silencing

Plants can be used as vehicles to express heterologous proteins stably or transiently even in industrial scale. Thus, it is possible to introduce a set of genes that guide the construction of the target molecules by means of stable transformation, virus vector or agroinfiltration (see below). Transient expres-sion systems have certain advantages over generation of stable transformants. The system is fairly easy to perform, it is fast, and not susceptible to chromo-somal effects, and it can be used for the high level gene expression prior to the activation of the RNA silencing mechanisms. Besides that, it allows gene expression in fully differentiated tissues.

The size and nature of transiently expressed gene defines the method of choice. Virus vector multiplies efficiently but may also spread systemically and eventually kill the plant, which is not desirable. There is also a size limit for the expressable gene, thus, genes larger than 2000 bp show instability and are poorly expressed from viral vectors. Agroinfiltration, on the other hand, is localized, and allows large fragments to be inserted and expressed. Agroinfiltration is also useful when post-translational modifications are required for protein function. The protein yield can also be increased by tar-geting the protein product into apoplast or cell organelles for easy extraction.

The drawback in using transient expression systems is that, as its name indicates, it is "transient"; gene expression peaks at 60–72 hours after which it rapidly declines. The decline is due to activation of RNA silencing. Some suppressors have been shown to be active in *N. benthamiana* and *A. thaliana*, but their use in crop plants awaits to be done.

Several plant genomics projects are on-going, and some [rice (*Oryza sativa*) and thale cress (*A. thaliana*)] have already been finished. The RNA silencing may prove as a useful tool in functional genomics. Functions of unknown genes and metabolic pathways are to be elucidated, and use of suppressor proteins that enable long-standing gene expression, may also be useful. The high-throughput analysis of orphan genes is facilitated by development of efficient methods to clone such genes for silencing assays. The system is functional in *O. sativa* and *A. thaliana* but not in species, such as wheat, where agroinfiltration is not functional. However, such plants are putative subjects for genomic analysis using viral vectors, if such exist. One of the advantages of virus-induced gene silencing (VIGS) is the avoidance of plant transformation. Thus, the phenotypic changes are observable within few weeks from infection. At least, *tobacco mosaic virus* (TMV; genus *Tobamovirus*), *potato virus X* (PVX; genus *Potexvirus*) and *tomato bushy stunt virus* (TBSV; genus *Tombusvirus*) have been used to develop VIGS vectors. The downside of virus vectors is that they also affect the phenotype of the plant, and many of them induce discolouration of leaves, growth retardation and some even kill the plant within weeks. Most plant viruses, and hence, viral vectors are also incapable of infecting whole plants, especially the meristems, which would allow propagation of the virus in subsequent generations. To overcome such effects, a *tobacco rattle virus* (TRV; genus *Tobravirus*)-based vector was developed, which induces mild symptoms, and is capable of infecting meristems due to its ability to be transmitted by seeds. The TRV does not encode a suppressor, which is also related to seed-transmission, and this is beneficial in RNA silencing since the spread of systemic signals are not prevented. Even though, TRV vectors are so far tested only in *N. benthamiana* and *A. thaliana*, they are likely to function in crop plants as well because the virus has a wide host range, and it infects more than 60 species, including tomato, potato and barley.

Techniques used to analyze viral suppressor activity

Various methods have been used to dissect the mechanism of plant viral suppressors in RNA silencing. This is reflected by the fact that various suppressor proteins have different modes of action, and they can be classified at least into three groups (see below). The suppressor can be introduced either into target cells subject to silencing together with the trigger or into silenced tissues to reverse silencing in a process called agroinfiltration. As a result, expression of a suppressor is localized and transient. In contrast, suppressor expression can be achieved by using virus vectors leading to persistent and systemic expression pattern or by stably integrating and expressing them as transgenes. The method of choice is the matter of the stage of suppression to be analyzed.

Virus infection and expression of a suppressor from a viral vector

By far the simplest experiment to test whether a virus encodes a suppressor is to inoculate *GFP*-silenced 16c plants, and follow the "recovery" or "reversal" of fluorescence. In similar assays, any other silenced plant can be used, if the result is easily measured. This method would be a fairly easy initial assay for choosing a virus for silencing assays and for isolation of the suppressor.

Expression of a suppressor from a viral vector is used mainly in silenced plants, to revert silencing at different stages of silencing cascade. Viral suppressor gene is cloned into viral vector, and *in vitro*-transcripts are inoculated onto silenced leaves. The appearance of green fluorescence is followed by using high-powered, long-wave ultraviolet lamp. The advantage of the use of viral vectors is the high level expression in multiple organelles of the suppressor, and the speed at which the assays can be accomplished compared to generation of transgenic plants. The drawbacks include the inability of PVX to accumulate in meristems, and the deleterious effect of target plants; infected plants die within 3–4 weeks.

Transient silencing suppressor assay

Most of the suppressors have been identified based on their ability to interfere with the local and systemic silencing of the green fluorescent protein gene (*GFP*) in *GFP*-transgenic *Nicotiana benthamiana* plants in an agroinfiltration assay. In such an assay, target gene is cloned into a T-DNA expression cassette driven by the CaMV 35S promoter in a binary vector in *Agrobacterium tumefaciens* strain. Cultivated agrobacteria are infiltrated into plant tissues using a syringe *via* a needle hole (but without a needle). The *GFP* is expressed transiently (over 2–5 days) in the infiltrated area (or patch) and *GFP* is visualized as green fluorescence under UV light. Decline in the fluorescence is attributed as induction of RNA silencing against the foreign gene expression. In the presence of suppressor, *GFP* expression remains high and is visualized as bright green fluorescence. This assay has the potential to identify suppressors that affect initiation and/or maintenance of RNA silencing since the silencing trigger is expressed simultaneously with its suppressor. Agroinfiltration is often accompanied by expression of the target gene from a viral vector (see above). Even if the silencing can be triggered in single cells, e.g. by introduction of a gene through biolistic means, it has rarely been used in silencing assays. Occasionally, modified virus genomes have been cloned to a binary vector to mediate high level, local expression of the putative silencer/suppressor.

Reversal assay

Most viral suppressors have been identified using the reversal of RNA silencing assay. In this approach, transgenic *N. benthamiana* (line 16c) expressing high levels of green fluorescent protein (*GFP*) is silenced by infiltrating the leaves with *A. tumefaciens* carrying a plasmid bearing a translatable *GFP* gene.

Agrobacterial infiltration and subsequent *GFP* expression remains localized but triggers a systemic silenced response in the whole plant. Target gene is agroinfiltrated into silenced *GFP*-plant that is red due to chloroplast autofluorescence, and the recovery of green fluorescence in infiltrated patch is followed. This assay allows not only to identify the suppressor activity on maintenance of silencing but also the site of activity (veins vs. tissues).

Grafting assay

Grafting has been used to analyze systemic movement of silencing signals. Wild-type scions are grafted into rootstocks bearing a transgene under study. The spread or suppression of silencing initiated in the rootstock is analyzed both in the leaves or rootstock as well as in the scion.

Selected Literature

Béclin C, Boutet S, Waterhouse P, Vaucheret H. A branched pathway for transgene-induced RNA silencing in plants. Curr Biol 2002; 12:684–688.

Goldbach R, Bucher E, Prins M. Resistance mechanisms to plant viruses: an overview. Virus Research 2003; 92:207–212.

Hannon GJ. RNA interference. Nature 2002; 418:244–251.

Li WX, Ding S-W. Viral suppressors of RNA silencing. Curr Opin Biotech 2001; 12:150–154.

Pickford AS, Cogoni C. RNA-mediated gene silencing. Cell Mol Life Sci 2003; 60:871–882.

Susi P, Hohkuri M, Wahlroos T, Kilby NJ. Characteristics of RNA silencing – similarities and differences across kingdoms. Plant Mol Biol 2004; 54:157–174.

Tang G, Reinhart BJ, Bartel D, Zamore PD. A biochemical framework for RNA silencing in plants. Genes Dev 2003; 17:49–63.

Tijsterman M, Ketting RF, Plasterk RHA. The genetics of RNA silencing. Annu Rev Genet 2002; 36:489–519.

Voinnet O. RNA silencing as a plant immune system against viruses. Trends Genet 2001; 17:449–459.

Voinnet O. RNA silencing: small RNAs as ubiquitous regulators of gene expression. Curr Opin Plant Biol 2002; 5:444-451.

Chapter 12

Virulence genes and their products

Yu. T. Dyakov, O. L. Ozeretskovskaya

The genes whose products cause a number of the defense responses described in earlier chapters in infected plants are called "Avr-genes," and proceeding from the same pattern, we will use the term "Vir-genes" for the genes whose products, on the contrary, provide pathogen protection from the immune response. Such protection can be provided by three ways:

1. Loss or structural change of the specific elicitors – the molecules recognized by plant receptors – which makes penetration of the parasite unnoticeable and plant–parasite relations compatible.
   In this connection, the recessive alleles of the avirulence genes are frequently called virulence genes. Since the elicitors, as was noted earlier, usually include the compounds either absolutely necessary for the parasite or at least those reducing its adaptivity, the complete loss of an elicitor is not always possible, and its modifications provide only a temporary effect due to high variability of the R-gene structure.
2. Suppression of defense responses at the pretranslational level. Theoretically, such suppression can be provided by:

   (a) modification of the plant cell receptors, with their subsequent inability to recognize the elicitors;
   (b) inactivation of the enzymes regulating separate stages of stress metabolism (PAL, pharensyl cyclase, etc.);
   (c) inhibition of the stages of basic cell metabolism (energy chains, membrane transport, etc.), which results in the loss of cell ability for immune responses.

   Consequently, the suppressors can be less toxic for the cells, without causing a visual damaging effect, or highly toxic, causing damage and death of plant cells. The former were called impedins, and latter, pathotoxins.
3. Suppression of defense responses at the post-translational level: chemical modification of the stress metabolism products toxic for the microorganism – phytoalexins (FAs) and phytoanticipidins – is carried out by the parasite enzymes that modify molecules of plant antibiotics.

Suppressors (impedins)

Unlike the toxins of necrotrophic pathogens that destroy the plant tissue, the suppressor molecules only prevent the manifestation of resistance by allowing pathogens to penetrate a plant and grow in it. The Japanese researchers Ouchi and Oka (1981) suggested the name "impedins" for such substances (in medicine this term is used to designate nontoxic substances of bacterial origin capable of suppressing anti-infection response mechanisms).

According to the compatibility concept suggested by K. Ouchi, previous inoculation with a pathogenic fungus makes a plant available for subsequent infection by an avirulent race of the same fungus or even by a nonpathogen. Such type of predisposition was investigated in a wide range of mildew fungi. Out of 51 pathogen isolates, 45 were found able to infect the leaves of barley which is not their host, if the leaves had been previously inoculated with the compatible race of the barley pathogen *Erysiphe graminis* f. sp. *hordei*. This effect was not limited to the haustorium-containing cells but propagated through diffusion to the adjacent cells. The active substance was isolated from the apoplast space of the affected cells and was found to be a low molecular weight substance.

Such experiments promoted a search of suppressors from other pathosystems (Table 12.1). As the researchers used different methods, no wonder that the suppressors were not always structurally homogeneous and had different properties.

Suppressors can be present both in the tissues of infected plants and in the pathogen cells. Usually, both the suppressor and the elicitor are contained in the same source: fungal mycelium, culture fluid, or apoplast fluid, which makes the suppressor isolation a difficult challenge. Under such circumstances, the suppressor can be fully or partially masked by high elicitor activity. For instance, the suppressor of synthesis of the pea FA pisatin is contained

Table 12.1.   Fungal suppressors of higher plant resistance (Beismann, Kogel, 1995)

Source	Origin	Chemical nature	Race specificity
Uromyces phaseoli/beans	Apoplast fluid	No data	No
C. fulvum/tomato	Same	Protein	No
Phytophthora infestans/potato	Spore exudate, mycelium	Glucan	Yes
P. megasperma/ soybean	Culture filtrate	Glycoprotein	Yes
Ascochyta rabiei/chick pea	Same	Same	No
M. pinodes/peas	Same	Same	No
Puccinia graminis/wheat	Apoplast fluid, spore exudate	No data	No

together with the elicitor in the fluid where the picnospores of *Mycosphaerella pinodes* germinated. The elicitor and suppressor of *P. infestans* are present in the fungus mycelium and spores. The suppressor of *A. rabiei* was found together with the elicitor in the fungal culture filtrate.

An optimum potential source for isolation of a suppressor (as well as a race-specific elicitor) is the apoplast fluid of infected plants. The ideal object for collecting the apoplast fluid – for the reasons explained earlier (p. 208–209) – is the causal agent of the tomato olive mold, the imperfect fungus *Cladosporium fulvum*. The active suppressors of protein origin, named EPS1 and EPS2, were received from the apoplast fluid of the tomato leaves inoculated with *C. fulvum*. The loss of these proteins due to inactivation of encoding genes resulted in strong inhibition of aggressiveness. The apoplast fluid isolated from the beans, sunflower, and wheat inoculated with the compatible strains of rust pathogens increased the bean susceptibility to the cowpea rust pathogen, avirulent for beans. The apoplast fluid from the tomato leaves inoculated with *C. fulvum* as well as from rust-affected beans prevented the formation of necrosis and deposition of callose in response to the treatment with nonspecific elicitors of the same fungi. A lignification suppressor was found in the apoplast fluid of the wheat leaves inoculated with stem rust, and it was separated from the elicitor using gel filtration.

To isolate a suppressor separated from the elicitor and to test its activity, several conditions should be met: (1) the elicitor should be used in a definite concentration so as not to mask the suppressor activity which may be not full; (2) the optimum time of the plant treatment with the elicitor as well as the suppressor should be determined experimentally; and (3) determination and assessment of activity should be easy and prompt, to enable the analysis of many samples.

Today, there is not enough data to show the structural determinant of the suppressor activity. For instance, invertase was isolated from two compatible races of *P. megasperma*, which inhibited accumulation of the phytoalexin glyceollin in soybean cotyledons. It was found to be mannanoprotein, its active part being carbohydrates. The suppressor from *M. pinodes* was a glycopeptide, the active part of which was discovered to be a protein, as pronase and proteinase had destroyed the suppressor activity. The active component of the suppressor from the intercellular space of the infected wheat leaves as well as of the fluid where the urediospores of *P. graminis* germinated was less than 5 kDa low-molecular weight compound. Periodate oxidation, proteolytic degradation, thermal treatment, or acidic hydrolysis did not influence the suppressor activity, thus indicating that neither carbohydrates nor proteins were an active part of the suppressor.

While most models of elicitor activity suggest that the primary elicitor site is a receptor localized on the cell surface, the suppressor activity can be more diverse. The suppressor is assumed to be able to interfere with the relations between the elicitors and the respective receptors, to interrupt signal transduction through inhibition of the secondary messenger formation, to suppress the gene activation as well as to inhibit the enzymes responsible for synthesis of the protective substances. For example, it was shown that the sections of potato tubers, even not possessing the R-genes, became resistant against all the tested races of *P. infestans* after the treatment with plasmolyzing solutions (KCl, saccharoses). The resistance was accompanied by a hypersensitive response (HR) and accumulation of the phytoalexins richitin and lubimin. Therefore, a conclusion was made that masking of the suppressor-binding sites had occurred in the plasmolyzed plasmalemma, i.e. the activity of the β-glucan suppressor of the late blight, probably, depends on the interaction of the elicitor and suppressor on the receptor site. If the suppressor was mixed with the microsomal fraction, which was assumed to contain the receptor, necrotization of potato tubers and production of phytoalexins were inhibited in response to the inoculation with an incompatible race or a nonspecific elicitor. The pea leaves treated with a mixture of a nonspecific elicitor and a suppressor of *M. pinodes* show a delay in the expression of phenylalanine ammonialyase and chalcone synthetases mPNA as compared with the leaves treated only with the elicitor. This is also indicative of the suppressor effect at the pretranslation stage of the response to the infection. The suppressor from the apoplast fluid of the tomato parasite *C. fulvum* is supposed to digest the nonspecific elicitor of the same fungus. This conclusion was made on the basis of the observation that heating inhibits the suppressor activity.

The specificity of the effect of the suppressor from *P. infestans*, i.e. its ability to suppress the potato immune response to inoculation with an incompatible race of the pathogen was discovered to be a property of only β-glucans isolated from the compatible races. The most convincing evidence of the suppressor specificity were the data received in inoculation of the potato cultivars with the late blight fungus races that have opposite relationship (potato cultivars with genes R1 and R4 inoculated with races 1 and 4). The response of the potato cultivars to inoculation with the parasite races was similar to the response of these cultivars to treatment with the glucans from the same races, which is quite a telling argument for glucan specificity. The data in Table 12.2 demonstrate specificity of the suppressor from *M. pinodes* at a species level. It can be seen that the pear parasite *Alternaria kekuchiana* does not attack the legumes, however, pretreatment of the leaves with the suppressor from *M. pinodes* confers susceptibility only to the species susceptible also to the suppressor producer. Therefore, in this case the suppressor is the primary factor restricting the range of host plants.

Thus, if interaction of the suppressor with the host-plant cells is highly specific, susceptibility of the tissue treated with the suppressor becomes nonspecific, i.e. it is manifested not only towards the pathogen that induces susceptibility but also towards the organisms that do not attack the given plant species (Table 12.3).

Mechanically wounded areas of potato tubers develop wound periderm that lies close to the wounded surface and has high protective properties. Treatment of potato tuber disks with the suppressor glucans isolated from the compatible races of the late blight pathogen suppresses the wound reparation process through interference with the phellogen deposition and resumption of

Table 12.2. Colonization of legumes (points) by *M. pinodes* and *A. kekuchiana* upon pretreatment with the suppressor from *M. pinodes* F5 and without pretreatment (Ouchi S, Oku H, 1982)

| Plant species | Degree of infectious hyphae formation (points) | | |
| | *M. pinodes* | *A. kekuchiana* | |
		Without F5	With F5
Pisum sativum	4	0	4
Trifolium pratense	1	0	1
T. repens	0	0	0
Lotus corniculatus	0	0	0
Milletia japonica	2	0	1
Lespedezia buergeri	2	0	2
Vigna sinensis	0	0	0
Medicago sativa	1	0	1
Glycine max	0–1	0	0
Vicia faba	0	0	0
Arachis hypogaea	0	0	0

meristematic activity in the tuber parenchyma cells (Figure 12.1). However, the suppression of the cell meristematic activity under a suppressor effect was only temporary. Several days after the treatment, the surface cells treated with the suppressor resumed the meristematic activity and began forming peridermal

Table 12.3. Effect of the suppressor glucans isolated from the races of *P. infestans* compatible and incompatible to the potato cultivar Lyubimets (R1), on the degree of tuber damage by different fungi (Medvedeva TE, 1985)

| Species of fungi | Number of darkened tuber cells (as percent of water-treated cells), preliminarily treated with glucans of | |
	Compatible race	Incompatible race
P. infestans	224	73
Colletotrichum atramentarium	224	125
Fusarium culmorum	138	61
Phoma exiqua	140	70
P. cyprena	144	113
Botrytis cinerea	270	118
C. cucumerinum	155	95
Magnoporthe grizea	192	89
Alternaria alternata	154	92

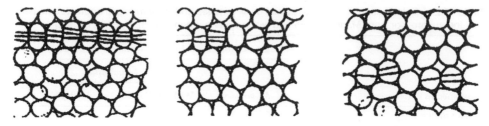

Figure 12.1.   Effect of β-glucans of *P. infestans* on formation of wound periderm in potato tubers. The tuber slide surface treated with (left to right): glucan of an incompatible race, water and glucan of a compatible race.

complexes. This resulted in formation of the abnormal multilayer periderm where one peridermal complex was as though heaped on another. Apparently, the loss of the ability to recognize wound signals and to resume meristematic activity was only a temporary effect caused by the suppressor in the cells on the wounding surface. The cells in the deeper parenchyma layers, where the suppressors had not penetrated or penetrated in low concentrations, began to divide and form the wound periderm. As soon as the suppressor activity stopped, the surface cells also started dividing and forming the wound periderm, which resulted in formation of the multilayer complexes. This fact suggests that the suppressor glucans, isolated from the parasite races compatible with particular potato cultivars, do not cause irreversible suppression of the defense responses but only their temporary delay. In light of the above, one can assume that the suppressor glucans either suppress the process of formation of the wound hormone responsible for wound reparation, or, more likely, they change the plant cell competence towards the hormone.

Pathotoxins

Any parasite toxins, including nonspecific vivotoxins (Chapter 5), are immunosuppressors, as they inhibit the plant cells ability for active resistance to infection through damaging the plant cells. For instance, alternaric acid, a toxin of the causal agent of the early leaf spot of potato *Alternaria solani*, causes depolarization of membranes and suppresses the hypersensitive response in the potato tubers of the late blight resistant cultivars in response to inoculation with *P. infestans*. As against nonspecific vivotoxins, we will further discuss the toxins that damage only a particular plant species or even plant cultivars. Some of them are primary pathogenicity factors (an organism cannot be a parasite without a toxin) and they were called "host-specific toxins", while others, like vivotoxins, are the secondary factors that intensify pathogenicity for particular plant species or cultivars (host-selective toxins). The general name of this group of compounds is "pathotoxins".

Pathotoxin producers and pathotoxin-caused diseases

The history of pathotoxin research is full of dramatic events in the destiny of several important agricultural crops.

In the 1950s, the oat cultivar Victoria from Uruguay was widely used in hybridization in the USA to develop new cultivars. Victoria possesses gene Pc2 that increased resistance against all North American races of the crown rust pathogen *P. coronata*. In different US states, the progeny of those hybrids took 80–100% of the oat areas. The new cultivars soon began dying from an unknown disease that caused root neck rot and leaf browning. The disease to be caused by a fungus, not described earlier, that was named *Helminthosporium victoriae* (the modern name of the anamorph is *Bipolaris victoriae*, and of the ascogenus stage, *Cochliobolus victoriae*). Its growth medium accumulated a toxin called victorin that damaged only the oat cultivars containing gene Pc2. After discarding these cultivars, the disease disappeared as rapidly as it had appeared, and its causal agent from time to time occurs in soil as a saprotroph or a weak parasite of dying or aging cereal plants – just as it was before the large-scale propagation of the cultivar Victoria progeny.

In 1970, an epidemic of southern helminthosporiosis of corn caused by a related fungus *Bipolaris maydis* (teleomorph of *C. heterostrophus*) broke out in the USA. The epidemic first attacked the corn cultivars with the Texas type of cytoplasmic male sterility (T-cms). These cultivars do not produce viable pollen and are not capable of self-pollination, therefore, they always have hybrid seeds. With T-cms in place, high-yielding heterosis hybrids of corn can be obtained, without the expensive mechanical breaking of male inflorescences in recipient plants. The mitochondrial gene T-cms was widely used in the selection programs: it was possessed by the cultivars covering about 85% of all the USA Corn Belt area. The southern helminthosporiosis of corn was well known to the American plant pathologists, but it never caused considerable losses. The epidemic of 1970 was caused by a new race of the fungus (race T) which unlike the previous race 0, turned out to be extremely toxic for the cultivars with T-cms. A toxin was isolated from the T-race growth medium that damages only the corn cultivars with T-cms cells. The American corn producers had to give up cultivation of corn with T-cytoplasm, which lead to the drop of T-race concentration in the populations. After the epidemic in Northern America, the T-race of *C. heterostrophus* has spread worldwide. It inflicts heavy losses on the maize in the southern regions of Russia.

In 1971, the agent of a severe sugarcane disease – eye spot (dark spots with a light halo developed on the leaf) – was identified in America. The pathogen was discovered to be the fungus *Bipolaris (Helminthosporium) sacchari*, its culture fluid also accumulated the toxin specifically damaging only the susceptible sugarcane cultivars. While in the New World, most fungi producing host-specific toxins belong to the formal genus *Helminthosporium* (though not only this genus: *Phyllosticta maydis* in maize and *Periconia circinata* in sorghum possess similar toxins), in Japan pathotoxins were found in the fungi from the genus *Alternaria*. These fungi and their toxins cause black spots on the leaves of susceptible cultivars of pear (*A. kekuchiana*), apple (*A. mali*), strawberry (*A. fragariae*), tomato (*A. lycopersici*), tangerine (*A. citri*), and other species. Morphologically, they all belong to one morphospecies, *A. alternata*, that is common in phylloplane of many plant species. In living plants *A. alternata* feeds on leaf exudates; in aging and weakened leaves it parasitizes with the formation of small black spots; after the leaf dying-off the parasite invades the plant remains. Mutations of individual genes are believed to result in the production of the host-specific toxins that transform a harmless resident of phylloplane into a highly pathogenic parasite. Pathogenic mutants were experimentally received from a saprotroph strain of *A. alternata* (Table 12.4). Therefore, the alternaria with host-specific toxins were assigned the status of pathotypes (or specialized forms) of the species *A. alternata* (however, the morphological and molecular studies of the late twentieth century – early twenty-first century bring evidence that some pathotypes of *A. alternata* should be set apart as independent species).

The characteristic features of all pathotoxins investigated so far include very low toxic concentrations towards the susceptible plants and high selectivity (the ratio of the minimum toxic concentrations for a resistant and susceptible cultivar) (Table 12.5).

Chemical composition

In terms of the chemical composition, pathotoxins are classified into several groups.

Cyclic peptides where the amino acids are connected by nonmatrix synthesis. They include victorin of *C. victoriae*, HC-toxin of *C. carbonum*, and AM-toxin of *A. alternata* f. sp. *mali* (Figure 12.2). All of them contain unusual amino acids, together with trivial ones. For instance, the group of pentacyclic peptides (victorins a, b, and c) contains 5,5-dichlorleucin-treo-β-oxylisine,

Table 12.4.   Toxin-producing mutants of *A. alternata* (Nishimura S, et al., 1982)

| Isolates | Number of lesions per 1 sq. cm of leaf | | Toxin production |
	Pears (cultivar Nijisseiki)	Apples (cultivar Red Gold)	
0-94 (initial)	0.0	0.0	–
0-94-P1	20.7	0.0	AK*
0-94-P2	27.5	0.0	AK
0-94-P33	12.6	0.0	AK
0-94-A3	0.6	13.8	AM**
0-94-A18	0.6	10.3	AM
0-94-A21	0.7	17.2	AM

* *toxin* A. kekuchiana; ** *toxin* A. mali.

Table 12.5. Pathotoxins of some phytopathogenic fungi

Fungus	Host	Toxin	Lethal dose for a susceptible cultivar	Selectivity
C. victoriae	Oat	Victorin (HV-toxin)	0.3 ng/ml	$>10^6$
C. carbonum	Maize	HC-toxins	0.5 μg/ml	10^1
B. sacchari	Sugarcane	HS-toxin	30 ng/ml	>10
A. alternata f. sp. *citri*	Tangerine	AC-toxin	30 ng/ml	10^4
A. alternata f. sp. *kekuchiana*	Pear	AK-toxin	10 ng/ml	$>10^4$
A. alternata f. sp. *Mali*	Apple	AM-toxin	2 ng/ml (application on leaves);	10^6
			0.1 ng/ml (petiole dipping)	10^5

erythro-β-oxyleucin, and α-amino-β-chloraryl acid. The specificity of reception seems to be connected with the glyoxylic acid residue.

The tetrapeptide HC-toxin contains, alongside proline and two alanines, 2-amino-8-oxo-9,10-epoxydecanoic acid, the latter appears to determine the peptide toxicity, for the rupture of the epoxy ring leads to the loss of toxicity. The structure of the inhibitor of histone deacetylation trapoxin is similar to HC-toxin, therefore, similar activity of HC-toxins on histones is supposed to lead to inhibition of expression of the immune response. In the genome of toxicogenic strains of *Cochliobolus carbonum*, a site was cloned containing a gene that controls synthesis (combination and ring closure of amino acids) of the enzyme toxin (HC-toxin synthetase), HTS-1. The synthesis and transmembrane transmission of HC-toxin also requires a complex locus Tox2, including the genes TOXA (encodes HC-toxin efflux pump), TOXC (encodes fatty acid synthetase); TOXF (aminotransferase that encodes branching of amino acid side chains); and TOXE (regulatory protein).

AM-toxin is represented by two forms possessing different toxicity; structurally, it is a cyclic tetrapeptide containing L-alanine, dehydroalanine, and isoforms of L-amino valerianic acid, closed in a ring by the lactone bond and connected to the *p*-metoxyphenyl group (Figure 12.2). The amino acid ring is closed by the enzyme cyclic peptide synthetase controlled by AMT gene.

Linear polyketols (Figure 12.2) are produced by two fungi: T-race of *C. heterostrophus* and *P.* (*Mycosphaerella*) *maydis* specifically pathogenic for maize with T-cms. Several homologous T-toxins differ in the chain length (from 39 to 41 carbon atoms). T-toxin synthesis is controlled by several genes. Gene FKS1 (Tox1A) is responsible for the synthesis of the enzyme polyketide synthetase which builds the polyketide chain, and gene DEC1 (Tox1B) is responsible for the synthesis of decarboxylase that eliminates terminal carboxyls. These genes are located on different chromosomes. PM-toxins have shorter chains (33–35 carbon atoms). To display toxicity, the total chain

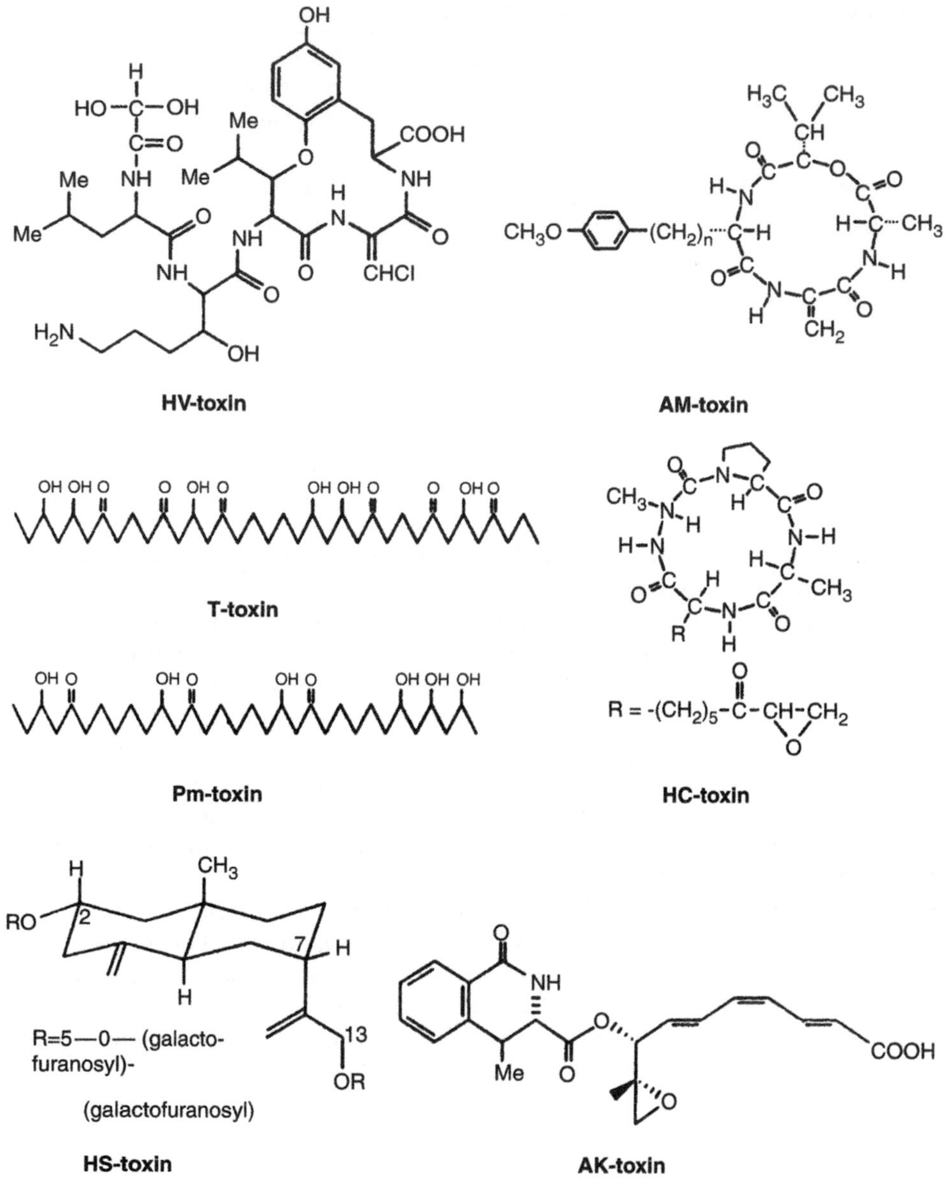

Figure 12.2.   Pathotoxins.

length and presence of not less than three oxidized groups (clusters =O–OH=O) is important.

Glycoside structure is characteristic of HS-toxins of *B. sacchari*, β-galactofuranosides, where two galactose molecules are connected by the glucoside bonds with sesquiterpene aglycone. Elimination of the sugars from the aglicon removes toxicity, but does not eliminate the ability to protect tissues from damage by the native toxin. Hence, the sesquiterpene part of the molecule is responsible for toxicity, and the galactofuranoside part, for binding by the receptor (specificity).

The toxins of *A. alternata* f. *kekuchiana* (AK), *A. alternata* f. *fragariae* (AF) and *A. alternata* f. *citri* (ACT) are the substituted products of 9-methyldecatrienic acid (Figure 12.2). Similarity of the structure determines cross pathogenicity: all three toxins, as well as their producers, cause necroses on pear leaves. Using the method of restriction enzyme-mediated integration (REMI), 15 toxin production-related genes were found in *A.alternata* f. *kekuchiana*. One of them (AKTS1) is unique for the pear parasite, and the others are also present in the genomes of *A.alternata* f. *fragaria* and *A.alternata* f. *citri*. Apparently, the common genes are involved in the synthesis of the common precursor of the three toxins.

Specificity and toxicity mechanisms

A low lethal dose and high selectivity point out to the receptor mechanism of specificity (resistant plants have no toxin-binding site). The first signs are observed only a few minutes after treatment of the susceptible plant cells with toxins show as leakage of metabolites and electrolytes, depolarization of the membrane potential, and membrane invagination. Therefore, the pathotoxin receptors are believed to be located on the membranes. AK-, HS-, AF-, ACT-toxins bind to the cytoplasmic membrane, T-, PM, ACR-toxins bind to the mitochondrial membrane, and AM-toxin, to the plasmalemma and membrane of the chloroplasts. In search of the receptors, the fungus culture is grown on the medium containing radioactive labeled salt, and the membrane fraction where the toxin is bound is determined by the radioactive label. To establish specificity of the reception, several conditions should be observed: (1) binding should be genetically specific (with the fraction of a susceptible but not resistant cultivar); (2) it should correlate with the biological activity (toxicity-reducing chemical modifications of the toxin also reduce the binding affinity); (3) binding should be ligand-specific (unlabeled toxin prevents binding the labeled toxin). For some toxins these conditions have been observed, plant binding sites have been isolated, and toxicity mechanisms have been explored.

Victorin labeled with 125I binds to two oat proteins: 100 kDa (P-component) and 15 kDa (H-component). The binding to the former receptor occurs only in planta and in combination with the membrane fractions of a susceptible cultivar, and binding to the latter receptor takes place both in a susceptible and resistant cultivars, *in planta* and *in vitro*. Functionally, both proteins are the forms of the enzyme glycine decarboxylase involved in photorespiration:

$$O_2 + \text{rubisco} \rightarrow \text{ribuloso-1, 5-biphosphate} \rightarrow$$

$$\rightarrow \text{phosphoglycolic acid} \rightarrow \text{glycolic acid}$$

The glycolic acid arrives in peroxisomes where it is oxidized by peroxidase to glyoxylic acid that is converted into glycine. In the mitochondria, glycine is decarboxylated by glycine decarboxylase and acetylated by serinoxymethyl transferase to CO_2, NH_3, NAD^+, and serine. This is a very important cycle in the functioning of plant cells, for mutations in the glycine decarboxylase gene are usually lethal. Mutations of this gene in humans cause a serious disease, hyperglycinemia. The enzyme per se is a P-protein, but in the presence of H-protein its activity grows 10,000-fold.

T-toxins and PM-toxins damage the mitochondrial membranes of the corn with cms. Their effect on the mitochondria results in the loss of Ca^{2+} and NAD^+, and dissociation of oxidative phosphorylation. The inner membrane of T-mitochondria contains a 13 kDa protein (Urf13), absent in normal mitochondria. It is responsible for the male sterility and susceptibility to toxins. Apparently, in combination with the toxin, protein Urf13 is polymerized and forms transmembrane channels that make pathways for leakage of electrolytes. The protein synthesis-controlling gene (T-urf13) was cloned and transferred in the model prokaryotes and eukaryotes (*E. coli*, yeast, drosophila, tobacco plant). All transgenic organisms have become susceptible to T-toxin.

HC-toxin is a structural analog of trapoxin that inhibits the mammalian enzyme histone deacetylase (HDAC). HC-toxin produces the same effect in tissues of plants, yeast, invertebrates, and vertebrates. It is not clear how HDAC inhibition leads to suppression of the defense responses, however, this enzyme was shown to be multifunctional and capable of deacetylation of not only histones but also other functionally important proteins (transcription activation factors, tubulins), as well as to act as positive regulators of transcription.

Plant resistance to pathotoxins is mostly determined by the absence of reception. However, cases are known when resistance was due to toxin degradation in a plant. For instance, the maize cultivars resistant to *C. carbonum* form the enzyme HC-toxinreductase that destroys HC-toxin.

Biological role of pathotoxins and ecology of producer

Pathotoxins of many fungi are the primary determinants of their pathogenic properties. This has been demonstrated by numerous experiments. For instance, the application of a drop of victorin solution on a leaf of the oat cultivar Victoria makes the treated leaf susceptible to inoculation not only with various phytopathogenic fungi but even with conditionally pathogenic species. The ascosporous progeny of the hybrids between the strains *C. victoriae* and *C. carbonum* was segregated into four classes: pathogenic for oat, pathogenic for maize, pathogenic for both crops, and nonpathogenic, with the class one strains producing victorins; class two, HC-toxins; class three, both toxins; and class four did not produce toxins. Thus, the pathotoxins completely paralyze the plant cell defenses and make the cells defenseless against an attack. Suppression of the defense responses of an infected plant by pathotoxins has been shown by direct experiments. For instance, the leaves of the lemon inoculated with a nonpathogenic strain of *A. alternata* induce synthesis of lipoxygenase (gene RlemLOX), chalcone synthase, hyperperoxide lyase (RlemHPL), propadiene oxide synthase, the inhibitor of polygalacturonase (RlemPGIP), and chitinase (RlemAchi). Upon inoculation with the ACR-toxin-producing strains, no expression of the above genes was observed. To understand the role of pathotoxins, the studies in pears inoculated with *A. alternata f. kekuchiana* are interesting. These authors observed two peaks of the toxin activity in the plants (by loss of electrolytes): 4 and 9 hours after the inoculation. The first peak was caused by the release of AK-toxin by the germinating spores. During 4 hours of germination the spore germ releases 10^{-6} μg of the toxin; this concentration does not kill the cells, however, it is sufficient to interfere with a number of functions (including the immune response) of about 100 cells in the hypha penetration area. The second peak is caused by the toxin released from the mycelium growing in the infected plant tissues. A considerably higher concentration of the mycelium-produced toxin kills the cells, which is manifested as the black spots along the veins appearing 11–12 hours after the inoculation. Thus, the biological concentrations of the toxins used by the fungi for infection are just sufficient for suppression of the cell defense potential, but they are not lethal, i.e. the pathotoxin producers interact with the plant as hemibiotrophs (parasite): they grow biotrophically at the first stages of the infection process and necrotrophically at the subsequent stages.

Pathotoxin producing fungi have caused a number of severe epidemics of agricultural crops. However, the cause of these epiphytotics was the human selection activity aimed at imparting to the plants of such unnatural properties, as, for example, male sterility. In natural phytocenoses, these fungi do not accumulate and have no epidemiological potential due to two circumstances:

1. Pathotoxin producers are highly pathogenic and frequently lethal for the susceptible plants. As the mutations resulting in resistance (receptor loss or structural change) occur frequently, and the selection pressure is very strong, the emergence of such parasite in a plant population rapidly leads to dying-out of susceptible individuals and accumulation of resistant individuals.

2. Pathotoxin producing strains have lower adaptivity and can withstand competition with the toxin-free mutants of the same species only in the presence of susceptible plants. When such plants are eliminated from the populations, concentration of the toxicogenic strains rapidly drops. Such was the case of the fungus *C. victoriae* after termination of cultivating the oat cultivar Victoria and its progeny, and T of *C. heterostrophus* after termination of production of the maize hybrids

with T-cytoplasm. In some fungi, the selection against toxicogenic strains is so strong that they occur very seldom even in susceptible crops.

For instance, in the gardens where *A. alternata* f. *kekuchiana*-susceptible pear cultivars were grown, from 510 isolated strains of *A. alternata* only 11 strains (2%) produced toxin, while the fungus populations isolated from the leaves of the resistant cultivars contained 0.1% of such strains. Among the strains of *P. circinata* isolated from the susceptible sorghum crops, only 25% produced toxin, with much more toxicogenic strains in the sorghum roots than in soil (34% against 13%). The drop in adaptivity of toxicogenic compared with nontoxicogenic strains is, probably, caused by high energy consumption in the toxin synthesis (mycelium of T-races of *C. heterostrophus* contains as much as 3% of toxin as dry weight), or by linking with lethal genes (Tox1 gene site of the same fungus is linked with reciprocal translocation).

Relationship between induction and suppression

Investigations of pathotoxins are interesting for studying the relationship between the two fundamental properties of parasites: induction and suppression of defense responses, in other words, between the genes with "plus" and "minus" functions. For example, victorin is toxic only for the oat cultivars that possess the resistance gene to crown rust Pc2 (causal agent *P. coronata*). Hybridization tests showed that resistance to rust (Pc2) and susceptibility to helminthosporiosis blight, i.e. susceptibility to victorin (Vb) are controlled by one locus. Consequently, the oat gene in question performs the resistance "plus" function towards *P. coronata* (its recessive allele imparts susceptibility to the plant), while the parasite virulence performs the "minus" function, which corresponds to Flor's "gene-for-gene" concept. In relation to *C. victoriae* (and to victorin) the same gene has the susceptibility "plus" function (the recessive allele imparts resistance), and the host–parasite relationship are shown as an inverted "gene-for-gene" rule (Table 12.6).

These contradictions turned reconcilable upon studying the quantitative relations between the parasite virulence factor (the toxin victorin) and oat resistance factor (the FA avenalumin) (Table 12.7).

Table 12.6.   Relationship of oat and two fungal parasites

Oat genes	Parasite genes			
	P. coronata		*C. victoriae*	
	A	A	Tox	Tox
Pc2 (Vb)	R	S	S	R
pc2 (vb)	S	S	R	R

Table 12.7. Induction of avenalumin (mg/g of raw tissue) by various elicitors (Mayama S, et al., 1986)

Elicitors	Isogenic oat lines	
	X-469 (Pc2 or Vb)	X-424 (pc2 or vb)
Water	Traces	Traces
Endogenous nonspecific elicitor	43	53
Mycolaminarine (β-glucan), 250 g/ml	38	34
Victorin, 10 pg/ml	86	Traces
10 ng/ml	Traces	Traces

It can be seen from the table that no difference was observed between the isogenic oat lines with and without gene Pc2 in relation to nonspecific elicitors. Victorin in very low (picogram) concentrations did not cause leaf blight in the oat line with gene Pc2 but induced synthesis of avenalumin, i.e. victorin behaved as a specific elicitor; victorin in higher concentration, on the contrary, did not induce synthesis of avenalumin but caused cell death in the cultivars with the same gene. Apparently, this is due to the fact that the living plant cells are necessary for activation of the immune response genes, and nanogram concentrations of victorin kill these cells.

Thus, the same plant gene functions as a resistance gene in response to a low concentration of the same fungal metabolite (victorin), and as a susceptibility gene in response to its higher concentration. Probably, these data have more fundamental importance: the same compound of the parasite in different concentrations and in different genotypes of the host-plant can serve as a specific elicitor or specific suppressor. In addition to the above example, this conclusion is also supported by other data.

1. The specific suppressors of *P. infestans* undermine the defense responses in the potato cultivars susceptible to the producer but enhance such responses in the resistant cultivars, i.e. they can be suppressors or elicitors depending on the host genotype.
2. Avr-genes of some bacterial species and vir-genes of other species or strains have homologous sequences in DNA.
3. Expression of avrD of *Pseudomonas syringae* pv. *tomato* and, accordingly, the elicitor activity of its product is different in different bacterial strains and species.
4. Products of bacterial Hrp-genes are necessary for the induction of hypersensitive response in nonhosts (elicitors) and pathogenicity for the hosts (immunosuppressors).
5. Protein NIP1 of *Rhinchosporium secalis* is simultaneously a specific elicitor for the plants possessing the resistance gene Rps1 and a factor of nonspecific pathogenicity, i.e. a suppressor (p. 135). A similar dual function is performed by protein Esp2 of *C. fulvum*, the loss of which does not only reduce the total pathogenicity but also confers

susceptibility on some tomato resistant lines; consequently, these lines possess the Esp2-specific receptor.

6. In *P. syringae* pv. *tomato*, avirulent to the tomatoes with the resistance gene Pto, two elicitor proteins, AvrPto and AvrPtoB, were found. Both proteins are bound with pPto in the yeast two-hybrid system and induce HR. However, their coexpression together with Pto in the tobacco *Nicotiana benthamiana* induces HR only in interaction with avrPto but not with avrProB. The latter inhibits the process of defense responses in the pathway downstream the binding with Pto, and it possesses a property of suppressing defense responses also in other host–parasite systems, i.e. it is a specific elicitor (in tomatoes with gene Pto) and a nonspecific suppressor.

These data support the "guard" model, according to which elicitors normally function as pathogenicity factors, i.e. as immunosuppressors, and, hence, they are sent by the pathogenic organisms to the plant tissue where they are recognized by the plant receptors and cause induction of a defense signal (Chapter 6).

At the same time, some researchers (Wolpert et al., 2002) believe that HR as a form of programmed cell death and the respective formation of the reactive oxygen species (ROS) plays a decisive role in protection against biotrophic parasites but promotes colonization of the infected plant by necrotrophic parasites.

Role of immunosuppression in pathogenesis

Immunosuppression as a biological property is widely used by biotrophic parasites. Necrotrophs suppress the defense potential of the living cell in a more coarse way – by killing it. As biotrophs feed on the contents of living cells, they need to use more subtle mechanisms of suppression: impedins and pathotoxins that suppress the defenses of the infected tissue in the concentrations below lethal. As the suppressed cells are partially or completely "disarmed", they become an easy prey not only because of a suppressor producing biotroph but also of the concomitant low pathogenic species unable to infect an intact plant. For instance, as it was shown earlier, the suppressor of *P. infestans* inhibits formation of the wound periderm in a potato tuber, which separates the damaged from healthy tissue. This makes the late blight affected tubers very susceptible to various putrefactive microorganisms (fungal and bacterial tuber rot pathogens). The lots of potatoes not damaged by late blight have better storage properties than the affected lots, though in normal storage conditions the causal organism *P. infestans* does not cause destruction of the infected tubers and does not transmit from infected to healthy tubers. Plant pathologists know well the complex flax disease "fusarial wilt on rust" where the cells around the rust pustule become highly susceptible to the necrotroph, fusarial wilt pathogen, due to the suppressive effect of that biotrophic parasite. Suppressivity is a key to understanding of the nature of concomitant plant diseases.

Enzymes degrading antimicrobial plant compounds

One more group of genes that provides parasite virulence towards particular plant species controls synthesis of the phytoanticipin and phytoalexin detoxifying enzymes. However, the finding of such an enzyme in the fungal mycelium or growth medium does not yet prove that this enzyme constitutes the primary virulence factor. It is necessary to show that such enzyme is present in the infected plants, that it is absent in the avirulent strains, and that inactivation of the encoding gene leads to the loss of virulence while its transformation into other microorganism species leads to their ability to degrade the antibiotic and even to acquire virulence towards the producer of that antibiotic. In some cases, such tests have been carried out.

Degradation of phytoanticipins

The structure of many phytoanticipins and the mechanism of their toxicity have been discussed in Chapter 6. In this section, some pathways of their degradation by phytopathogens will be discussed.

The tulip parasite *B. tulipae* prevents the ring closure of the tulip aglicon glucoside, tuliposide – which is the case upon infection by the nonspecialized parasite *B. cinerea*. Therefore, the resulting product is not highly toxic lactone but much less toxic acid (Figure 12.3).

It was written earlier that the saponin avenacin localized in the oat roots protects the roots from infection by the root rot pathogen *Geumannomyces graminis*.

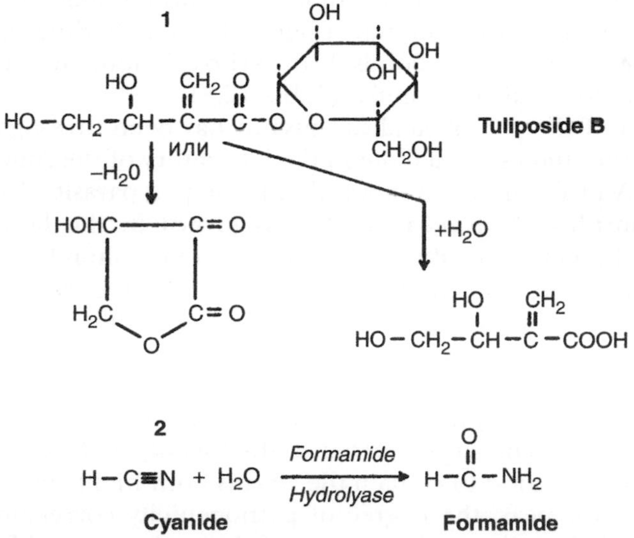

Figure 12.3.   Phytoanticipin degradation.
(1) Tuliposide degradation by *Botrytis* species. Metabolites of *B. cinerea* induce production of highly toxic lactone (left); *B. tulipae* induce production of slightly toxic acid (right). (2) Cyanide degradation by *Stemphylium loti*.

The oat-attacking strains of this fungus (*G. graminis* var. *avenae*) have a specific enzyme – glucoside hydrolase – avenacinase that eliminates sugar residues and transforms avenacin in an unsolvable, less toxic form. The encoding gene was cloned using polyclonal antibodies to the purified avenacinase as a probe. Its transformation in the genome of the saprotroph fungus *Neurospora crassa* made the latter unsusceptible to avenacin. Inactivation of the gene by transposon mutagenesis in the strain *G. graminis* var. *avenae* resulted in the loss of the ability to degrade avenicin and infect oat, while retaining pathogenicity to wheat that does not produce avenacin.

The steroid glycolalkaloid of tomato tomatin is hydrolyzed by tomatinase (β-glucosidase) of the strains of *A. solani* isolated from the infected tomatoes, but it is not hydrolyzed by the nonpathogenic strains. In another tomato parasite, *Septoria lycopersici*, the gene of tomatinase was cloned, however, its transformation in the genome of the saprotrophic *fungus Aspergillus nidulans* did not enhance tolerance of the transformed strain to tomatin *in vitro*. Tomatinase of *S. lycopersici* eliminates glucose from α-tomatin and thus does not only transfer α-tomatin to less toxic compound, but the latter, in turn, suppresses induction of the defense response, as it interferes with the fundamental pathways of signal transduction leading to resistance.

Pathogens of the cyanogenic plants sorghum *Gloeosporium sorghi* and bird's foot trefoil *S. loti* possess the enzyme cyanide – hydrolyase, which converts cyanide in less toxic formamide (Figure 12.3).

Degradation of phenol phytoalexins

Fungi can modify the pterocarpan molecule (most phenol FAs have the pterocarpane carcass) by various ways (Figure 12.4), including ring hydroxylation, rupture of the lactone bonds, hydroxyl oxidation, etc. The resulting compounds are less toxic than initial (Table 12.8).

Degradation of the pea phytoalexin pisatin has been investigated in fine detail, due to the studies carried out in the laboratory of the American plant pathologist G.Van Etten. It was shown that if the pea parasite *Nectria haematococca* (anamorph – *F. solani* f. sp. *pisi*) is grown in the liquid nutrient medium containing pisatin, after some lag period mycelium begins growing, which is accompanied by the drop of pisatin content in the medium and accumulation of its derivative 3,6α-dioxy-8,9-methylene dihydroxypterocarpan (Figure 12.5). Consequently, in the presence of pisatin the fungus accumulates the enzyme that eliminates the methyl group from C-3, and the lag period shows that the enzyme formation is induced by the pisatin present in the growth medium. The large number of the investigated natural strains of *N. haematococca* was found to contain the strains both highly pathogenic and nonpathogenic for peas, the degree of pathogenicity correlating with the ability to demethylate pisatin. In crossing of the pathogenic with nonpathogenic strains, the hybrid monoascosporous isolates were distributed into two groups: pathogenic and nonpathogenic, and only the pathogenic group possessed the ability to demethylate pisatin (Figure 12.6).

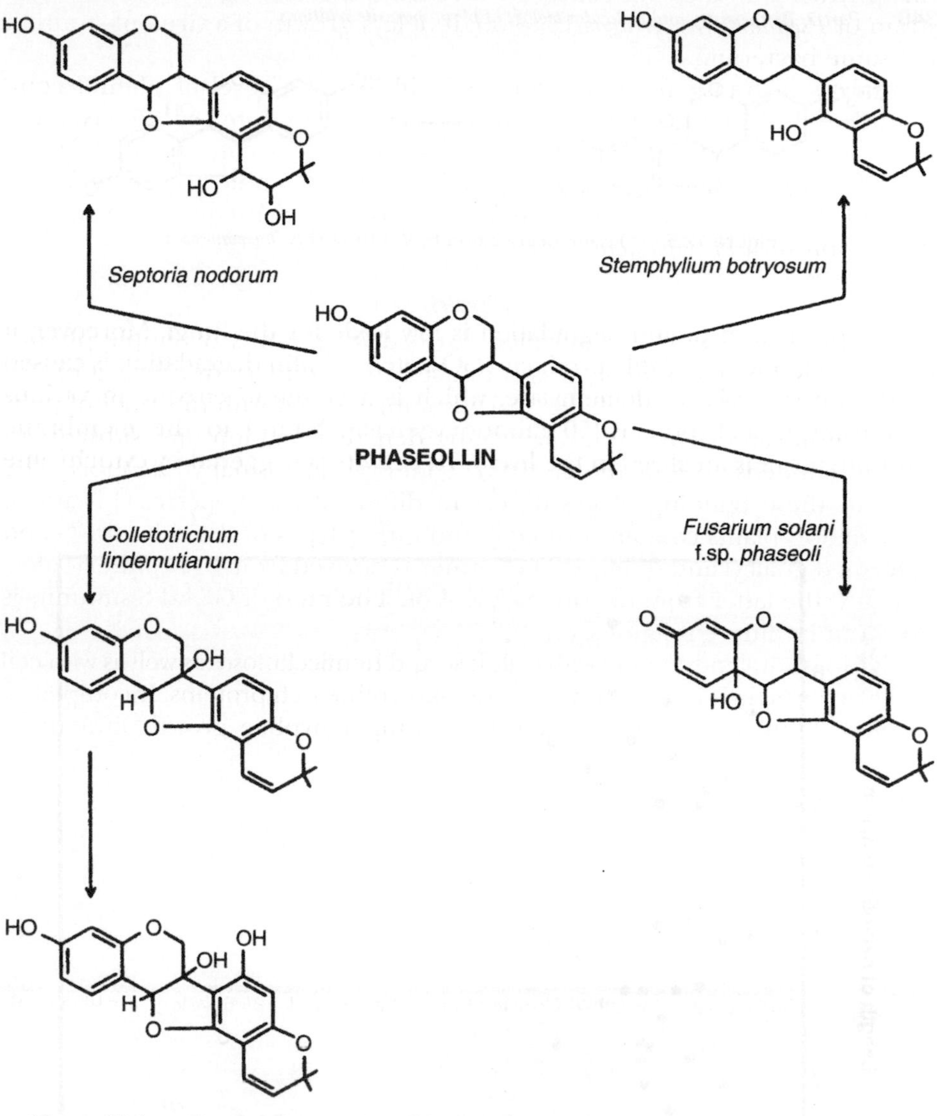

Figure 12.4. The pathways of phaseollin degradation by different fungi. (Fuchs A, 1982.)

Table 12.8. Toxicity of the FA pisatin and the product of its degradation (Barz W, et al.,1980)

Fungus species	Percent of growth inhibition	
	Pisatin	3-O-demethylpisatin
H. turcicim	67	6
N. crassa	78	28
F. solani f.sp. *phaseoli*	67	6
Penicillium expansum	36	6
Rhizopus stoloniferum	74	6
S. botryosum	66	3

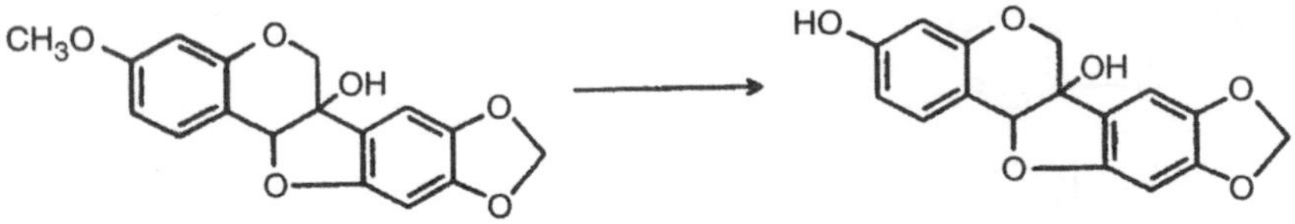

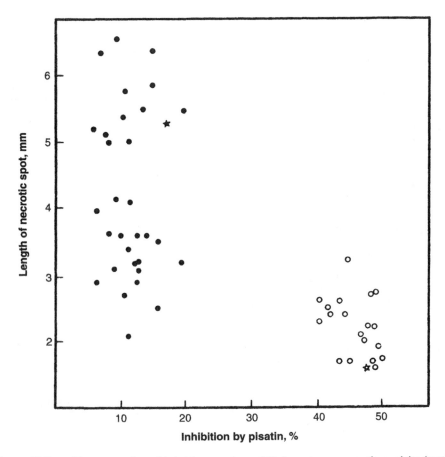

Figure 12.5.   Pisatin degradation by the fungus *N. haematococca.*

The product of pisatin degradation is low toxic for the fungi. Moreover, it further degrades to volatile products (CO_2, etc.). Pisatin degradation is caused by the enzyme pisatin demethylase, which is a common enzyme in various organisms, cytochrome E-450 monooxygenase, bound to the membrane (in humans, it is localized in the liver). Its activity is triggered by cytochrome

Figure 12.6.   The segregation of hybrid progenium of *N. haematococca* on pathogenicity (vertical scale) and pisatin sensibility (horizontal scale). (Tegtmeier KJ, Van Etten HD, 1982.) Each circle is descendant of one ascospore. Only black circles has enzyme pisatin demethylase. The parent strain has marking by starlets.

P-450 and NADPH-cytochrome P-450 reductase. Expression of the enzyme is induced by pisatin and repressed by glucose and amino acids, i.e. it is subject to catabolite repression.

Several genes that control synthesis of pisatin demethylase were found in the genome of *N. haematococca*. Gene Pda1 provides high activity, Pda2 and Pda3, low activity, and Pda6, intermediate activity of the enzyme. The recessive alleles of these genes are deletions. As fungal chromosomes are much shorter than chromosomes of the higher eukaryotes, they can be separated in electric field using a special procedure (pulse electrophoresis). In separation of *N. haematococca* chromosomes it was found that, in addition to the chromosomes available in all strains, many strains contain small chromosomes of variable number (dispensable or B-chromosomes). These chromosomes are not essential for the fungus, but Pda-genes are located on these very chromosomes. The strains that have lost the B-chromosomes can exist only as saprotrophs.

B-chromosomes also contain the toxin-formation genes in the pathogenic forms of *A. alternata*. Such chromosomes are lacking in the saprotrophic strains. Apparently, the loss of B-chromosomes causes a fast drop in the rate of occurrence of pathogenic strains after the termination of cultivation of the susceptible plant cultivars and species.

Gene Pda6 was cloned and transferred in the other fungi: *C. heterostrophus* (maize parasite) and *A. nidulans* (saprotroph). Both transformants acquired the ability to destroy pisatin *in vitro*, which points out to the gene expression in the new organisms, however, only the former became infectious for peas. This experiment does not only show the role of the enzyme pisatin demethylase in pathogenicity to peas but also the differences between the basic pathogenicity and specific virulence. The phytopathogenic species *C. heterostrophus* possesses the basic pathogenicity which seems to be determined by many genes (Chapter 0); therefore, acquisition of just one more additional gene of virulence to peas was sufficient to provide the ability to infect a new host. The saprotroph *A. nidulans* has no basic pathogenicity, therefore, the virulence gene cannot provide its ability to parasitism. Finally, this experiment shows the fundamental differences between the prokaryotic and eukariotic parasites. It has been shown (Chapter 5) that the transfer of only one pectatelyase gene from the genome of the phytopathogenic bacteria *Erwinia chrysanthemii* to the *E. coli* genome rendered transgenic strains pathogenic for plants. Apparently, the transfer from saprotrophic to parasitic life is considerably simpler in prokaryotes than in eukaryotes, and it can be provided just by a few genes. This explains the virtual absence of obligate bacterial animal or plant parasites and discovery of some bacterial strains capable of parasitizing both on plants and animals. For a parasitic mode of life, the lower eukaryotes (fungi) need to possess a system of basic genes; a considerable percent of phytopathogenic fungi is made by obligate parasites. Finally, in even more differentiated organisms, invertebrate animals and higher plants, parasitism is almost always obligate, and the divergence between parasites and nonparasites has reached an irreversible level.

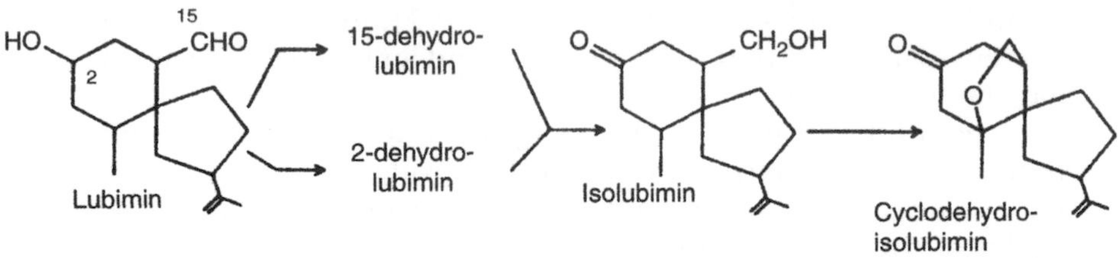

Figure 12.7.   Lubimin degradation by *Gibberella pulcaris* (Desjardins AE, Gardner HW, Plattner RD, 1989).

Table 12.9.   Relationship between lubimin degradation and pathogenicity of the strains of *Gibberella pulcaris* to potato tubers (Desjardins A, et al.,1989)

Strain origin	Number of strains	Average tolerance to lubimin (growth % vs. control)	Lubimin in medium after fungus growth (μg)	Average pathogenicity (% of the most pathogenic strain)
Potato	8	87	0	55
Nonpotato (soil, pine, maize, cactus, carnation)	12	33	54	5

Degradation of terpenoid phytoalexins

The pathogen of the dry rot of potato *G. pulcaris* (anamorph – *F. sambucinum*) decomposes the main potato phytoalexins richitin and lubimin to less toxic metabolites (Figure 12.7). The strain ability for FA metabolization correlates with their pathogenicity (Table 12.9).

Hybridological analysis of the strains has shown that the ability to degrade terpenoid FAs is controlled by several loci, however, high pathogenicity for potato is caused by only one of them (Rim1).

Selected Literature

Bart W, et al. Ann Phytopathol 1980; 20:435–452.

Beismann B, Kogel KH. Identification and characterization of suppressors. Ann Rev Phytopathol 1995; 33:259–275.

Desjardins AE, Gardner HW, Plattner RD. Phytochemistry 1989; 28(2):431–437.

Desjardins A, et al. Molec Plant Microbe Interactions 1989; 2:26–34.

Fuchs A. In: Active defense mechanism in plants. Wood RKS (Ed.), Plenum Press 1982; 331.

Mayama S, et al. Physical And Molec Plant Pathol 1986; 29:1–18.

Medvedeva TE. Specific suppression of potato defense reaction by glucans of *Phytophthora infestans*. Ph.D. Thesis 1985 (In Russian).

Morrissey JP, Osbourn AE. Fungal resistance to plan antibiotics as a mechanism of pathogenesis. Microbial Mol Biol Rev 1999; 63:708–724.

Nishimura S, Kohmoto K, Ptani H., Ramachandran P., Tamura F. Pathological and epidemiological aspects of *Alternaria alternata* infection depending on a host-specific toxin. In: Plant Infection. Asada Y, Bushnell WR, Ouchi S, Vance (Eds), Jap Sci Soc Press & Springer-Verlag 1982; pp. 199–214.

Oliver R, Osbourn A. Molecular dissection of fungal phytopathogenicity. Microbiology 1995; 141:1–9.

Ouchi S, Oku H, Physiological basis of susceptibility induced by pathogens. In: Plant infection. Asada Y, Bushnell WR, Ouchi S, Vance CP (Eds), Jap Sci Soc Press & Springer-Verlag, 1982; pp. 117–136.

Tegtmeier KJ, Van Etten HD. Phytopathology 1982; 72:608–612.

Van Etten H, Mattws D, Mayers P. Phytoalexin detoxification: importance for pathogenicity and practic implications. Ann Rev Phytopathol 1980; 27:143–164.

Walton JD, Panoccione DG. Host-selective toxins and disease specificity. Perspectives and progress. Ann Rev Phtropathol 1993; 31:275–303.

Wolpert TJ, Dunkle LD, Ciuffetti LM. Host-selective toxins and avirulence determinants: what's in a name? Ann Rev Phytopathol 2002; 40:251–285.

Chapter 13

General and specific aspects of plant and animal immunity

Yu. T. Dyakov

Historically, plant resistance studies, on the one hand, and investigation of human and animal resistance, on the other hand, evolved by different ways. In phytopathology, attention was concentrated on identification of the causal agents of disease, the parasitic fungi, due to the work of the outstanding botanists of the 19th century the Tulasne brothers, de Bary, Kuen, Woronin, Marshall Ward, etc. The studies of Ivanovsky and Bejerink allowed revealing the possibility of viral infection for plants much earlier than the same was discovered for animals and humans. However, these studies were not accompanied by any serious plant prevention and treatment work. In human and veterinary medicine, on the contrary, the primary attention was focused on the ways of infection, prevention and patient treatment. In those days, when nothing was known about the black smallpox pathogen, the English physician Jenner developed a method of vaccination to protect against smallpox infection. The well-known Russian immunologist and microbiologist Ilya Mechnikov wrote about it: "When people were still groping in the dark regarding the diseases of humans and higher animals, the plant pathology had been already investigated in detail, and etiology of a great number of plant diseases had been well-defined. Nevertheless, the question of immunity in botany remained in the background, so we don't have any special publications on this subject." (I.I. Mechnikov, "Immunity Against Communicable Diseases", 1903, p. 29). The first intensive studies on plant resistance against diseases began after the work of the medical immunologists: Pasteur, Mechnikov, Ehrlich, and due to the influence of their success. It was not surprising that most efforts were aimed at plant vaccination and search for plant antibodies. The results of these studies were summarized in the reports of D. Carbone and C. Arnaudi "L'immunita nello piane" (1930) K.S. Chester "The problem of acquired physiological immunity in plants" (1933) and N.I. Vavilov "Theory of Plant Immunity To Infectious Diseases" (1935). However, the search for antibodies ended in a failure, the practical results of plant vaccination were insignificant, and eventually the interest for such work petered out, to restart 50 years later. In 1928, C. Owens wrote: "No progress has yet been made in this direction and it is doubtful if this ever can be done with plants as it is done in the animal kingdom. The structure of plants is so different from that of animals, especially as regards a circulatory system, that it is does not seem likely that much success can ever be attained in that direction".(Owens C.E., "Principles of Plant Pathology", 1928, p. 125). The immune system of vertebrate animals is the result of a long evolution that lead to development of the special organs and cells carrying the immune functions and the humoral system delivering the immune cells to the infected site. Plants lack all these things. These differences were brilliantly foreseen in 1902 by Erwin Smith, who wrote: "To my mind, the problem of problems in pathology, both animal and vegetable, during the next fifty years will be the varying nature of the host plant or host animal as related to the parasite. This is the burning question. Why is that some individuals are so very susceptible to disease and others so resistant"? (Erwin Smith, Plant Pathology: Retrospect and Prospect. Science, 1902, p. 611). Therefore, unlike the human medicine where the major success was achieved in development of acquired immunity, in phytopathology since the beginning of the 20th century the extensive studies on the use of intra-species and inter-species hereditary differences in resistance, genetic analysis of resistance, development of remote hybridization methods, and transmission of resistance genes between plants have been carried out. The Russian plant pathologist A. A. Yachevsky wrote that the resistant cultivars make a basis of plant protection against disease. All other measures should be viewed as palliative. This does not mean that vertebrate animals have no hereditary differences in resistance to infectious diseases. Many examples of such differences are provided in the books by Hutt F.B., "Genetic Resistance to Disease in Domestic Animals" (1958) and by Efroimson V.P., "Immunogenetics" (in Russia, 1971). However, a veterinarian, and especially a physician have only limited possibility to make use of hereditary differences in resistance to disease. Only a phytopathologist can artificially inoculate huge plant populations, select resistant individuals, cross them with the susceptible individuals, reject all affected progeny through artificial inoculation, carry out mutagen treatment, and choose the resistant mutants. This pathway of hybridization and selection has brought the most impressive results of enhancing plant immunity against disease. "The aim of human medicine is to preserve the individual; the phytopathologists, on the other hand, as a rule has little interest in the individual, his main objective being the preservation of the population. Moreover while much of the activity of the physician is devoted to cure or therapy, that of the phytopathologist is more directly concerned with prevention or prophylaxis" (Chester K.S. Quarterly Rev. Biol., 1933, V.8, p. 314). Thus, the disease resistance of vertebrate animals and plants differs both in biochemical mechanisms and methodological approaches. The different strategies of enhancing plant and human resistance have led to separation of the studies conducted by phytopathologists and physicians, with no mutual enrichment of the ideas, methods, and results.

The situation has changed in the past decades when the molecular methods of research came into common practice not only in the medical but also phytopathological studies. With the growing knowledge of the nature of disease resistance, especially due to the use of the recombinant DNA, the differences in immunity mechanisms of plants and vertebrate animals have come to appear less striking. Today, more and more data is collected on their similarity. Some data on such similarity have been discussed earlier in this book. They will be presented here in a summarized form.

Immunomodulators of pathogenic microorganisms

Animal and plant pathogenic microorganisms release their metabolites in the infected tissues and cells that influence the immune responses (immunomodulators). One type of them – inducers or elicitors – are used by the host for recognition of the foreign compounds and induction of defense responses in the host organism. These substances enhance both the resistance to the elicitor-producing bacteria and to the subsequent infection by other pathogenic microorganisms, i.e. they constitute one of the factors of induced acquired immunity. On the contrary, other immunomodulators–immunosuppressors or impedins–suppress the defense responses and reduce the host immune properties. Both types of immunomodulators of phyto- and zoopathogens are similar in the structure and mechanism of action on their hosts.

Fungal immunomodulators
Polysaccharides
Tables 13.1 and 13.2 contain the data on the fungal carbohydrates that possess immunomodulating effect on the plant and animal cells.

It can be seen from the table that most polysaccharide immunomodulators are linear or branched β-glucans or mannans.

Table 13.1.   Carbohydrates, plant immunomodulators

Name	Structure	Producer
Hepta-β-glucoside	β-3,β-6-glucoside with ≥ 5 glucose residues	*Phytophthora soyae*
β-glucan	β-(1-3),β-(1-6) branched glucan with 13–15 glucose residues	*P. infestans*
Oligochitin	Linear polymer of (1→4)-N-acetyl glucosamine; ≥ 6 units	Many fungi
Glycopeptide	Active oligosaccharide side chain enriched with mannose	*Puccinia graminis*
Peptidoglucan	Active side α-(1–6) chain of 10–20 mannose residues	*Saccharomyces cerevisiae*

Note: Depending on the chain length, its configuration, and the plant genotype, most of the above polysaccharides can be elicitors or suppressors of the defense responses.

Table 13.2. Immunomodulating fungal polysaccharides possessing pharmacological properties (Kashkina MA, Elinov NP. 1985.)

Name	Producer	Structure
Aubasidan	*Aureobasidium pullulans*	β-1,6-, α-1,4-, β-1,3- branched glucan
Glucan	*Saccharomyces cerevisiae*	β-1,3-, β-1,6- branched glucan
Chrestin (coriolan)	*Coriolus versicolor*	β-1,4-, β-1,3-,β-1,6 glucan (25–38% of protein)
Cryptococcal heteropolysaccharides	*Cryptococcus* spp.	Branched polyuronides
Lentinan	*Lentinus edodes*	β-1,3-, β-1,6- branched glucan
Mannan	*Sporobolomyces* sp.	β-1,3-, β-1,6- linear mannan
Pachiman	*Poria coccos*	β-1,3-, β-1,6- linear glucan
Rodeksman	*Rhodotorula rubra*	β-1,3-, β-1,4-linear mannan
Chitin	Many fungi	β-1,4-N-Ac-glucosamine (linear)
Scizophylanl	*Scizophyllum commune*	β-1,3-, β-1,6-glucan (branched)

Note: Depending on the dose, many of the above compounds can stimulate or suppress formation of antibodies.

Lipids

A non-specific elicitor of *Phytophthora infestans* arachidonic acid (AA) (Figure 10.8, G) in potato cells is included in phosphotides, oxidized by lipoxygenase to eicosanoids, it activates the oxidative burst, synthesis of phenols and phytoalexins, and induces apoptosis (HR). In animal cells AA is oxidized by lipoxygenase to leucotrienes, and by cyclooxygenase to prostaglandins, thrombaxanes, prostacyclines – the inflammation mediators that regulate cAMP synthesis, aggregation of thrombocytes, and other immune response related reactions.

The sphingolipid pathotoxins of phytopathogenic fungi: AAL-toxin of *Alternaria alternata f.* sp. *lycopersici* and fumonisin of the maize parasite *Fusarium moniliforme* (Figure 13.1) inhibit the enzyme ceramide synthetase necessary for synthesis of ceramide from two sphinganin molecules (Figure 13.1). The ceramide phosphorylates retinoblastoma, a regulator of the cell cycle (transfer of G1 to S) and apoptosis (Figure 9.6). Fumonisin-mediated protoplast apoptosis of tomato and *Arabidopsis* needs involvement of the SA and JA signalling pathways. In the infected plants, sphingolipid toxins suppress the defense responses, inhibit sugar transport, and cause neoplastic growth and necroses. Feeding livestock animals with the maize contaminated by fumonisin causes severe, frequently lethal poisoning, with concomitant hepatotoxicoses, various forms of neoplastoses, and cell death.

Peptides

Fungal cyclic peptides are powerful immunosuppressors of both plants and animals. The cyclic pentapeptide HV-toxin (victorin) (see Figure 12.2) suppresses

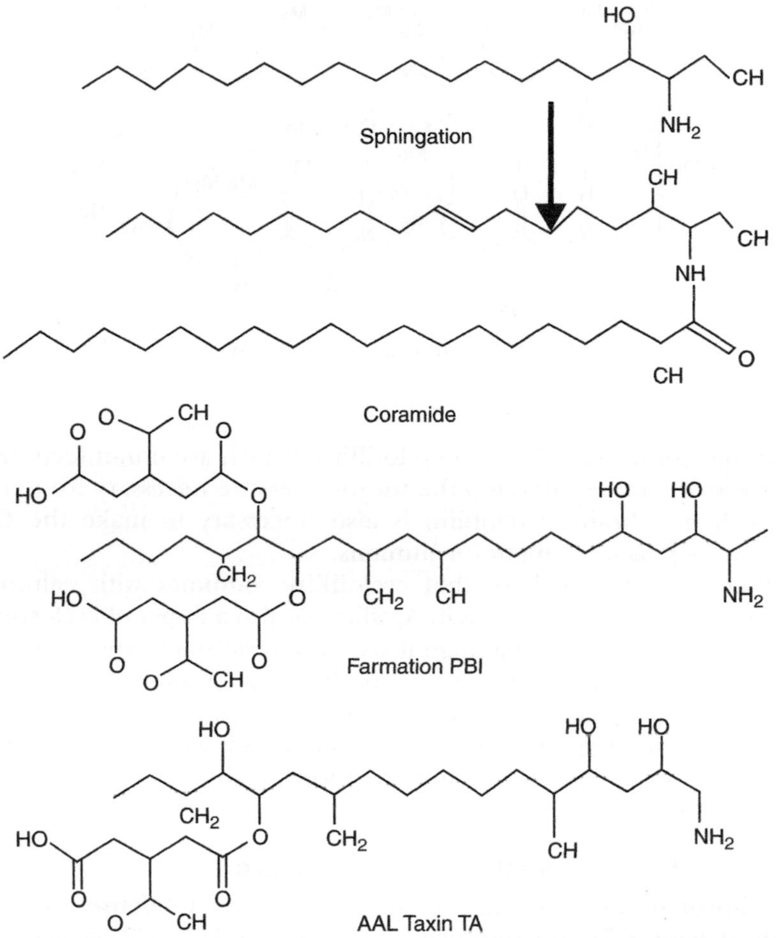

Figure 13.1. *Top*: Synthesis of ceramide from two sphingane molecules using the enzyme ceramide synthetase. *Bottom*: sphingolipid fungal toxins, ceramide synthetase inhibitors.

the immune response of the susceptible plant cells. Its receptor glycine decarboxylase is a central enzyme in the nitrogen catabolism of plants and animals. The suppressors of immune responses of the susceptible plants, the cyclic tetrapeptides HC- and AM-toxins (see Figure 12.2), are homologues of trapoxin, the histone inhibitor in animal cells. The cyclic undecapeptide cyclosporine (Figure 13.2), produced by many fungi, is an immunosuppressor of vertebrate animals and humans. Its targets are the intracellular proteins cyclofillins found in bacteria, fungi, plants and animals. Cyclofillins possess the active enzymes peptidyl-prolyl-isomerases involved in many processes, including stress response, regulation of the cell cycle, Ca^{2+} signalling system, and control of the transcription repression due to isomerization of the peptide bonds. The mutants of the fungus *Magnoporthe grizea* in

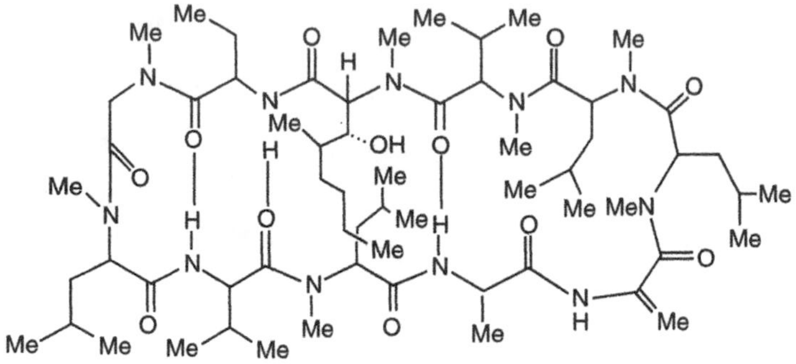

Figure 13.2.   Structure of cyclosporine A.

the gene that controls synthesis of cyclofillins, CYP1, are non-infectious, and their appressoria do not develop the turgor pressure necessary for penetrating the cell membrane. Cyclofillin is also necessary to make the fungus *Cryptococcus neoformans* virulent for humans.

Viand et al. (2002) believe that cyclofillin combines with calmodulin-dependent phosphatase, calcinevrin A, which is also a target of cyclosporines. Fungi need calcinevrin for pathogenicity, and vertebrates, for gene expression in T-cells. In particular, calcinevrin binding by cyclosporine inhibits expression of the gene IL-2 that controls synthesis of IL–2. Cyclosporine obtained from the fungi of the genus *Tolypocladium* is widely used in pharmacology as a suppressor of immune responses for rejection of foreign transplanted organs.

Viral immunomodulators

Structural proteins of some plant viruses are elicitors of defense responses in the hosts (Chapter 7). For instance, the 17.5 kDa TMV CP induces HR in resistant tobacco species and cultivars. It possesses the elicitor only in the form of crystalline aggregates. A 17 kDa cytokine (cytokines are the protein molecules involved in transmission of the immune signal between the immune cells) the tumor necrosis factor (TNF) produced by T-lymphocytes and macrophages, causes development of the classical inflammation signs: swelling, redness, pain, and fever. It causes cell necrosis. The three-molecule aggregate is also active. The TNF protein is homologous to the structural protein of the TNV satellite virus.

Protein RepA of plant heminiviruses as well as proteins EA and E7 of animal viruses suppress the retinoblastoma gene, the role of which was described earlier.

Bacterial immunomodulators

Many taxonomically related bacteria cause plant and animal disease. For example, the genus *Pseudomonas*, alongside numerous biotrophic and necrotrophic plant parasites, also contains a widely known non-specific animal and human wound parasite *P. aerugenosa* that is also able to cause disease in some plants. The pathogen of sour skin of onion *Burkholderia cepatia* causes dangerous diseases in the people suffering from immunodeficiency. Bacteria from the genus *Erwinia* (soft rot pathogens) can cause severe human poisoning. This is explained by the presence of common virulence factors towards plants and animals in these and other bacteria.

Factors responsible for the primary relations with the hosts. The external membrane components of Gram-negative bacteria, lipopolysaccharides (LPS), play an important role in the pathogen attachment to their hosts. The mutants of *Pseudomonas aeruginosa* and *Ralstonia solanacearum* with modified LPS composition lose their virulence towards animals and plants, respectively. Another component of the bacterial cell surface, pili type IV, were found in the plant and animal pathogens *Pseudomonas syringae* pv. tomato and *P. aeruginosa.* They are necessary for adhesion to the cells of plants and animals. The protein of bacterial flagellae flagellin is also necessary for surface adhesion. Resistant plants and animals recognize flagellin as an elicitor. On the other hand, mutants of the phytopathogenic bacteria *P. fluorescens, P. avenae, Erwinia carotovora* subsp. *artroseptica,* and the human pathogen *Shigella flexneri* lose their virulence if they are stripped of the protein flagellin.

Factors responsible for avirulence and virulence. The product of the avirulence gene of avrRxv *Xanthomonas vesicatoria,* which attacks tomatoes and pepper, is homologous to the virulence factor yaupj of the bacteria *Yersinia pseudotuberculosum* that causes intestinal infections. These and other proteins are transported through the bacterial membranes by special proteins (products of hrp-genes) of the type III secretory system; these proteins are homologous in phytopathogenic bacteria and human parasites from the genera *Yersinia, Salmonella,* and *Shigella.* The bacterial peptides flagellins are the ligands bound by the immune molecules of plants (R-protein FLS2 is kinase with a LRR-receptor motif) and animals (Toll-like receptor that also has the LRR-motiff, TLR5).

In the causal agent of tuberculosis *Mycobacterium tuberculosum,* hypervirulence is caused by phenol glycolipid, which is synthesized using polyketide synthetase. PGL inhibits the response of the innate resistance system to infection (formation of cytokines: TNF, interleukins 6 and 12, etc).

Host immune molecules

Antibodies (immunoglobulins) of the mammals have a characteristic structure with two polypeptide chains: the heavy chain H (molecular weight (50 kDa) and the light chain L (molecular weight (25 kDa) connected by disulfide bonds. Each chain contains the alternating constant (C) and variable (V) sites. A great diversity of antibodies in a mammal body that enables recognition of numerous foreign antigens is provided by the unique genetic control of their synthesis (Figure 13.3).

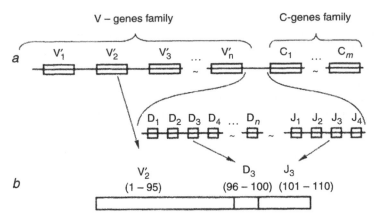

Figure 13.3.   Simplified diagram of (a) antibody family, and (b) assembly of the gene that controls the variable site of the antibody (Abilev, 1996).
(a) – Basic diagram of the chromosome site containing the families of variable (V′) and constant (C) genes. The number of V′ genes varies, depending on the type of chains and animal species, from 50 to 1000, and C genes, from 1 to 9; (b) – Assembly of V gene from the fragments. V′ is the gene fragment that controls the first 95 amino acids of the V region; D is the short fragment that controls from 2 to 10 amino acids (the number of D genes can be up to 20); J is the short fragment that controls from 5 to 15 amino acids of the V-gene region.

Plant R-proteins (Chapter 8) include some structures that provide, on the one hand, interaction with the ligand, and on the other hand, signal transduction to the molecules of the intracellular signalling systems. Various combinations of R-protein fragments provide resistance to viruses, bacteria, fungi, and nematodes of the plants from different taxons. Consequently, these structures emerged a very long time ago, before the divergence of the forerunning plants into the modern classes and orders, and they used to perform other, more general functions involved in the excitation signal reception and transmission to the genome. Their immune function is secondary.

Resistance genes in many plants form series of alleles or closely linked clusters. Formation of the functionally linked clusters and structurally similar genes is caused by intra- and intergenic exchanges of the DNA sites with direct or inverted repeats. R-proteins possess such repeats in the LRR-site that continuously generates new specificities (Figure 13.4). For example, five susceptible to *Cladosporium fulvum* variants of tomato selected from Cf4/Cf9 heterozygotes were all generated by intergenic crossovers. Thus, similar to numerous immunoglobulin genes that provide synthesis of antibodies in mammals, plants possess the proteins which are encoded by a linked gene family. High variability of their gene products enables prompt response to infection by new virulent species and races of parasites.

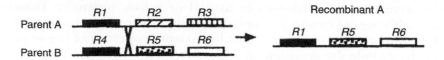

B. Creating a novel variant *R* gene from a heterozygote

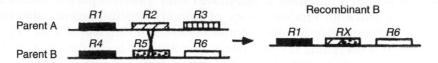

C. Creating a novel variant *R* gene from a heterozygote by unequal crossover

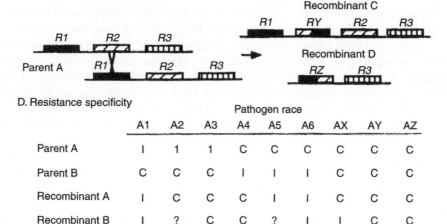

D. Resistance specificity

| | \multicolumn{9}{c}{Pathogen race} | | | | | | | | |
	A1	A2	A3	A4	A5	A6	AX	AY	AZ
Parent A	I	1	1	C	C	C	C	C	C
Parent B	C	C	C	I	I	I	C	C	C
Recombinant A	I	C	C	C	I	I	C	C	C
Recombinant B	I	?	C	C	?	I	I	C	C
Recombinant C	I	?	I	C	C	C	C	I	C
Recombinant D	?	?	I	C	C	C	C	C	I

Figure 13.4. Recombination between LRR sequences of R-proteins and generation of new race specificities in plants (Hammond-Kozak KE, Jones J, Rev. Plant Physiol. Plant Molec. Biol. 1977; 48: 576-607 (Figure 5)).
A–C – intra- and intergenic ectopic recombinations; D – generation of new vertical resistance genes.
I – incompatibility, C – compatibility.

Signal transduction and immune response

Signal transduction is a stage between the receptor excitation and immune response gene activation, during which the signal is repeatedly multiplied. The process is carried out by the signalling systems (usually common for all the cells) involved in regulation of various aspects of the organism's life activity. Some signalling systems function in infected or sick cells and tissues of plants and animals.

It was written in the beginning of this chapter that plants do not produce antibodies: therefore, their disease resistance was evolving along the line of exploitation of the mechanisms of natural or innate immunity. However, animals also possess natural immunity where it performs the important functions of the first line of defense. The important properties of the acquired immunity include the provision of: a) the subtle specificity mechanisms due to clonal selection of antibodies and T-cell receptors; and, b) the ability to fast response to a recurrent appearance of the antibody in the organism. However, acquired immunity is manifested after a certain period; therefore, it does not make an efficient factor of primary defense against infection. The natural resistance factors become involved in defense as soon as the pathogen has overcome the integument or mucosa and taken root in the internal environment of the body. "The clonal response retains the immune memory of the individual, but the innate response retains the evolutionary memory of the species" (Fearon, 1997). The key function of the innate immunity mechanisms is recognition of the molecular patterns released by the pathogen and activation of the synthesis of the immune molecules (cytokines, etc). One of the important mechanisms of such recognition is the Toll-like receptors. In drosophila, Toll is a receptor protein with extracytoplasmatic domain LRR that controls embryo polarity and synthesis of the antifungal peptide drosomycine in adult flies. Through the small ligand Spatzle, the Toll-protein activates the cytoplasmic protein Tube that passes excitation to the serine/threonine kinase Pelle that phosphorylates the protein complex Cactus/Dorsal, which causes Cactus degradation and release of Dorsal that constitutes the transcription regulation factor of the Rel/NF-kB family (Figure 13.5). Rel proteins function in the form of dimmers; they belong to the family of the transcription factors that regulate the gene expression level.

The transcription factor NF-kB in mammals is the central regulator of inflammation and immunity, and it is also involved in the processes of apoptosis, carcinogenesis, growth, and differentiation. Thus, the Toll-like IL-1 reception system is an important factor of human innate immunity. This system is responsible for the inflammatory and immune response to microbial infection. The family of Toll-like receptors is activated when combined with various microbial ligands: lipopolysaccharide (ligand TLR4), peptidoglucan (TLR2), bacterial lipoprotein (TLR1/TLR2), flagellin (TLR5), etc. All these molecules are microbial elicitors that induce defense responses in plants. In macrophages, activation of the Toll-like receptor of IL-1 is followed by activation of protein kinase IRAK (Pelle homologue) that, through a series of MAP-kinases, phosphorylates the IKB/NF-kB complex and thus releases the NF-kB transcription factor from the inhibitor protein, which leads to immune response expression by the gene (Figure 13.5).

The structure of many plant proteins (N, M, L6, PRF, and FLS2) is similar to Toll. Tomato Pto kinase and the kinase domain of rice protein Xa21 are homologous to Pelle and IRAK. *Arabidopsis* also possesses the transcription factor NPR1 the release of which activates expression of the

Toll-like signalling in:

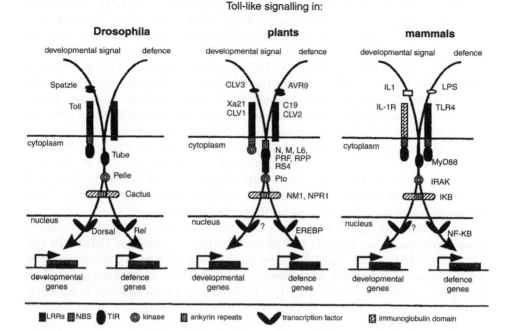

Figure 13.5. Toll signalling pathways in *Drosophila*, plants and mammals responsible for induction of genes that play role in development or defense (Takken FLW, Joosten HAJ, Europ. J. Plant Pathol. 2004;135:1113-1128 (Figure 1)).

A small ligand, which can be either a small protein like Spätzle, AVR9 or CLAVATA3 (CLV3), or alternatively a lipopolysaccharide (LPS) or interleukin 1 (IL1), binds to a Toll-IL like receptor. The transmembrane receptor contains either a leucine rich repeat (LRR) domain or an immunoglobulin domain and can be linked to a nucleotide binding site (NBS) or a kinase domain. Binding to the Toll receptor or homologues. Like Xa21, Cf, CLV, the interleukin 1 receptor or Toll-like receptor 4 (TLR4), activates the signalling cascade/the TIR domain activates Tube, which recruits Pelle (or Pto or IL1 receptor associated kinase (IRAK.) Pelle then posphorylates Cactus (or NIM1, NPR1 or *IKB*) which is degraded, thereby releasing the transcription factor Dorsal (or EREBP or NF-kB). Dorsal moves as a dimer to the nucleus and regulates nuclear gene expression.

immune response genes. Other plant proteins also have a similar structure (protein kinase with the LRR-receptor domain):

- CLAVATA1 (CLV1) regulates stem cells morphogenesis in the apical meristem in response to ligand (secreted peptide CLV3);
- receptor (BAκ1) associated with brassinosteroids that is responsible for the activity of brassinosteroid hormones;
- bacterial flagellin receptor FLS2; and
- determinant of self-incompatibility (S-alleles) in the cruciferous SPK.

In macrophages, the transcription factor NF-kB regulates the oxidative burst. Plant infection is also accompanied by the oxidative burst caused by the arrival of the reactive oxygen species. An important role in this process is played by the NADPH-oxidase system of the cytoplasmic membrane. In plants, it is similar to those in mammal macrophages and neutrophils. The immune sera with the key component of the animal NADPH-oxidase complex interact with plant proteins of the respective size. Rice genes homologous to the gene of the membrane complex of the NADPH-oxidase component of animal neutrophils have been cloned. However, formation of the reactive oxygen species in plant cells also involves cell wall peroxidase and oxalate oxidase. These alternative pathways of generation of the reactive oxygen species are not suppressed by the specific inhibitors of oxidative burst in animals.

An active component of the plant signalling system is salicylic acid. Salicylic acid and its acetylated form, aspirin, also play a significant role in defense response regulation in mammals. It inhibits synthesis of prostaglandins and activity of the tumor necrosis factor gene, and limits production of IL-1, thus performing anti-inflammatory and antipyretic functions.

The research of the recent years has revealed many common features between apoptosis in animals and hypersensitive response in plants. In the cells of infected plants, the DNA free 3′ ends are released, Ca^{2+}-dependent endonuclease is activated, and the ca. 50 mb DNA fragments as well as oligonucleosome fragments emerge. Residual apoptotic bodies have been found that migrate towards the cells periphery. In animals, apoptotic vesicles are absorbed by the neighbouring or specialized cells, whereas in plants phagocytosis is prevented by the cell wall. In hypersensitive response, periderm is produced around the lesion, as the healthy cells acquire meristematic activity. The same microorganism immunomodulators can be inductors of apoptosis in animals and plants. AAL-toxin and fumonisin cause apoptosis in plants and animals. Finally, many plant R-proteins have the sites (NBS) homologous to the animal apoptosis regulators (Ced4, Apaf1). Probably, these domains are also functionally involved in the regulation of apoptosis.

Conclusion

The molecular studies have revealed that there is more in common than specific between the simplest (primitive elementary) eukaryotes, yeast, and the crown of creation, man, regarding the cell life supporting processes and the cell interaction with the environment. Therefore, there is no surprise in availability of the common mechanisms of natural or innate resistance to infectious diseases in such dissimilar taxons as higher plants and vertebrate animals. Stepping up from the primitive plant immunity mechanisms towards the perfect mechanisms characteristic of animals, probably, consists in the separation of functions.

In plants, every cell is able to synthesize the molecules that recognize foreign matter, to transduce signal to the nuclear parts, to produce toxic action on the parasite, etc. Signal molecules function endocellularly and provide local immune response. Though the response to infection can result in systemic acquired resistance, it is not absolute and shows only as decreased susceptibility to the repeated infection of the uninfected areas.

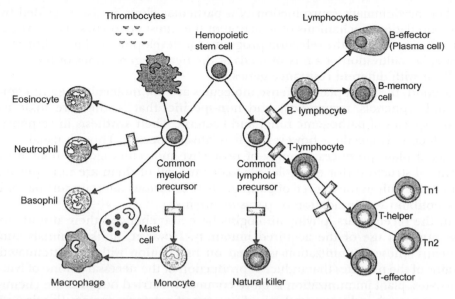

Figure 13.6.   Cells involved in immunity of vertebrate animals and humans (Haitov R, Allergy, Asthma, and Clinical Immunology 1999 N1:1–16).

In animals, the immune functions are divided between a number of cells: T- and B-lymphocytes, macrophages, etc. Signal molecules transfer information between the cells, which provide signal enhancement and its systemic transmission. This mechanism determined the evolution of a family of signal molecules, cytokines (interleukins, TNF, interferon, etc.) and their receptors on the surface of the immune cells (Figure 13.6).

The evolving of such mechanism in plants is also unlikely due to structural reasons: presence of the polysaccharide cell membranes interfering with intercellular metabolism.

Emergence of the functionally different cells capable of independent reproduction and systemic spread entails another key difference between the immune systems of plants and animals.

The key defense molecules of vertebrate animals, antibodies, are highly specific to particular species or even strains of the pathogenic microorganisms. Such specificity requires, first, a great diversity in the structure of the immune molecules, which can be provided only by proteins, and, second, availability of the mechanism of preferential reproduction of only the clone which produces the necessary antibody. Obviously, every plant cell that carries all immune functions cannot provide such a sophisticated defense system. The recognition function in plants is performed by LRR-domains of the constitutional R-proteins whose variability cannot be so prompt as variability of antibodies. Therefore, first, genetic diversity of resistance genes in plants develops at a population level. Natural populations of many investigated plant species are highly heterogenic to the resistance genes.

The predominant reproduction of a particular clone is also provided by the population mechanisms of selection of the fittest genotypes. These regularities were used in the selection programs for development of multiline cultivars, i.e. cultivation of a mix of seeds of the near isogenic lines of the same cultivar with different resistance genes.

Second, the main plant defense molecules are low molecular weight compounds (phenols, terpenoids) and non-specific, that is, they are toxic to many species of pathogenic fungi and bacteria. Their synthesis in response to infection is specific – however, only relatively specific. Though usually an infected plant produces a family of protective molecules closely related in terms of structure (for example, phytoalexins), all of them are non-specific and differ only in the degree of pathogenicity to various microorganisms and susceptibility to degradation by parasite enzymes.

In the recent years, phytopathologists have also changed their attitude to the practical use of the acquired immunity. However, unlike animals and humans whose immunization is based on inoculation with the attenuated strains of the parasites that induce reproduction of the necessary clone of lymphocytes, plant immunization is more commonly carried out with the chemical compounds inducing synthesis of non-specific defense factors (biotic and abiotic elicitors), which will be discussed in the following chapter.

Selected Literature

Asai T, Stone JM, Heard JE, et al. Fumonisin B1-induced cell death in Arabidopsis protoplasts required jasmonate-, ethylene-, and salicylate-dependent signalling pathways. Plant Cell 2000; 12:1823–1835.

Cao Hui, Baldini RL, Rahme LG. Common mechanisms for pathogens of plants and animals. Ann Rev Phytopathol 2001; 39:259–284.

Dushay MS, Eldon ED. Drosophila immune responses as models for human immunity. Am J Human Genet 1998; 62:10–14.

Fearon D.T. Seeking wisdom in innate immunity. Nature 1997; 338:323-324.

Hacker J, Kaper JB. Pathogenicity islands and the evolution of microbes. Ann Rev Microbiol 2000; 54:641–679.

Haitov R, Allergy, asthma, and clinical immunology 1999; N1:1–16.

Hammond-Kozak KE, Jones J. Plant disease resistance genes. Ann Rev Plant Physiol Plant Molec Biol 1997; 48:575–607.

Kashkina MA, Elinov NP. Mycol & Phytopathol 1985; 1:345.

Nurnberger T, Brunner F, Kemmerling D, Piater L. Innate immunity in plants and animals: striking similarities and differences. Immunol Rev 2004; 198:249–266.

Parnicske N, Jones JDG. Recombination between diverged clusters of the tomato Cf-9 plant disease resistance gene family. Proc Natl Acad Sci USA 1999; 96:5850–5855.

Schreiber SL, Crabtree GR. The mechanism of action of cyclosporin A and FK506. Immunol Today 1992; 13:136–141.

Spaink HP. A receptor in symbiotic dialogue. Nature 2002; 417:910–911.

Takken FLW, Joosten HAJ. Plant resistance genes: their structure, function and evolution. Europ J Plant Pathol 2000; 106:705.

Chapter 14

Beneficial plant–microbe
interactions

I. A. Tikhonovich, N. A. Provorov

The biological impacts of plant–microbe interactions are not restricted
to pathogenesis: many bacteria and fungi improve the plant growth due
to intensive supply of nutrients, to defense from pathogens and pests, and for
the regulation of development (Table 14.1). The best studied beneficial
plant–microbe system is the nitrogen-fixing symbiosis formed by leguminous
plants and nodule bacteria (rhizobia). Analysis of this system provides the
genetic methodology and molecular approaches that may be used to study many
other symbiotic systems, including mycorrhizae, endophytic and epiphytic
associations. A range of commonalities between the pathogenic and mutualistic plant–microbe interactions were revealed from the analysis of these
associations.

Nitrogen-fixing symbioses

Being the principal sources of fixed carbon in biosphere, plants usually suffer
from a deficiency of nitrogen. This deficiency looks like a paradox since the
plants are surrounded by nitrogenous compounds: atmosphere is composed
of 78% N_2 while the soil organic matter (humus, remains of dead organisms)
contains sufficient amounts of N. However, plants cannot use these nitrogen
sources because:

(i) plants (as well as all other known eukaryotes) are devoid of the nitrogenase
     enzyme required for N_2 fixation: distribution of this enzyme (and of *nif*
     genes encoding for its synthesis) is restricted to prokaryotes;
(ii) plant roots can assimilate only the soluble inorganic N compounds
     (ammonia and nitrate ions) which are deficient in the majority of soils.

This is why many plants try to balance their C/N ratio by cooperating with
different microbes, namely with: (i) N_2-fixing prokaryotes (to be discussed in
this Section) or (ii) mycorrhizal fungi that are extracting N from soil compounds (see further Sections).

Symbioses between plants and N_2-fixing prokaryotes are very diverse with
respect to taxonomy of partners and to the location of microbes within the host
(Table 14.2). This diversity suggests that symbioses with N_2-fixing microbes
have been evolved many times in the plant evolution under the selective

Table 14.1.   Major examples of beneficial plant–microbe Interactions

Impacts on the plant life	Mechanisms	Microbes involved
Mineral nutrition	N₂ fixation	Rhizobia (*Azorhizobium, Bradyrhizobium, Mesorhizobium, Rhizobium, Sinorhizobium*), actinomycetes (*Frankia*), cyanobacteria (*Nostoc, Anabaena*), endophytes (*Acetobacter, Azoarcus*), rhizospheric bacteria (*Azospirillum*)
	Mobilization of nutrients from soils	Mycorrhizal fungi, phosphate-solubilising bacteria
Defense from pathogens	Suppression of pathogen development (synthesis of antibiotics, competition for nutrients), activation of plant defense systems	Various endosymbiotic and epiphytic bacteria and fungi
Defense from pests and herbivores	Synthesis of toxins	Endophytic fungi (Clavicepitaceae) and bacteria (*Clavibacter*)
Developmental regulation	Synthesis of phytohormones	Plant growth-promoting rhizobacteria (PGPR), some mycorrhizal and endophytic fungi
	Control of embryogenesis	Orchid mycorrhizal fungi

Table 14.2.   Symbioses between plants and N₂-fixing prokaryotes

Location of microbes in plants	Symbioses formed by		
	Gram-negative bacteria	Cyanobacteria	Actinomycetes
Intracellular (involving *de novo* development of special symbiotic compartments)	Rhizobia with legumes (Fabaceae) and some non-legumes (*Parasponia, Ulmaceae*)	*Nostoc* with angiosperms (*Gunnera*)	*Frankia* with dicots (Rosid 1 clade)
Endophytic (within intercellular spaces, tissue cavities or vascular system)	*Azoarcus, Acetobacter* with cereals	*Nostoc* or *Anabaena* with pteridophytes (*Azolla*), liverworths (*Anthoceros*) and gymnosperms (*Cycas*)	Not known
Epiphytic (on outer surfaces)	*Azospirillum, Enterobacter* in rhizospheres of different plants	*Nostoc* with mosses (*Sphagnum*)	Not known

pressures imposed by nitrogen starvation. From the physiological viewpoint, N₂-fixing organisms (either symbiotic or free-living) should cope with two problems arising from: (1) high energy cost of nitrogenase reaction and (2) its incompatibility with aerobic processes required for the intensive generation of ATP. Therefore, very restrictive control on N₂ fixation is required and the nitrogenase reaction is switched on only under severe limitation of combined N, when the life without N_2 fixation becomes really impossible.

Legume–rhizobia symbiosis

Due to the integration of leguminous plants and nodule bacteria into the symbiotic system, a range of novel structures and metabolic capacities are acquired by both the partners. As a result of these innovations, the plant adaptive potential may be improved greatly. Being highly diverse (up to 19,000 species distributed all over the world), legumes (family Fabaceae) in cooperation with rhizobia make the critical invest greatly in global biological nitrogen fixation. Inoculation of leguminous crops with effective rhizobial strains is of great importance for constructing the sustainable agricultural systems.

Genetic programs for the nodule development

The interaction of legumes with rhizobia results in the formation of special organs – nodules that provide ecological niches for the microsymbionts (Figure 14.1A). The legume–rhizobia interaction represents a multi-step process based on tightly coordinated expression of the numerous symbiotic genes in both plants and bacteria. Nodulation results in the formation of sub-cellular compartments containing endosymbionts. In legume–rhizobial symbioses, the plant tissues and cells harbouring these compartments develop *de novo* in the root cortex.

Nodule ontogeny. Let us consider the development of symbiotic structures in the best-studied indeterminate nodules formed by evolutionary advanced papilionoid legumes (e.g., *Pisum, Medicago, Trifolium* from the tribe Trifolieae). The nodule development is initiated during the pre-infection stage when rhizobia recognise the appropriate host plants and colonise their root surfaces. Nodule morphogenesis in clover, pea and alfalfa starts from rhizobial infection of root hairs, which are strongly deformed and curled (stage Hac – Hair curling). Within curls, the hair wall is weakened and the plasmalemma invaginates ensuring active uptake of bacteria by the host. The bacteria are encapsulated in a special tunnel – the infection thread (IT) in which the walls are built up from the material of plant cells, while the internal space contains a matrix synthesised by both the partners (stage Itf – Infection thread formation). Although ITs develop either between or within the plant cells, rhizobia inside the IT always have an intercellular location.

Synchronous with IT development the nodule primordium starts from the mitotic reactivation, dedifferentiation and proliferation of the cortical cells (stage Ccd – Cortical cell division). The growing nodule primordium is subjected to the histogenesis, which results in the formation of cortical, vascular and central N_2-fixing tissues (stage Ntd – Nodule tissue differentiation). A few days after inoculation, the bacteria-containing ITs reach the cortex and colonize the emerging primordium by further growing and branching.

A key stage of endosymbiosis is represented by the bacterial "release" from infection droplets into the plant cytoplasm *via* endocytosis-like process

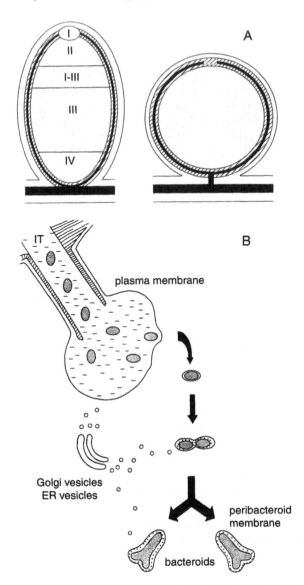

Figure 14.1.   Structural organization of the legume–rhizobia symbioses (Spaink et al., 1998). (A) Anatomy of the major types of nodules (left – indeterminate, right – determinate). I – meristem, II – prefixation zone (wherein the infection of plant cells by rhizobia occur), II–III – interzone (start of N_2 fixation); III – N_2 fixation zone; and IV – senescence zone; (B) Development of intracellular symbiotic compartments in the legume nodules. IT – infection thread and ID – infection droplet.

(stage Bar – <u>Ba</u>cterial <u>r</u>elease, Figure 14.1B). Infection droplets are unwalled regions of the ITs (they usually arise at the growing tips of ITs) at which the rhizobial cells come into close contact with the plant cell plasma membrane. Inside the plant cytoplasm, the bacteria are surrounded by special peribacteroid membranes (PBMs) that are derived from the plasmalemma of infection droplets and then are formed mainly from endoplasmic reticulum and Golgi vesicles. The bacterial cell(s) surrounded by PBM represents a major intracellular symbiotic compartment – the symbiosome. Within symbiosomes the rhizobia differentiate into bacteroids (stage Bad – Bacteroid differentiation). They are several times larger than free-living bacteria and are altered in shape (in pea nodules the mature bacteroids are X- or Y-like).

A pronounced differentiation is typical of rhizobia-infected plant cells. They possess an increased amount of internal membrane structures participating in PBM formation and biosynthetic processes. The plant cellular differentiation is followed by *de novo* synthesis of a range of proteins responsible for the energy supply to nitrogenase and for the assimilation of fixed nitrogen.

The resulted nodules belong to the morphologically most complex indeterminate type characterized by stable apical meristem and histologically zonated central part. A persistent system of symbiotic ultrastructures providing bacteria migration from intercellular compartments (ITs) to intracellular ones (symbiosomes) is developed within these nodules. However, some other papilionoid legumes (soybean, pole bean – tribe Phaseoleae; trefoils – tribe Loteae) form the determinate nodules that may be considered morphologically more simple with respect to indeterminate ones (Figure 14.1A). In a determinate nodule the meristem exists for a few days only; after its disappearance the nodule growth and additions to the infected tissue are stopped. An important distinction is that, in indeterminate nodules, the apical meristem remains uninfected, whereas at the early stages of development in determinate nodules the rhizobia colonise cells that are still mitotically active. In the central part of a determinate nodule the histological zonation is absent, however infected cells (where N_2 is fixed) are often interspersed with uninfected ones (where the transport forms of fixed nitrogen are synthesized).

Plant control over nodule development. The plant genes involved in symbiosis may be divided into two groups: *Sym* genes and nodulin genes. These genes were identified using different approaches: *Sym* genes by formal genetic analysis (selection of plant mutants defective in nodule structures); nodulin genes by molecular isolation (identification of proteins or RNAs-nodulins-synthesized *de novo* in nodules). Nodulin and *sym* genes play different roles in the processes, which may be referred as "management of bacteria" inside plant roots.

Most of the nodulin genes were already cloned and sequenced but functions for many of them remain unclear. There are several examples demonstrating that nodulins are the structural elements of newly constructed temporary organelles (ITs, symbiosomes) formed during symbiosis. For many nodule-specific proteins, the sub-cellular locations (cytoplasm, PBM, IT wall) or enzymatic activities were determined. Some nodulins are involved in the formation of the symbiotic structures. For example, "early nodulin" ENOD2 is actively synthesized in the nodule parenchyma, while ENOD5 and ENOD12 are accumulated in the IT walls. Nodulin 26 is synthesized during endocytosis as a component of PBM, which may be required for the transport of signals or nutrients between partners. ENOD40 is involved in balancing the hormonal status of the developing nodules. This protein was also revealed in non-legumes and its function may be related to the control of auxin: cytokinin ratio that is altered greatly after inoculation and plays an important role in nodule initiation.

The regulatory role for the realization of symbiotic programs is assigned to *Sym* genes. These genes are usually not induced specifically in nodules and there are many examples of the high functional and sequence homologies between them in different legumes. Up to now more than 50 *Sym* genes have been revealed and their number is thought to be close to saturation.

The sequential order of action for *sym* genes has been revealed by using the double mutant line approach which is based on the pairwise combinations of recessive mutations in the same genotype subjected to the backcrosses with the initial single mutant lines. The proposed scheme covers the whole developmental process of nodule ontogeny but for the early events of symbiosis it was detailed on the basis of mutations in numerous plant genes (Table 14.3). Nowadays the functional order of *Sym* genes was detailed to a great extent for the early events in symbiosis, especially in the model legumes (*Medicago truncatula, Lotus japonicus*). For the late stages of interaction the most detailed genetic dissection is available for pea (*Pisum sativum*).

One of the genes involved in the IT formation is *NIN* (No Infection) the structure of which was described in the lab of Stougaard in Denmark (Schauser et al., 1999) in *Lotus japonicus* as the first example of *Sym* genes sequence. Its ortholog – *Sym35* has been revealed in *P. sativum* using the syntheny and microsyntheny between the large pea genome and the much smaller genome of model legume *L. japonicus*. The phenotypes of *NIN* or *sym35* mutants are characterized by excessive root hair deformation in response to bacterial infection, lack of IT formation and of cortical cell divisions. Analysis of the *NIN/Sym35* DNA sequences shows that their protein products have a complex domain structure. For some domains the reasonable functions implemented in the nodule development may be proposed. The motif of transcriptional regulator and the homology with the *MID* locus involved in the adaptation to nitrogen starvation in *Chlamydomonas* may indicate the possible role of *NIN/Sym35* as a trigger for the establishment of symbiosis under the conditions where plant should use the "biological" nitrogen.

The set of mutations in the plant genes allows us to study the impacts of plant genes on the expression of bacterial genes during symbiosis. The expression of *nodA* (involved in the early signal exchange) and *dctA* (controls the dicarboxylate utilization by symbiotic bacteria, see further) were studied in mutant nodules arrested at the IT, infection droplet and bacteroid differentiation.

With the use of reporter fusions to these bacterial genes it was shown that *nodA* and *dctA* are expressed at all stages with a pattern similar to that of constitutive, symbiosis-unrelated genes. These genes seem to be subjected to gradual down-regulation in nodule bacteria correlating with the stage of bacteroid differentiation. In contrast, expression of oxygen-dependent gene *fixN* is up-regulated in bacteroids.

Table 14.3. Examples of the legume genes controlling the nodule development

Developmental stages (steps)	Codes*	Controlling genes (legume species)**
Pre-infection		
Curling of root hairs	Hac	*sym8, sym9, sym10, sym19, sym30* (Ps); *rj₁*=nod1 (Gm); *rn1* (Ca); *nn₁, nn₂* (Ms); *sym3* (Ma)
Induction of the cortical cell divisions	Ccd	*sym5* (Ps)
Infection of plants and formation of nodules		
Development of the infection threads	Itf Iti	*sym7, sym14, sym35* (Ps); *sym1, sym5* (Ma); *r, t* (Tp)
	Ith	*sym2, sym36* (Ps)
	Itr	*sym5, sym34* (Ps)
Autoregulation of nodulation	AON	*nod3, sym28, sym29* (Ps); *sym5* (Vf); *nod* (Pv); *nts1=nod2* (Gm)
Differentiation of nodule tissues	Ntd	*sym33* (Ps)
Bacteria release (endocytosis) into the plant cells	Bar Itn	*sym33* (Ps); *d, I₁, i_c* (Tp)
	Idd	*sym40* (Ps)
Differentiation of bacteroids	Bad	*sym31, sym32* (Ps); *in₁, in₂, in₄, in₅* (Ms)
Functioning of nodules		
Nitrogen fixation	Nif	Plant mutants defective in N_2 fixation are impaired in different steps of nodule development
Nodule persistence	Nop	*sym13, sym25, sym26, sym27* (Ps)

*Phenotypic codes: Hac – root hair curling, Ccd – cortical cell division, Itf – infection thread (IT) formation (Iti – IT initiation, Ith – IT growth in root hair; Itr – IT development in root cortex; Itn – IT development in the juvenile nodule tissue); AON – autoregulation of nodulation, Ntd – nodule tissue development, Idd – Infection droplet differentiation, Bar – bacteria release from infection threads to plant cell cytoplasm, Bad – bacteroid differentiation, Nif – nitrogen fixation and Nop – nodule persistence.

**Ca – *Cicer arietinum*, Gm – *Glycine max*, Ma – *Melilotus albus*, Ms – *Medicago sativa*, Ps – *Pisum sativum*, Pv – *Phaseolus vulgaris*, Tp – *Trifolium pratense*, and Vf – *Vicia faba*.

A marked repression of the bacterial genes involved in early events depends on the degree of bacteroid differentiation and fullest repression is achieved only in N_2-fixing bacteroids. Within this process, bacteroid differentiation could resemble an adjustable process of bacterial adaptation to the symbiosome interior that is basically stressful for free-living rhizobia. This adaptive process results in the mature, actively nitrogen-fixing bacteroid, which lost the reproductive activity (transition to an organelle-like state) and the capacity to regain the free-living state. The deviation from the normal developmental process caused by plant mutations leads to the inability of bacteria to be released into host cell or to the rapid degradation of bacteroids. In all cases there is no N_2 fixation and the phenotypes of such plant mutants are designated as Fix⁻.

Several plant genes – *Sym31, Sym33, Sym40* – are of great interest because they control bacteroid formation involving the deep reorganizations of bacterial cells. At the same time *Sym31* is involved in the regulation of nodule number by the combined nitrogen (see further). Therefore, analysis of the plant gene action clearly shows that the fate of symbiotic bacteria is controlled by the set of plant genes providing very effective regulation of the bacterial gene activity and the differentiation of bacteroids.

During the nodule development, the plant *Sym* genes are also involved in the process of nodule tissue development which is induced in parallel with the infection. Some pea mutants lack even the earliest steps of nodule tissue development including cortical cell division (Ccd⁻ phenotype) but retain the root hair curling. In pea, the expression of at least eight genes is required for the induction of cortical cell division but they are not necessary for root hair curling and colonization. The phenotypic characterization of a large set of mutants allowed us to classify the symbiotic genes according to the stage of nodule development that they control. The comparison of two programs – infection by rhizobia and nodule development – leads to the conclusion that many *Sym* genes are involved in the two parallel processes. Their differences are clear from the fact that the orders of action for the same *Sym* genes are not identical (Figure 14.2). The mechanisms providing for a differential involvement of a particular plant gene in either infection or developmental programs remain obscure.

Systemic control over nodulation.   One of the most important functions of host during symbiosis is to provide the systemic regulation of the number of nodules the formation and functioning of which represent the energy consuming processes. The nodule number in legumes is regulated by numerous factors including the genotype of bacterial partner, environmental factors (temperature, drought, salinity, N and P nutrition, environmental pollutions), etc. However, the crucial role in this control is implemented by the endogenous plant factors including phytohormones, receptors of symbiotic signals and factors for autoregulation of nodulation (AON).

As it was demonstrated by Gresshoff and his co-workers (2004), several effects result from the arrest of AON by the plant mutations: (1) number and mass of nodules per plant increase markedly; (2) in spite of increasing the total nitrogenase activity the plant mass decreases due to over-expenditure of energy for supporting the excessive nodules; and (3) nodulation becomes tolerant to the inhibitory levels of nitrate, suggesting that the external regulation of nodulation and AON are interactive. These features demonstrate that AON represents the plant capacity to reach the optimal number of nodules in accordance with the nitrogen content in a soil. Evidently, without AON the N_2-fixing symbiosis becomes non-adaptive.

The fact that the shoot is involved in AON is clear from two types of experiments. In the first one, root systems were split to allow the inoculation of the two parts of root at different times. In wild-type soybean plants, the difference of seven days results in 100% suppression of nodulation on the part of root that was inoculated later. However, in AON-deficient mutants the suppression of nodulation is reduced and only a weak reduction of nodule number is achieved in the part of the root, which was inoculated later.

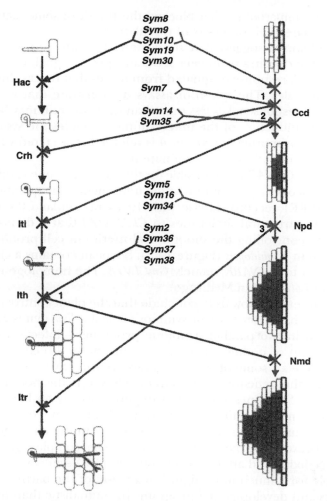

Figure 14.2. Sequential functioning of pea (*Pisum sativum*) symbiotic genes involved in control over the infection process and nodule development (Tsyganov et al., 2002).
Phenotypic codes for developmental stages are given in the footnote to Table 14.3; Crh – bacterial colonization of pocket in the curled root hair. Npd – nodule primordium development and Nmd – nodule meristem development. Arrows indicate the impacts of the plant genes on developmental stages or the interdependencies between different stages.

The second type of experiments is represented by reciprocal grafts of the hypocotyl or epicotyl between mutant and wild-type lines. The grafting experiments suggest that the nodule number in the root (stock) is usually controlled by the genotype of shoot (scion). These data confirmed further the importance of the shoot/root regulatory circuit in nodulation control. In the upper part of the plant, the leaves appeared to be the major source of autoregulation. It was shown that the wild-type stem segments grafted in between the mutant root neither blocked the transit of some factor nor produced the shoot-derived inhibitors of nodulation.

The data obtained suggest that soon after the induction of the nodule primordium, the dividing plant cells send an upward signal, which elicit the negative feedback response originated from leaves that blocks the formation of excessive nodules. The split-root assays demonstrate that in clover and alfalfa the AON response takes less time than in soybean. It may be supposed that during the induction of the indeterminate nodules the systemic signal is released from primordia quicker or it is released more rapidly from shoots than during the induction of determinate nodules.

The first gene (*GmNARK*) responsible for the AON was cloned and sequenced in soybean. It encodes a protein with single trans-membrane spanning, leucine-rich repeat (LRR) receptor kinase with 19 extra-cellular LRR spans. This gene shows the structural similarities to *CLAVATA1* that is demonstrated to be involved in controlling the shoot apical meristem cell proliferation and differentiation in *Arabidopsis* (Figure 14.3). Soybean contains a close relative (90% identical) to *GmNARK*, namely *GmCLV1A*. The homologous *har1* gene was cloned in *Lotus* and in Medicago.

The data presented allow us to conclude that the plant control over nodule development is based on a set of symbiosis-specific *Sym* genes and nodulin genes involved in two parallel symbiotic programs – infection by rhizobia and development of nodules. The data on promotor structures allow us to suppose that at least some of the *Sym* genes are directly triggered by microbial signals in the gene-for-gene manner. However, the nodulins seem to be regulated by *Sym* genes and do not depend directly on bacterial signals. This hierarchy may constitute a sufficient distinction between beneficial and deleterious interactions since in the latter group the direct gene-for-gene interactions between partners are prevailing. The complex gene hierarchy revealed in N_2-fixing and arbuscular mycorrhiza (AM) symbioses (see further) is responsible for a much more tight integration of the partners' biochemical pathways and developmental programs in mutualistic than in antagonistic systems. Since nearly all symbiotic genes have close homologues among the genes involved in various plant traits (defense from pathogens, development, stress adaptations), it is logical to suppose that both groups of genes were recruited during the plant evolution into symbiotic programs from other developmental programs.

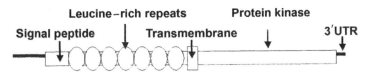

Figure 14.3. A generalized structure of the NARK-type LRR receptor kinase in soybean (Gresshoff et al., 2004).

Specificity and signaling

Interaction between legumes and rhizobia is highly specific: the symbiotic nodules are induced by rhizobial strains only within the strictly defined ranges of plant species. This specificity was first revealed in the late 1890s when it was demonstrated that alfalfa, soybean, pea and clover plants need different bacterial strains to form nodules. Therefore, the host specificity is defined as a spectrum of plant species or genotypes that can respond to inoculation with a particular rhizobial strain was addressed as its key symbiotic property.

During the last 20 years, spectacular success has been achieved in the elucidation of the molecular mechanisms involved in the symbiotic specificity. Now it is clear that specificity should be addressed in terms of partners' recognition leading to the signal exchange between plants and bacteria. When we consider the signal transduction we are concerned with how external influences, particularly the presence of specific messengers, can determine the events that happen inside the target organisms. To do this, the signal should interact with the complementary receptor in cooperation with which a united signaling pathway is formed.

An important prerequisite for the analysis of specificity, recognition and signaling is constituted by the fact that any change in symbiotic phenotype (non-nodulation, inefficient nodulation, decreased or enhanced nodulation or nitrogen fixation, as well as a changed symbiotic specificity) may be obtained by the mutation or recombination events occurring in either bacteria or host (Figure 14.4). Several groups of genes responsible for nodulation and symbiotic nitrogen fixation were identified in rhizobia (Figure 14.5).

Taxonomic basis of specificity. The first insights into the mechanisms of specificity were made *via* analyzing the taxonomic regularities responsible for variation of this property in both partners. Nitrogen-fixing nodules are

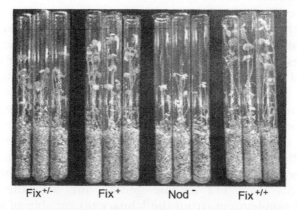

Fix$^{+/-}$ Fix^{+} Nod^{-} Fix$^{+/+}$

Figure 14.4.   Selection of symbiotic mutants in alfalfa rhizobia (*Sinorhizobium meliloti*) in aseptic tube tests. Alfalfa (*Medicago varia* Mart.) seedlings are grown on vermiculite amended with N-free mineral medium wherein the plant development is limited by intensity of N$_2$ fixation. Phenotypes: Nod – nodule formation and Fix – nitrogen fixation (experimental details are given in: Provorov and Simarov, 1990).

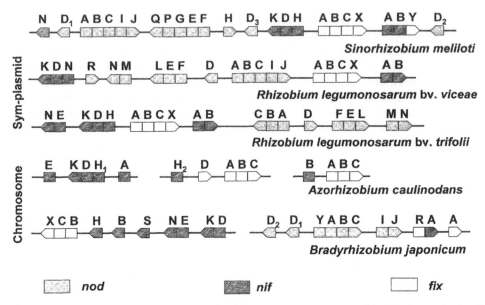

Figure 14.5. Major symbiotic genes in rhizobia. *nod* genes are responsible for the synthesis of Nod-factor, *nif* and *fix* genes for nitrogen fixation (for more details on organization of *nif/fix* genes see: Fisher, 1994).

formed in more than 90% of the legumes (family Fabaceae). Some legumes were classified into groups within which the cross-inoculation occurs readily while the species from different groups do not cross-inoculate. The best-studied examples of this classification are represented by clover (*Trifolium*), alfalfa (*Medicago, Melilotus, Trigonella*), pea (*Pisum, Vicia, Lathyrus, Lens*) and goats' rue (*Galega*) cross-inoculation groups (CIGs) for which separate groups of rhizobia are specific (Table 14.4).

However, such strictly defined specificity is limited to the herbage legumes growing in temperate zones and representing the "galegoid complex" to which the tribes Trifolieae, Vicieae and Galegeae belong. For many tropical and subtropical legumes a strict separation into CIGs is not typical. Instead we observe a wide variation in the degree of symbiotic promiscuity. For instance, the species of tribe Pheseoleae (*Phaseolus vulgaris, Vigna unguiculata, Glycine max*) may be inoculated with rhizobia isolated from the taxonomically distant legume species representing different tribes and subfamilies of the Fabaceae. Moreover, many tropical rhizobia can inoculate very wide spectra of hosts. For example, strain NGR234 (*Sinorhizobium fredii*) inoculates plants from 112 legume genera and also the *Parasponia* – an exceptional non-legume (from the Ulmaceae) capable of symbiosis with rhizobia.

Rhizobia, in contrast to their plant hosts, do not form a monophyletic group. The majority of rhizobia belong to the α-proteobacteria assigned to the family

Table 14.4.   Host specificity in the major rhizobia groups

Genera	Species	Major hosts	Location of major symbiotic genes	N$_2$ fixation *ex planta*
Rhizobium	*R. leguminosarum*		On *Sym*-plasmids (300–500 kb)	Absent
	bv. *viceae*	*Pisum, Vicia, Lathyrus, Lens**		
	bv. *trifolii*	*Trifolium**		
	bv. *phaseoli*	*Phaseolus*		
	R. galegae	*Galega**		
Sinorhizobium	*S. meliloti*	*Medicago, Melilotus, Trigonella**	On *Sym*-plasmids (1300–1700 kb)	Absent
	S. fredii	*Glycine*		
Mesorhizobium	*M. loti*	*Lotus, Lupinus*	In chromosomes	Absent
Bradyrhizobium	*B. japonicum*	*Glycine, Vigna*	In chromosomes	Low, not sufficient for diazotrophy
Azorhizobium	*A. caulinodans*	*Sesbania, Aeshynomene*	In chromosomes	High, provides diazotrophic growth

*These plants comprise the cross-inoculation groups (CIGs) that are separated strictly from each other and from the rest of the legume species (CIGs' structure and evolution are summarized in Provorov, 1994).

Rhizobiaceae (Figure 14.6) solely on the basis of their ability to nodulate the legumes. Not less than five genera (*Azorhizobium, Bradyrhizobium, Mesorhizobium, Rhizobium, Sinorhizobium*) are distinguished among these bacteria. Moreover, some β-proteobacteria were recently discovered that can form N$_2$-fixing nodules with legumes. Therefore, rhizobia represent very diverse group of Gram-negative bacteria the molecular phylogeny of which demonstrates that they have been diverged from a hypothetical common ancestor much earlier than the host plants appeared.

From the rhizobial side, correlation between specificity and taxonomy is even more questionable than from the plant side. For example, bacteria inoculating the plants of clover and pea CIGs belong to the same species (*Rhizobium leguminosarum*) and differences between the biovars (bv. *trifolii* and bv. *viceae*) pertain only to few genes responsible for the host specificity. Moreover, some legumes may be inoculated by taxonomically distant bacteria (soybean – by *Sinorhizobium fredii* and *Bradyrhizobum japonicum*; trefoil – by *Mesorhizobium loti* and *Bradyrhizobium* sp.), which share the genes required to inoculate a common host.

In spite of the observed inconsistencies with phylogenetic data, the symbiotic specificity constitutes a stable property for both partners: each plant species possesses a genetically defined range of potential symbionts; each strain of rhizobia has a fixed spectrum of plant host species. Under adverse environmental conditions (e.g., high temperature, flooding, heavy salinity) the symbiosis development may be arrested, however specificity does not change.

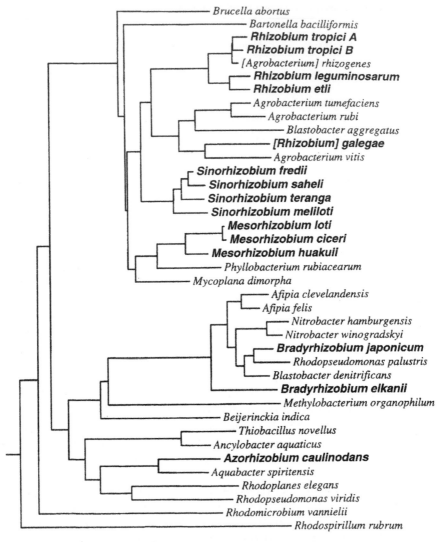

Figure 14.6.   Phylogeny of the rhizobia and related α-proteobacteria (from: Young and Haukka, 1996). Analysis of diversity in rDNA regions demonstrated that rhizobia do not comprise a coherent group. Rather, they are represented by several genera included in the Rhizobiaceae family only because of their ability to nodulate legumes. The evolutionary implications of this phylogeny assume that the symbiotic properties were acquired by diverse rhizobia either independently or due to horizontal transfer of symbiotic genes. (For more details see: Provorov, 1998; van Berkum and Eardly, 1998.)

Bacterial genes for host-specific nodulation.   It is usually considered that full expression of rhizobial host specificity involves the formation of normally developed, nitrogen-fixing nodules in particular plant species. However, nodulation itself is a highly specific process and this is why nodulation genetics contributed greatly to the understanding of the mechanisms of specificity.

The regular genetic analysis of host specificity was greatly facilitated when the extra-chromosomal location of symbiotic genes was established in fast-growing rhizobia. The symbiotic plasmids (p*Syms*) were visualized in the late 1970s although the extra-chromosomal location of symbiotic genes was pre-dicted much earlier, proceeding from the facultative dependency of rhizobia on the symbiotic interactions. It was demonstrated that the majority of *Rhizobium* and *Sinorhizobium* strains harbor high molecular weight plasmids (300–1700 kb), one of which can be hybridized with the genetic probes for *nif* genes encoding the nitrogenase proteins. Curing the rhizobia of these plas-mids by elevated temperature or by acridine orange dye usually results in the loss of nodulation on all potential hosts.

The involvement of p*Syms* in the control of host specificity was proved by Beringer and co-workers, (1980) (UK), who transferred p*Syms* between the *R. leguminosarum* bv. *viceae* strains contrasting in its ability to nodulate the "Afghan" peas. Resistance of these peas to infection by the European rhizobia was at first demonstrated in 1928 by Govorov (USSR) and more than forty years later, Lie, (1978) (The Netherlands) isolated from the Middle East soils the strain TOM capable of nodulation of "Afghan" peas. A high molecular weight plasmid was found in this strain the conjugation of which into the European *R.l.* bv. *viceae* strains resulted in the transfer of the "Afghan" pea nodulation ability.

An extended genetic analysis of the rhizobial host specificity was facilitated by Tn5-*mob* system constructed in the lab of Pühler at Germany (Simon et al., 1983). It ensured the mobilization of diverse p*Syms*, the great majority of which are devoid of self-conjugative capacities. Analysis of the results obtained demonstrated that the rhizobial host specificity may be expressed in different degrees i.e., intra-species, inter-species and inter-genera transfer. For example, transfer of p*Syms* between biovars of *R. leguminosarum* (*viceae, tri-folii, phaseoli*) usually lead to the full expression of specificity. As a result, bv. *tri-folii* may be converted into bv. *viceae* by the substitution of their p*Syms*: the recombinants which acquired p*Syms* from the bv. *viceae* strains elicited N_2-fixing nodules with the pea and vetch, instead of the clover plants. Moreover, the recombinant strains with the extended host specificities may be obtained by combining the p*Syms* of different origins within the single strain. However, transfer of p*Syms* from *R. leguminosarum* to *S. meliloti* leads only to a partial expression of specificity: recombinants may elicit only abnormal non-N_2-fixing nodules with the host of the donor strain. These results suggested that the methodology to search for genes of host specificity should be based on transfer and expression of the bacteria nodulation genes within and between the rhizobia inoculating the legumes from different CIGs.

Application of the traditional methods of bacterial genetics for the studies on symbiosis was limited by the lack of expression of many symbiotic genes in the free-living cells. This limitation was avoided by using transposons (mainly, Tn5 and its derivatives), which confers the drug resistance on the recipient and the selection of those bacteria which maintain this resistance stably. Transposons inactivate the target genes into which they are inserted and usually a single mutagenic lesion is introduced into the bacterial genome.

Another important feature of Tn5 is that the site of mutation is labeled by the drug resistance enabling the researcher to trace the mutant allele in genetic experiments. Moreover, Tn5 provides the possibility to identify the tagged gene using a specific probe facilitating the subsequent isolation of the wild-type allele by DNA hybridization and PCR amplification.

The most definite data concerning the control of specificity were obtained for genes the inactivation of which results in: (a) the absence of nodules (Nod⁻) or in a delayed nodulation (Nod^d) with the host from initial CIG, (b) the changes of host range (inoculation of a host from other CIGs). These genes were called *nod*, *nol* and *noe* and their mutations usually lead to the alteration in early symbiotic events such as the inability to curl the root hairs (Hac⁻) or to elicit the formation of infection threads (Itf⁻).

The data suggest that the rhizobial nodulation genes are subjected to common regulation at the transcriptional level. This was first suggested in the lab of Kondorosi at France (Rostas et al., 1986). They revealed the presence of highly conserved "*nod* box" sequence upstream the coding regions of all identified genes which participate in nodulation. This *nod* box was originally defined in *Sinorhizobium meliloti* as a 47-bp consensus required for *nod* gene induction and the same motif was revealed in other rhizobia. These data suggest that in each rhizobial strain the family of *nod*/*nol*/*noe* operons represents the regulon induced in the host-specific manner. The conservation of *nod* boxes among rhizobia provided a strategy to search the *nod* genes based on the promoter hybridization to the *nod* box probe.

Comparison of the coding regions from different *nod* genes revealed that they should be divided into two groups: "common" *nod* genes and the "host specific" ones. Typical common *nodABC* genes have been found in all rhizobia studied so far. These genes are very similar not only in their structures but also in functions: for instance, interchange of genes between *Rhizobium* and *Bradyrhizobium* may restore nodulation activity in the mutant, usually without any alteration in the host range.

However, the majority of *nod* genes are not conserved among rhizobia. In contrast to *nodABC*, presence of these genes is restricted to the particular rhizobia species, or they are present in different species but cannot functionally complement each other. These genes have been named the host-specific *nod* genes since they satisfy at least one of the following criteria: (a) mutations in these genes cannot be fully complemented by DNA from other rhizobial species; (b) these mutations result in the narrowing or extending the host range; and (c) transfer of these genes into the other rhizobia may result in the expression of the donors' specificity in the recombinant strains.

Regulation of nod gene activity.   Bacterial nodulation genes may be distinguished not only based on their structures but also on their expression in symbiotic and free-living states. All nodulation genes are transcriptionally silent in the absence of suitable plant hosts with only one exception of *nodD* gene, which exists in all rhizobia. Several lines of genetic evidence confirmed that *nodD* genes are the determinants for host specificity. Some point mutations in these genes cannot be complemented by *nodD* alleles from other rhizobial species and such mutations may also result in an altered host range. Different *nodD* alleles are conserved in their nucleotide sequences and share homologies with genes from the *lysR-araC* family of transcription regulators. Specifically, the N-terminal domain of NodD contains a helix-turn-helix motif characteristic typical for DNA binding proteins. It was shown that NodD proteins bind to the *nod* box due to this motif.

In order to elucidate a plant signal recognized by NodD protein, it was suggested to examine the root exudates because it was demonstrated that the treatment of free-living rhizobia with these exudates enables them to excrete a factor that elicits the early symbiotic responses in the sterile roots. These responses include: root hair curling, induction of nodule primordia or even development of the "empty" (devoid of bacterial infection) nodule-like swellings. In order to develop a trap system for detecting the plant-inducing compound(s), Lugtenberg (The Netherlands) proposed a transcriptional fusion between *nod* gene promoters and the structural gene for β-galactosidase (*lacZ*). Three components have been shown to be important for the induction of the *nod* genes (Figure 14.7):

– promoter including the *nod* box fused with the coding part of *lacZ*;
– *nodD* genes of different origins sensing the hypothetical plant signals; and
– inducing substances produced by plants from different CIGs.

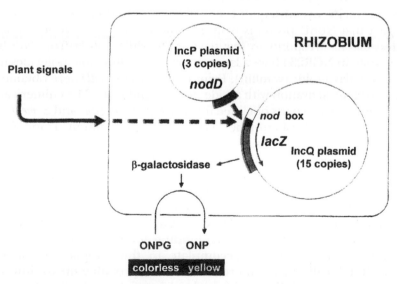

Figure 14.7. Trap system for the identification of plant signals activating nodulation genes in rhizobia (details in the text). Kindly presented by Ben Lugtenberg (Leiden University).

Plant-specific inducers of nod genes.   By using the above-described trap system, it was demonstrated that the principal legume inducers sensed by NodD proteins are represented by flavonoids. Flavonoids are widely distributed throughout the plant kingdom and are abundant in many plant organs. On the basis of modifications in the three carbon rings more than 4000 flavonoids discovered to date can be classified into several groups, including flavonols, flavones, isoflavones and antocyanine pigments. Besides the plant–microbe signaling, the non-pigmented flavonoid compounds play a central role in the signal transduction pathways required for pollination, hormone metabolism, protection from stresses and for the ecological interactions: defense from pathogens (phytoalexins) and allelopathy.

Identification of the first nod gene inducer luteolin in alfalfa root exudates has been reported by Long (USA) for the induction of *Sinorhizobium meliloti nod* genes. Using synthetic or natural compounds of known structures, the features of flavonoids required for *nod* gene induction by any NodD protein were determined. The core structure of different flavonoids is common to different legumes while the positions of $-OH$ or $-OCH_3$ groups vary in six different sites of the core. Such heterogeneity provides enough opportunities for nodule bacteria to differentiate their hosts. The spatial–temporal distribution of flavonoids in the rhizosphere can be directly observed on different parts of the root system. Nodules generally appear in the zone of maximum induction of *nod-lacZ* activity corresponding to the emerging root hairs.

These data provide the strong evidence that *nodD* gene is a transcriptional regulator of the *nod/nol/noe* regulon and that the NodD proteins perform the role of receptors recognizing the plant signals with the following induction of nodulation gene expression. The role of flavonoid–NodD interaction in the specificity of subsequent signaling and nodulation was suggested by the fact that NodD proteins from narrow-host-range rhizobia (*S. meliloti, R. leguminosarum*) respond to few flavonoids while NodD from the broad-host-range strain NGR234 have a larger spectrum of inducing molecules including even non-flavonoids (vanillin). Point mutations in *nodD* were obtained that result in *nod* gene activation with a wide range of inducers. The chimerical *nodD* allele consisting of two parts obtained from *R. leguminosarum-* and *S. meliloti-*born alleles provides FITA phenotype (Flavonoid Independent Transcriptional Activation) in free-living cells.

However, the correlation of NodD induction with the origin of this protein and of the inducing compounds is poor (Table 14.5) demonstrating that the differences in symbiotic specificity between the major CIGs are not restricted to the "flavonoid–NodD" interaction. Moreover, up to now the stable complexes of flavonoids with NodD have not been visualized in spite of the extended genetic data demonstrating that such binding should occur.

Flavonoids were proved to accumulate in the cytoplasmic membrane where NodD is localized, and therefore the most probable site for interaction between receptor and flavonoid is the bacterial inner membrane. The C-terminal domain of NodD protein is thought to be the sensor of different

Table 14.5. Induction of β-galactosidase activity (modified Miller units) in the trap system *nod* box – *lacZ* + *nodD* (Figure 14.7)

Flavonoid effector	Origin of the effector	The rhizobial species used as the source of *nodD* allele		
		Sinorhizobium meliloti	*Rhizobium leguminosarum* bv. *viceae*	*Rhizobium leguminosarum* bv. *trifolii*
Luteolin	Alfalfa	6.0	19	22
Apigenin	Pea	2.6	24	29
7-Hydroxyflavone	Clover	0.55	7.6	29

inducers since its structure varies among the rhizobial species. Experiments with *R. leguminosarum* and *R. meliloti* NodD have shown that binding of these proteins to *nod* box does not require the presence of flavonoids although the stability of this binding may be enhanced by flavonoid treatments. The precise mechanism of *nod* box activation upon NodD binding is not clear and the acceptable idea is that regulation by flavonoids involves the changes in conformation of "NodD–promoter" complex, which activates the *nod* gene transcription.

The *nod* gene induction process is complicated by the fact that each legume species releases several types of flavonoid molecules, which display the opposite effects on the activity of target genes. For the fast-growing rhizobia, flavones are usually the inducers while isoflavones are the repressors of *nod* genes. In contrast, for the slow-growing rhizobia inducers are represented by isoflavones, whereas repressors are the flavones. The anti-inducers often have chemical structures close to those of the inducers suggesting a competition for the same binding site on NodD protein. The resulting activities of *nod* box promoters seem to depend on a fine balance between different types of flavonoid effectors present in the root vicinities.

In addition to the one-component receptor NodD sensing the host signals and activating the *nod* genes, some rhizobia regulate these genes *via* the two-component transcriptional activators. For example, *Bradyrhizobium japonicum* strains have *nodV/nodW* genes, the first one is coding for trans-membrane sensor of flavonoids that is autophosphorylated upon contact with these signals. Later, it transduces the phosphate group on the receptor that activates the promoters of *nod* genes. Importantly, such additional regulators are found in the broad-host-range rhizobia and apparently enable them to cope with the diverse hosts.

Since the *nod* gene induction by plant signals is a low specific process, while nodulation is highly specific, it was supposed that the specificity in legume–rhizobia symbiosis is determined mainly by a bacterial signal encoded by *nod/nol/noe* genes (Figure 14.8). This hypothetical signal was called "Nod factor".

Genetic control of Nod factors synthesis. From the previous sections, it is clear that the function of rhizobial nodulation genes should be addressed

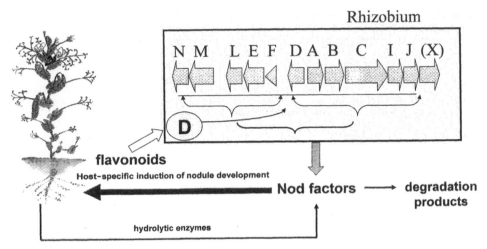

Figure 14.8. Signal exchange between rhizobia and the legume plants. (Details in the text.) Kindly presented by Ben Lugtenberg (Leiden University).

in terms of production of the specific feedback signal, distinguished as Nod factor and that this signal provides the start for the developmental program resulting in the nodule morphogenesis. Despite the analysis of a Nod factor structure requires chemical analyses, the genetic approaches provided important insights into this structure. Importantly, the genetic data demonstrated the differentiation of nodulation genes into the "common" and "host-specific" groups. This differentiation suggests that the signal(s) encoded by *nod* genes should consist of a core part that undergoes various modifications required for the nodulation of different legume hosts.

The biochemical experiments have shown that the Nod factors induce the very early reactions in nodule initiation. Filtration through molecular sieves indicates that the Nod factor has relatively low molecular mass, heat stable, partially hydrophobic and resistant to proteases. Highly important was the fact that Nod factor treatment displays the effects having the same host specificity as it is observed for the nodulation by living bacteria. This has been shown by Denarie group (France) using the bioassays with vetch (nodulated by *Rhizobium leguminosarum* bv. *viceae*) and alfalfa (nodulated by *Sinorhizobium meliloti*), where the action of a Nod factor was detected upon root hair deformations (Denarie et al., 1992). Inactivation of *nodABC* genes suppresses the activity of a Nod factor completely for both hosts. However, *S. meliloti nodH* mutants have an altered host range: they cannot infect and nodulate the homologous host – alfalfa, but can infect vetch, a heterologous host. When the *S. meliloti* genes *nodFEG*, *nodH* and *nodPQ* were transferred into *Rhizobium leguminosarum* bv. *viceae* background, recipient cells gained the ability to nodulate alfalfa and retained a reduced but detectable nodulation on vetch.

These data suggest that synthesis of Nod factor core part is controlled by common *nodABC* genes while its modifications responsible for affinities to different CIGs are controlled by host-specific genes. Since *S. meliloti* has additional *nod* genes in comparison to *R. leguminosarum* it might be expected that the former species possesses extra chemical groups in their to Nod factors in comparison to the latter species.

The performed analyses of chemical structure generally confirmed the genetic data. Nod factors produced by different rhizobial species were proved to be the acylated lipo-chito-oligosaccharides (LCOs) i.e. oligomers of β-1,4-*N*-acetyl-D-glucosamine (usually 3–6 GlcNAc units), carrying a fatty acyl chain at the non-reducing terminus. Synthesis of the core structure is controlled by common *nod* genes. The polymerization reaction is performed by the chito-oligosaccharide synthase NodC resulting in the formation of short chains composed of two to six GlcNAc residues. NodB protein specifically deacetylates the *N*-acetyl group at the non-reducing terminus. Finally, NodA attaches the fatty acid (16–20 carbon atoms) to the *N*-deacetylated precursor (the fatty acid is available in the acyl form). The glucoasamine synthase NodM contributes to the synthesis of the precursor GlcNAc (Figure 14.9A).

It is important to note that host specificity is based on modifications of either this core structure (called "minimal Nod factor") or on its decorations with additional chemical groups. For example, variability in the length of oligo-chitin chains may be essential for the Nod factor recognition that takes place in some combinations between symbiotic partners. Special alleles of *nodC* gene may confer on the *S. meliloti* strains the ability to nodulate *Medicago laciniata* – the unique medic species that cannot be nodulated by the majority of *S. meliloti* strains.

The acyl (fatty acid) moiety at the non-reducing terminus may be represented by two forms: either relatively common, saturated or mono-unsaturated fatty acid, or by more specific highly unsaturated (2–4 double C=C bounds) fatty acid. Nod factors of the second type are produced by *R. leguminosarum* and *S. meliloti* and their synthesis requires the ketoacetylsynthase NodE and acyltransferase NodF. In *R. leguminosarum*, the length of fatty acid and the degree of its unsaturation are controlled by the allelic variation of *nodEF* genes. Transfer of these genes between *R. leguminosarum* and *S. meliloti* may exchange their host specificities.

The most pronounced modifications of host specificity are achieved *via* decorations of core Nod factor structure with the additional chemical groups (Figure 14.9B). The reducing terminus of the oligo-chitin backbone is the target for action of the broad group of bacterial enzymes. For example, genes *nodH, noeE* and *nodPQ* are involved in the decoration of the reducing terminus by a sulfate group. NodH and NoeE are sulfotranferases that modify Nod factor with a sulfate group at the reducing terminus. The special donor of sulfate group is synthesized by enzymes encoded by the *nodPQ* genes. These genes provide the differences between two Nod factors – a rather simple one required for the nodule induction in vetch and a more complex one (sulfated) – for the alfalfa nodulation.

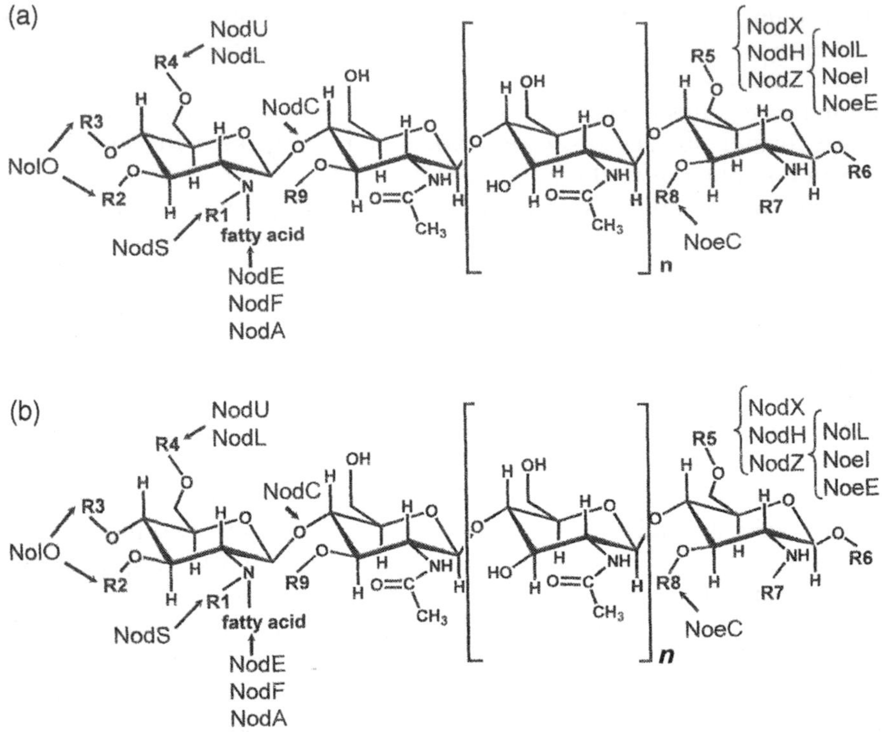

Figure 14.9. Nod factors – principal determinants of rhizobial host specificity.
(A) Biosynthesis of "minimal" Nod factor (from: Spaink and Lugtenberg, 1994; details are given in the text); (B) Major host-specific modifications in the core Nod factor structure implemented by the products of *nod/ nol/ noe* genes (Ovtsyna and Staehelin, 2005). The encoded substitutions are: acetyl (NodX, NolL, NodL), fucosyl (NodZ), sulphate (NodH, NoeE), arabinosyl (NoeC), carbamoyl (NodU, NolO), methyl (NoeI, NodS). NodA, NodE, NodF are responsible for synthesis and attachment of acyl group, NodC – for oligomerisation of β-1,4-*N*-acetyl-D-glucosamine residues.

The other well-documented example of host-specific decorations of the Nod factor core is represented by *R. leguminosarum* bv. *viceae* strains nodulating the "Afghan" pea. The host-specific gene *nodX* was identified in the strain TOM, which encodes for the acetyl transferase responsible for the synthesis of a double acetylated Nod factor. Transfer of this gene to the other *R. l.* bv. *viceae* strains them with the ability to nodulate "Afghan" pea demonstrating that additional acetylation of Nod factor is essential for extending the host specificity.

Therefore, host specificity of nodulation is provided by the signal exchange between legume plants and nodule bacteria and the Nod factor represents the key determinant in this signaling circuit. For demonstrating the complexity of the Nod factor functions, it is necessary to note that each rhizobial strain releases a variety of LCO molecules, which may contribute to the host range phenotypes. For example, the broad-host-range NGR234 strain secretes up to 18 different LCOs varying in nearly all host-specific parameters in their chemical structures. *B. japonicum* releases a mixture of pentameric and tetrameric

Nod factors. The first one is responsible only for triggering the host receptor, the second one is also for autoregulation of the *nod* gene activities.

Plant genes involved in the perception of Nod factors. Realization of the symbiotic functions of Nod factors occurs at the level of host that senses the bacterial signals and switches on the developmental program regulated by the internal signaling cascade. The application of purified Nod factors initiates two kinds of processes in the legume plants: (1) preparation of the epidermal root cells to the rhizobial entry; and (2) formation of nodule primordia in the root cortex. In epidermal cells, Nod factors induce the curling of root hairs and the re-initiation of their polar tip growth. These processes are accompanied by depolarization of plasma membrane, changes in H^+ and Ca^{2+} fluxes, regular oscillations of intracellular Ca^{2+} concentrations (calcium spiking), cytoplasm streaming and re-arrangements of cytoskeleton preparing the root hair cells for the formation of the ITs. Within the cortical cells, Nod factors induce the development of the pre-ITs. They are represented by intracellular channels surrounded by cytoskeleton structures, which facilitate the passing through of the ITs filled with rhizobia.

Nod factors also induce their own degradation by plant enzymes *via* the significantly increased hydrolytic activities. In certain hosts, Nod factors induce the additional synthesis and secretion of flavonoids resulting in an increased nodulation gene transcription. The Nod factor effects are similar to the effects induced by plant hormones and there is increasing evidence that Nod factors change the phytohormone balance within the root, for example, cytokinin:auxin ratio is increased during the nodulation.

The major role in the signal transduction within the root after treatment with Nod factor is implemented by the plant *sym* genes that may be identified using the plant mutants with Nod$^-$ phenotypes (Table 14.3). Up to now a number of legume *sym* genes have been cloned and we may propose the functions which they play in the Nod factor recognition and subsequent signal transduction in three model legumes – pea, diploid medic *Medicago truncatula* and trefoil *Lotus japonicum* (Table 14.6). The phenotypes of the relevant mutants reflect the early events in symbioses formed by legumes with either nodule bacteria or with the arbuscular mycorrhiza fungi. Among the products of identified genes are: receptor-like protein kinases and transcriptional regulators as well as Ca^{2+}/calmodulin-dependent kinase.

Two genes in *L. japonicus* (*LjNFR1*, *LjNFR5*), their orthologs in pea (*Sym10*, *Sym2*) and in *M. truncatula* (*NFP*, *Lyk3*) that were predicted to be involved in Nod factor perception encode receptor-like kinases (Figure 14.10A). Their protein products harbor LysM domains, which are present in the *Escherichia coli* MltD protein that binds peptidoglycans. LysM domains seem to bind also the *N*-acetyl-glucosamine-*N*-acetylmuramic acid backbone because these domains are present in proteins that are known to bind chitin, which is chemically similar to the Nod factor backbone. Chitin oligomers can induce Ca^{2+} spiking in legumes however, binding of Nod factor to these LysM-receptor-like kinases has yet to be shown in the direct biochemical assays.

Table 14.6.   Examples of the sequenced legume *Sym* genes

Genes	Mutant phenotypes	Gene products
LjSYMRK = MsNORK = MtDMI2 = PsSym19	Nod⁻ (Hac⁻ Ccd⁻) Myc⁻	Receptor-like protein kinase
LjHar1 = PsSym29 = GmNARK	Nod⁺⁺ Myc⁺⁺	Receptor like protein kinase
LjNFR5 = PsSym10	Nod⁻ (Hac⁻Ccd⁻) Myc⁺	Receptor-like protein kinase
LjNFR1 ~ MtLYK3 (~PsSym2)	Nod⁻ (~ Ith⁻ Cmd⁻) Myc⁺	Receptor-like protein kinase
LjNin = PsSym35	Nod⁻ (Hac⁺⁺ Iti⁻ Ccd⁻) Myc⁺	Transcriptional activator
LjASTRAY	*Nod⁺⁺ Myc⁺*	Transcriptional activator
MtDMI1 ~ PsSym8	Nod⁻ (Hac⁻ Ccd⁻) Myc⁻	Ligand-operated cation channel
MtDMI3 = PsSym9	Nod⁻ (Hac⁻ Ccd⁻) Myc⁻	Ca²⁺/calmoduline-dependent kinase

Orthologous genes are linked by "=", presumable orthologous genes by "~". The gene origins are: *Lj – Lotus japonicus, Ms – Medicago sativa, Mt – M. truncatula, Ps – Pisum sativum* and *Gm – Glycine max*. Phenotypes: Nod – nodulation (Hac – root hair curling, Iti – initiation of infection thread, Ith – development of infection thread in root hairs, Ccd – cortical cell divisions and primordia induction, Cmd – cyclic meristem in young nodule) and Myc – arbuscular mycorrhiza formation.

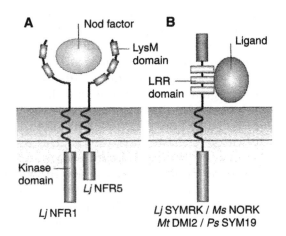

Figure 14.10.   Structures of putative components of symbiotic signaling cascade in model legumes (Oldroyd and Downie 2004).
(A) Receptors for Nod factors in *Lotus japonicus*; (B) Leucine-rich-repeat (LRR) receptor-like kinases in *L. japonicus* (Lj) and *Medicago truncatula* (Mt) (genes are specified in Table 14.6 and in the text).

The available model of recognition process is based on a set of plant genes capable of accepting the extra-cellular bacterial signals and transmiting them for regulation of the downstream signal transduction. According to the current knowledge, Nod factor receptor in *L. japonicus* might be a heterodimer that comprises two LysM-receptor-like serine/threonine kinases. Mutations in either genes cause defects in early symbiotic responses, and this indicates that both are equally essential for Nod factor perception. *Lj*NFR5 lacks a kinase-activation loop, and so it probably forms a heteromer with another protein, *Lj*NFR1, which harbors the complete kinase-activation domain (Figure 14.10A). From the data presented, we might suppose that the recognition of Nod factor is a one step process.

The responses to stimulation of plant receptors by Nod factors are eventually mediated by protein phosphorylation and dephosphorylation. They represent the major devices by which the external signaling information encoded by the ligand binding to LRR domain is transformed into an adaptively valuable cellular response (Figure 14.10B). This transformation is implemented by the specific kinases or phosphatases involved directly or indirectly in modifications of target proteins. With a few exceptions, phosphorylation due to these secondary messengers occurs on serine and threonine residues.

Impacts of the host defense reactions.   The early stages of partners' interactions at which the recognition, signaling and infection processes occur, involve the dynamical dialogue between rhizobia and the host factors responsible for the homeostasis in plant cells and tissues. Many of these factors are common to the plant defense from pathogens. The best-documented examples of the legume nodulation functions/genes common to pathogenic interactions include:

1. flavonoid biosynthesis the products of which are required for the induction of rhizobial *nod* genes;
2. chitinase-like proteins participating in processing the Nod factors;
3. receptor-like protein kinases which are broadly represented in the signaling cascades involved in nodule development;
4. early nodulins many of which have close homologues among cysteine-rich PRPs (Pathogen-Regulated Proteins);
5. extensins accumulated in ITs;
6. cytoskeleton rearrangements in epidermal and cortical cells hosting the ITs;
7. *PHAN* and *KNOX* genes induced in nodule primordia that are similar to genes induced in giant cells by nematodes; and
8. oxidative burst in rhizobia-infected tissues.

From this list one can conclude that the reactions exhibited by plant at the entry of rhizobia resemble greatly the hypersensitive response induced by the entry of incompatible pathogens. However, in the case of rhizobia–legume symbiosis the defense-like response does not result in the suppression or killing of the microbe, but in its hosting and in regulation of its performance within the plant organism.

From the rhizobial side, the nodulation is also controlled by a range of genes common to pathogenic bacteria. The impacts of principal rhizobial signal, the Nod factor, on a root have many similarities with the action of elicitors which induce the plant defense reactions. Functional similarities between Nod factors and elicitors are responsible for the above-presented commonalities between nodule development and hypersensitive responses. Moreover, the oligosaccharide moiety of Nod factors is similar to chitin the fragments of which can act as elicitors of plant defense reactions in legumes and non-legumes. Binding sites, which might represent receptors for chitin oligomers have been described in various plants. These reactions include syntheses of phytoalexins, reactive oxygen species and hydrolytic enzymes.

The data presented allow us to speculate that: (1) rhizobia evolved Nod factors to mimic the more ancient plant symbionts which synthesize the chitin-like molecules (e.g., pathogenic or AM fungi); and (2) host-specific modifications in Nod factor allow their differentiation from chitin oligomers thereby avoiding or attenuating the plant defense responses.

In order to avoid the hyper-induction of the defense-like host responses, rhizobia synthesize a range of suppressors. Due to their operation, the rhizobial entry or the treatment with purified Nod factors do not lead to the full-scale hypersensitive response. Highly important role in the dialogue with the host defense systems is implemented by the bacterial surface exopolysaccharides (EPS) and lipopolysaccharides (LPS). The mutations blocking synthesis of EPS1 (acid EPS or succinoglucan) in *S. meliloti* lead to the formation of non-fixing "empty" nodules (Inf⁻Nod⁺). This phenotype is due to high activities of defense reactions that exclude bacteria from host tissues. However, many stages of nodule histogenesis may be stimulated by Nod factors the synthesis of which is retained by the EPS1-defective mutants.

The data on molecular similarities between the mutualistic and pathogenic interactions represent the convincing evidence on the validity of the broad approach to plant–microbe interactions suggested by Anton de Bary more than 120 years ago. His definition of symbiosis as of "prolonged living together of differently named organisms" did not suggest the principal contrasts between the pathogenic and mutualistic interactions. Rather, Anton de Bary preferred to address them as the closely related phenomena, based on similar mechanisms and developed from the common evolutionary roots. The fine analysis of different plant–microbe systems provides us an exciting opportunity to address the background for the ideas forwarded many decades before the birth of molecular biology.

Metabolic integration

In the legume–rhizobial system, as in other symbioses between plants and nitrogen-fixers, two fundamental processes are combined: fixation of N_2 and of CO_2. However, it does not mean that biochemistry of nodule should be reduced to a simple exchange of N and C compounds between plant and bacterial cells. Rather, integrated system of metabolic pathways is formed which produces C and N compounds distributed between the partners. In this joint venture, partners fulfill a range of complementary functions (Table 14.7) responsible for highly efficient cooperation.

The central role in the biochemical machinery of nodule is implemented by nitrogenase enzyme eliciting the reaction:

$$N_2 + 8H^+ + 8e^- + 16\,ATP = 2NH_3 + H_2 + 16\,ADP + 16\,Pi$$

Table 14.7. Key biochemical functions responsible for symbiotic nitrogen fixation in legumes

	Impacts of partners	
Functions	Rhizobia	Legumes
Synthesis of nitrogenase	Encode nitrogenase proteins (NifH, NifD, NifK) and cofactors (*nif* genes)	Possibly, regulate *nif* genes
Defense of nitrogenase from O_2	Allow *nif* gene expression only under anaerobic conditions	Synthesize leghemoglobin (Lb) and construct the diffusion barrier in nodules
Energy supply	Uptake C4-dicarboxylates (mainly, malate)	Donate photosynthates and process them into C4-dicarboxylates with the nodule-specific C-metabolic enzymes
Utilization of N_2 fixation products	Export NH_4^+ into plant cells	Assimilate NH_4^+ with the nodule-specific isoforms of N-metabolic enzymes

Nitrogenase is the huge molecular complex composed of six polypeptides and two cofactors. In spite of its complexity, nitrogenase does not possess a strict substrate specificity since it can reduce triple chemical bonds not only in N_2 but also in cyanides, azides and acetylene. The latter reaction is of great importance since it is broadly used has a fast test for nitrogenase activity.

Oxygen regulation. The basic property of nitrogenase is its oxygen sensitivity: N_2 can be reduced to NH_3 only under anaerobic conditions (<50 nM of O_2). This property constitutes a range of problems for N_2- fixers since nitrogenase reaction is energy consuming and an intensive N_2 fixation is possible only under sufficient energy supply. It may be provided by either respiration or photosynthesis, which are correlated with the consumption or elution of O_2. The resulting "Oxygen Paradox" is resolved using different mechanisms.

In the legume nodules, the most important role in resolving the "Oxygen Paradox" is implemented by the host. It creates within the nodules two barriers protecting N_2-fixing bacteroids from external O_2: structural and biochemical (Figure 14.11). The structural barrier, called the diffusion barrier, consists of a tightly packed layer(s) of thick-walled cells in nodule outer tissues that physically isolate the infected central zone of nodule from the atmosphere. The biochemical barrier consists of the leghemoglobin (Lb) – the protein similar to hemo- and myo-globins. Lb may constitute up to 50% of total proteins in the plant cells containing N_2-fixing symbiosomes and both parts of Lb (polypeptide and heme) are encoded by plant genes. During symbiotic N_2 fixation, Lb implements two functions: it binds O_2 molecules and transports them to symbiosomes.

Functioning of the plant O_2 barriers is reflected in the bacterial regulatory cascade responsible for the transcription of *nif* genes in bacteroids. This cascade involves two steps at which external O_2 tension is monitored: two-component FixL/FixJ system that is induced at moderate decrease of O_2 tension

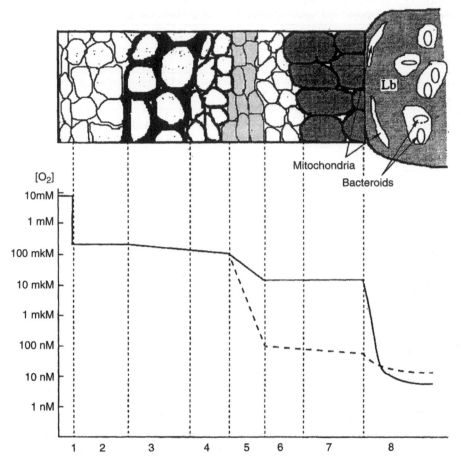

Figure 14.11.   Barriers protecting the nitrogenase from O_2 in the legume nodules (from: Topunov et al., 1998).
Histological zones of nodules are: 1 – epidermis, 2 – outer cortex, 3 – entodermis, 4 – middle cortex, 5, 6 – inner cortex and 7, 8 – central zone with infected cells containing bacteroids.

(5–10 times) and NifA transcriptional regulator that completes the activation upon a sharp decrease of O_2 tension (100–1000 times).

C- and N-metabolism.   Consider the integrated plant–microbe biochemical machinery working in the best-studied indeterminate nodules (Figure 14.12). In order to fuel this machinery, the host should spend up to 30% of products of its photosynthesis, which are consumed by the nodules (that usually constitute less than 1% of the plant biomass). About a half of these products are lost irreversibly in respiration that supply ATP for nitrogenase reaction and C atoms to build the symbiotic structures. Another half is a reversible investment used to assimilate the fixed N and to transport it to shoots.

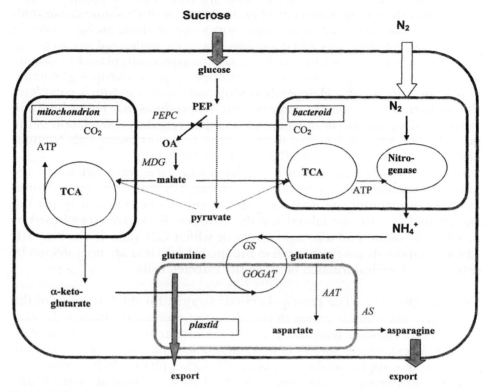

Figure 14.12.   Integration of C- and N-metabolic pathways in the indeterminate legume nodule. TCA – tricarboxylic acid cycle; PEPC – phosphoenol pyruvate carboxylase, MDG – malate dehydrogenase, GS – glutamine synthetase, GOGAT – NADP•H-dependent glutamate synthase, AS – asparagine synthetase, AAT – aspartate amino transferase, PEP – phosphoenol pyruvate and OA – oxaloacetate.

The photosynthesis products supplied to the infected nodule cells are fermented in an unusual way, since its major product is not pyruvate but malate that can be included immediately into the tricarboxylic acid (TCA) cycle. Malate together with some other C4-dicarboxylic acids (succinate, fumarate) are supplied directly into bacteroids which, in contrast to free-living rhizobial cells, do not metabolize hexoses. Among C-metabolic pathways, the TCA cycle dominates in bacteroids and nearly all ATP produced is used to supply energy to nitrogenase. The immediate product of its activity, ammonium (NH_4^+) is excreted into the host cytoplasm since the own enzymes for its assimilation are repressed in bacteroids.

Assimilation of ammonium in nodules is implemented with the active participation of mitochondria and plastids in rhizobia-infected plant cells. This process is characterized by tight cooperation of different metabolic systems and by a strict compartmentalization of the relevant biochemical reactions. The C-skeletons which acquire the amino groups are represented by α-ketoglutarate (α-Kg) excreted by mitochondria as an intermediate of

TCA cycle. α-Kg is transported into plastids where it receives amino group from glutamine (Glu) resulting in its conversion into glutamate (α-Kg + Gln → 2Glu). Later, Glu migrates into cytoplasm to acquire additional amino group (Glu + NH_4^+ → Gln) and the resulting Gln migrates into plastid to donate amino group to α-Kg. This cycle is controlled by plant enzymes glutamine synthetase and NADP · H-dependent glutamate synthase, which completes the primary assimilation of fixed nitrogen. Its products may either be transported to the plant shoots or be involved in the secondary assimilation of the fixed N. The latter is required to enrich the transport forms with N atoms and therefore to minimize the C price of N assimilation.

In the indeterminate nodules (Figure 14.12), the transport forms are mainly glutamine and asparagine that contain 2 N atoms per 4–5 C atoms. However, in determinate nodules (e.g., in soybean or common bean) the assimilation is more specialized and the transport forms are represented by ureids (allontoine and allontoic acids) in which C:N ratio is close to 1:1. These compounds are synthesized in the special cells that are not infected by rhizobia and are intercalated between the infected cells.

Genetic regulation.   The principal genetic strategy for the formation of the integrated biochemical system in nodule is represented by the strictly coordinated expression of symbiotic genes in the interacting plant and rhizobial cells. This expression was at first visualized as synthesis of multiple "late nodulins" that appear in the cytoplasm of infected plant cells just before the onset of nitrogenase reaction. The majority of these nodulins are represented by nodule-specific forms of C- and N-metabolic enzymes that are adapted to the unusual conditions inside symbiotic plant cells: strict anaerobiosis and alkaline pH caused by the excess of ammonium.

An important role in switching on the synthesis of late nodulins is usually assigned to the decreased O_2 tension since these proteins appear after completing the oxygen barriers in nodules. However, it looks highly probable that this coordinated expression is elicited also by some incoming signals from the bacterial cells. An attempt to reveal the mechanisms for such cross-regulation was made using the model of Lb encoding genes in tropical legume *Sesbania rostrata*. In the promoter zone of the gene *Lb3*, two sites were found to be potentially specific for the bacterial gene regulators (Figure 14.13). In *Azorhizobium caulinodans* (the *S. rostrata* symbiont) the proteins which bind these sequences were identified. Moreover, the bacterial mutations that knock-out these proteins lead to a decreased Lb synthesis and N_2-fixing activity.

May a reverse flow of signals from host to bacterial cells be involved directly in the regulation of the biochemical machinery of nodules? Theoretically, it looks probable but the experimental evidence is still poor. Some preliminary indications for such signaling were obtained for bacterial *dct* genes controlling the uptake of C4-dicarboxylates by bacteroids. These genes encode for succinate permease DctA and a two-component regulatory system DctB/DctD responsible for transcriptional activation of *dctA* gene. In the free-living rhizobia, this system is activated by C4-dicarboxylates and the disruptions

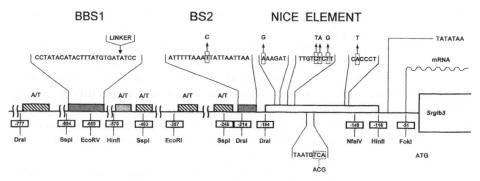

Figure 14.13.   Complex structure of promoter of the leghemoglobin (*Lb3*) gene in *Sesbania rostrata* (from: de Bruijn et al., 1994). BBS1 and BS2 – sites for binding the potentially regulatory proteins from bacterial partners; NICE (Nodule Infected Cell Expression) element shares CACCC motif with the promoter region of mammalian β-globin genes.

of regulatory genes (*dctB* or *dctD*) result in the complete loss of DctA synthesis. However, under symbiotic conditions, inactivation of these regulators results only in a partial loss of DctA synthesis suggesting that some alternative activator(s) of *dctA* expression are used in nodules. Involvement of plant signals in this activation is highly probable and their nature awaits its researchers.

Symbioses with N₂-fixing cyanobacteria

Some nitrogen-fixing cyanobacteria (*Nostoc, Anabaena*) form with various plants the symbioses that differ in many respects from the legume nodules. Many of these differences are caused by the ability of cyanobacteria to fix N_2 in free-living state and to support this fixation with products of their own photosynthesis. The mechanisms of free-living N_2 fixation are used in symbioses, but here it is controlled by the hosts due to the localization of cyanobacteria in special symbiotic organs. In the water fern *Azolla*, cyanobacteria are located within the leaf cavities, in the gymnosperm *Cycas* – within the intercellular spaces of coralloid roots, in the gymnosperm *Gunnera* – within the cells in the glands at the bases of leaf petioles.

Microbe ontogeny

When transferred into N-free medium, 5–10% of cells in the *Nostoc* or *Anabaena* filaments are differentiated into heterocysts that, like the rhizobial bacteroids, are devoid of reproductive ability. Heterocysts are enlarged and have thick walls blocking the oxygen diffusion inside the cells. Due to the arrest of photosynthesis, microaerobic conditions are reached inside the heterocysts permitting the nitrogenase synthesis. Another pathway of cellular differentiation in *Nostoc* is represented by motile hormogonia – filaments consisting of small cells with gas vacuoles.

These variants of cellular differentiation are also used by cyanobacteria during symbioses with plants, but herein the microbial ontogeny is combined with the development of specialized plant structures and with very high N_2-fixing activity required to supply plants with nitrogen. During the pre-infection contacts, massive formation of motile hormogonia is induced in *Nostoc* that colonize actively the interiors of *Gunnera* glands. This process may be elicited by host: active differentiation of hormogonia was observed in *Nostoc* treated with matrix extracted from the *Gunnera* glands and containing polysaccharides and proteins. Within these glands cyanobacteria propagate actively and infect the plant cells. In the points of tight cellular contact, the plant cell walls are lysed and after the entry of cyanobacteria their integrity is restored. Within plant cytoplasm the cyanobacteria are encapsulated into the symbiosome membranes and later up to 80% of cells are converted into N_2-fixing heterocysts.

Biochemical integration

The cyanobacteria are biochemically versatile organisms capable of using either autotrophicaly fixed or combined forms of C and N. This versatility is very useful for differentiation of symbiosis (Figure 14.14). In heterocysts of free-living cyanobacteria, glutamine synthetase (GS) is expressed along with nitrogenase and the product of primary assimilation of ammonium (glutamine) is exported to the vegetative cells where the amino groups are donated to α-Kg under the control of GOGAT. The major source of ATP for nitrogenase reaction is the photosystem I that retains its activity in heterocysts, while the photosystem II and Calvin cycle are repressed.

The heterocysts formed in cyanobacteria within the host plants, differ from free-living heterocysts. In symbiosis with *Gunnera* the GS activity is lost

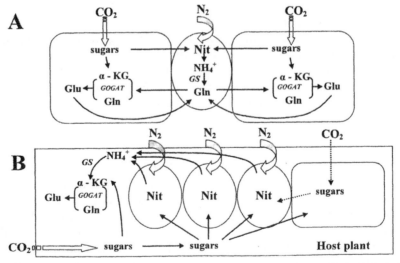

Figure 14.14. Integration of C- and N-metabolic pathways in *Nostoc* under free-living conditions (A) and in symbiosis with *Gunnera* (B). Nit – nitrogenase, Gln – glutamine, Glu – glutamate, α-KG – α-ketoglutarate, GS – glutamine synthetase and GOGAT – glutamate synthase (completed by E. Dolgikh).

by the *Nostoc* heterocysts and the ammonium is exported into the plant cells wherein it is assimilated *via* α-Kg. The principal sources of C and energy for heterocyst are the sugars imported from the plant cells since own photosynthetic activity in symbiotic cyanobacteria is decreased sharply (Figure 14.14B).

A unique property of regulation of N_2 fixation in cyanobacteria is constituted by the programmed rearrangements of *nif* gene structure in heterocycts. In the vegetative *Nostoc* or *Anabaena* cells one of the nitrogenase structural genes, *nifD* is interrupted by 11–23 kb DNA fragments, and integrity of this gene is restored during the heterocyst differentiation due to precise excision of these segments. This process is controlled by endonuclease XisA encoded by the excised fragment that may be found in the heterocyst cytoplasm in the plasmid form. It is interesting to note that the mutations arresting this rearrangement resulted in zero nitrogenase synthesis but do not affect the heterocyst formation. Moreover, the mutants defective in heterocyst differentiation can fix N_2 under anaerobic conditions, under a good supply of sugars. Therefore, processes of cellular differentiation and the induction of nitrogenase activity may be uncoupled in these cyanobacteria mutants. However, in wild-type cyanobacteria these processes are well coordinated: transcription activator NtcA was identified that switches on the genes for both heterocyst differentiation and nitrogenase synthesis.

A range of differences between legume–rhizobia and plant–cyanobacteria symbioses were revealed with respect to structural and functional modifications of partners' cells (Table 14.8). However, some general principles of their operation are evident too. In both, symbioses N_2 fixation is combined with photosynthesis resulting in integrated partners' C/N biochemical pathways.

Table 14.8. Comparison of two types of N_2-fixing plant–microbe symbiosis

	Symbioses between	
Properties	Rhizobia and legumes	Cyanobacteria and plants
Specificity of interaction	Narrow: great majority of rhizobia inoculate only the restricted groups of legumes	Wide: *Nostoc* inoculates different types of plants
Action of plant signals on microbes	Activation of *nod* genes (by plant flavonoids)	Massive formation of motile hormogonia
Differentiation of nitrogen-fixing cells	Only in plants (bacteroids)	In plants and in free-living state (heterocysts)
Rearrangements of *nif* genes prior to N_2 fixation onset	Not shown	Typical
Source of ATP for nitrogenase reaction	Respiratory chain	Photosystem I
Protection of nitrogenase from oxygen	Mainly by plants (leghemoglobin and diffusion barrier in nodules)	Mainly by bacteria (cell walls of heterocysts)

In both systems, energetic supply of the microbial N_2-fixing machinery as well as assimilation of its products are implemented mainly by the host plants. Deep and irreversible cellular differentiation before the onset of N_2 fixation occurs in rhizobia (bacteroids) and in cyanobacteria (heterocysts). A pronounced allocation of major functions occurs between the partners' cells: synthesis of C skeletons accepting fixed N is separated from N_2 fixation and from primary N assimilation. Due to high intensity of N_2 fixation, microbial cells minimize the set of fulfilled functions and excrete ammonium that is a first free product of nitrogenase reaction.

Mycorrhizae

Mycorrhizae represent the most broadly distributed strategy of plant–microbe symbiosis. They are formed by plants and the root-colonizing fungi wherein a part of the fungal partner (mycobiont) is inside the root while the other part is outside of it. Based on the anatomy of intra-radical portion of mycobiont, the mycorrhizae are classified into endomycorrhizae (fungal hyphae penetrate the plant cells forming the specialized sub-cellular structures) and ectomycorrhizae (EcM) (distribution of hyphae is restricted to the intercellular spaces). The universal form of endomycorrhizae is represented by arbuscular mycorrhiza (AM) formed by the majority (75–90%) of terrestrial plants. The EcM is restricted to trees and shrubs representing angiosperms (mainly, dicots) and gymnosperms (Table 14.9).

Separation of the mycobiont into inter-radical and extra-radical parts reflects the basic function of mycorrhizae as an intermediate between the plant and the soil. This function is of global importance since only some plants (mainly short-living herbs and aquatic forms) are able to cover their demands of mineral nutrients and water without the aid of a mycobiont. Under artificial conditions, majority of plants can develop normally without a fungal partner, however under field conditions severe competition for nutrients occurs within the plant communities. Under these conditions, the independent survival of plants is hardly possible leading to their ecologically obligatory dependency on the mycobionts. In the orchids, this dependency is even more deep since many of these plants cannot start their ontogeny without the aid of a specific fungus which is indispensable for the embryogenesis and seed germination.

Development

Being multi-cellular eukaryotes, the mycorrhizal fungi undergo the symbiosis-specific morphogeneses acquiring the structures that cannot be developed by free-living fungi. The plants' input into building the symbiotic structures is represented by organization of specialized sub-cellular contacts (AM) or modifications in root anatomy (EcM). The most specialized symbiotic development is undergone by orchids, in some of which the roots are reorganized into the coralloid structures in which the assimilatory function is substituted by the function of hosting the mycobiont.

Table 14.9. Comparative analysis of the major types of mycorrhiza

Partners	Characters	Arbuscular mycorrhiza	Ectomycorrhiza
Host plants	Diversity	75–90% of terrestrial plants (including about 225 000 angiosperm species)	Dicots and gymnosperms (about 5000 species, mainly trees and shrubs)
	Distribution	Worldwide	Mainly, temperate zones
	Nutrients obtained from fungi	Mainly phosphorous, accompanied by nitrogen, potassium and calcium	Mainly nitrogen, accompanied by phosphorous, potassium and calcium
Mycobiont	Diversity	Glomeromycota (family Zygomycetes, order Glomales), not more than 200 species. The best-studied genera: *Glomus, Gigaspora*	Basidiomycetes ($\approx$ 5000 species), Ascomycetes (400–500 species). The best-studied genera: *Amanita, Cortinarius, Hebeloma, Pisolithus*
	Dependency on symbiosis	Genetically obligatory: cannot complete living cycle independent of the host	Facultative or ecologically obligatory: some species implement sexual processes (formation of dikaryons, meiosis, development of fruiting bodies) only in symbiosis
	Sexual processes	Reduced	Usually present
	Major structures formed during symbiosis	Appressoria (attachment of germination tube to the roots), infective hyphae (intercellular growth in outer and inner root cortex), arbuscules (sub-cellular structures within the cortical cells), external hyphae with the spores	Mantle (multiple layers of hyphae covering the root), Hartig net (dense network of hyphae colonizing the outer cortex), extra-matrical mycelium with fruiting bodies

Arbuscular mycorrhiza (AM)

Ontogeny. AM development involves several processes: (1) pre-infection, (2) formation of intercellular (intra-radical) mycelium, (3) development of intracellular symbiotic structures and (4) development of sporulating extra-radical mycelium (Figure 14.15). It starts from the germination of the fungal spores and braching of the growing germ tube elicited by the plant exudates. According to recent data, the "branching factor" secreted by *Lotus japonicus* roots to elicit the pre-infection branching in *Glomus* germ tube is idenitified as 5-deoxy-strigol. Interestingly, it belongs to the group of root exometabolites involved in eliciting the seed germination in parasitic plants (*Striga, Orobanche*) giving one more example of the universal regulatory factors involved in deferent types of plant symbiotic interactions. A rapid growth of the germ tube from the spore toward the root is caused by the active chemotaxis and is finalized by the attachment of appressoria to the root surface. However, if the germ tube does not reach a root within 5–7 days after germination, process may be reversed: germ tube aborts and nutrients returned

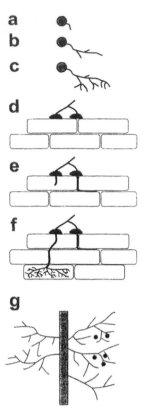

Figure 14.15. Development of arbuscular mycorrhiza (AM) (Marsh and Schultze, 2001). Developmental stages: (a) spore germination; (b) pre-infection growth; (c) pre-infection branching; (d) appressorium formation; (e) penetration and intra-radical growth; (f) arbuscule formation and (g) growth of external hyphae and spore formation.

into the spore. The huge size and high metabolic potential of spores (that may be 0.5 mm in diameter and contain several thousands of nuclei) permit them to repeat several times their attempts to reach the root.

After attachment to the root, the infective hyphae start to grow from the appressorium into the cortex colonizing its outer and inner layers, but never crossing the pericycle. After establishing a network of intercellular hyphae, fungus starts to develop two types of structures: intracellular and extra-radical. The fully developed highly branched arbuscules fill nearly the whole volume in infected cortical cells.

A pronounced differentiation is characteristic of the plant cells containing arbuscules. The vacuole is reduced or degraded, amount of α-tubulin is increased, the nucleus is deformed (sometimes it becomes surrounded by the branches of arbuscule) and chromatin is decondensed due to a high transcriptional activity. The amount of endoplasmic reticulum and Golgi vesicles is greatly increased ensuring the biogenesis of periarbuscular membranes. The latter is separated from the fungus surface by a structurally differentiated interface that contains polysaccharides and enzymes of the mixed plant and fungal origins. The basic feature of the arbuscules is to sustain the physical connection with intercellular hyphae. However, arbuscules are short-living structures: every 4–7 days they are digested by plant cells and the novel arbuscule may be developed from the nearby hyphae.

The final developmental stage in AM is represented by the formation of the system of extra-radical hyphae which are of crucial importance for both the partners. For the host, the trophic functions of extra-radical mycelium (assimilation of soil nutrients) are most important, while for the mycobiont the reproductive function (formation of spores) is indispensable. The extra-radical hyphae are variable in their morphology: the more thick (10–20 mkM in diameter) first-order (runner) hyphae exit immediately from the roots and the more thin (2–5 mkM) second-order (absorptive) hyphae branch from the runner hyphae. The extended system of extra-radical hyphae (70–80 m per 1 m of root) permits the mycobiont to explore the soil within several millimeter around the roots actively absorbing the soil nutrients. An important property of the extra-radical hyphae is represented by their ability to penetrate the neighboring roots and to form an underground network of hyphae connecting different members of a plant community.

The development of AM fungi within the root is strictly controlled by the host and the mycorrhizal development should be considered as a result of complex dialogue between the fungus and the host defense systems. The defence-like reactions induced within the root cortex during AM development include: modification of the cell walls, synthesis of phytoalexins, accumulation of

Table 14.10. Defence-like reactions induced by the development of symbiotic fungi in plant roots (from: Provorov et al., 2002)

Plant reactions	AM fungi (*Glomus*)	Root pathogens (*Rhizoctonia*)
Modification of the cell walls	Weak thickening (without a pronounced structural modification); phenolics are not accumulated	Strong thickening with the papillae formation; phenolics are accumulated
Activity of phenylpropanoid pathway*	Low; PAL, CHI and CHS are active during root colonisation, PAL and CHS - during arbuscule formation; no activation of IFR	High; synchronic induction of PAL, CHS, CHI and IFR
Synthesis and accumulation of callose	Absent at early stages, low level during the formation of arbuscules	Intensive at the early and late stages of infection
Synthesis of peroxidases, chitinases, glucanases	At early stages only	At all stages of infection
Synthesis of pathogen-regulated protein PR-1	Low level in the vicinity of arbuscules only	High level without a specific sub-cellular localization

*PAL – phenylalanine ammonia-lyase, CHI – chalcone isomerase and CHS – chalcone synthase and IFR – isoflavone reductase.

callose and of some pathogen-regulated (PR) proteins, including peroxidases and lytic enzymes (Table 14.10). However, the plant defence from fungal pathogens differs significantly from the reactions for AM development. During *Glomus* growth the intensity of plant reactions inside roots is low, they are less prolonged and highly differentiated in time and space compared to pathogenesis. It looks like the plant defence reactions are repressed by some fungal signals required for the stable co-existence of partners.

Developmental genetics.   The analyses of plant control over AM development started in the late 1980s when it was demonstrated that the legume mutants defective in early nodule development are often arrested in AM formation. This discovery enabled the researchers to enter the area of AM genetics that was previously closed because: (1) the direct isolation of the plant mutants defective in AM development is complicated by the absence of appropriate selective techniques; and (2) genetic analysis of AM fungi is hardly possible since they cannot grow of *ex planta* and are devoid of sexual reproduction.

The analysis of the numerous legume nodulation mutants (Table 14.3) allowed researchers to identify a range of *Sym* gene alleles with double (Nod⁻ Myc⁻) phenotypic effects. The mutations arresting AM development segregate into two groups: Myc^{-1} and Myc^{-2}. The first one is composed of mutants in which AM development is blocked after the formation of appressoria and the infective hyphae are aborted immediately after penetrating the epidermis. Phenotype Myc^{-1} was revealed in the majority of Hac⁻ mutants (defective in the root hair curling occurring after the rhizobial inoculation). The Myc^{-2} mutants form appressoria and intercellular mycelium after *Glomus* inoculation, however arbuscules are not developed. Myc^{-2} mutants were identified in pea among the Itf⁻ mutants arrested at the IT growth: it was also found in some Bar⁻ mutants defective in the rhizobial endocytosis.

A detailed classification of the Myc^{-2} mutants was suggested by Marsh and Schultze (2001) using the model legume *Lotus japonicus*. They differentiated these mutant phenotypes into Coi⁻ (infective hyphae initiated from appressoria pass the epidermis but cannot enter the outer cortex), Ici⁻ (infective hyphae cannot enter inner cortex) and Ard⁻ (arbuscule differentiation starts but only the truncated arbuscules can be formed). The Fix⁻ pea (*Pisum sativum*) mutants forming the histologically normal nodules but defective in N_2 fixation are usually not arrested in AM formation, although a pronounced decrease in the amount of infective hyphae or arbuscules ($Myc^{+/-}$ phenotype) may be revealed.

Further similarities between AM and nodulation were revealed at the molecular level. Mycorrhization stimulates *de novo* synthesis of a range of proteins (mycorrhizins), which are absent in non-inoculated roots. These products comprise 4–5% of the total root proteins, and some of them are common for AM and nodule development. They include proteins of PBM and periarbuscular membranes, early nodulins (ENOD2, ENOD11, ENOD12, ENOD40) and even leghemoglobin (LB), which is weakly synthesised in the plant cells containing arbuscules. A range of common genes is involved in signalling cascades regulating root nodulation and development of AM in legumes (Figure 14.16).

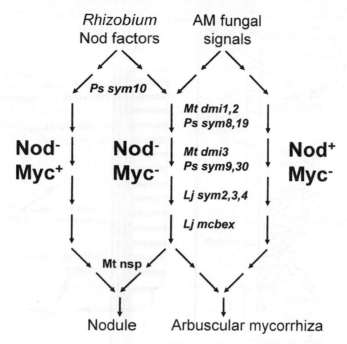

Figure 14.16. Common stages in the plant signal cascades regulating the development of root nodules and arbuscular mycorrhiza in legumes (Marsh and Schultze, 2001). Phenotypes: Nod – nodule formation and Myc – formation of arbuscular mycorrhiza. Symbiotic genes originated from the model legumes: Ps – *Pisum sativum*, Lj – *Lotus japonicus* and Mt – *Medicago truncatula* (genes are specified in Tables 14.3 and 14.6).

The genetic and molecular commonalities between nodules and AM suggest that in the course of legume evolution some genes have been recruited into the nodulation program from the much more ancient AM developmental program. This hypothesis looks probable since according to the paleontological data AM appeared together with the first land plants (400–500 million years ago) while the nodulation appeared much later, with the origination of Fabaceae family (60–70 million years ago).

Ectomycorrhiza (EcM)

In contrast to AM fungi, the majority of EcM fungi are able to grow without hosts following the free-living (saprotrophic) strategy. However, in cooperation with the tree hosts, EcM fungi implement the symbiotic development that improves the adaptive potential of both the partners (Figure 14.17). This development starts from the pre-infection processes that may involve either induction of dormant spore germination by root exudates or the root-directed growth of hyphae from the soil mycelium. There were several attempts to identify the plants' signals, which induce the pre-infection specifically. Using the model system "*Eucalyptus* (host)–*Pisolithus* (mycobiont)", the candidate compounds were suggested including zeatin (which induced

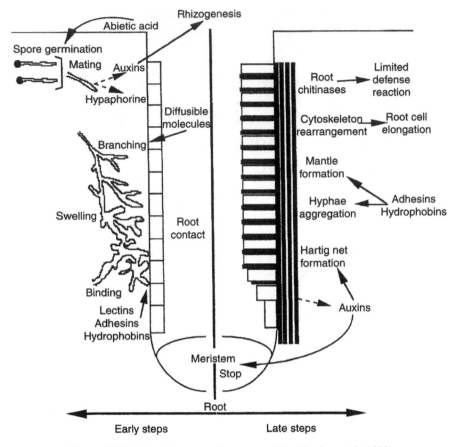

Figure 14.17.   Development of ectomycorrhiza (Barker et al., 1998).

a specific branching of fungal hyphae near the root) and flavonol rutin that can induce the hyphae growth in very low concentrations.

The hyphae attracted by the host roots grow and branch actively on the root surface and as a result a multilayer mantle is formed covering the root. From mantle, two novel structures are started: extra-matrical mycelium and the intra-radical mycelium called Hartig net. The extra-matrical hyphae are functionally analogous to the extra-radical hyphae of AM fungi in implementing trophical and reproductive functions. However, the EcM extra-matrical hyphae are much more extended: they run for several meters from roots and may form very dense mycelium or even mats in the surrounding soil. The reproduction implemented by EcM fungi usually involves sexual processes often occurring within specialized fruiting bodies (known as mushrooms). In many EcM fungi, the sexual reproduction is implemented only in association with a host.

Although the tree hosts do not form special organs in response to the penetration by EcM fungi, a pronounced modification of root structures usually occurs. They include formation of multiple thick and short lateral roots (that increase the surface available for fungal colonization) and reduction of root hairs (which are functionally substituted by extra-mantle mycelium). As a result, in many trees the own assimilatory activity of root is decreased and substituted by the functions of hosting of and metabolic exchange with the mycobiont.

The molecular basis for the EcM-induced root modifications is still not clear, however it is widely accepted that a sufficient role belongs to phytohormones synthesized by fungi. The EcM-fungi actively synthesize the auxin-like compounds and their production is increased in association with the hosts. In the fungus *Hebeloma cylindrosporum,* a mutant with an enhanced IAA production was isolated that is characterized by severalfold increase in the rate of colonization of the *Pinus pinaster* roots.

During interaction, both EcM partners synthesize a range of novel proteins demonstrating that the differential gene expression is induced by the signal exchange, which underlies the symbiosis formation. Among the symbiosis specific fungal proteins, the best studied are hydrophobins – small cysteine-rich proteins that are accumulated near the contacts with plant cells and in the fruiting bodies. An important role in the cell-to-cell interactions is supposed for these proteins. The other group of fungal proteins up-regulated in the EcM-fungi are represented by symbiosis-regulated acid proteins (SRAPs) for which the role in surface interactions is suggested due to the presence of cell wall signal peptides with the Arg-Gly-Asp (RDG) motif specific for many animal and microbial adhesins. Structural genes for hydrophobins and SRAPs are transcriptionally activated during EcM formation, however, no impact on this activation was detected after the treatment of fungi with purified root exudates.

As in the case of AM, a range of localized defence-like reactions were detected in ECM roots including accumulation of phenolics and flavonoid phytoalexins, and the synthesis of lytic enzymes. However, these reactions are weak and do not result in the suppression of fungal growth, but in its organization and in balancing the co-ordination of partners' developmental processes. The phylogenetic studies made by Hibbett and co-workers (2000) did not demonstrate any close relationships between EcM fungi and pathogenic Basidiomycetes, while close relationships of EcM fungi with different wood- and litter-decomposing saprotrophs were evident.

Metabolism

The adaptive gains obtained by plants and fungi from AM and EcM formation are based on the active exchange of nutrients. Due to the ability for intensive osmotrophic nutrition, mycobionts collect the soil compounds containing the whole spectrum of macro- and micro-elements and provide them to the host. The latter covers the energy and carbon demands of the mycobionts due to the export of the products of photosynthesis. Fungi usually receive these products in the form of hexoses the catabolism of which provides energy for osmotrophy and for the development of symbiotic structures

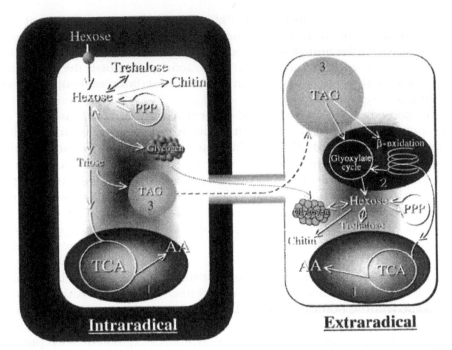

Figure 14.18.   Major stages of carbon metabolism in arbuscular mycorrhizal fungi (Bago et al., 2000). 1 – mitochondria, 2 – glyoxysome, 3 – lipid bodies, PPP – polyphosphates, TCA – tricarboxylic acid cycle, AA – aspartic acid, TAG – triacylglycerides.

as well. Two common features in AM and EcM metabolic systems should be noted.

The first feature is the marked differentiation of the biochemical activities among extra- and intra-radical parts of mycobiont (Figure 14.18). Specifically, the sites of metabolic exchange with the host may be located many meters aside from the sites of extraction of nutrients from the soil. This fact assumes that the special mechanisms for bidirectional transport should be involved to organize the flows of nutrients along the hyphae.

The second feature is the pronounced separation of the fungal and plant cells by the interface that involves the plasma membranes and cell walls from both the partners. The symbiosis-specific mechanisms for traffic of metabolites through these borders require the tight coordination of enzymatic activities in the contacting partners' cells.

Arbuscular mycorrhiza (AM)

Inoculation by AM fungi improves the plant provision with the whole spectrum of soil-borne nutrients, however the most pronounced effect pertains to the phosphorous (P) nutrition. This function is very important for the plants because the major part of P pool is represented by unsoluble compounds that cannot be assimilated by roots. The additional amounts of P obtained by plants due to inoculation with *Glomus* are several times more than additional amounts of N and K. Therefore, special mechanisms for uptake from soil and for translocation into host are available for phosphorous with respect to other nutrients.

Uptake of P from soils occurs in soluble forms (mainly $H_2PO_4^-$ usually addressed as P_i) and involves several transport systems some of which are characterized by their extremely high affinity for P_i. After its transportation inside the hyphae, P_i is included into polyphosphates represented by linear polymers of P_i (catalyzed by the polyphosphate kinase). The granules rich in polyphosphates and in the P-containing esters are formed and move along the hyphae. For an organized movement, these granules are often packed into cylindrical vacuoles the transport of which is facilitated by tubulin fibrils.

After reaching the arbuscules, P compounds are hydrolyzed by the active phosphatases and the released P_i crosses the partners' interface. The crucial role of arbuscules in donating P_i to the host was proved by the dependence of the arbuscule development on the soil P status. After the inoculation of the plant with AM fungus, arbuscules are formed actively under the limited or moderate P_i supply, while the absence or excess of P_i block the arbuscule formation.

In order to support the mycobiont activity, the reverse flow of C compounds from host occurs mainly *via* interface between plant cells and intercellular hyphae. The major part of carbon pool obtained by the fungus is included into the lipids, glycogen and trehalose (Figure 14.18). Lipid and glycogen are formed as granules that are suitable to be transported from intra- to extra-radical mycelium. The major part of plant-derived C compounds is used to provide energy for osmotrophic processes and for developing the fungus, as well as for assimilating N compounds absorbed from the soil.

An interesting hypothesis concerning the role of AM in the plant N nutrition was suggested from the data demonstrating that the majority of AM fungi harbor obligatory endosymbiotic bacteria closely related to the genus *Burkholderia*. In these bacteria, the *nif* genes encoding for nitrogenase structure were found and the *in planta* expression of these genes was demonstrated that may result in the fixation of molecular nitrogen. These data may confirm the old observations suggesting the N_2-fixing activity of some AM fungi that was later rejected due to the inability to find *nif* genes in the eukaryotic genomes.

Ectomycorrhiza (EcM)

In contrast to AM, the basic output of EcM formation is to improve the host nitrogen nutrition. The major biochemical activity of the extra-matrical hyphae is represented by the uptake of ammonium implemented by the high affinity transporters. The fungus not only absorbs free ammonium (that is usually limited in soils) but also extracts it from organic soil compounds destroyed with the help of powerful extra-cellular enzymes. With the example of soil peptides, it was demonstrated that mobilization of soil N is a multi-step process including hydrolysis of soil polymers, uptake of small molecules by fungus, assimilation of the obtained N and its transfer through the interface to plant cells. A sufficient role in the decomposition of soil matter by EcM fungi is implemented by extra-cellular peptidases, fatty acid esterases (degradation of cuticles), tyrosinase, peroxidase and polyphenol oxidase (oxidation of phenolic acids and tannins), lignase and different enzymes for degradation of plant cell wall components. After the nutrients are absorbed from the soil, they are accumulated in the mantle and are transported into the Hartig net. From there, the nutrients move to the sites of metabolic exchange by using vesicles, vacuoles and tubules, the transportation of which is facilitated by a reduction of septation within the internal part of mycobiont.

The efficiency of partners' metabolic exchange is improved due to invaginated and extended fungal plasma membranes that amplify the contact zone with the root cells. The plant photosynthates reach the fungus-infected root in the form of sucrose that is hydrolysed into hexoses by the plant-encoded acid invertase located at the interface. Later, the hexoses are transported into the hyphae by special fungal transporters. Absence of such transporters in the contacting plant cell makes the C flow unidirectional. Products of plant photosynthesis are used to supply energy and nutrients for the mycobiont growth and to bind the assimilated N that is donated to the host mainly in the form of amino acids.

Orchid mycorrhiza

Since its discovery more than a century ago, orchid mycorrhiza still remains a mysterious biological phenomenon that awaits its investigators. Nearly more than 20000 orchids are obligate mycotrophs in which the embryogenesis and the seed germination cannot be implemented without a fungal partner. A unique strategy of plant–fungal metabolic relationships is represented by orchid mycorrhizae wherein the carbon flow from fungus to plants was registered using the isotopic techniques. A sufficient part of this carbon may have originated from other plants since the majority of orchid mycobionts are forming ectomycorrhizal or even pathogenic associations with the nearby non-orchid plants.

In some orchids, the carbon flow from fungi is restricted to the early developmental stages, while the adult plants are capable of photosynthesis and receive C independently of mycobiont. However, many groups of orchids are known (for example, *Cephalanthera, Corallorhiza, Epipogium, Neottia*) that have lost irreversibly the ability to photosynthesize and receive their carbon from the mycobiont throughout their life. An intensive digestion of the fungal hyphae inside the plant cells was described in these endomycorrhizal systems. This is why some authors address the orchids as fungi-eating plants or as the parasites of their mycobionts. Interestingly, the non-photosynthetic orchids may be huge in size and widely distributed: for example, *Galeola* – a liana grown in India, Indonesian islands and Northern Australia has a stem length up to 40 m.

Defensive symbioses

One of the basic strategies in plant–microbe interactions is to protect the plants against their natural enemies: pathogenic microbes, pests and herbivores. These interactions differ in their mechanisms from N_2-fixing symbioses and mycorrhizae: the defensive symbioses do not involve the bilateral metabolic exchange since the microbes usually do not donate nutrients to the plant metabolism. Rather, the host covers trophic and energetic costs of the microbes, which implement a range of functions resulting in a decreased damage to the plants from biotic stresses. A potential to be used in the biocontrol of these plant enemies attracts a great attention towards the defensive symbioses.

Epiphytic strategy: plant growth-promoting rhizobacteria

Very efficient defense of plants from pathogens may be implemented by the microbes colonizing the hosts' outer surfaces. The best-studied examples are Plant Growth-Promoting Rhizobacteria (PGPR), especially *Pseudomonas* (*P. fluorescens, P. chlororaphis, P. putida*), *Serratia* (*S. marcescens*) and *Bacillus* (*B. cereus*) species. Many of these bacteria are capable of preventing the plant attack by pathogenic fungi (*Fusarium, Trichoderma, Verticillium* and others) very effectively. Using wide spectrum of approaches, Lugtenberg and co-workers (2001) revealed several mechanisms that may be involved in host defense.

The best-studied mechanisms are represented by the direct suppression of pathogens by the bacterial antibiotics. For example, many *Pseudomonas* strains produce phenazins active against *Fusarium*; the immediate impacts of phenazine-1-carboxamide (PCN) on the biocontrol activities were proved genetically, biochemically and microbiologically. Indeed, the bacterial mutations leading to PCN$^-$ phenotype result in the loss of biocontrol activity. Genes for PCN synthesis pathway are transcriptionally activated in the host rhizosphere. In addition to antibiotics, PGPR can excrete the toxic volatile metabolites, for example the involvement of hydrogen cyanide in the antifungal activities was demonstrated *via* biochemical assays implemented during the mixed PGPR–*Fusarium* cultivation.

The direct suppression of pathogens by PGPR is best expressed when bacteria exhibit a high root-colonizing activity. Due to active root colonization, PGPR may exclude the pathogens from the plant rhizosphere: being colonized by high number of PGPR cells, roots become less-accessible to the pathogens that do not find enough space and nutrients to develop an infective inoculum. This is why the genes encoding the bacterial surface components required for the root adhesion (for example, lipopolysaccharides and flagella) are very important in this type of defense. Inactivation of these genes usually results in the loss of plant-protecting activity by PGPR strains.

The microscopic observations demonstrated that the suppression of *Fusarium* may be correlated to *P. chlororaphis* attachment not only to the root surfaces but also on the pathogen hyphae. Due to this attachment, some

PGPR strains may implement their biocontrol functions acting as hyper-parasites or even as the predators of the phytopathogenic fungi. This mechanism of suppression may be correlated to the production of bacterial enzymes destroying the pathogen hyphae. For example, some *Serratia, Pseudomonas* and *Bacillus* strains produce the extra-cellular chitinases that inhibit the pathogen at different stages of its development including the conidia germination and development of hyphae on the roots.

However, sometimes the biocontrol activities of PGPR do not correlate with the colonization of the host roots, and plant protection is implemented by the local bacterial colonization. In this case PGPR inoculation can lead to Induced Systemic Resistance (ISR) that makes the plant non-accessible to root pathogens. The ISR effects may be elicited by the PGPR cells attached to roots or penetrated their outer tissues. Some molecules produced by PGPR (LPS, flagella components, siderophores) may elicit the ISR response even in the absence of living bacteria. In contrast to the standard Systemic Acquired Resistance (SAR) reaction, ISR induction does not need a hypersensitive response resulting from the pathogen invasion. However, after switching on the ISR many components of SAR reactions (including salicylate and jasmonate) are operating in plants.

The activities of plant-protecting PGPR are implemented under the direct control of the host. First, plant provides nutrients and energy required for the PGPR development in rhizosphere. It was demonstrated that the intensive growth and expression of the biocontrol-related genes are induced by root-excreted organic acids (succinate, malate, fumarate), while mono- and disaccharides are not optimal for the host-beneficial activities of PGPR.

Under field conditions, the biocontrol activities are expressed by PGPR strains in close association with many other rhizospheric microbes, which may benefit the plant growth due to mechanisms not involved directly in the antagonism with pathogens (Figure 14.19). For example, many PGPR strains improve the plant growth by providing additional nutrients. A good example is N_2 fixation by *Azospirillum* and *Herbaspirillum* strains isolated by Dobereiner and Day (1976). The dissolution of phosphates by some *Bacillus* strains may contribute sufficiently to the growth of plants in the P-deficient soils, especially in the absence of appropriate mycorrhizal fungi. Moreover, development of AM and EcM may be improved by some PGPR resulting in the formation of the tripartite "plants–fungi–bacteria" symbioses.

The improvement in plant nutrition may also be achieved by the bacteria-mediated synthesis of phytohormones, mainly of auxins (IAA). It was revealed in a range of rhizospheric bacteria belonging to *Azospirillum, Pseudomonas, Enterobacter,* and the direct involvement of bacterial auxins in the stimulation of root growth was demonstrated using the IAA-deficient *A. brasilense* mutants. It is often suggested that the major substrate used by PGPR to synthesize IAA is represented by the tryptophan secreted by host roots. For example, very active IAA synthesis was detected in *Pseudomonas* strains grown in rhizosphere of radish plants that excrete 30–100 times more tryptophan than wheat or tomato.

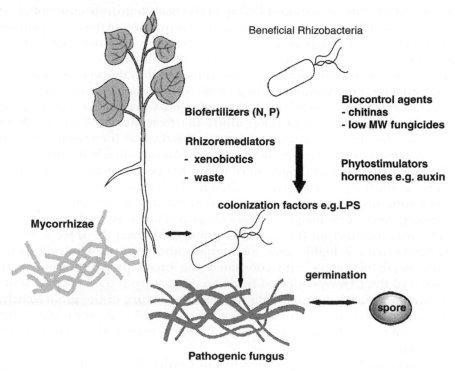

Beneficial Rhizobacteria

Biocontrol agents
- chitinas
- low MW fungicides

Biofertilizers (N, P)

Rhizoremediators
- xenobiotics
- waste

Phytostimulators
hormones e.g. auxin

colonization factors e.g.LPS

Mycorrhizae

germination

spore

Pathogenic fungus

Figure 14.19. Beneficial functions of plant growth-promoting rhizobacteria (Lugtenberg, 2004).

However, some *A. brasilense* strains may produce IAA effectively in the tryptophan-deficient cereal rhizospheres by using the alternative biochemical pathways.

Endophytic strategy: toxin-producing fungi

Many plants support in their tissues a variety of endophytic microbes some of which are symptom-less endophytes, while others fulfill different beneficial functions. They include N_2 fixation (*Azoarcus* and *Acetobacter*), synthesis of phytohormones (*Gibberella*) and of plant-protective substances (Clavicipitaceae fungi, belonging to Ascomycota). The latter are of great agricultural interest since they are living in cereals including grain (wheat, rye, millet) and fodder (ryegrass, fescue) crops. The fundamental interest in this group of fungi is caused by the mechanisms of symbioses, which involve tightly interconnected elements of mutualism and antagonism.

The widely known plant-protective endophytes are ergot fungi (genus *Claviceps*) that are sometimes considered as severe plant pathogens. This consideration is mainly based on the cellular mechanisms of symbiosis between the fungus and its host. The interaction involves the active penetration of fungus into the host ovary in which the fungal germ tube develops from a spore following the pathway of pollen tube. Within the ovary, fungus implements the necrotrophic development resulting in the complete destruction of plant tissues replaced by the fruiting body (sclerotium). The nutrients and energy for this development are obtained from the plant vascular bundle feeding the tiller: special trophic hyphae are developed to sink nutrients from the bundle. After completing its development, sclerotium falls into the soil and later the sexual processes are executed leading to the formation of ascospores.

Many mechanisms typical of plant–pathogen interactions are involved in the development of ergot fungi inside the host including the synthesis of enzymes for the degradation of plant cell walls and for the inactivation of the reactive oxygen species. Nevertheless, *Claviceps* should not be considered as an actual phytopathogen since under field conditions, the reduction of plant productivity after infection is often negligible (percentages of infected plants within the population and of destroyed ovaries within an infected tiller are usually low). The image of *Claviceps* as a dangerous pathogen comes mainly from its effect on the quality of grain yield: even the trace amounts of sclerotia make it highly toxic for humans and animals. The reports on massive death of people who consumed the bread prepared from grain collected in the *Claviceps*-infected fields are known from the Medieval times up to the nineteenth century, until the infective nature of the grain toxicity was revealed. It is caused by derivatives of lysergic acid (e.g., ergovaline) that lead to heavy neurotic and gangrene symptoms (some of these compounds are of medical use).

The other well-studied representatives of Clavicipitaceae can be found in the genera *Neotyphodium* (formerly *Acremonium*) (anamorph – *Epichloë*). They implement the complex living cycles including asexual and sexual reproduction, regular changes of hosts and a very interesting phenomenon of obligatory involvement of insects for the conidia dispersal (Figure 14.20). For *Neotyphodium* genotypes, the vertical transmission through subsequent host generations was demonstrated – a feature typical for many insect symbionts but rare in the plant symbionts. During tiller development, the endophytes reach the seeds and infect them without any necrotic effects. As a result, more than 90% of the daughter plants contain the *Neotyphodium* endophyte if they originated from the infected mother plant.

Infections by endophytic fungi may result in diverse effects on the plant fitness. Interactions with *Neotyphodium* are predominantly mutualistic: endophyte induces vegetative and generative development and improves resistance to some abiotic stresses, for example to drought. In addition, endophyte synthesizes alkaloids, lolins and peramins that make the plants toxic to some pests (aphids, weevils). Unfortunately many *Neotyphodium* strains also produce alkaloids, loliterm B and ergovaline deleterious for cattle and the application of strains not producing these compounds is of high practical interest. The impacts of *Epichloë* are more complicated since the plant protection from herbivores may be accompanied by the phytopathogenic effects. The latter are represented by the "choke disease" of ryegrass: intensive development of mycelium on the plant surface leads to inhibition of florets development, although vegetative growth may be enhanced.

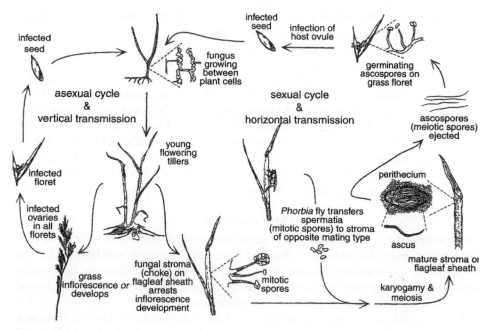

Figure 14.20. Living cycles in the Clavicepitaceae fungi implementing the defensive symbioses with cereal plants (Schardl et al., 1997).

Asexual living cycles are implemented by the mutualistic *Neotyphodium* and *Epichloë* species and sexual cycles by the pathogenic *Epichloë* species.

The Clavicipitaceae fungi represent only a part of diversity in the plant-protective ascomycetes. The other well-known examples are p-endophytes of cereals belonging to the genera *Gliocladium* and *Phialophora*. These systemic endophytes can confer on hosts the resistance to fungal pathogens due to the synthesis of antibiotics. For example, *Gliocladium* produces antibiotic gliovirin active against *Pythium* or gliotoxin active against *Rhizoctonia*. A specific group of plant protectors is represented by the genus *Cordyceps* – the epiphytes which cause lethal infections in phytophagous insects. These infections are initiated after the insects feed on epiphyte-harboring plants and resulted in the development of mycelia and fruiting bodies on the killed pests.

Ecological and agricultural impacts of beneficial plant–microbe interactions

Intensive analysis of plant–microbe interactions leads to a sufficient change in the view of plant organism as a chemical machinery, converting inorganic substances into organic ones using the solar energy. Instead of this mechanistic view, plant appears to be an organizer of symbiotic community in which many vital (trophic, defensive, regulatory) host functions are delegated to the microsymbionts.

It is often considered that the role of beneficial microbes in plant adaptability is based on improving their nutrition. Due to the wide application of isotopic (^{15}N, ^{14}C) techniques, metabolic exchange among the partners becomes available for precise evaluation. However, restriction of the plant–microbe interactions to this exchange would not be correct. For example, in defensive symbioses the microbes do not donate any nutrients to the host. Even if the transfer of nutrients to host is intensive (N_2-fixing symbioses, mycorrhizae), it comprises only the immediate effect of interaction. At the level of plant community, microbial-based nutrition is usually accompanied by improving competitiveness in symbiotic plants that may greatly change the ecosystem structure. Based on the whole spectrum of effects involved, microbial-based strategies of sustainable agriculture may be proposed.

Plant nutrition

Improvement in plant nutrition is implemented by the most widely distributed plant–microbe systems – mycorrhizae and N_2-fixing symbioses. In the majority of present-day terrestrial plants, N and P nutrition is essentially symbiotrophic. In the case of AM or EcM symbioses, up to 80% of mineral nutrients are obtained by plants from mycobionts.

Any mutualistic interaction involves a fine balance between the benefits, which plants may obtain from microbes and the costs, which should be paid for these benefits. This balance is based on the energy supply of symbiosis: 20–30% of plant photosynthetic products are invested in the N_2-fixing or mycorrhizal symbioses. However, these expenses may be returned to the host rather quickly. For example, it is more profitable for plants to obtain P through mycorrhiza than through their own root hairs (that are many times more thick and have much lower surface volume ratio than the hyphae). Moreover, the C cost may be decreased greatly since photosynthesis is stimulated by the symbioses formation. Such stimulation was demonstrated for N_2-fixing symbioses and mycorrhizae demonstrating a great self-regulatory potential of symbiotic systems.

Diversity of plant communities

Formation of the plant-microbe symbioses usually leads to an increased diversity of plant communities because the inter- and intra-species competition for the natural resources becomes less severe. The well-documented examples are represented by pasture legumes which become more competitive in the mixture with cereals after inoculation with rhizobia, since two components of the grass mixture use different sources of N the legume consumes the fixed nitrogen, while the cereal uses combined nitrogen assimilated from soil or fertilizer. Similarly, mycorrhization of the grassland plant communities with AM fungi may greatly increase its diversity as it improves competitiveness in the AM-responsive species and reduces the competitiveness in the non-responsive species that may dominate in the absence of AM (Table 14.11).

Increase in the diversity of plant communities resulting from the plant–microbe symbioses may be due to changes in the movement of macro-elements

Table 14.11. Effect of inoculation with arbuscular mycorrhizae (AM) fungus *Glomus* on the success of different plant species within a grassland community (adapted from: Smith and Read, 1997)

Plants	Plant mass (mg)		% Survival (after 6 months of vegetation)	
	AM⁻	AM⁺	AM⁻	AM⁺
Mycorrhizal species				
Centaurum erythraea	0.23	7.08	2	64
Galium verum	1.87	9.39	11	58
Hieracium pilosella	0.93	7.63	6	49
Leontondon hispidus	0.83	3.72	13	42
Plantago lanceolata	3.62	33.96	10	71
Sanguisorba minor	5.06	17.14	6	53
Scabiosa columbaria	2.56	10.19	16	84
Non-mycorrhizal species				
Arabis hirsuta	0.26	0.13	42	8
Rumex *acetosa*	9.72	8.77	60	11

in the ecosystems. Among these changes, the most evident is the increase in overall rates of P and N turnover due to mobilization of these macro-elements from chemically inert soil compounds. The other change is represented by the allocation of the symbiotically acquired nutrients within the ecosystems. Very often it is based on the donation of nutrients from symbiotic systems to the other components of biocenosis. An economically important example is represented by *Azolla–Nostoc* N_2-fixing symbiosis that excretes combined N which may be assimilated by the nearby growing rice plants. Mycorrhizal fungi can elicit the flow of fixed C between different plants representing the example of direct transfer of symbiotically consumed nutrients (Table 14.12). Crop rotations and inter-cropping of cereals with legumes are used for centuries as the efficient strategy to allocate the fixed N into the non-symbiotic components of field ecosystem.

Microbial-based sustainable agriculture

During the prolonged co-evolution with cultured plants, humans learned to implement many vital plant functions using the agrochemicals – mainly, mineral fertilizers and plant-protecting dyes. The wide use of these agrochemicals underlies "intensive agriculture", however, this resulted in a high ecological price namely a global pollution of the environment. A sufficient part of it is due to the losses of N and P from fertilizers that accumulate in soil and water leading to the adverse ecological changes. The plant-protective agrochemicals and the products of their degradation are the potent sources of xenobiotics;

Table 14.12.   Transfer of carbon from *Plantago lanceolata* (exposed to $^{14}CO_2$ and grown at light) to *Festuca avina* induced by AM formation (adapted from: Smith and Read, 1997)

Regime for *F. avina* growth	Parts of *F. avina* plants	% ^{14}C (in the total C) in *F. avina* plants	
		AM⁻*	AM⁺
Light	Roots	0.08	2.8
	Shoots	0.10	0.44
Shade	Roots	0.03	36.4
	Shoots	0.04	0.32

*The figures demonstrate the background level of ^{14}C since *F. avina* was not exposed to $^{14}CO_2$.

many of them may be transformed into toxic or mutagenic substances that reach humans through food chains. The real alternatives to the agrochemicals are by microbial preparations without which no ecologically friendly agriculture is possible. Of course, inoculants will not substitute the fertilizers or pesticides totally but may decrease by manyfold the levels of their application.

Microbial inoculants have been used for more than a century starting with the inoculation of legume crops with rhizobia in the late 1890s. At present, PGPR are increasingly used for the inoculation of various non-legume crops including cereals and vegetables. Inoculation with AM fungi is very promising for a broad spectrum of crops, however preparation of inoculants is complicated by the inability of fungi to propagate independent of plants.

As a result of more than a century of research, rich collections of beneficial plant microbes were established in many institutions and diverse technologies for manufacturing and application of inoculants were worked out. The great experience accumulated about using the microbial preparations and several lessons from it should be drawn for the future work.

For example, it is clear that many of the widely used crops differ in their response to microbial inoculation. For example, many "old" crops with a prolonged history of cultivation (alfalfa, pea, soybean, pole bean) grow better under the combined N supply than after inoculation even with most effective rhizobial strains. In contrast, some "young" crops (Kura clover, hairy vetch, goats' rue) usually prefer to use symbiotically fixed N_2 (Figure 14.21). Moreover, in the "old" crops, the wild-growing genotypes and local varieties exceed the agronomically advanced cultivars in the nitrogen-fixing activity. Similar difference between wild-growing and cultivated plant genotypes were reported for AM symbiosis.

These data suggest that during cultivation and breeding, plants lost a significant part of their symbiotic potential, and the resulted crops grow better using mineral fertilizers than using the appropriate microbial symbionts. Therefore, special efforts from the plant breeders are required to restore (or even to increase!) the symbiotic potential of cultured plants. A validity of this approach was demonstrated in pole bean: introduction of genes for

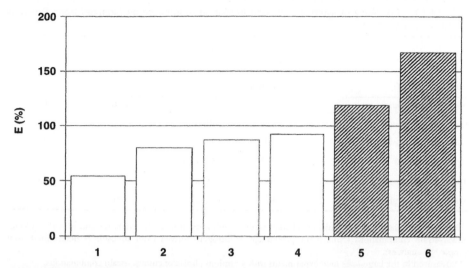

Figure 14.21. Symbiotic efficiency in some legume crops. Bars represent the coefficients of symbiotic efficiency (E) calculated as: $E = M_s/M_c$, where M_s and M_c are the indices of plant productivity determined under the conditions optimal for symbiotrophic nitrogen nutrition and combined nitrogen nutrition, respectively.
1 – pea (*Pisum sativum*); 2 – alfalfa (*Medicago sativa*); 3 – pole bean (*Phaseolus vulgaris*); 4 – soybean (*Glycine max*); 5 – hairy vetch (*Vicia villosa*) and 6 – Kura clover (*Trifolium ambiguum*). (Data are summarized in: Provorov and Tikhonovich, 2003.)

high N_2-fixing activity from wild-growing (climbing) varieties into agronomically advanced (bush) cultivars created the commercially attractive genotypes which retain the beneficial traits of "cultured" phenotype in combination with high N_2-fixing activity.

An important prerequisite for improving the symbiotic efficiency is the coordination of genetic modifications in the plants and their microsymbionts. Two-factor analysis of variance in the productivity of symbiotic systems suggests that symbiotic efficiency is controlled equally by genotypes of both the partners. Therefore, coordinated breeding should be aimed at creating the optimal (specific) combinations of partners' genotypes. A great potential of this work may be illustrated by 3– to 4-fold differences in the amounts of N_2 fixed by legume crops under optimal (experimental) and real (field) conditions (Table 14.13).

In the practical sense, coordinated breeding means improvement of plant affinities towards the specially selected or genetically engineered highly effective strains. These affinities are often assessed on the basis of competitiveness of introduced microbes (effective symbionts) that is expressed against the local (usually, ineffective) symbionts. An important approach for this improvement would be to use the plant genes that block symbioses with non-effective strains. For example, in soybean fields at USA the less effective and highly competitive serotype 123 of *Bradyrhizobium japonicum* dominates. In order to exclude it from soybean nodulation, the genes restricting susceptibility to this

Table 14.13.   Efficiency of symbiotic nitrogen fixation in legume crops (adapted from: Provorov and Tikhonovich, 2003)

Crops	Fixed nitrogen (kg/ha per one season)		Ndfa*	Yield increase** (%)
	Potential*	Real**		
Pea (*Pisum sativum*)	135	40–60	0.66	14
Vetch (*Vicia sativa*)	157	40–65	0.70	18
Soybean (*Glycine max*)	390	60–90	0.88	24
Lupine (*Lupinus angustifolius*)	220	80–120	0.81	22
Alfalfa (*Medicago sativa*)	550	140–210	0.88	25
Goats' rue (*Galega orientalis*)	480	130–220	0.91	35

*Measured in glasshouse and micro-field trials under conditions optimal for symbiosis and with the selected cultivar–strain combinations (Ndfa is the percentage of nitrogen derived from atmosphere determined using isotopic ^{15}N analyses).

**Measured in the large-scale field experiments with a random choise of cultivar–strain combinations.

serotype have been introduced into commercial *Glycine max* cultivars from wild-growing *G. soja* lines.

A similar approach may be used in pea the inoculation of which is often less-effective in Europe and North America. This is due to numerous *Rhizobium leguminosarum* bv. *viceae* local populations composed of the symbionts from wild-growing *Vicia* and *Lathyrus* species (that cross-inoculate readily with *Pisum sativum*). To avoid competition between inoculant and local pea symbionts, the plant *sym2* allele which arrests nodulation with the great majority of local *R. leguminosarum* bv. *viceae* strains was introduced into the commercial pea varieties from the "Afghan" peas. Only those *R. leguminosarum* bv. *viceae* strains which harbor *nodX* gene can overcome this resistance, however, they are very rare in European populations. Engineering of *nodX* gene into the commercial *R. leguminosarum* bv. *viceae* strains may be a good approach to ensure their exclusive penetration into the roots of *sym2/sym2* pea lines (obtained by introgression of *sym2* allele into commercial varieties).

Another important goal of coordinated plant–microbe breeding is to avoid the limiting (bottleneck) stages in the metabolic exchange in symbiosis. In the N_2-fixing nodules, transfer of energy from host to bacteroids is usually considered to be among the most narrow bottlenecks. To overcome this, several groups of researchers engineered the additional copies of *dctABD* genes (responsible for the uptake of dicarboxylates by bacteroids) into highly effective alfalfa symbionts (*Sinorhizobium meliloti*) and obtained a sufficient (two times or more) increase in the N_2-fixation activity. However, the yield of plants inoculated by these recombinants is improved poorly. The biochemical analysis of alfalfa (*Medicago sativa*) inoculated with these recombinant *S. meliloti* strains suggests that the plant cannot assimilate the excessive N the concentration of which may be increased greatly. In order to avoid such as bottleneck, Barnes suggested the selection of alfalfa for the activities of key nodule-specific enzymes controlling nodule metabolism (PEPC, GOGAT; see Figure 14.12) which were demonstrated to limit the efficiency of N_2-fixing machinery.

In this chapter, we presented only the general ideas concerning the ecological and practical prospects for the application of beneficial plant–microbe interactions. In order to use them in constructing novel sustainable agro-systems, a cooperation of specialists from different fields of microbial and plant research is required. Organization of such cooperation is an important task in the twenty-first century that should be fulfilled to maintain the ecological balance in agricultural systems.

Selected Literature

Bago B, Pfeffer PE, Shacher Hill Y. 2000. Carbon metabolism and transport in arbusclar mycorrhizas. Plant Physiology. 124: 949-957.

Barker SJ, Tagu D, Delp G. 1998. Regulation of root and fungal morphogenesis in mycorrhizal symbiosis. Plant Physiol. 116: 1201-1207.

Barnes DK, Heichel GH, Vance CP, Ellis WR. 1984. A multiple-trait breeding program for improving the symbiosis for N2 fixation between Medicago sativa L. and Rhizobium meliloti. Plant and Soil. 32: 303-314.

Beringer JE, Brewin NJ, Johnston AWB. 1980. The genetic analysis of Rhizobium in relation to symbiotic nitrogen fixation. Heredity. 45: 161-186.

de Bary A. 1879. Die Erscheinung der Symbiose. Strassburg, Verlag Von Karl J. Trübner, p. 30.

de Bruijn F, Chen R, Fujimoto SY, Pinaev A, Silver D, Szczyglowski K. 1994. Regulation of nodulin gene expression. Plant and Soil. 161: 59-68.

Denarie J, Debelle F, Rosenberg C. 1992. Signaling and host range variation in nodulation. Annu. Rev. Microbiol. 46: 497-531.

Dobereiner J, Day JM. 1976. Associative symbiosis in tropical grasses: characterization of microorganisms and dinitrogen fixing sites. In: Proc. 1-st Intern. Symp. Nitrog. Fixat. Eds. Newton W, Nyman CJ. Washington State Univ. Press. Pulman. p. 518.

Fisher HM. 1994. Genetic regulation of nitrogen fixation in rhizobia. Microbiol. Rev. 58: 352-386.

Govorov LI. 1928. The peas of Afghanistan. Bull. Appl. Botany, Genet. Plant Breed. (in Russian). 19: 497-522.

Gresshoff PM, Buzas DM, Laniya T, Men A, Jiang O, Schenk PM, Hayward A, Kam J, Li D, Miyahara A, Nontachaiyapoom S, Indrasumunar A, Brcich T, Gualtieri G, Davis P, Carroll B. 2004. Systemic regulation of nodulation by a leaf-controlled LRR-receptor kinase. In: Biology of Plant-Microbe Interactions. V. 4. Eds. Tikhonovich IA, Lugtenberg BJJ, Provorov N.A. IS-MPMI. St.-Petersburg, Russia. pp. 369-372.

Hibbett DS, Gilbert LB, Donoghue MJ. 2000. Evolutionary instability of ectomycorrhizal symbioses in basidiomycetes. Nature. 407: 506-508.

Kaminski PA, Batut J, Boistard P. 1998. A survey of symbiotic nitrogen fixation by rhizobia. In: The Rhizobiaceae. Molecular Biology of Model Plant-Associated Bacteria. Eds. Spaink HP, Kondorosi A, Hooykaas PJJ. Dordrecht, Boston, London: Kluwer Acad. Publ. pp. 431-460.

Lie TA. 1978. Symbiotic specialization in pea plants: the requirement of specific Rhizobium strains for peas from Afghanistan. Ann. Appl. Biol. 88: 462-465.

Long SR. 1989. Rhizobium-legume nodulation: life together in the underground. Cell. 56: 203-214.

Lugtenberg B. 2004. Molecular aspects of biocontrol traits. In: Biology of Plant-Microbe Interactions. V. 4. Eds. Tikhonovich I.A., Lugtenberg B., Provorov N.A. St.-Petersburg, Russia. Biont. pp. 310-311.

Marsh J, Schultze M. 2001. Analysis of arbuscular mycorrhizas using symbiosis-defective plant mutants. New Phytologist. 150: 525-532.

Oldroyd GED, Downie JA. 2004. Calcium, kinases and nodulation signaling in legumes. Nature Rev. 5: 566-570.

Ovtsyna AO, Staehelin C. 2005. Bacterial signals required for the Rhizobium-legume symbiosis. Recent Res. Develop. Microbiol. 7: 631-648.

Provorov NA. 1994. The interdependence between taxonomy of legumes and specificity of their interaction with rhizobia in relation to evolution of the symbiosis. Symbiosis. 17: 183-200.

Provorov NA. 1998. Coevolution of rhizobia with legumes: facts and hypotheses. Symbiosis. 24: 337-367.

Provorov NA, Borisov AY, Tikhonovich IA. 2002. Developmental genetics and evolution of symbiotic structures in nitrogen-fixing nodules and arbuscular mycorrhiza. J. Theor. Biol. 214: 215-232.

Provorov NA, Simarov BV. 1990. Genetic variation in alfalfa, sweet clover and fenugreek for the activity of symbiosis with Rhizobium meliloti. Plant Breeding. 105: 300-310.

Provorov NA, Tikhonovich IA. 2003. Genetic resources for improving nitrogen fixation in legume-rhizobia symbiosis. Genetic Res. Crop Evol. 50: 89-99.

Rostas K, Kondorosi E, Horvath B, Simoncsits A, Kondorosi A. 1986. Conservation and extended promoter regions of nodulation genes in Rhizobium. Proc. Natl. Acad. Sci. USA. 83: 1757-1761.

Schardl CL, Leuchtmann A, Chung KR, Penny D, Siegel MR. 1997. Coevolution by common descent of fungal symbionts (Epichloe spp.) and grass hosts. Molec. Biol. Evol. 14: 133-143.

Schauser L, Roussis A, Stiller J, Stougaard J. 1999. A plant regulator controlling development of symbiotic root nodules. Nature. 402: 191-195.

Simon R, Priefer U, Pühler A. 1983. A broad host range mobilization system for in vivo genetic engineering: transposon mutagenesis in gram-negative bacteria. Biotechnology. 1: 784-791.

Smith SE, Read DJ. 1997. Mycorrhizal Symbiosis (second edition). San Diego, London, New York, Boston, Sydney, Tokyo, Toronto: Academic Press, pp. 590.

Spaink HP, Lugtenberg B. 1994. Role of rhizobial lipo-chitin oligosaccharide signal molecules in root nodule organogenesis. Plant Molec. Biol. 26: 1413-1422.

Topunov AF, Rozov FN, Petrova NE. 1998. How the entry of oxygen into legume root nodules is regulated? Four types of regulation. Russian J. Plant Physiol. 45: 935-941.

Tsyganov VE, Voroshilova VA, Priefer UB, Borisov AY, Tikhonovich IA. 2002. Genetic dissection of the initiation of the infection process and nodule tissue development in the Rhizobium-pea (Pisum satrivum L.) symbiosis. Ann. Botany. 89: 357-366.

Van Berkum P, Eardly B. 1998. Molecular evolutionary systematics of the Rhizobiaceae. In: The Rhizobiaceae. Molecular Biology of Model Plant-Associated Bacteria. Eds. Spaink HP, Kondorosi A, Hooykaas PJJ. Dordrecht, Boston, London: Kluwer Acad. Publ. pp. 1-24.

Young JPW, Haukka KE. 1996. Diversity and phylogeny of rhizobia. New Phytol. 133: 87-94.

Part 3

Practical use of molecular studies

The results of investigation on the mechanisms of pathogenicity and resistance in infectious plant diseases are primarily important as a fundamental basis for the development of highly efficient plant disease and pest control measures. These investigations have not changed the plant protection strategy (and they could not have changed it): similar to the conventional systems, there are two prevailing approaches: chemical protection and genome modification to develop cultivars resistant against pathogens. However, both the methodologies and capabilities of these approaches have cardinally changed.

Part 3

Practical use of molecular studies

Chapter 15

Molecular basis of plant immunization

Yu. T. Dyakov, V. G. Dzhavakhiya, T. Korpela

Plant resistance to disease can be enhanced through inoculation with nonpathogenic microorganisms or strains of pathogenic microorganisms (vaccination), as well as through treatment with chemical products: abiogenic or biogenic elicitors (immunization).

Vaccination

As against human health, vaccination has limited practical application as a way of plant protection. There are several major areas of practical use of plant vaccination.

1. A method of protection against viral diseases. Plant inoculation with a "mild" strain protects plants from "severe" strains of a relative virus.
   It was said earlier that plants can respond to pathogen penetration by the activation of various own defense mechanisms. If adequate defenses are available in response to the challenge, the plant can become more resistant against repeated infection by the same or other pathogen.
   The phenomenon of developing resistance to the pathogen resulting from a previous inoculation with some pathogenic or nonpathogenic microorganism is called cross-protection. Resistance arising in such cases frequently occurs in naturally growing plants, as in natural conditions plants are continuously exposed to various pathogens. It is systemic resistance and it is ensured by natural defenses.

The principal ideas of studying immunization are based on the experimental data showing that treatment of plants with attenuated strains of phytopathogenic microorganisms can induce resistance to more pathogenic strains. For a number of reasons, practical application of this method of plant biological protection is still an open question. One of the key reasons limiting the practical implementation of cross-protection is a highly possible mutation of a slightly pathogenic strain towards a highly pathogenic genotype. This can lead to disastrous consequences – taking into account high doses of the inoculum required to treat large planted areas. Besides, production of the necessary amount of the inoculum in some cases may be unprofitable. Moreover, the avirulent immunizing strain may be inadvertently transferred to other plant species where it may be virulent. Studies on the propagation and harmfulness of the populations of virus pathogens showed that an effect of simultaneous inoculation with two or more strains may be more deleterious for the plants.

Possible development of a certain pathogenicity synergism upon combined inoculation of plants with two viral strains as well as the synergism of their recombinant progeny was described by Zhou and collaborators in1997. These scientists discovered that the cassava epidemics in Uganda had been caused by simultaneous infestation of the plants by a hybrid of the African cassava mosaic virus and the East African cassava mosaic virus, and by one of the parent strains. These viruses are transmitted by the whitefly. Emergence of synergism between the hybrid and the parent strains also caused significant economic losses in cassava in Kenya, Tanzania, Sudan, and Congo. It should be noted that in South America, cassava homeland, these viruses are unknown. That is, cassava, brought to Africa in the sixteenth century, was infected with the mosaic virus transferred by the whitefly from other endemic African plant species.

Thus, it was found in practice that treatment with a nonpathogenic strain may lead to the emergence of undesirable synergism between this strain and already naturally occurring virulent strain of the same or taxonomically close pathogen.

Nevertheless, certain cases may lead to the situations when the cross-protection method is economically justified. In particular, such a situation may arise in the case of a high rate of propagation of an endemic disease, with its stopping unfeasible, where the predicted losses obviously exceed the possible losses from an immunizing strain. An example is the use of a mutant strain of the papaya ring spot virus (PRSV) for the cross-protection of papaya in Hawaii. As initially the low pathogenic strains of this virus could not be isolated, the mutants of a virulent PRSV strain were obtained by treatment with nitrous acid solutions. One of the resulting mutants was found to be actually capable of protecting papaya from the natural strain of the virus. However, the mutant itself caused the emergence of peculiar signs on the leaves and fruits of some papaya cultivars.

The necessity of additional treatments as well as unwillingness of the farmers to treat plants with a virus prevented broad application of this method.

A similar case of practical use of cross-protection against the *Citrus tristeza* virus in Brazil is known. It should be noted that in the case of PRSV the virus protein coat gene expression leads to resistance of transgenic papaya plants to this virus.

In some countries, greenhouse tomatoes are inoculated with attenuated tobacco mosaic virus (TMV) strains to provide protection against infestation with the severe strains that cause tomato streak disease. A relative safety of this method is based, first, on the isolation of the greenhouse plants from the plants grown in the open ground, and, second, inability of TMV to spread by insects.

2. Prevention of tree infection by root rot causal agents. The Basidiomycetes root rot (*Heterobasidium annosum*) and autumn honey agaric (*Armillaria mellea sensu lato*) expand the damaged area by infecting the roots of the healthy trees through penetration of mycelium and rhizomorphs released from the roots of the affected trees. The new foci are made by the basidiospores germinating through recently stripped wood. Fresh stubs in wood-cutting areas get infected very rapidly. For this reason, methods have been developed to inoculate fresh stumps with the preparations containing spores and mycelium of saprotroph basidiomycetes (*Poria giganthea*, *Pleurotus* spp., etc.) that promptly grow and occupy wood in wood-cutting areas.

3. Treatment of the chestnuts infected by the fungus *Criphonrectria parasitica*. The causal agent of chestnut canker *C. parasitica* arrived to Northern America and Europe from East Asia where it grows in the local tolerant species without causing much damage. The European and, especially, American species turned out to be very susceptible, and their disease has led to large-scale death of the chestnut forests. In 1951, low virulent strains were found in Italy that were growing slowly and did not damage an infected tree. These "hypovirulent" strains were found to contain a viral dsRNA, the infection of which resulted in the inhibition of colony growth rate in vitro, weakening of pigmentation, decrease in conidia production, and loss of the sexual stage and pathogenic properties. In the presence of viral RNA or protein, production of CPG-1 protein, which constitutes an α-subunit of the multipurpose regulator of G-protein signal transduction, was inhibited. The result is reduction of cycloadenylate and Ca-phosphoinositol signalling pathways that provide expression of the key enzymes necessary for morphogenesis and pathogenesis of LAC-1 (laccase), CBH-1 (cellobiohydrolase), CRP (hyrophobin cryparin), and Vir1 and Vir2 (sexual pheromones). As the hypovirus is able to transfer the viral RNA from hypovirulent to virulent strains in common growth through anastomoses, the Ministry of Agriculture of France developed a process of chestnut inoculation with the paste containing a mixture of hypovirulent strains. Though now in Europe about half of the natural strains are infected by viruses, efficiency of biological control is questionable. Natural ousting of the virus-free strains by the infected strains due to their infestation is more likely. In America, such a way of chestnut protection was inefficient because the *C. parasitica* populations turned out to be highly polymorphous in the vegetative incompatibility genes that prevent migration of the hypovirus from the infected to healthy strains.

4. Treatment of seeds with rhizospheric bacteria. Seed soaking in the suspension of some *Pseudomonas* spp. strains provided leaf protection in cucumber, tomato, and carnation against leaf pathogens and Fusarium wilt.

The biological control is provided by various mechanisms, the most important being the following:

1. Antibiosis due to the production, by the antagonist bacteria, of antibiotic (antifungal and antibacterial) compounds, including chitinase.
2. Competitive colonization of the root surface, as well as competition for iron sources in soil.
3. Induction of systemic acquired resistance (SAR) by the antagonist nonspecific elicitors. In particular, bacterial lipopolysaccharides are SAR inducers in the treated plants. Many strains of pseudomonads used for biological control produce l-aminocyclopropane-l-carboxylate (ACC) desaminase, a regulator of the level of the plant hormone ethylene.
4. Direct parasitism of the biological control agents, such as *Trichoderma* strains, towards the pathogenic fungi.

Immunization

Chemical protection of plants has been traditionally based on the use of pesticides toxic for pathogenic organisms. The use of fungicides has undergone three stages in its history. The first and second generation fungicides (inorganic and organic contact products) are nonspecific, they possess a broad range of effects on various fungi and do not damage plants just because they are actively absorbed by the fungal cells, but they do not penetrate the plant tissues through the cuticle. These features impede the tactics of chemical treatment, for it is necessary, first, to coat the surface of the plants being treated very carefully, and, second, to repeat the treatment a number of times. The third generation fungicides are systemic organic products specific for particular groups of fungi and not toxic for plants. They penetrate plants, spread systemically in its tissues, and kill the parasites not only on the surface but also inside the plant. Therefore, their application is much more convenient than application of contact fungicides. However, their site-specificity (effect on particular specific structures and stages of metabolism) leads to a situation where modification of a susceptible site can bring partial or complete loss of susceptibility to the fungicide. Therefore, long-term use of systemic products usually results in the loss of their efficacy due to the accumulation of the parasite's resistant strains.

The fourth generation of plant protecting products is represented by the substances designed not to destroy pathogens but rather to undermine their pathogenicity and enhance plant defenses. Such agents possess a number of advantages compared with the true fungicides:

1. Less hazard to people, nontarget organisms, and the environment due to high specificity and absence of biocidity in the agents.
2. Enhanced resistance to pathogenic agents in the host plants possessing no resistance genes allows using the agronomically better cultivars.

3. Less hazard of accumulation of the pathogen strains resistant to fungicides.
4. Resistance induction at the horizontal level (which is possible) will provide a longer term plant protection compared with the application of fungicides and will require lower concentrations of the host-specific substances.
5. Most resistance inducers are multifunctional, i.e. they reduce plant susceptibility to fungi, bacteria, viruses, and sometimes also to nematodes and insects.

Two areas of such research (they are actually more numerous) are considered.

Agents affecting parasite pathogenicity

Many chemical compounds do not possess a distinct antifungal effect, but rather inhibit pathogenic properties and thus slow down the intensity of the course of disease. Some of them are listed below.

Inhibitors of melanin synthesis. In a previous section, the importance of melanin in the cell walls of many fungi for penetration in the mesophyll of a host plant was pointed out. Melanin is required for spore germination, formation of appressoria, penetration in plant tissue and growth in the plant, and protection against solar UV, fungitoxic leaf diffusates, and destructive effect of the plant hydrolytic enzymes (Dzhavakhiya et al., 1990). Many important plant parasites from the genera *Colletotrichum, Pyricularia, Verticillium, Phyllaphora, Polystigma, Guignardia*, etc. form pigmented appressoria. Some benzene derivatives inhibit the synthesis of melanin, thus preventing pathogen penetration in the plant. Their protective effect correlates with the degree of melanization inhibition (Table 15.1).

The most active compound in the above list is tricyclasol, widely used for the protection of rice against the rice blast disease. This agent has no effect on the linear growth of mycelium in vitro but inhibits melanin synthesis,

Table 15.1. Effect of benzene derivatives on melanization of *Pyricularia oryzae* (*Magnoporthe grizea*) mycelium and rice protection against infection (Inoke et al., 1984)

Compound	Concentration (μg) required for melanin inhibition	ED_{90}* (μg)
Tetrachlorophthalide	50.0	45.9
Pentachlorobenzene	100.0	94.7
alcohol	1.0	15.6
Chlorobenthiazone	0.5	7.9
Tricyclasol	1.0	19.7
Pyroquinol		

* Concentration required to reduce rice infestation by 90%.

which makes the colonies light pink instead of dark grey. Tricyclasol also slows down the formation of the resting structures – sclerotia *Botrytis cinerea* and *Sclerotinia sclerotiorum*, reduces their cell wall melanization, and accelerates germination.

Inhibitors of parasite enzyme synthesis. Many *organophosphorous pesticides* (the insecticides phospirate and diisopropyl orthophosphate, the fungicides ketacin and quinosan) and benzimidazoles (the fungicide benomyl), in the concentrations nontoxic for parasites, inhibit the fungal cutinase necessary for cell wall penetration and thus reduce infectivity of the fungi of the genera *Fusarium, Botrytis* and others.

Agents enhancing the release of nonspecific elicitors to plant tissue. Some compounds that seem to have an influence on the structure and function of the fungal membranes (they increase the loss of proteins by the germinating spores) are able to enhance elicitor release to the environment, thus improving protective properties of the infected tissue. This effect is characteristic of the phosphorus-containing insecticides and fungicides, synthetic pyrethroids, and the antibiotic polyoxin (Table 15.2).

As can be seen from the table, the phosphorus-containing insecticides (cidial, phthalophos, phasolon, sevin) do not only enhance the release of richitin elicitors from *P. infestans* spore germs but also themselves induce richitin formation in potato tubers, while the pyrethroid neopinamine itself does not induce richitin but enhances its formation in response to infection.

The phenylamide agent metalaxyl, efficient against oomycetes, in concentrations below fungicidal ones induced an immune response to the inoculation of the late blight-susceptible soybean cultivars with *Phytophthora megasperma*.

Table 15.2. Effect of insecticides on growth and pathogenicity of *Phytophthora infestans* (Mustafa and Dyakov, 1979)

Agent concentration (µg/ml)	Inhibition of spore germination (%)	Content of the phytoalexin richitin (µg/ml) in diffusates of the tubers treated with		Tomato leaf damage index*
		Agent	Agent + spores of virulent race	
Control**				10.9
Cidial (100)	0	35	45	0.3
Phasolon (40)	0	10	35	3.2
Phthalophos (100)	0	11	22	4.8
Sevin (100)	0	12	31	1.1
Neopinamine (100)	0	2	28	1.8

*Infestation index: (number of lesions per 50 infected leaves × diameter of sporulent spot)/100.
**Rishitin content in diffusates of the tubers infected with a compatible race: 2 – 3 µg/ml.

It has been shown that metalaxyl-induced phytoalexin accumulation occurs upon plant inoculation with metalaxyl-susceptible but not metalaxyl-resistant strains. Metalaxyl induced release of a nonspecific elicitor to a growth medium in vitro also only for susceptible but not resistant strains.

The agents influencing plants resistance

The ways of plant resistance enhancement against the disease induced by chemical treatment are dealt briefly.

1. Building of a protecting film on leaf surface. Spraying of bean leaves with water emulsions of polyacrylic acid derivatives provides a thin film on the leaf surface that reduced water loss but did not affect the photosynthesis. The film protected the beans from penetration of *Uromyces phaseoli*: its spores stuck to the film and could not germinate.
2. Enhancement of cell wall resistance against pectolytic enzyme attack. The phytohormone naphthylacetic acid reduces the degree of pectate methoxylation in the septum, which increases the number of Ca-pectate bridges and protects the septum from the attack by the parasite pectolytic enzymes. In particular, a protective effect of naphthylacetic acid has been demonstrated in tomatoes against the Fusarium wilt pathogen. Glucan – a nonspecific elicitor of *Pseudomonas syringae* – insolublizes the soybean cell wall coat proteins, thus enhancing the cell wall strength, which was demonstrated by the cell resistance to protoplasting (Table 15.3).
3. Activation of phenol metabolism, oxidative enzymes, and phytoalexins. Carboxyacids of dichlorcyclopropane (carboxyacid 2,2-dichlor-3, 3-dimethylcyclopropane (DCP)) protect rice from blast disease without being toxic to *P. oryzae*. The treated plants increase the level of peroxidase, the content of phenols, lignin, brown pigments, and the phytoalexin momilacton. However, DCPs have a low antifungal effect against the fungus *P. oryzae* and inhibit melanin synthesis, therefore, the primary mechanism of the protective effect of these compounds is not clear. The herbicide simazin modifies many plant metabolic processes

Table 15.3. Percent of soybean mesophyll protoplast release caused by lytic enzymes (Brisson et al., 1994)

Exposure (h)	Lytic solution	Lytic solution + elicitor
0.5	0-5	0-5
1.0	45	10
2.0	65	12
5.0	80	60
18.0	95	95

and influences their disease resistance. For instance, in wheat treated with simazin the nitrogen content increases almost by one-third, and sugar content drops almost by half. The content of phenol glucoside 2,4 - dihydroxy-7-metoxy-1, 4 -benzoxasin-3 - one (DIMBOA) grows by 33% and its aglicon, by 50% (see Chapter 5), which improves resistance to mildew and Cercosporella root rot. The fungicide carbendazim, in nontoxic concentrations for the fungi, protected the treated plants due to enhanced activity of phenylalanyl ammonia lyase and phenol oxidation, as well as ROS reactive oxygen species accumulation.

A considerable effect on biochemical pathways involved in plant resistance is rendered by retardants. Chlorcholinchloride (CCC) enhances wheat susceptibility to *Septoria, Fusarium,* and *Alternaria,* and reduces susceptibility to *Cercosporella,* enhances tomato and cotton resistance to *Fusarium* and *Verticillium,* and cucumbers, to *Cladosporium cucumerinum.* Tomato resistance is caused by the formation of tylose, and cotton accumulates the phytoalexin hemigossipol. The prospects of practical use of the above-described protection principles are not obvious for the following reasons:

(i)   The effect of immunization is usually calculated by the change in the number and size of lesions in inoculated immunized and non-immunized plant, but much less often by the change in yield of the infected plants. Sometimes immunization at slight epidemics can even result in the yield drop, as the immunization-induced protective processes consume energy. For instance, in response to barley inoculation with mildew, respiration of a resistant plant strengthens by 80% within 16–24 h, which results in a drop in the number of grains by 7% and grain weight by 4% (inoculation of a susceptible plant leads to a 26% drop in the number of grains and 11% drop in grain weight).

(ii) The antifungal plant compounds phytoalexins and phytoanticipins are toxic for animals and humans, and their accumulation in plants may lead to poisoning. Synthetic inducers, as against biogenic, inhibit the processes of natural decomposition of phytoalexins in plant tissue, which increases the hazard of toxicity of the treated plant tissues.

(iii) Plant sensibilization, i.e. an effect where defense responses occur only in response to infection. Such response of a host plant is observed in the treatment of tomatoes and cotton with dinitroaniline herbicides (trifluraline, etc.) The hypocotyl extract of the tomatoes treated with the agent and inoculated with *Fusarium oxysporum* f.sp. *lycopersici* is toxic for the fungus, while the hypocotyl extract of the treated but not infected plants is not toxic.

Aluminium phosetyl (Al-tris (*o*-ethyl phosphonate)), without being toxic for oomycetes in vitro, protects potatoes from *P. infestans,* grapes from *Plasmopara viticola,* and tobacco and pepper from *Phytophthora nicotianae.* The phytoalexin capsidiol is accumulated in the fruits of pepper and stems of tobacco only after treatment and inoculation (capsidiol is not accumulated without inoculation). In the mutants of *Arabidopsis thaliana* NahG and nim1, incapable of immune response, the fungicides aluminium phosetyl, metalaxyl and $Cu(OH)_2$ are considerably less effective than in nonmutant plants. This indicates that in addition to direct antimicrobial effect these fungicides are involved in the induction of the immune response.

The described defense mechanism is very promising, as it imitates, in the parasite-susceptible plants, the responses typical of the resistant forms therefore, this mechanism is dealt in more detail.

Biogenic elicitors

The studies headed by O. L. Ozeretskovskaya showed that the treatment of potato tubers with a high concentration (100 µg/ml) of the biogenic elicitor from the mycelium of *P. infestans* resulted in intensive but short-term defense response (necrotization and FA accumulation). Treatment with the elicitors in low concentration (5 µg/ml) did not produce any visible signs or accumulation of stressful metabolites but sensitized the plant cells which became resistant against the subsequent inoculation with the pathogen, the resistance covering all the tuber cells and lasting for a long time (up to 3 months). The sensibilization seems to consist of the cell preparation for a prompt response to the subsequent infection (increase in the volume of agranular endoplasmatic reticulum, number of mitochondria, and cAMP concentration).

The authors of the studies postulated the rules that need to be observed for the induction of plant resistance to disease using biotic elicitors:

1. Induction of resistance can be achieved using the biogenic elicitors produced by the pathogens.
2. The elicitors should be used in certain concentrations that do not cause FA accumulation in the treated tissue.
3. The elicitors should be separated (if possible) from the suppressors.

Field experiments show that plant treatment with biogenic elicitors is very efficient and does not suffer from the disadvantages characteristic of fungicides. It is distinguished by:
 – better environmental safety as it is based on the activation of natural resistance mechanisms in plants;
 – systemic effect and long duration of the protective action;
 – low concentration of the active substance;
 – involvement of numerous defense systems in resistance operation, which reduces the probability of accumulation of the parasite-resistant strains;
 – integrated protective effect against various phytopathogenic fungi, bacteria, nematodes, and, probably, viruses;
 – absence of a toxic effect on the nontargets organisms;
 – absence of toxic chemicals in the harvested plants;
 – in some cases, stimulation of growth processes in the treated plants; and
 – intensification of wound reparation processes.

Some promising biogenic elicitors for plant immunization are considered here.

Lipids. It was shown earlier (see Chapter 7) that unsaturated fatty acids possess elicitor properties for potatoes. In low concentrations that do not damage the treated leaves they induce systemic resistance against *P. infestans.* Their protective properties decreased with the decrease in the number of double bonds in the molecule: 20:5 > 20:4 > 18:2 > 18:3 > 18:1. The plant protection was best 5 days after the induction and lasted two weeks. Arachidonic acid (AA) in potato leaves is metabolized by 5-lipoxygenase to 5-*S*-hydroperoxyeicosatetraenoic acid (5-*S*-HPETE), which in a concentration of 1 μg per tuber section induced higher protective effect than AA in a 20 μg concentration. In the AA-treated potato tubers, resistance to *P. infestans* was revealed a day later and lasted for a long time (up to several months). The immunized tubers were found to contain PR-proteins, peroxidase, polyphenoloxidase, and lipoxygenase (LOG).

Amino sugars. Among the biogenic elicitors, the most popular immunizers are the agents produced from chitin, chitosan (deacetylated chitin) possessing a greater induction activity than chitin. Chitin and chitosan cause rapid membrane depolarization and induce local and systemic resistance to fungi, bacteria, viruses, and nematodes. Regarding the phytopathogenic viruses, chitin and chitosan reduce the number of necroses in the infection of super-susceptible plants and inhibit the spread of systemic viral infections over the plant. Chitosan binds to the anion components of the plant membranes and cell walls (phospholipids, pectin), which impedes virus adsorption on the wall and enhances the synthesis of callose that impedes long-range transport of the viruses. In tobacco protoplasts, chitosan induces the synthesis of 130 kDa protein, an RNA-dependent RNA polymerase that causes the post-translational gene silence, including the viral genes, as well as the synthesis of 23 kDa ribonuclease that degrades viral RNAs. Chitosan, through the formation of complexes with anion peroxidases, causes thickening and lignification of the cell walls, and, through its combination with nucleic acids, it acts as a transcription regulator. Chitosan sets off induction of the synthesis of phytoalexins, PR-proteins, including chitinase and -glucanase, proteinase inhibitors, reactive oxygen species, as well as lectin (hemagglutinin) accumulation. Chitosan activates LOG, an important intermediate product of signal transduction in the cell. LOG is involved in the signalling systems induced by salicyclic acid (SA) and jasmonic acid (JA). Different signalling pathways seem to involve different LOG isoenzymes. It has been shown that wheat inoculation with *Puccinia graminis* f.sp. *tritici* and wheat treatment with the glycopeptide elicitor isolated from the germ tubes of this fungus lead to 92 kDa LOG (LOG-92) accumulation, and treatment with chitosan or Me-JA brings LOG-100 accumulation, i.e. chitosan and rust inoculation induces different pathways of signal transduction.

Chitin oligomers of a size less than 5 molecules of acetylglucosamine do not induce defense responses in plants. Chitosan hexamers–octamers possessed the greatest induction ability and protective properties in various plants.

Plant spraying and application of solutions and suspensions of chitin and chitosan to soil have an effect on microbe diversity of the phylloplane and rhizosphere, increasing the number of chitinolytic microorganisms (bacillus, actinomycetes, and fungi), which are important antifungal and antinematode agents.

The water-soluble 5-kDa chitosan is produced from sea crab shells or as a by-product in the production of citric acid from *Aspergillus niger*.

The possible practical applications of some biogenic elicitors for the induction of plant resistance to phytonematodes were considered in the systems: tomatoes – root-knot nematode *M. incognita*, cucumbers – root-knot nematode *M. incognita*, and potato – *Globodera rostochiensis*. AA (C20 unsaturated fatty acid) and chitosan with a molecular weight 5 kDa and 85% deacylation were used as the elicitors. Both these compounds are intrinsic in parasitic phytonematodes, i.e. to a certain extent they can be the compounds that induce a phytoimmume response upon the plant invasion. As elicitors possess a property to induce systemic resistance, they offer a unique opportunity to treat not only the plants but also the seeds, as the plants grown from these seeds are immunized. Analysis of the immune potential of the susceptible plants, the seeds of which had been treated with the elicitors showed that their defense response to the invasion becomes similar to the response of the resistant cultivars. The enhancement of the immune potential has led to a noticeable improvement in the plant growth and development, less root damage by nematodes, as well as to the change of morpho-physiological and population parameters of the parasite (smaller size, drop in fertility, and change of age and sex structure of the nematode population). The investigated elicitors had the best effect in combination with the signal molecules, methyl jasmonate and SA.

On the basis of these elicitors, plant protection products have been developed, and they are already used at present to control phytoparasitic nematodes.

Harpins. Commercial use of the bacterial elicitors has also begun. One of them, Messenger™, contains the bacterial protein harpin EA from *Erwinia amylovora* as the active substance. The experiments with *A. thaliana* mutants showed that harpin induces SAR genes PR-1 and PR-2. However, in the transgenic plants containing gene nahG and unable to accumulate SA, harpin loses the ability to induce resistance. The same effect is also observed in nim1 mutants defective in the ability to signalling transduction in the SA pathway. However, the mutations blocking Me-JA and ethylene signalling pathways have no effect on the inductive properties of harpin; consequently, harpin induces resistance in the signal SA transduction pathway through the nim1 mediator. Harpin triggers active biochemical processes as soon as 5–10 min after a leaf treatment, and it is recommended for the protection of vegetables and grapes against mildew.

Based on harpin EA, the product Messenger has been developed which contains 3% of harpin and 97% of environmentally inert biological material. This biopesticide works at low doses, and it completely and rapidly breaks down in natural conditions into environmentally safe components. Messenger(r) is produced as water-soluble pellets and can be used for spraying the above-ground parts of plants using ordinary commercial sprayers, or by treating seeds or soaking sprout roots.

The US Environmental Protection Agency categorized Messenger as a biochemical pesticide. Messenger induces plant defense responses without a direct action on the pathogen.

Field tests have shown that the agent possesses an ability to induce the natural mechanisms responsible for growth and resistance to pathogens in more than 40 species of agricultural crops, including citrus, tomato, pepper, cucumber, and strawberry; traditional field crops, such as cotton, rice, and corn; and also rose.

Cold shock proteins. In addition to harpins, other elicitors of protein origin are also known. For instance, a low molecular weight (7.2 kDa) thermostable protein factor MF2 (Microbial Factor 2) has been isolated from *Bacillus thuringiensis*. The MF2 protein was shown to be able to induce plant resistance in the following pathogen–host pairs: TMV – tobacco, potato X-virus – tobacco, late blight – potato, rice blast–rice (Djavakhia et al., 2001). The MF2 protein can be classified among nonspecific elicitors. A gene of MF2 protein has been cloned, sequenced, and registered in the GenBank as a cold shock protein CspD (cold shock protein; hereinafter the designation CspD is used instead of MF2).

Transgenic tobacco plants have been obtained that carry the bacterial gene CspD, and enhanced resistance of such plants to TMV and to the pathogenic fungus *Alternaria longipes* has been shown.

The data regarding elicitor activity of the bacterial CSP have been confirmed by the Swiss researchers from the Botanisches Institut, Universitat Basel, who studied elicitor activity of the bacteria *Micrococcus lysodeikticus* extracts and discovered that the active substance in the extracts was a CspD (Felix and Boller, 2003). It should be noted that these authors studied the protein elicitor activity by adding it in cell culture suspension and recorded indirect parameters such as pH variation of the media (induction of "oxidative burst") and increase in intensity of ethylene synthesis. Thus, an artificial system without the pathogen has supported the data on elicitor properties of bacterial CspDs.

It was also shown that a peptide of 15 amino acid residues (csp15 – VKWF-NAEKGFGFITP) is sufficient for elicitor activity of CspD. The peptide csp15 constitutes a consensus sequence of bacterial CSP. This peptide has also been synthesized in Russia, and its protective activity was shown in natural pathogen – host systems: *Septoria nodorum* – wheat and TMV – *Nicotiana tabacum.*

As the CspD genes in one or other structural variation are found in genomes of many living organisms, the bacterial gene CspD is not completely foreign to plants. In addition, toxicological tests in mice show that CspD protein is not toxic for warm-blooded animals. In this case, the mechanism of action of CspD protein is likely to interact with some intracellular receptors triggering the plant natural defense responses against, at least, viral and fungal pathogens.

All these factors taken together suggest possible practical uses of cold shock bacterial protein genes in plant protection against disease. At the same time, the most important theoretical aspect is the question on the nature of intracellular receptors interacting with CspD.

Abiogenic elicitors

It has been shown that, along with biogenic elicitors, abiogenic (synthesized) elicitors are also very promising. They possess a broad range of effects on plant defense systems. Certainly, the initial attention was focused on the natural signalling molecules, SA and JA. However, they have a number of disadvantages that prevented their commercial use. SA is phytotoxic and, in exogenous application, it rapidly metabolizes into β-glycosides incapable of phloem transport. The protective effect of JA and its methyl ester is provided not by the induction of SAR so much but rather by their toxicity for fungal parasites. Therefore, the central focus today is made on SAR-inducing synthetic agents.

Some noncanonic amino acids possess a certain immunomodulating activity. Aminobutyric acid induces the accumulation of PR-proteins in the treated tobacco leaves, and *n*-amino benzoic acid enhances tulip and gladiolus resistance to white rot. The β-aminobutyric acid (BABA) was found to be an efficient inducer of disease resistance, as it increases the level of PR-proteins in tomato leaves and protects them from late blight.

Among these acids, 2, 6-dichlorisonicotinic acid and its methyl ester designated as INA are very interesting. Similar to SA, they induce SAR to bacterial, fungal, and viral diseases, induce synthesis of PR-proteins, but they do not increase SA concentration. Moreover, they induce SAR even in transgenic tobacco plants that produce the enzyme salicylate dehydrogenase and are not capable of accumulating SA.

In 1997, the Swiss company Siba produced the first commercial synthetic elicitor Bion based on INA, which showed high efficacy against fungal, bacterial, and viral diseases of wheat, cucumber, tobacco, beans, sugar beet, rose, barley, and rice. It induces local and systemic resistance of tobacco plants against TMV, *Cercospora nicotianae*, *Peronospora tabacina*, *P. nicotianae*, and *P. syringae* pv. *tabaci*. Induction of resistance with dichlorisonicotinic acid occurs only after inoculation with the pathogen or treatment with its elicitor in low concentration. The INA possesses a dual effect: it induces protective potential of a noninfected plant and sensitizes it after inoculation. As INA is an active LOG inducer, this compound may be involved in the ocadenoid signalling pathway, where the central element is JA. Studying gene expression in rice treated with INA, JA, or infected with compatible or incompatible pathogens (*P. oryzae* and *Pseudomonas syringae* pv. *syringae*, respectively) showed that some genes (PR-genes) express at any exposure, but there are genes expressed only with INA or only upon infection (Figure 15.1). Schwerizer et al. (1997) believe, that INA, being an elicitor and plant sensibilizer, activates various signalling pathways: the pathway leading, similar to infection, to PR-gene expression, and the pathway involved in the expression

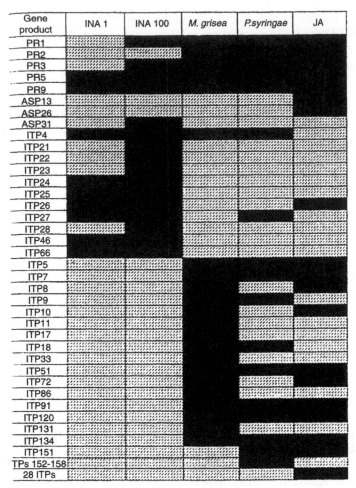

Figure 15.1. Gene-expression patterns in INA-treated versus bacterial (*P. syringae*) and fungal (*Magnoporthe grisea*) pathogen-attacked or jasmonate-treated leaves of rice (Schwerizer et al., 1997, Figure 6).

of other genes, probably, also responsible for resistance. The stress caused by the treatment with high concentrations of INA leads to the accumulation of JA, which activates both pathways.

Derivative INA *N*-cyanomethylamide-2-chlorisonicotinic acid induces rice resistance to *P. oryzae*, which is also accompanied by an increase in LOG activity.

Pyrasocarboxy acids possess the effect similar to INA (Figure 15.2). The advantages of 3-chlorine-1-methyl-1-*N*-pyrasol-5-carboxy acid (CMPA) are good water solubility and complete absence of phytotoxicity. A representative of this group of compounds probenazole (PBZ) and its active substance BIT

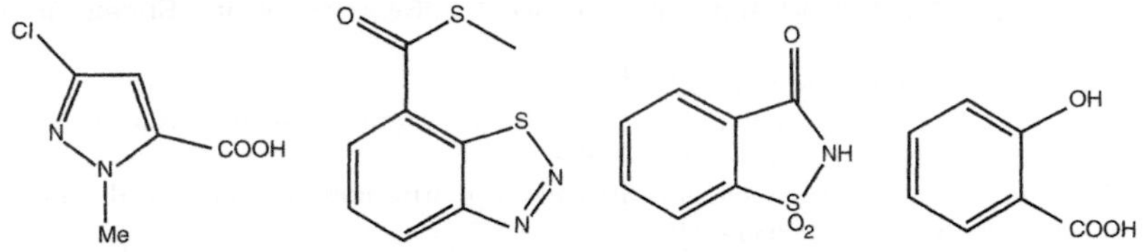

Figure 15.2.   Structure of chemicals that induced disease resistance (Nakashita et al., 2003). CMPA – 3-chloro-1-methyl-1H-pyrazole-5-carboxyl acid; BTH – benzo(1,2,3) thiadiazole-7-carbothioic acid-*S*-methylester; BIT – benzisothiazole; and SA – salicylic acid.

(Figure 15.2), used for the protection of rice against rice blast, induce SAR in the SA-signalling pathway and SA accumulation in plants. A great hope is pinned on benzo-(1,2,3)-thiadiazole-7-carbothioic acid-*S*-methylester (benzothiadiazole, BTH; the commercial agent Acibenzolar). In *A. thaliana* it causes accumulation of SAR-related mRNA genes (PR-1, PR-2, and PR-5) and induces SAR against viral, bacterial, and fungal infections, the immune answer being observed in the mutants unsusceptible to ethylene and JA, as well as in the plants transgenic in nahG (SA degrading). However, the BTH effect is not observed in noninducible mutants nim1, which is indicative of the activation of the BTH signalling pathway of SAR transduction. In wheat, BTH induces gene expression in LOG and sulfur-rich proteins. BTH is a more active inhibitor of catalase and ascorbate peroxidase than SA, therefore, BTH can increase the H_2O_2 level, and, hence, promote the accumulation of ROS that cause activation of the NF-kB transcription factor and induction of apoptosis. This mechanism is an important factor of the BTH immunizing effect, as pretreatment with antioxidants suppresses PR-1 protein accumulation under the BTH or SA influence. The BTH stirs a great commercial interest for the protection of fleshy fruits in storage. Treatment of harvested strawberries with BTH prevents the development of grey rot within 2 days at the storage temperature 5°C. This is equivalent to a 15–20% increase in the fruit storage life.

Selected Literature

Brisson L.F., Battchikova N.A., Korpela T. and Khomutov R.M. Plant cell 1994; 6:1703–1712.

Djavakhia VG, Nikolaev ON, Voinova TM, et al. DNA sequence of gene and amino acid sequence of protein from *Bacillus thuringiensis*, which induces non-specific resistance of plants to viral and fungal diseases. J Russ Phytopathol Soc 2000; 1:75–81.

Dzhavakhiya VG, Averianov AA, Minaev VI, Ermolinski B.S., Voinova T.M., Lapikova V.P., Petelina G.G. and Vavilova N.A. Structure and function of cell wall melanin of rice blast causal agent *Pyricularia oryzae* Cav. J Gen Biol 1990; 51:528–535.

Felix G, Boller T. The highly conserved RNA-binding motif RNP-1 of bacterial cold shock proteins is recognized as an elicitor signal in tobacco. J Biol Chem 2003; 278:6201–6208.

Fulton RW. Practices and precautions in the use of cross protection for plant virus disease control. Ann Rev Phytopathol 1986; 24:67–81.

Hammerschmidt R., Kuc J. (eds) Induced resistance to disease in plants. Kluwer Academic; Amsterdam 1995.

Inoke S. et al. J Pesticide Sci 1984; 9:731–736.

Kessman H, Staub T, Hofmann C, et al. Induction of systemic acquired disease resistance in plants by chemicals. Ann Rev Phytopathol 1994; 32:439–459.

Lyon GD, Newton AC. Do resistance elicitors offer new opportunities in integrated disease control strategies? Plant Pathol 1997; 46:636–641.

Mustafa HM, Dyakov YuT. Effect of pesticides on the late blight development on potato and tomato leaves. Chemistry in Agriculture 1979; 1:56–59, in Russian.

Nakashita H., Yamaguchi I, Yoshida S. Pyrazolecarboxylic acid derivative induced acquired resistance in tobacco. Biosci Biotechnol Biochem 2003; 67(12):2615.

Peng J-L, Dong H-S, Dong H-P, et al. Harpin-elicited hypersensitive cell death and pathogen resistance require the NDR1 and EDS1 genes. Physiol Mol Plant Pathol 2003; 62:317–326.

Schwerizer P, Buchala A, Metraux, J-P. Gene-expression patterns and levels of jasmonic acid in rice treated with the resistance inducer 2,6-dichloroisonicotinic acid. Plant Physiol 1997; 115:67.

Sequeira L. Mechanisms of induced resistance in plants. Ann Rev Microbiol 1983; 37:51–79.

van Loon LC, Bakker PAHM, Pieterse CMJ. Systemic resistance induced by rhizosphere bacteria. Ann Rev Phytopathol 1998; 36:453–483.

Yasuda M, Nishioka M, Nakashita H, et al. Pyrazolcarboxylic acid derivates induced systemic acquired resistance in tobacco. Biosci Biotechnol Biochem 2003; 67:2614–2620.

Chapter 16

Creation of disease-resistant plants by gene engineering

V. G. Dzhavakhiya, L. A. Shcherbakova

All-round introduction of intensive technologies in crop growing involves a number of negative processes, such as contamination of the environment and agricultural produce with xenobiotics, high economic and energy cost. One of the alternative ways of development of this sector of agronomy can be a maximum use of the biological potential of agricultural crops.

Certain hopes here are pinned on gene engineering, approaches and methods to alter the plant genome construction for directional modification of their genetic properties. Gene engineering is based on the technology enabling construction of genetic structures in the form of the so-called recombinant (hybrid) molecules (DNA or RNA) in laboratory conditions. The main point of this technology is that a certain DNA fragment is "cut out" using the specific enzymes, restrictases, and is connected (using the enzyme ligase) to another DNA molecule (vector) that retains the ability for replication upon insertion in the recipient plant cell. This technology basically allows isolation of a gene, interesting to the investigator, from any organism, and, with a suitable vector available, transfer the foreign gene to a desirable plant and try to obtain its expression in that plant. Recombinant nucleic acid technology makes the process of obtaining new plant forms more targeted and this considerably expands the opportunities of manipulation with a plant genome, simultaneously cutting the total time input for development of new cultivars of agricultural crops. The advantages of gene engineering technologies compared with traditional hybridization are:

1. Hybridization impoverishes the whole genomes of the two parents, therefore, the progeny receives, alongside one or several desirable (target) genes, a lot of undesirable genes from the parent, and getting rid of the latter genes requires long-lasting expensive procedures (back-crosses). Gene engineering manipulations import only one target gene to the genome of the cultivar being improved.
2. Hybridization enables transgenosis only from closely related plants (cultivars of the same species, less often, species of the same genus) that has no reproduction barriers. In gene engineering manipulations, it is possible to transfer genes from virtually any organisms to plants, which dramatically increases the opportunities of constructing desirable metabolism.

Methods of transgenic plant creation

Use of pTi plasmid

Efficient methods of transformation (inclusion of a foreign gene in the hereditary system of the recipient plant) have been designed for most dicotyledon and some monocotyledon plants, such as rice, wheat, and corn. The transformation process is carried out either by making use of the mechanisms of the natural exchange of genetic material during interaction of bacteria of the genus *Agrobacterium* with the plant, or by direct insertion of recombinant DNA in plant cells (Figure 16.1).

Development of vector systems for gene transfer in plants began from elucidation of the molecular mechanism of tumour formation in plants upon inoculation with the phytopathogenic bacterium *Agrobacterium tumefaciens*. It was found that these bacteria, capable of infecting most dicotyledon plants, possess a plasmid (extrachromosomal self-replicating genetic item) that can naturally penetrate the host cells and build a particular DNA fragment (T-DNA) in a plant genome. The tumour-inducing plasmid of *A. tumefaciens* was respectively called pTi plasmid (tumor inducing). Integration of the bacterial plasmid genes in the host plant genome and their expression result in a considerable altering of the cell metabolism towards formation of tumour.

Genetic analysis has enabled identification of the genes that control the plasmid ability to induce tumour formation, as well as of the genes encoding synthesis of unique amino acids, opines, in the T-segment of pTi-plasmid. The process of T-DNA transfer and integration in a plant is controlled by the virulence genes (vir-genes) located in pTi plasmid outside the T-DNA segment. An essential structural feature of this segment is direct repeats of 25 base pairs at both ends of the segment (Figure 16.2, LB and RB). Inclusion of any DNA fragment between these two nucleotide sequences results in its physical transfer to the plant chromosome. To obtain expression of a foreign gene integrated in a plant genome, vector systems have been developed using modified Agrobacteria strains that make it possible to insert practically any genes in a genome, obtain their expression, and regenerate whole plants from the transformed cells.

Plasmid pTi is a unique phenomenon. There are grounds to believe that it is a natural chimaera, as it contains two gene patterns: one pattern expresses in plants, the other in a bacterial cell. Regulatory elements of the genes in the T-DNA segment are intended for functioning in a plant cell, whereas the other genes of pTi plasmid are controlled by bacterial promoters.

Even at the early stages of research it became clear that manipulations with a whole pTi plasmid are difficult due to the absence of the unique restriction sites in this large molecule. This problem was addressed by various ways.

Use of intermediate vectors

The point of this approach is that first T-DNA are cloned in *E. coli* using plasmid pBR322. Then the DNA fragments containing the target gene are built

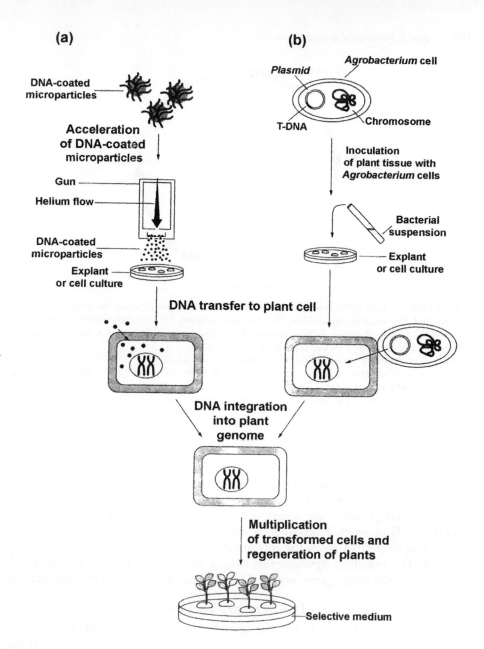

Figure 16.1. There are two main approaches for transferring genes into plant cells.
(a) Direct gene transfer delivers DNA directly to the nuclear or plastid genome of the plant cell through various techniques such as electroporation or chemical treatment, which stimulate the passive uptake of DNA through protoplast membranes, and biolistics, which uses acceleration of DNA-coated microparticles to carry DNA directly into plant cells. (b) Indirect gene transfer is based on a gene-transfer mechanism mediated by *Agrobacterium*. During natural infection, these phytopathogenic bacteria transfer and integrate oncogenes into the plant genome. The transferred DNA from *A. tumefaciens* is a discrete segment of DNA from a small plasmid (Ti plasmid) resident in the cell. This Ti plasmid has been genetically engineered to produce efficient non-pathogenic vectors for plant-cell transformation. Given the low transformation efficiencies of both direct and indirect methods, the selection of transgenic plants is achieved through the combined transfer of a selectable marker gene together with the gene of interest.

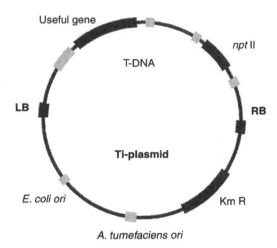

Figure 16.2. A large (200 kbp) circular DNA plasmid is responsible for the production of the tumour and the transforming abilities. This tumour inducing (Ti) plasmid transfers and inserts a region of the plasmid, called T-DNA, into the host chromosome. The T-DNA transfer genes are located outside the T-DNA on the Ti plasmid.
Diagram of some major regions of the Ti plasmid of *A. tumefaciens*:
T-DNA = transferred DNA fragment of Ti plasmid; **LB** = left border of T-DNA; **RB** = right border of T-DNA; ***nptII*** = neomycinphosphotransferase II marker gene; ***E. coli ori*** = origin of replication of the plasmid in *E. coli*; ***A. tumefaciens ori*** = origin of replication of the plasmid in *A. tumifaciens*; **Km R** = kanamycin resistance marker gene.

in particular places of the T-DNA of the previously isolated plasmid, and the resulting chimaeric plasmids are cloned in *E. coli* cells (Figure 16.3).

The cloned plasmids are inserted in the cells of *A. tumefaciens* that have a wild-type pTi plasmid. Homologous recombination results in the build-in of the chimaeric T-DNA from the intermediate vector to an intact wild-type pTi plasmid. The resulting *Agrobacterium* clones, which carry the recombinant pTi plasmid with the target gene, are used in experiments on plant transformation.

Use of binary vectors

It was already mentioned that the genes from the vir-region and T-DNA-restricting direct repeats are necessary for inclusion of pTi plasmid. It was found that for successful transformation these two groups of genetic elements can be situated in different plasmids of the same bacteria, i.e. in this case no homologous recombination is required.

Improvement of the vector systems brought the hybrid plasmids which constitute complex structures. The created structures are considerably smaller compared with wild-type pTi plasmids, and they are more convenient for genetic manipulations. The oncogenes responsible for tumour formation and impeding tissue differentiation have been removed from the T-DNA, which made plant regeneration from the infected tissue culture possible. At the same time, the right and left terminal repeats limiting the

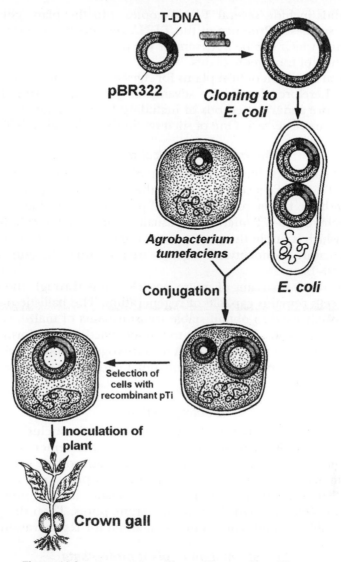

Figure 16.3.   Construction and use of intermediate pTi-vector.

T-DNA region are retained, and the target gene being transferred to the plant is built in between them. The Ti-plasmid region responsible for virulence (*vir*-genes) is also preserved in the transforming plasmids. In the binary vector systems this function is provided by the assistant plasmid with built-in vir-region.

pRi-plasmids

Another type of plasmids was found in *A. rhizogenes* and was named pRi plasmids (root inducing; they cause rootlet proliferation in plants). At plant inoculation, these plasmids include several T-DNA copies into the plant genome and assist opine synthesis. Presence of the T-DNA renders no harmful effect on the host plant, which can be a reason for the preferable use of pRi plasmids in construction of transgenic plants.

The natural spectrum of host plants for *Agrobacterium* is limited by dicotyledon plants; therefore, despite the advance of gene engineering due to this approach, more general methods of including foreign genetic information in plant cells are necessary. One of such methods is biological ballistics.

Method of biological ballistics

The method consists of "firing" at recipient plant cells from a special vacuum gun by metal micro particles coated with the vector carrying an appropriate construct with the target gene. The metal particles (for instance, tungsten) with a diameter of 0.6–1.2 microns are usually applied on the cellophane substrate film, which is put in the gun opposite the gun mouth. Cell suspension, callus or immature monocotyledon germs are put under the gun mouth at a distance of 10–15 cm.

After the "shot", a stream of micro particles runs through the cells, and part of the cells remains capable of regeneration. The ballistic method has been successfully used to obtain stable transformants of maize, rice, wheat, barley, and other crops. In addition to monocotyledon transformation, this method enables direct transfer of DNA in embryogenous pollen with the subsequent prompt obtaining of transgenic dihaploid plants suitable for subsequent use in selection.

Electroporation

Another method using physical effect on plant cells for inclusion of foreign DNA is electroporation. The method is based on enhancement of permeability of biological membranes under the effect of high-voltage electric pulses. A short-term action on the mixture of plant protoplasts and transforming DNA (250–300 v during fractions of a second) results in DNA molecule penetration in the cells through the cell membrane pores. Then the protoplast suspension is diluted and planted in the medium for regeneration.

Transformation of plant protoplasts

In treatment of plant protoplasts with chemical fusion inducers, capture of DNA molecules from solution by the fusing cells is possible, against a background of high concentration of calcium ions, which results in protoplast transformation. Despite the low efficacy of transformation compared with the previously described methods, this approach has certain advantages, based on the fact that the method does not require special mechanisms for vector inclusion in a plant cell and integration into the plant genome.

The foreign fragment of DNA gets stably included in the structure of the nuclear DNA, and, provided that the appropriate regulatory regions are available, expresses.

Microinjections

The solution of merely technical problem of protoplast fixation on glass using polylysine made possible the injection of foreign DNA preparations into cells through micro needles with the external diameter 2 microns. The approximate efficacy of transformation here varies within 10–20%. As the species-related limitations for this method are yet unknown, it can be regarded as an all-purpose technique.

Marker genes for transformant selection

As frequency of transformation is usually low, marker genes are inserted in hybrid plasmids for convenience of the transformant selection at selective media. As marker genes, antibiotic resistance genes from *E. coli* are used (kanamycin, chloramphenicol, etc.), to which the plant cells are susceptible. In addition to the antibiotic resistance genes, the genes encoding biosynthesis of the enzyme luciferase, isolated from bacteria of the genus *Vibrio*, are also used as marker genes. In this case the transformants are distinct against a general background by the ability for luminescence. A convenient marker is also the gene of bacterial β-glucuronidase (GUS) that allows identification of the transformants by specific staining of the selective media.

For expression of bacterial marker genes, the appropriate signal sequences are introduced in plant cells for the start and termination of the gene transcription process. The following promoters were used as regulatory elements:

– 35S-promoter of the cauliflower mosaic virus (CMV);
– promoters of pTi plasmid spin genes;
– promoter of the gene of the small subunit of the enzyme ribulosediphosphate-carboxylase;
– promoter of the chlorophyll-binding protein gene, etc.

Transformant regeneration

The bottleneck of transgenosis technology is regeneration of the whole plants from transformants. Regeneration is easier in multicellular structures. The most common method is leaf disks. Leaf tissue disks, cut by a standard punch, are incubated in suspension of Agrobacterial cells, the plasmid of which contains an insertion with the target gene. Then the pieces are transferred to the selective regeneration medium containing the antibiotics inhibiting growth of Agrobacteria (carbencillin), as well as the antibiotics inhibiting growth of untransformed tissues (kanamycin, neomycin, chloramfenicol, metatrextate). The selective medium enables selection of only those leaf tissues, which contain marker genes, i.e. tissues transformed by T-DNA. Presence of the target gene is further tested by genetic and molecular analysis.

Vectors based on the genome of cauliflower mosaic virus (CMV)

In addition to pTi- and pRi-plasmids, others vector and transformation systems are also used in transgenic plant creation technologies.

CMV is one of few DNA-containing viruses whose genome is used to transfer foreign genes to plants. A relatively small size of the circular virion DNA (8 Kbp) allows easy manipulation with it under *in vitro* conditions, similar to an ordinary plasmid, and insertion of this DNA in plant cells by mechanical inoculation of the leaves. As CMV infection spreads in a plant systemically, the inoculation of a small leaf area leads to prompt propagation of the recombinant DNA over the entire plant. Another characteristic feature of this DNA is its large copy number (up to 50,000 per cell). However, the greatest advantage of the CMV genome is its promoter, which is widely used in gene engineering constructs due to its high activity, absence of tissue specificity, and ability to induce expression of foreign genes not only in Cruciferae but also in plants from other families.

Shuttle vectors based on chloroplast and mitochondrial DNA

The chloroplast genome is a circular DNA containing some number of genes necessary for functioning of the chloroplasts and maintaining their structure. These DNA also contain the genes of ribosomal and transport RNA. The size of a chloroplast DNA varies within 120–180 tbp. Each chloroplast contains tens of copies of the circular DNA. The complete nucleotide sequence of the tobacco and rice chloroplast DNA has been discovered.

An essential property of the chloroplast genome is its ability for independent replication and transcription, which allows the use of chloroplast DNA to construct vectors. Combination of a bacterial plasmid with the chloroplast DNA fragment, which contains regulatory elements for replication and transcription in a plant cell, can lead to obtaining a multipurpose vector. Such vectors are called shuttle vectors. They are notable for the ability for independent replication in bacterial cells as well as expression of foreign genes in eukaryotic cells.

Mitochondrial DNA is also a circular duplex with the length varying between 1 and 50 thousand base pairs. It can be used for construction of shuttle vectors, as it possesses all necessary regulatory elements for independent replication and transcription in a plant cell.

Development of transgenic plants resistant to viral disease

Resistance provided by expression of the viral coat protein in a plant gene
Cross-protection of plants against viral infection

The phenomenon of cross-protection against viruses in plants was first described more than 50 years ago. In a nutshell, it is the inability of a viral pathogen to infect the plants which are previously inoculated with another virus. The mechanisms of cross-resistance were studied using a very common approach: mutants defective in some gene were obtained, and the role of these gene products in the process was studied. For instance, when TMV mutants unable to synthesize a full viral coat protein (VCP) were used for primary inoculation, it was shown that these mutants can induce tobacco plant defenses. However, plant treatment with the viral strains possessing a normal VCP (wild-type genome) resulted in a much greater protective effect.

A TMV VCP mutant, unlike other mutants in this gene, induces non-specific plant resistance – as far as resistance to turnip mosaic virus (TuMV) that belongs to another taxonomic group, poty viruses (see Chapter 1), while the plant treatment with a wild-type TMV strain induces resistance to other TMV strains in a much greater degree than to TuMV. This suggests the existence of at least several cross-resistance mechanisms.

The above described approach to studying cross-resistance mechanisms has a number of limitations, first of all related to the fact that in some viral genes mutations cannot be expressed phenotypically. For instance, defects in the replicase gene will result in a "lethal" outcome for the viruses.

A considerable progress in this area was made after the discovery of the transfer of individual viral genome fragments to plant cells, with subsequent expression of the transferred viral genes. This process is called transgenosis, and the resulting plants are called transgenic.

Transgenosis of viral protein

One of the first viral genes transferred to tobacco plants was the TMV coat protein (VCP) gene. The gene was expressed in tobacco plant cells using heterologic promoters. In these experiments transgenic tobacco plants with a high level of TMV coat protein expression were found to be more resistant to inoculation with TMV suspension than to inoculation with TMV RNA preparations. An assumption was suggested that in this case one of the possible resistance mechanisms could be blocking of dissociation of the coat protein of TMV virions, which is necessary for the release of the viral RNA and to start of the initial stages of the infection process. Successful infection of transgenic plants with TMV RNA preparations in this case is explained, in addition to the absence of the stage of dissociation, also by the fact that intercellular transport of the viral RNA in this system does not need the coat protein.

The blocking of the primary stages of the infection process was confirmed in the experiments on inoculation of protoplasts of different types of tobacco cells. The expression of the TMV coat protein gene in the protoplasts of epidermal cells leads to development of resistance of these protoplasts to inoculation with TMV, while the use of the promoters specifically directing the VCP gene expression in mesophyllic cells does not bring resistance of the protoplasts of these cells to inoculation with TMV.

The hypothesis concerning the mechanism of VCP-induced resistance of transgenic plants to TMV was directly supported in the experiments with β-glucuronidase (GUS) mRNA encapsulated in TMV coat protein. Such TMV-like particles (pseudovirions) were used to inoculate protoplasts in transgenic and non-transgenic tobacco. The level of expression of GUS mRNA in the protoplasts of the transgenic lines of tobacco possessing the TMV coat protein gene was much lower than the expression in the protoplasts of the non-transgenic plants. The only explanation is the blocking of

VCP separation from GUS mRNA in the cells synthesizing a large amount of TMV coat protein.

The further studies of the mechanisms of plant resistance to viruses provided by the VCP gene expression in transgenic [VCP (+) plants] showed the presence of multilevel defenses. Plant inoculation with preparations of the viruses causing leaf spot in all events showed a decrease in the average specific number of infection spots per one leaf in VCP(+) transgenic plants compared with VCP(−) lines of the same cultivar. Therefore, one of the possible sites of antiviral defenses can be the barrier set up at the initial stages of development of the infection process. Thus, the expression of VCP genes in the cells of transgenic plants seems to stir VCP interaction with the plant tissue structures responsible for resistance to penetration of viral infection.

A comparative study of the subsequent stages in development of the infection process in VCP(+) and VCP(−) tobacco plants was carried out as follows. TMV concentrations in the inoculums were chosen in such a way that they caused about the same specific number of necroses per leaf of a VCP(+) or VCP(−) plant. The analysis of the average number of viruses in each infectious spot showed that there is no significant difference in this parameter between VCP(+) and VCP(−) lines. However, the immunoassay of the leaf segments cut out from the areas between the infectious spots showed that in case of VCP(+) plants the virus accumulation in these parts of leaves is reliably smaller than in the VCP(−) plants. This suggests a conclusion that expression of the VCP-encoding TMV genes induces development of a second defense barrier in tobacco plants, at the level of the virus spread in the leaf tissue.

To verify the VCP effect on TMV transport in the vascular system, a grafting experiment was carried out. A fragment of a transgenic plant containing a VCP was grafted on a TMV-susceptible non-transgenic plant. It was found that the TMV transport through the grafted segment was suppressed, which confirms the assumption regarding the blocking of the mechanisms of virus transmission over the vascular system of the transgenic plants expressing the VCP gene. However, the TMV movement over the vessels is probably accompanied by the alternating processes of virions assimilation and dissimilation. In this case the blocking of the virus transport over the vascular system may be a secondary consequence of virion dissimilation inhibition.

An additional evidence for the hypothesis of the blocking activity of VCP genes at the stage of virion dissimilation is based on the results of experiments using the alfalfa mosaic virus (AMV). However, the resistance mechanisms in this case are more complex than in case of TMV resistance. In the lines with a low level of expression of the VCP-encoding AMV genes, resistance was observed only at inoculation with AMV virions. In contrast, in the transgenic lines expressing a mutant VCP or many wild-type VCP, resistance to AMV occurred at inoculation with both RNA preparations and virions. Similar to TMV, the dependence of resistance on the type of inoculum suggests that AMV resistance occurs at least at two levels. One level can include inhibition of virion dissimilation at a low level of VCP expression. The second level of resistance occurs in the lines with mutant VCP or with a high level of expression of a wild-type VCP gene, and it can include any VCP interactions necessary for the infection cycle of the virus. These can be interactions with the viral RNA, necessary for the virion assembly or virus replication, or inter- actions with the VCP receptors of the host plant.

Resistance to the potato X-virus (PXV) was found to be efficient at inocu- lation with both RNA inoculum and virion inoculum. Similar to AMV resist- ance, blocking of all stages of the infection process necessary for PXV replication and transport in the infected plant can also take place here. There are grounds to believe that interaction of the VCP synthesized by a transgenic plant with the site of initiation of PXV assembly can result in development of resistance. In potex viruses, the assembly begins at the 5′ end of the viral genome; therefore, interaction with VCP can block translation of the viral RNA-dependent RNA polymerase which is encoded at the 5′ end of the open reading frame. However, resistance to PXV may be also caused by inhibition of intercellular transport for which the VCP is a necessary factor.

Summarizing the previous data, the following conclusions can be drawn regarding the defense mechanisms implemented in the transgenic plants that express a virus coat protein gene:

1) In VCP(+) plants infection blocking develops at a stage preceding the synthesis of the minus chain of the viral RNA.
2) The level of defense of the VCP(+) tobacco plants is much higher against TMV than against TMV RNA. This protection is induced by the synthesis of TMV coat protein in the cells of the transgenic plants and is not connected with the synthesis of VCP m-RNA transcripts. A similar situation occurs in case of transgenic plants carrying the alfalfa mosaic virus coat protein.
3) The level of defense of VCP(+) transgenic plants is determined not only by the coat protein gene expression but also by the ability of that protein to produce intracellular virus-like aggregates.
4) VCP(+) plant stem engrafting on the shoots of VCP(−) tobacco plants develops induction of resistance to the systemic TMV infection, without preventing development of the primary stages of infection.

It would be logical to assume that in addition to the viral coat proteins there are also other proteins the gene expression of which in plants could result in development of resistance to viral infections. Consider the proper- ties of a number of such proteins that are interesting both theoretically, since they can be used as a tool to study virus resistance mechanisms, and practi- cally, since the resulting transgenic agricultural crops can stir a concrete practical interest.

Resistance of transgenic plants induced by expression of transport proteins

In the majority of the earlier examples, resistance of transgenic plants is induced by the expression of wild-type VCP genes. However, genetic engineering can produce pathogen-induced resistance using negative mutant forms of the dominant viral genes. Efficiency of this approach was demonstrated in transgenic expression of viral transport proteins (TP), with resistance observed only when transgenic plants express a defective TP gene. Transgenic expression of normally functioning wild-type TP genes either has no effect on viral infection or increases the plant susceptibility.

Resistance induced by TMV TP gene expression results from competition for the linking sites in the cytoplasmic membrane between a mutant TP and a wild-type TP inserted in the viral cell.

A particular feature of such type of resistance is its relative non-specificity. For instance, TMV TP gene expression results in development of plant resistance to cucumoviruses, potexviruses, tobraviruses, and, certainly, tobamoviruses. Similarly, expression of a defective TP of the Brom Mosaic Virus induces resistance to TMV (Malyshenko et al., 1993). These facts suggest a conclusion that transport proteins of phytoviruses belonging to different taxonomic groups can interact with the same receptors of the cytoplasmic membrane, which promises unquestionable good future prospects for this area of research both in a practical and theoretical aspect.

Resistance of transgenic plants induced by expression of antiviral antibodies

Unlike vertebrates, plants have no immune system capable of producing antibodies that can recognize and inactivate pathogens and parasites penetrated into the host organism. However, due to the achievements of gene engineering and the technologies designed for development of transgenic plants, immunoglobulin-encoding genes can be introduced into a plant genome, with their expression resulting in the formation of antibodies or their fragments capable of functioning in a plant. This approach was used to develop transgenic plants resistant against viral infection.

If the genes, the expression of which results in the accumulation of those antibodies or their fragments that are able to bind the antigens playing an important role in pathogenesis (for instance, toxins, enzymes, transport proteins), are introduced into a transgenic plant, such technology can be used as a method of disease control.

Today, the structure of immunoglobulins (antibodies) has been investigated very well. Hypervariable sites responsible for specific binding with an antigen have been revealed. The genes encoding the sequence of amino acid residues in polypeptide chains of immunoglobulins and their functional fragments are known. Methods of antibody gene engineering have been developed enabling construction of the genes the expression of which leads to formation of the entire antibody molecule or its antigen-binding fragments, such as Fab-fragment and single-chained Fv-fragment (scFv).

Antibody expression in plants

Gene engineering achievements have opened an opportunity to develop transgenic plants where molecules of antibodies or their functional fragments are expressed. One of the ways to assure expression of antibodies in a plant is to first obtain the plants where the heavy and light polypeptide chains of immunoglobulins express separately, then the whole antibody molecules are obtained in the progeny from the crossing of such plants. Using another approach, both genes are cloned in the same expression cassette either with tandem or divergent promoter. Introduction of this construct into plants results in formation of the plant functional antibodies. To obtain expression of the whole antibody molecules in plants, it is important to include a signal peptide sequence in front of the genes of both heavy and light chains, to assure chain self-assembly in the endoplasmic reticulum. This purpose is achieved through the use of secretory proteins of various organisms (plants and animals). Without signal peptides, it is impossible to obtain successful expression of the whole antibody molecules in a plant. One of the possible reasons may be interference of the reduction properties of the cytoplasm content in formation of disulphide bonds. The single-chained Fv (scFv) which lacks signal sequences can express in cytoplasm and specifically bind the antigen, however, data is available that inclusion of N-terminal and C-terminal signal peptides enhances scFv expression and its stability in the plant. Due to a relatively small size, this fragment possesses good tissue permeability. It does not need self-assembly for its functioning. In addition, it can be fairly easily produced by gene engineering methods. All these factors make scFv attractive for development of transgenic plants producing antibodies responsible for disease resistance. The use of this fragment may enable avoiding such problems involved in providing expression of whole antibodies as glycosylation and formation of intermolecular disulphide bonds.

Antibodies against plant pathogens

As soon as the first plants with functioning antibodies were obtained, the idea arose that this approach could be used for plant protection against diseases (Hiatt et al., 1989). Then, the next steps in application of this strategy were made to design a gene engineering construct of scFv obtained from monoclonal antibodies possessing high affinity to virus coat protein that causes mottled crinkle of artichoke. Transgenic tobacco plants where successful expression of gene scFv was obtained were less often affected by the virus, and they showed milder signs of the disease.

Later, cloning was carried out of cDNA of heavy and light chains of the antibodies with high affinity to the epitope present only on the surface of an intact TMV. Simultaneous expression of the heavy and light polypeptide chains resulted in self-assembly of functional antibodies in the tobacco plants. The number of lesions in the inoculated transgenic plants went down. This reduction correlated with the number of antibodies produced by the plant.

Recently, the genes responsible for the synthesis of the heavy and light polypeptide chains of monoclonal antibodies against root-knot nematode were separately cloned. Crossing of the transgenic plants gave the progeny with expression of the whole antibody molecules. However, antibody expression in these plants did not affect the parasite. Probably, there was no interaction between the antibodies and nematode secretory antigen, as they were spatially separated in the plant.

An attempt to develop transgenic corn plants producing antibodies against the corn stunt pathogen has been undertaken in the USA. The causal microorganism, spiroplasma, falls into the group of plant pathogenic phytoplasmas and spiroplasmas, the prokaryotes lacking the cell wall. These microorganisms cause more than 300 world known plant diseases. These diseases are detrimental for crop yield, and sometimes they completely destroy some agricultural crop. Only very few controls are efficient against these diseases. Monoclonal antibodies, which actively inhibit the spiroplasma growth, were produced by hybridome technique. Then, using gene engineering technology, the gene encoding scFv of these antibodies was introduced into plant cells and its expression was obtained in the plant cell culture. However, the plants regenerated from the transgenic cells and grown in greenhouse did not show distinct resistance to the pathogen. The reasons probably were the change of affinity or functional capability of scFv *in vivo*, as well as the fact that the expressing fragment and the penetrated pathogen could be localized in different places of the plant cell.

Development of transgenic plants resistant to fungal diseases

There are a number of various plant defenses against fungal pathogens, including:

- morphological barriers
- local death of the infected cells
- biosynthesis of phytoalexins
- synthesis of the enzymes degrading pathogen cell walls
- synthesis of the ribosome-activating proteins

Consider the possible uses of various host or pathogen genes, the products of expression of which influence their interaction, in development of transgenic plants resistant to fungal diseases.

Plant resistance to fungi as a consequence of foreign phytoalexin gene expression

Plant resistance to fungal phytopathogens can be provided by a range of defenses, both constitutive and induced. The inducible synthesis by the plant of phytoalexins – low-molecular-weight secondary antimicrobial compounds – is considered one of the most important defenses. An advantage of such a defense is a prompt triggering of phytoalexin synthesis in response to inoculation of the host plant tissue with a pathogen, stress, or treatment with some natural or synthetic compounds. The genes that control biosynthesis of some phytoalexins have been studied. The gene engineering methods were used to transfer these genes to the recipient plants of other species, genus, or even family, which allowed demonstration of the role of phytoalexins in defense responses of the plants where the built-in genes are foreign. In addition to theoretical importance of these transformation experiments, they suggest practical uses of the gene engineering technologies for development of transgenic varieties of agricultural crops with the resistance to fungal phytopathogens induced by the gene expression of highly efficient foreign phytoalexins. This section discusses the experimental results in this area.

The peanut and grape genes encode stilbene synthetase (STS), which is responsible for synthesis of the stilbene-like phytoalexin resveratrol. When these genes are transferred to the tobacco plant genome, they express resveratrol in response to inoculation with a fungal pathogen. The STS is a key enzyme involved in formation of stilbene at transformation of one molecule of p-cumarol-CoA and three molecules of malonyl-CoA to 3,4,5-trihydroxystilbene (resveratrol). This evidence points to a general possibility of expression of the phytoalexin resveratrol in a foreign host plant; however, the level of STS gene expression in tobacco is low.

After treatment with an elicitor preparation from *Phytophthora megasperma*, the final number of STS mRNA accumulated in the calluses of transgenic tobaccos totals only approximately 0.5% of the STS mRNA accumulated in a homologous system, the peanut callus. The reason can be the normal presence of at least 6 or 8 different STS genes in the peanut cells, while in these experiments only one STS gene was used. Greenhouse phytopathology tests with a transgenic tobacco containing the peanut gene showed a relatively higher resistance of the transgenic plant to inoculation with *P. megasperma.*

The level of expression of the STS genes was improved by the transfer of two grape STS genes to tobacco plants. The STS mRNA synthesis in the tobacco calluses was 5–10% of the amount induced by STS mRNA elicitor synthesis in the grape calluses. Similar to the grape, the STS synthesis in transgenic tobacco was also found to be inducible.

In addition to producing a large amount of the phytoalexin resveratrol, the resulting transgenic tobacco showed a high degree of resistance to *B. cinerea.* The level of the disease suppression varied in the range of 52 – 82% compared with the original non-transgenic tobacco.

Improvement of resistance to *P. infestans* and *Fusarium sulfureum* was obtained in experiments with transgenic tomato and potato plants that express two grape STS genes.

In the analysis of these experimental data, one important aspect should be emphasized. To improve disease resistance, a particular pathogen should induce biosynthesis of phytoalexins in the host plant. The molecular analysis of STS gene expression showed that resistance can be induced in a heterologic plant, at least by necrotrophic fungi. For instance, *B. cinerea* or *Alternaria longipees* can induce synthesis of resveratrol in tobacco. From the perspective of the practical agronomy, it would be important to find out if biosynthesis of foreign phytoalexins can also be induced by biotrophic fungal pathogens (such as *Erysiphe graminis*, *Bremia lactuca*, *Plasmopara viticola* or *Puccinia* spp.) in the respective host plants. The experimental results in the pathogen–host pair *P. viticola-Vitis vinifera* showed a positive correlation between the induced synthesis of the phytoalexin and the host plant resistance. This significant factor allows a hope for an actual use of the above approaches to develop economically important sources of resistance and their further use in plant selection activities.

One of the promising areas of the search for the ways to improve plant resistance to fungal diseases by gene engineering is the detailed investigation of the mechanisms of regulation and the induction of certain key genes involved in the biosynthesis of phytoalexins. Casben synthetase from the castor-oil plant – the terminal enzyme involved in the biosynthesis chain resulting in formation of the diterpentoid phytoalexin casben, can be taken as an example. Synthesis of casben synthetase is induced in plant tissues in response to inoculation with the fungi, treatment with a fungal protein or other elicitors. Today, kDNA of this enzyme has been cloned, which opens a way to development of the transgenic plants with possible inducible resistance to fungal diseases. Since casben synthetase catalyzes only one final stage, unique in the chain of the casben biosynthesis, the transfer of the casben synthetase gene to other plants would help to identify the role of casben within the defenses against fungal pathogens.

Another possible approach to the improvement of plant resistance to fungal pathogens could be incorporation in plant genomes of the genes, the products of which can modify the structure of phytoalexins, making them either more fungitoxic or more resistant against the activity of fungal detoxicants. For instance, different fungal pathogens of peas are known to possess different susceptibility to (+) or (−) stereoisomers of the phytoalexin pisatin. In addition, these isomers differ in susceptibility to the degradation factors of the same pathogen. There are the enzymes isomerases (in this case, isoflavon oxidoreductase) that could be used for preferential synthesis of a (+) or (−) stereoisomer of the phytoalexin pisatin, depending on the pathogen against which the host plant resistance needs to be improved. At present, the enzyme isoflavon oxidoreductase has been isolated and purified, and also the gene responsible for synthesis of this enzyme has been cloned. Implementation of this approach could be an example of development of transgenic plants with artificially increased activity of phytoalexins protected from the degradation factors of the fungal pathogens.

There are also other ways to improve fungitoxicity of phytoalexins. For instance, the fungitoxic properties of stilbenes are known to be connected with the hydrophoby of these compounds. Therefore, their biological activity can be probably increased by methylation. In this connection, the transfer of the genes, similar to resveratrol methyltransferase or pinosylvin monomethyltransferase, to new host plants can result in formation of more fungitoxic monomethyl ethers pterostilbene or pinosylvin, respectively.

Pinosylvin synthetase (PSS) – stilbene synthetase from pine – forms 3,5-dihydroxy stilbene (pinosylvin), which is more hydrophobic than resveratrol. The PSS gene has been isolated and cloned, and the gene was transferred from a common pine to a tobacco plant. the PSS expression was controlled by 35S RNA promoter and STS promoter from grape. Activity of this enzyme was found in the freshly isolated extracts of tobacco tissues. Phytopathologic experiments in greenhouse showed improvement in resistance of these transgenic plants at inoculation with *B. cinerea*. These data enable an opportunity for synthesis of other foreign phytoalexins, in addition to resveratrol, thus improving disease resistance and enlarging the range of antifungal activity of new transgenic plants.

A possible candidate for such studies could be the enzyme bibenzyl synthase (BBS) that catalyzes a stage in the biosynthesis of the phenol component of phenanthrene phytoalexins. These phytoalexins are produced in the tissues of the orchid *Bletilla striata*. Since the gene corresponding to this enzyme has been isolated and cloned, it could be transferred to foreign plant species to induce synthesis of dihydrophenanthrene phytoalexins.

In addition to expression of the ribosome activating proteins, chitinases, glucanases, and other PR-proteins, there is a certain promise in the approaches based on pathogen-inducible synthesis of known phytoalexins or their derivatives resulting from the transfer of the appropriate genes to foreign host plants by the gene engineering methods. It should be stressed that so far none of the mentioned approaches has brought the development of a commercial product.

Thus, the use of new endogenous antimicrobial compounds is one of the most promising strategies in the molecular selection of plants for resistance to fungal pathogens. The major limitation of this approach so far has been the small number of the genes responsible for synthesis of phytoalexins or their modification, available for such experiments. As the knowledge about the molecular mechanisms of the fungus–plant interaction is gained, the opportunities of application of the above-described biotechnology approaches and methods in agricultural practice will grow.

Host plant peroxidases

Induction of peroxidase synthesis is one of the early plant responses to interaction with a microbial pathogen. Peroxidases are involved in several plant defenses, including lignification, cell wall protein binding, wound healing, and generation of antimicrobial active radicals. Fungal phytopathogens can specifically induce synthesis of new peroxidase isoforms in host plant tissues. Despite of the presence of a numerous peroxidase gene family in the plant genome, the individual peroxidase genes (Shpx6a and Shpx6b) were isolated from the tropical leguminous plant *Stylosanthes humilis* using a modified polymerase chain reaction (PCR). The transcripts of these genes are found at the earliest stages of plant inoculation with the fungal pathogen *Colletotrihum gloeosporioides*, as well as after the plant treatment with the defense response regulator methyl jasmonate (MeJA).

The bioengineering constructs built on the basis of these genes were transferred to tobacco plants. Inoculation of the leaves of transgenic tobacco with the conidia of the phytopathogenic fungus *Cercospora nicotianae* as well as inoculation of these plants with *Phytophthora parasitica* var. *nicotianae* induced expression of the genes Shpx6a and Shpx6b. A similar effect was also observed after treatment of the tobacco leaves with MeJA. These results are fundamentally important, as it was the first demonstration that the synthesis of the particular plant peroxidases can be induced through inoculation with a pathogen and treatment with a chemical inductor. This provides an opportunity to study the signalling processes accompanying the fungal infection in the given host–pathogen systems. A practical value of the data obtained is showing a real way of using peroxidase genes to develop transgenic plants resistant to fungal diseases.

Elicitins

Proteins from the species *Phytophthora*, called elicitins, differ among themselves in the amino acid sequences as well as in the ability to induce necroses and plant defense responses. The proteins from the subfamily of β-elicitins (with alkaline properties) possess a greater activity in induction of necroses than the proteins from the α-subfamily (with acid properties). Expression in *Escherichia coli* of the synthetic gene encoding β-cryptogenins results in production of a protein capable of inducing necroses in tobacco leaves. The replacement of the only amino acid (lysine in position 13 by valine) reduces the necrosis formation activity to the level of a similar activity of β-elicitins. The quarternary structure of β-cryptogenin as well as the active centres of these proteins responsible for induction of necrosis formation and induction of the host plant defense responses have been identified.

The phytopathogenic fungus *P. parasitica* var. *nicotianae* can efficiently infect tobacco plants as it produces an elicitin not recognizable by tobacco. The gene of β-cryptogenin, expressed under the control of the 35 CaMV virus promoter, was transferred to these plants, and the resulting transformants included the clones with high resistance to inoculation with *P. parasitica* var. *nicotianae*. An impressive evidence that the resistance is provided by the synthesis of foreign β-cryptogenin is reduction of resistance in the transformants that express a gene of the same elicitin, but with replacement of lysine by valine in position 13. The results of these experiments place the genes of pathogen-produced proteins – elicitins – among the promising candidates for development of transgenic plants with resistance to fungal pathogens.

"Barnase" and "Bastar" genes

Concentration on the biotechnological approaches to improve plant resistance to fungal phytopathogens using the genes that control the synthesis of chitinases, gluconases, or ribosome-activating proteins, as well as using the low-molecular-weight fungitoxic compounds such as phytoalexins can result in real development of highly resistant cultivars. However, at least some foreseeable shortcomings of such approach should be taken into account. A high level of resistance of the transgenic plants produced by such a way is based on the specificity of the mechanisms used. Selective resistance to a narrow taxonomic group of phytopathogens is not advantageous from the agronomic point of view.

Theoretically, the pathogen more easily overcomes the highly specific protective barriers of the host plant due to, for instance, one or several mutations in the fungus genome. In this connection, it would seem the most acceptable, in terms of agricultural practice, to develop transgenic plants with multiple resistance to fungal diseases using the genes responsible for natural non-specific defenses, for instance, the hypersensitive response. Theoretically, the overcoming of such type of defense should be much more difficult and take the pathogen more time compared with highly specific disease resistance mechanisms.

One of the possible approaches to the artificially produced hypersensitive response, accompanied by a fast cell death of a susceptible cultivar, is synthesis of cytotoxic compounds in the host tissues. The success of such approach depends on the possibility to contain the cytotoxic effect within the infected cells, which can be achieved by the expression of cytotoxin genes under the control of a promoter that would be specifically activated in response to introduction of the pathogen in the plant tissues. Another possible way to ensure such type of protection is suppression of cytotoxin synthesis in all uninfected cells.

The promoter of the potato gene prp1-1 has a regulatory site which is normally inactive but can promptly and locally activate the transcription of the controlled gene in response to introduction of a fungal pathogen. This promoter site does not respond to abiotic factors and is inactive in any uninfected tissue.

The barnase gene, isolated from Bacillus amyloliquefaciens, encodes RNAase that possesses high cytotoxic activity. A gene engineering construct has been developed where the barnase gene was placed under the control of the mentioned prp1-1 fragment of the promoter (see Figure 16.4). To minimize the destructive effect of barnase in uninfected tissues, a chimaeric gene construct – a barstar gene controlled by 35S promoter of the cauliflower mosaic virus (CMV) – was introduced in the genome of the same clones. Barstar is a barnase suppressor, the barstar gene was also isolated from *B. amyloliquefaciens*. The resulting construct was used for transformation of the potato cultivars susceptible to late blight. The resulting transgenic plants were found to be more resistant to late blight compared with the original non-transgenic plants, which was manifested in delayed spore-bearing and reduced production of the sporangia developed on the leaf surface.

Some transgenic clones showed expressly pronounced morphological changes in the leaves – to the point of the leaf tissue destruction unrelated to the fungal infection. These lines were not used in further work. The appearance of such variants among the transgenic plants shows that efficiency of the protection provided by a two-component system depends on the balance between the expression of two target genes and that this balance in different lines can be shifted to either side.

To check the expression of the target gene of barnase, transgenic potato lines with normal leaf morphology were treated with ethylene (chemical inducer of prp1-1 promoter activation). The effect of the ethylene treatment

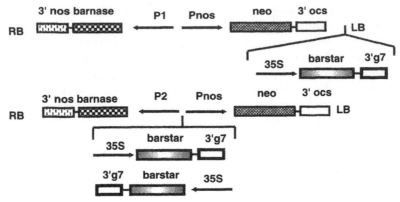

Figure 16.4. Chimaeric gene constructs used for transformation. In plasmids pTCV13 and pTCV14 the barnase gene was fused with the *prp1-1* promoter regions covering positions – 402 to +31 (P1) or – 402 to –130 combined with the CaMV35S TATA box region – 46 to +8 (P2), respectively. Plasmids pTCV15, pTCV16 and pTCV17 contain the barstar gene under control of the CaMV 35S promoter (35S), in addition to one of the prpM/barnase constructs; pTCV16 and pTCV17 differ in the orientation of the CaMV 35S-barstar gene construct with respect to the the *prp1-1* promoter fragment. All plasmids harbour the neomycin phosphotransferase II gene (neo) under control of the *nos* promoter (Pnos), as a selectable marker. 3′g7, 3′nos and 3′ocs, polyadenylation signals of the *A. tumefaciens* T-DNA gene 7, nos and *ocs*, respectively; LB and RB, left and right T-DNA border repeats, respectively.

was a large-scale destruction of the leaf tissue of the transgenic lines; hence, the built-in barnase gene can really express in the cells of these plants in response to the induction of prp1-1 promoter.

In this case, the low resistance of the transgenic plants can be possibly explained either by a high level of constitutive expression of the barstar gene or low inducing activity of the prp1-1 promoter in response to the fungal infection. For an optimum balance between the expression of the two target genes, it is necessary either to replace the CMV 35S promoter with a less active promoter or to find a more susceptible promoter for the barnase gene.

An alternative model to generate a prompt and efficient protective barrier in the form of drying cells in the pathogen entry area can be the simultaneous transformation of plants by the avirulence genes of a fungal pathogen and the respective resistance gene of the host plant.

Enhancement of plant resistance to bacterial diseases by genetic engineering

Bacterial diseases have a major economic impact on many agricultural crops, as they inflict great yield losses in cereals, vegetables and fruit. In most cases, the use of protective agrochemical technologies is not sufficient for efficient control of bacterial diseases. Use of chemical control agents has major limitations because of their potentially harmful effect on the environment; therefore, control of bacterial diseases frequently does not go beyond prevention. Existence of natural sources of resistance to bacterial diseases enables routine selection of resistant cultivars (hybridization). Classical hybridization is confined to intra-species crosses (or crosses between closely related species), when sources of potentially efficient factors of resistance are available. Resistance gene transfer from wild species to commercial cultivars usually requires a great deal of effort and time. The present day accomplishments in development of plant transformation methods and advances in studying plant – pathogen interactions enable the use of genetic engineering for creation of disease resistant plants. Unlike classical hybridization, genetic engineering allows modification or introduction of one or more resistance traits in susceptible cultivars from non-related sources.

During the last decade, many various genetic approaches for obtaining plants resistant to bacterial diseases using genetic engineering have been suggested, including production of antibacterial proteins of non-vegetable origin, inhibition of bacterial pathogenicity and virulence factors, enhancement of plant natural defenses, and artificial induction of programmed cell death at the infection site. These approaches are based on the knowledge of the mechanisms of action of antibacterial agents and successive stages of "plant–bacterium" relationship. This section of the book describes various approaches and discusses their opportunities and limitations in creation of transgenic plants resistant against bacterial diseases.

Development of non-plant antibacterial proteins

Antibacterial proteins are important components of antimicrobial defense mechanisms in many groups of animals. As they work synergistically, they render bactericidal effect on a broad range of Gram-negative and Gram-positive bacteria. Genes that encode some antibacterial proteins have been cloned, and the possibility of expression of these genes in plants has been shown. Attempts are being made to obtain transgenic plants resistant against bacterial diseases.

Lytic peptides of insects. Lytic peptides are low-molecular-weight proteins that produce pores in bacterial membranes. They include cecropines the genes of which were isolated from the silkworm, and their synthetic analogues Shiva-1 and SB-37 were expressed in transgenic potato and tobacco plants. The results regarding acquiring resistance were contradictory. The progeny of transgenic tobacco plants that express the gene Shiva-1 showed delay in disease sign development and drop in percent of the plants destroyed after inoculation with *Ralstonia solanacearum* (bacterial wilt pathogen). Resistance to *R. solanacearum* and *Pseudomonas syringae* pv. *tabaci* (pathogen of tobacco fire wilt) has not been found in the transgenic tobacco plants that express the cecropine gene, because of cecropine degradation by plant proteases. However, resistance occurs in the case of expression of a stable analogue of cecropine (MB 39) in tobacco. Progeny of these transgenic plants is also resistant against leaf inoculation with *P. syringae* pv. *tabaci*.

Attacines are another class of lytic peptides isolated from the silkworm, though their mechanism of action has not been completely investigated. The attacine gene E was introduced in apple plants, and some transgenic clones showed drop in susceptibility to *Erwinia amylovora,* the bacterial blight pathogen. One of these lines showed a 50% decrease in the number of disease signs in conditions of field experiment in artificial infection background.

Lysozymes.   Lysozymes are widespread enzymes with specific hydrolytic activity against bacterial peptoglycan of cell walls. Three various lysozyme genes (of ovalbumin, bacteriophage T4, and human lysozyme) expressed in plants. Cell-free extracts from transgenic tobacco plants that produce chicken ovalbumin lysozyme inhibited growth of some bacterial species. Experiments in greenhouse and *in vitro* showed partial resistance to *E. carotovora atroseptica* in the transgenic potato plants producing lysozyme of bacteriophage – T4, and a small decrease in the signs of the disease, caused by *P. syringae* pv. *tabaci,* in the tobacco plants producing human lysozyme.

Other antibacterial peptides.   Lactoferrin is an iron-bound glycoprotein with antibacterial properties. Expression of the human lactoferrin gene resulted in a 5 to 25 day delay of development of the signs of tobacco plant infection with *R. solanacearum.* This resistance develops due to digestion of lactoferrin resulting in formation of a short peptide with high antibacterial activity. The gene of tachilepsin, which encodes a lytic peptide isolated from the Kamchatka crab, also expressed in transgenic potato, bringing down the number of tubers affected with rot caused by *E. carotovora.*

Efficiency of these approaches can be enhanced by concurrent expression of several genes of the lytic peptides, which provides for synergistic effects. Today, active search of antibacterial proteins from different sources is under way. Discovery of new families of protein molecules will open new opportunities to genetic engineering in terms of creation of transgenic plants resistant against bacterial diseases. Studying of the structure and functions of these proteins can lead to construction of synthetic molecules with improved expression and stability in plant tissues.

Mechanisms responsible for susceptibility or resistance in plant–bacterium relationship
General resistance

Most plants are resistant to potentially pathogenic bacteria due to the resistance named general resistance or non-specific resistance. This resistance is genetically complex, non-specific, and durable. It includes a number of constant factors (cuticle, features of cell wall structure, presence of phenol compounds) and inducible defense mechanisms. The latter supposes recognition of an incoming pathogen by means of multiple plant cell receptors, and availability of signalling pathways, which results in activation of the defense mechanisms.

Susceptibility

Phytopathogenic bacteria develop various biochemical "weapons" to over-come non-specific resistance, and they cause disease in those plants that con-stitute their host range. These "weapons" are known as pathogenicity factors if they play a leading role in pathogenesis, or virulence factors when they only determine the degree of disease. In case of successful course of disease, the permanent defenses are ineffective. The inducible defenses are either not activated or are activated too late, and the bacteria can reproduce and infect new tissues.

Specific resistance

Host plant populations possess disease resistance genes. Specific resistance can be qualitative (usually controlled by single dominant R-genes) or quan-titative (controlled by several genes). The molecular mechanisms and action of qualitative resistance are better investigated than those of quantitative resistance. R-gene products recognize the specific bacterial signals encoded by avirulence genes (avr). This recognition triggers the earlier mentioned inducible plant defenses, including the hypersensitive response. Unlike basic resistance or specific quantitative resistance, qualitative resistance is often non-durable: selection pressure of the resistant host plants can lead to single mutations of the bacterial avr-genes, which can result in a new race of the pathogen unrecognizable by the host plant, and, therefore, capable of caus-ing disease. In some cases avr-genes encode essential functions, which gives selective advantage to the bacteria (for instance, gene avrBs2 of *Xanthomonas campestris* pv. *vesicatoria* or gene avrRpm of *P. syringae* pv. *maculicola*). These avr-genes are stable, and the respective resistance is durable.

Inhibition of bacterial pathogenicity factors

Expression in a plant of any mechanism that inhibits the bacterial pathogenic-ity factors should, theoretically, result in resistance or lowered susceptibility. Though many of these factors (toxins, pectic enzymes, exopolysaccharides and hormones) are well investigated at many bacterial species, this knowledge has not been widely used for development of gene engineering approaches to creation of crops resistant against bacterial diseases.

Inhibition of bacterial toxins

Toxins produced by various pathotypes of *P. syringae* cause signs as chlorotic spots on leaves of the host plant. Due to antimicrobial activity of the toxins these bacteria have advanced self-defense mechanisms. In *P. syringae* pv. *tabaci*, gene ttr (resistance to tab-toxin) encodes the enzyme that inactivates the tab-toxin, inhibitor of glutamine synthetase. A cloned gene of this toxin was expressed in tobacco plant, and the transgenic plants were found to be absolutely resistant against the bacteria. The basis of the second mechanism is synthesis of a target insusceptible to the toxin: *P. syringae* pv. *phaseolicola* synthesizes ornithine carbo-moyl transferase (involved in biosynthesis of arginine), which is known to be resistant to its own phaseolotoxin. The respective gene (arg K) has been cloned and used for transformation of the leguminous plants (soyabean), which as discovered to be resistant against *P. syringae* pv. *phaseolicola* infection.

Other possible approaches

Expression of monoclonal antibodies in transgenic plants is an alternative approach used already successfully for protection of plants against viral infection. Single-chained antibody fragments are specially suitable for this purpose as their synthesis is not too complex, and they can reach the target in various divisions of the cell. For successful protection of plants, antibodies specific for one of the bacterial pathogenicity or virulence factors should be used. A procedure for obtaining antibodies against pectolytic enzymes of *E. chrysanthemi* and *E. carotovora* has been described.

Another possible approach is based on the shortage of iron in plant tissues. *E. chrysanthemi* that produce pectolytic enzymes, causing soft rot synthesizes siderophore (absorbing iron) called chrysobactin, provides bacteria with iron in iron-deficient conditions prevailing in plants. Iron is necessary for bacteria to provoke systematic signs in plants. Plant transformation by the transferrin-encoding genes (transferrins are strong absorbers of iron in vertebrates) to deprive the bacteria of iron during infection has been suggested as a method of disease resistance enhancement.

Enhancement of natural plant defenses

Molecular interactions resulting in activation of plant defenses are intensively studied, and the latest accomplishments open a number of opportunities to enhance crop resistance to diseases and pests by genetic engineering.

Enhanced production of elicitors

Elicitors are pathogen signal metabolites, recognized by plant cells, which trigger plant defenses. They are produced either by the pathogen or by plant cell components, such as cell wall, upon hydrolyzing action of the pathogen. Production of elicitors in plants using recombinant DNA technologies is an attractive approach to enhancement of plant resistance to disease, and attempts have already been made to apply it in potato. Transgenic tubers producing the pectolytic enzyme of *E.carotovora* (known to release oligomeric saccharides, working as elicitors, from cell wall) were much less susceptible to bacteria than the tubers of non-transgenic plants. In this case, the enzyme was produced intracellularly and the elicitors were released only upon infection, as a result of the enzyme release from the tissues destroyed during infection.

Expression of cloned resistance genes

In the last decade, the molecular basis of the host plant and pathogen interaction has been investigated and avirulence genes (avr) and the respective resistance genes (R) have been cloned. These developments have already been used to protect rice against bacterial leaf spot caused by *X. oryzae* pv. *oryzae*. Transformation of a susceptible rice cultivar by R-gene

Xa-21, isolated from a resistant line, yielded rice plants highly resistant to many isolates of *X. oryzae*.

Signalling pathways in the plant kingdom seem to be similar, at least in closely related species. Infact,, the transgenic tobacco that expresses R-gene Pto (isolated from tomato and providing resistance to *P.syringae* pv.*tomato*) showed resistance to *P.syringae* pv.*tabaci*. This opens opportunities for interspecific and intergeneric transfer of functional R-genes. R-gene products that provide resistance to bacteria or fungi have similar structural features, for instance, leucine-rich repeating sequences which may be involved in recognition of avr-signals of avirulence genes. An opportunity to create synthetic R-genes, capable of recognizing a broad range of pathogens and / or designed for stable Avr-signals of these pathogens, looks promising.

Enhanced production of reactive oxygen species

Early response which occurs immediately after plant infection by bacterial pathogens is local, transient, but massive generation of reactive oxygen species (ROS) called oxidative burst. Accumulation of hydrogen peroxide and other ROS seems to be responsible for a number of functions in plant defenses: (1) triggering the local death of hypersensitive cells; (2) intensification of direct antimicrobial activity; and (3) strengthening of plant cell wall. In addition, hydrogen peroxide can serve as a signal for induction of the genes that control defenses in surrounding tissues.

Expression of gluco-oxidase gene (GO) from *Aspergillus niger* induced production of a large amount of hydrogen peroxide in tissues of transgenic potato plants, which resulted in enhanced resistance to *E. carotovora*. Considerable increase in concentration of hydrogen peroxide in plant tissues was found only at bacterial infection. Bacterial pectinolytic enzymes probably released the intracellular glucose which, in turn, served as substrate for GO.

Expression of plant defense genes

Expression of heterologic plant defense genes that results in synthesis of the phytoalexins involved in pathogenesis of proteins (PR) or antimicrobial peptides was mostly used to obtain transgenic plants resistant against fungi; only one group of antimicrobial peptides, thionines, was used against bacteria. Their activity is explained by electrostatic interaction with the membrane phospholipids, leading to formation of pores. *In vitro*, thionines are toxic for a number of plant bacterial pathogens. However, the transgenic tobacco plants expressing thionine genes of barley had different resistance to *P. syringae* pv. *tabaci*. In one case, three days after the leaf inoculation with the pathogen, very low levels of necrotic damage was observed, while in another case no change of resistance was observed, in spite of the fact that transgenic plants synthesized a large amount of thionines. This negative result was put down due to lack of thionine secretion in intercellular spaces where bacteria usually occur.

Transfer of individual defense genes usually provides only partial resistance to the transgenic plants. This may be explained by the fact that defense mechanisms in the plant work against the pathogen as a system. Therefore, expression of a combination of heterologic defense genes should be more effective in terms of enhancing resistance of transgenic plants. Actually, it has already been shown by the example of several plant–fungus systems where synergistic action of chitinase and glucanase increased plant resistance to fungi.

Another strategy can be the controlled activation of the systemic acquired resistance (SAR) resulting in plant immunity. Today, many studies are focused on identification of the genes involved in induction of this event, and they provide opportunities to obtain the plants resistant against a broad range of pathogens, including bacteria, by means of genetic manipulations.

Artificially induced programmed cell death at the infection site

Fast and localized cell death in the primary infection site is a frequently occurring event in resistance, and is interpreted as hypersensitive response (HR). Several systems have been suggested to obtain HR in transgenic plants; however, to avoid a destructive effect of the generalized hypersensitive response, induction of the transgenic plant should be controlled directly at the infection site. Therefore, one of the key success factors of such strategy is the choice of the specific promoter. The artificial induction systems of fast cell death have been suggested and they are discussed in the following sections.

Bacterial opins gene

One more approach is the use of the bacterial gene (b0) that encodes the proton pump (bacterial opsin protein). The mechanism of action is not yet exactly known. Expression of b0-gene in transgenic tobacco plants resulted in enhanced resistance to some viruses and full resistance to *P. syringae* pv. *tabaci*. The transgenic plants also accumulated a large amount of salicylic acid, which is a key chemical signal of the systemic acquired resistance induced by the pathogen. However, expression of b0-gene in transgenic potato has lead to the change of phenotype and lowered viability.

Possible limitations of commercial use of bioengineering methods in creation of plants resistant against bacterial diseases

The earlier described approaches to obtain plants resistant against disease by means of genetic engineering are based on several investigated plant – pathogen interaction models. Expansion of this research, involving new plant and bacterial species, and scale-up to commercial production will require stringent adherence to certain criteria, such as efficiency, reliability, absence of toxicity, and low impact on the environment.

It is difficult to assess efficiency of a strategy before the plant transformation experiments. In addition to successful integration of the target gene in the recipient plant genome, efficiency of this gene depends on the level of its expression, which in some cases can be influenced by the gene silence caused by homology of nucleotide sequences between the bacterial and introduced gene. It can become a serious problem in development of the procedures using homologous sequences (R-genes, thionines, phytoalexins). Moreover, the proteins should be synthesized, exported from the cell and move to the necessary place rapidly enough to avoid serious modifications in process of transport; they should also be stable in their destination and not be degraded by the plant proteases.

Predicting the duration of resistance is even more difficult, as it depends on ability of the bacteria to overcome the new type of resistance. As the action of antibacterial peptides on bacterial cell walls or membranes is not connected to specific receptors, only very considerable changes in the structure of these peptides can make bacteria capable of resisting the lytic mechanisms, and this is unlikely. On the other hand, it is well known that avr-gene mutations are frequent; therefore, the resistance caused by integration of the appropriate R-gene can be overcome. Combination of several various methodological approaches in handling a particular plant genotype will reduce the risk of multiple mutations of the pathogen that can lead to loss of resistance.

Gene engineering selection of plants draws attention to possible toxicity or allergic hazard in consumption of transgenic plants for humans and animals, as well as the impact of these plants on the environment. The former question regards the methods dealing with the peptides not included in human or animal diet: in this case complete and comprehensive assessment of toxicity is necessary. Such peptides as lactoferrin, or peptides isolated from edible plants can hardly be hazardous. The basic question regarding the effect on the environment is possible impact on natural micro flora (endophytic and epiphytic bacterial populations, rhizospheric microflora) due to expression of foreign genes in transgenic plants. It is especially important with regard to the methods using antibacterial peptides that possess a broad range of activity against various pathogens. The complete solution of this question requires development of the adapted procedures. In addition, there is a hazard of transferring genes in closely related plant species, where the consequences are difficult to predict. Finally, an absolutely unexplored field is pathogen evolution against a background of interaction with the transgenic plants resistant to this pathogen.

Selected Literature

Baulcombe DC. Mechanisms of pathogen-derived resistance to virus in transgenic plants. Plant Cell 1996; 8:1833–1844.

Cornelissen DJC, Melchers LS. Strategies for control of fungal diseases with transgenic plants. Plant Physiol 1993; 101:709–712.

Djavakhia VG, Nikolaev ON, Voinova TM. DNA sequence of gene and amino acid sequence of protein from *Bacillus thuringiensis*, which induced non-specific resistance of plants to viral and fungal diseases. J Rus Phytopathol Soc 2000; 1:12–23.

Fisher R, Hain R. Plant disease resistance resulting from the expression of foreign phytoalexins. Curr Opinion in Biotechnol 1994; 5:125–130.

Fitchen JH, Beachy RN. Genetically engineering protection against viruses in transgenic plants. Ann Rev Microbiol 1993; 47:739–764.

Fokunand CN, Beynon JL, Watson KA, et al. Advancement in genetic modification technologies towards disease resistance and food crop production. Biotechnol 2004; 3:1–20.

Malyshenko SI, Kondakova OA, Nazarova JV, et al. Reduction of tobacco mosaic virus accumulation in transgenic plants producing non-functionl viral transport protein. J General Virology 1993; 74:1149–1156.

Michelmore R. Molecular approaches to manipulation of disease resistance genes. Ann Rev Phytopathol 1995; 33:393–427.

Mourgues F, Brisset M-N, Chevreau E. Strategies to improve plant resistance to bacterial diseases through genetic engineering. TibTech 1998; 16:203–210.

Strittmatter G, Janssens J, Opsomer C, Botterman J. Inhibition of fungal disease development in plants by engineering controlled cell death. Biotechnol 1995; 13:1085–1089.

Conclusion

We start this book with the remark: "Organisms in ecosystems can interact with each other in different ways as shown in Table 0.1. Apparently, two organisms in direct contact, or in chemical contact through extracellular metabolites cannot avoid the interaction. Zeroes in Table 0.1 only show that the interaction is so weak that it cannot be detected by the usual methods." Speaking generally, the abilities to very tight and specific interactions constitute the fundamental property of all organisms, and these interactions have much more profound and diverse impacts on their individual development and the natural histories than the influences of the abiotic environment.

Plant research in the last few decades gives strong support to these ideas. Rapid development of research in Molecular Plant–Microbe Interactions (MPMI) has changed our vision of plant life tremendously. Due to this development, it is becoming clearer that the plant is not only a chemical machine that, by using the solar energy, converts inorganic compounds to organics. Rather, the plant is an organizer of complex communities of microbes, which are colonizing the surfaces, tissues and intracellular compartments and are forming the "Microbiome" (Lederberg and McCray, 2001) responsible for enormous genetic and adaptive potential of the higher plants. Having a fastened lifestyle and being restricted in their behavioral adaptations, plants need for their survival many more functions than can be encoded in their own genomes. To reach these adaptations, plants broadly use the biochemical diversity of microbes which retain many of the molecular functions that have been lost during the progressive evolution of higher eukaryotes.

At present, research on Molecular Plant–Microbe Interactions enters a new stage of development. Broad application of transcriptomics, proteomics and metabolomics enable us to implement a comprehensive analysis of symbiotic systems: now we can trace nearly all molecules synthesized during the interaction. The results of this development is intriguing: it turns out that many molecular mechanisms are common for different types of interaction. The same or closely related gene networks are involved in the formation of various pathosystems, in beneficial interactions, in defense against herbivores and even in adaptations to abiotic stresses.

This means that MPMI research returns to its historical roots since more than 120 years ago, Anton de Bary, Professor of Botany at Strassburg University (Figure 1) coined the term "Symbiosis" that he defined as a prolonged living together of distant organisms. Being the initiator of phytopathology, Anton de Bary did not separate deleterious and beneficial interactions, rather he preferred to address them as different manifestations of a common biological phenomenon.

This approach seemed to be forgotten in the twentieth century when scientists preferred to compare mutualism and parasitism as two contrasting types of interactions. Curiously, genetic studies were started simultaneously for beneficial (Nutman, 1946) and pathogenic (Flor, 1946) plant–microbe symbiosis. However, for several decades, their genetics have been developing independently. The restoration of the broad approach to symbiosis is based on essentially new vision of Plant–Microbe Interactions as the continuum of biotic interactions which constitute an essential part of plant development. Plants appeared to have very old systems for monitoring the microbial performance: many important elements in these systems are ancient and are essentially universal for vascular plants, mosses and algae. At the same time, the symbiotic systems in plants are very plastic: they may be reorganized readily to meet new challenges from the rapidly evolving microbial world.

Many recent data demonstrate that the inter-conversion of the mutualism and parasitism is an ordinary event not only in history, but today also. So *Armillaria* species in natural forests form ectotrophic mycorrhizae but in forests, disturbed by anthropogenic factors, they form an aggressive parasite. *Ceratobasidium* fungi (anamorph *Rhizoctonia*) in some plant species form mycorrhizae but they are root-rot agents on other plant species. There are many examples of conversion from symptomless endophytes or benign plant pathogens into the severe ones initiated by changes in the genetic and physiological properties of crops with respect to their wild-growing relatives as well as by large-scale cultivation of genetically uniform cultivar populations.

Weeding out the population of their hosts, parasites create the conditions for formation of multicomponent phytocenosis, which is ecologically more stable. For example, root rot caused by *Stachybotrys chartarum* usually kill 20% of the seedlings of very prolific *Anemone nemorosa* before they reach soil surface. Abscission of leaves, infected by rust *Gimnosporangium juniperinum* in rowan-tree, allow juniper, the second host, to develop under the tree, because it suffers from the disease less than the rowan-tree (see Zadoks, 1987). In this way heteroecious parasite regulate a ratio between its two hosts.

Endophytic fungi – ascomycetes placed among *Clavicipitales* order and basidiomycetes placed among *Ustilaginales* order being the cause of sexual sterility change the ratio of sexual and vegetative reproduction of their hosts (Clay, 1991). On the other hand, rust *Uromycea aritriphylli* hardly infect tubercles of its host *Arisaema triphyllum*, favoring its seed reproduction (see Bierzychudek, 1988). Their actions lead to interconversion of plant populations from outcrossing to clonality and *vice versa*. There are many more such examples.

Of course, the presented data do not mean that the beneficial and deleterious plant–microbe interactions are identical at the cellular or molecular levels. For example in legumes the systemic autoregulation of nodulation (AON) is implemented by genes completely different from those involved in systemic acquired resistance to pathogens (SAR). The physiological mechanisms of beneficial symbioses (integrated system of C and N metabolism in the legume nodules; tightly coordinated bilateral exchange of nutrient through the mycorrhizal interface) differ dramatically from the mechanisms for exploitation of the host metabolism by biotrophic pathogens. Nevertheless, a lot of commonalities revealed among the beneficial and deleterious interactions at the molecular level suggest that these types of interaction represent an evolutionary continuum and the common methodologies may be used to research and to use practically these interactions.

Broad prospects for application of molecular knowledge in sustainable agriculture are opened through the molecular research of Plant–Microbe Interactions. Along with the traditional approaches to defend crops from pathogens and to use the inoculants of beneficial microbes to improve the crop production, a range of novel ideas have been forwarded based on the increasing knowledge about the plant–microbe genetic systems controlling their interactions. The nitrogen-fixing or biocontrol microbes exploit the same regulatory circuits and occupy the same habitats inside and outside the plants as the pathogens. The evidence on plasticity of systems responsible for symbiotic interactions inspires the researchers to manipulate these systems artificially – to combat pathogens more efficiently or to provide the plants with new useful functions. Through a combination of the novel molecular tools and traditional breeding methods, several successful attempts have been made to engineer plants harboring the microbial genes responsible for beneficial traits.

Moreover, in order to construct novel sustainable agro-systems we should move from engineering particular genotypes to the construction of novel genetically integrated plant–microbe communities. Manipulations with the signaling processes and metabolic interactions in plant–microbe associations should be used to create the controlled rhizospheres, as well as the tripartite and multipartite symbioses. The horizons for this ecologically oriented evolution will be extended in parallel to accumulation of molecular knowledge in Plant–Microbe Interactions. We shall be very happy if this book can provide useful contributions to this existing development.

References

Bierzychudek P. Trends in Ecol Evolut 1988; 3:6–7.
Clay K. Trends in Ecol Evol 1991; 6:162–166.
Flor HH. Journ Agric Res 1946; 73:335–357.
Lederberg J, McCray AT. Scientist 2001; 15:8.
Nutman PS. Nature 1946; 157:463–465.
Zadoks JC. Causes, Effects, and Processes. In: Disturbance in Grasses The Netherlands: Dortrecht; 1987:201–207.

Index

AAL-toxin 249, 354, 362
Abscisic acid 155, 234, 262, 282, 301, 309
Abiotic elicitors 187, 364
Abscission 155, 468
Acer pseudoplantanoides 143, 193
AC2 protein 321
Acetabularia 312
Acetate malonate 268, 269
Acetate mevalonate 268, 269
Acetosyringone 141
Acquired resistance 5, 178, 286, 298
Actinomycetes 25, 28, 205, 366, 433
Actinomycin D 178, 269, 271
Additivity 121
Adhesion 139, 140, 208, 212, 260, 261,
    357, 409
Aecidiospores 130
Aflatoxin 107
African Cassava Mosaic Virus 321, 424
Agarose gel 98, 99, 249
Aggressiveness 16, 89, 128, 129,
    329, 468
Agrobacterium 27, 87, 440–442, 444
Agrobacterium tumefaciens 27, 141, 156, 324,
    440–442, 458
Agrobacterium rhizogenes 27, 156, 443
Agroinfiltration 322–324
Albugo candida 25
Alfalfa 14, 204, 267, 273, 282, 286, 367,
    374–376, 382–385, 416–419
Aluminium phosetyl 430, 431
Alternaria 14, 24, 88, 152, 333, 334, 430
Alternaria alternata 14, 152, 249, 331, 334, 335,
    337, 339, 340, 347, 354
Alternaria brassicola 142, 241, 301
Alternaria citri 334
Alternaria fragariae 334, 337
Alternaria kekuchiana 330, 331, 334, 335,
    337, 339, 340
Alternaria longipes 434
Alternaria lycopersici 334, 354
Alternaria mali 334, 335
Aminobutyric acid 435

Ammonium 152, 176, 284, 302, 393, 394,
    396–398, 407
Amplicon 99, 103, 104
Amplification 16, 96, 98–103, 108, 211, 245,
    317, 380
AM-toxin 335–337, 355
Anabaena 366, 395, 397
Anastomosis groups 12–14, 91
Anthocyanin biosynthesis 315
Antibacterial peptides 460, 465
Antibody 76–80, 82, 83, 85, 92, 93, 95, 102,
    358, 360, 363, 450, 451, 462
Antibody expression 450, 451
Antigen(s) 76–88, 90–95, 97, 102, 110, 112,
    222, 254, 357, 450–452
Antisense-RNA-mediated 316, 317
AP (Alkaline Phosphatase) 81, 94, 97, 108
Aphids 35, 91, 101, 162, 172, 294, 412
Apoplast 62, 63, 72, 199, 209, 210, 239, 275,
    322, 328–330
AP-PCR (Arbitrary Primed Polymerase Chain
    Reaction) 108
Apple 4, 14, 24, 26, 27, 87, 334, 335, 460
Appressorium 5, 6, 50–55, 59, 62, 63, 126,
    140, 400
Arabidopsis 52, 54, 58, 62–71, 150, 165, 218,
    220, 223, 225, 227, 231, 237–242, 252,
    254, 258, 259, 274, 285–287, 291, 300,
    354, 360, 374
Arabidopsis thaliana 51, 52, 64, 149, 219, 225,
    226, 291, 319, 323, 431, 433, 437
Arabinase 7
Arabinogalactane 288
Arabinose 143, 288
Arachidonic acid (AA) 185, 189, 192, 193,
    301, 306, 310, 311, 312, 354, 432
Arachis hypogaeca 331
Arbuscular mycorrhiza 374, 387, 390, 398, 399,
    400–403, 405–407, 410, 414–416
Arbuscule 399–402, 407
Arg K gene 461
Armillaria mellea 22, 128, 425
Ascochyta rabiei 328

Ascomycetes 22, 24, 25, 130, 399,
    413, 468
Ascospores 21, 412
Ascus 413
Aspergillus 66, 100, 107, 146, 280
Aspergillus nidulans 210, 344, 347
Aspergillus niger 66, 146, 199, 433, 463
Aspirin 241, 302, 362
ATP 7, 60, 109, 202, 206, 207, 211, 222,
    228, 230, 262, 318, 366, 390, 392, 393,
    396, 397
ATP-dependent RNase 318
Attacines 460
Aureobasidium pullulans 354
Aurone 285
Auxin 148, 149, 155–157, 195, 287, 291, 370,
    387, 405, 410
Avenacin 173, 343, 344
Avenalumin 267, 340, 341
Avenacinase 344
Avidin 82, 97, 100
Avirulence 181–184, 186, 187, 190, 200, 204,
    208–211, 214, 220, 254, 292, 294, 295,
    327, 357, 458, 463
Avirulence alleles 133, 135, 182
Avirulence genes (Avr) 132–135, 181–184, 186,
    187, 190, 200, 208, 210, 214, 220, 294, 327,
    458, 461–463
    Avr4 210, 221
    Avr9 209–212, 214, 221, 361
    avrBs2 204, 461
    avrBs3 204
    avrD 205, 224, 341
    avrPto 205, 212, 221, 223, 225, 342
Avr-signals 463
Azolla 366, 395, 415
Azorhizobium 366, 377
Azospirillum 366, 410

Bacillus 25, 26, 409, 410, 433
Bacillus amyloliquefaciens 457
Bacillus megathericus 26
Bacterial pathogenicity factors 207,
    225, 459, 461, 462
Bacteriophage T4 460
Bacteroids 139, 369, 370–372, 391–395, 397,
    398, 418
Ballistics 444
Banana 87, 88, 296
Bark 22, 141, 171
Barley 6, 12, 32, 86, 153, 163, 210, 211,
    217–220, 224, 239, 266, 274, 281, 323,
    328, 430, 435, 444, 463
Barnase 457, 458

Barnase gene 456–458
Barnase suppressor 457
Barstar gene 457, 458
Basidiospores 56, 57, 130, 131,
    184, 425
Bean 8, 14, 63, 65, 87, 90, 146, 189, 190,
    191, 195, 199, 249, 266, 282, 289, 290,
    296, 300, 328, 329, 394, 429, 435
Beet 13, 14, 22, 23, 87, 90, 141, 173, 175,
    199, 294, 312, 435
Benomyl 428
Benzimidazole 428
Benzothiadiazole 304, 437
Beta- tubulin genes 108
Bibensyl synthase (BBS) 455
Binary vectors 442
Biochemical markers (for studying
    the plant colonization by
    pathogens) 110, 111
Biocontrol 89, 298, 409, 410, 469
Biological ballistics 444
Biotin 82, 96, 100, 163
Biotrophs 4, 7, 8, 10–12, 15, 22–24, 49, 50,
    52, 59, 62, 72, 75, 156, 342
Bipolaris maydis 333
Bipolaris sacchari 333, 335, 337
Bipolaris victoriae 333
Blasicidine 271
Bletilla striata 455
Blight 13, 14, 25–27, 49, 88–91, 110, 113,
    114, 117, 123–125, 127, 141, 152, 161,
    162, 164, 190, 193, 198, 265, 269, 306,
    307, 310–312, 330, 332, 340–342, 428,
    434, 435, 457, 460
Blotting as a means of plant pathogen
    detection:
    immunoblotting 77, 93, 94
    Southern blotting 98
    Western blotting 93, 94
Blueberry 87
Blunt-ended dsRNAs 318
b0-gene 464
Botrytis 14, 52, 53, 57, 65, 66,
    343, 428
Botrytis allii 14, 88
Botrytis cinerea 14, 53, 54, 57, 58, 65–68, 88,
    146, 172, 173, 191, 286, 301, 331, 343,
    428, 453, 455
Botrytis fabae 14
Botrytis tulipae 14, 343
Bradyrhizobium 366, 377, 380
Brassinosteroids 361
Bremia lactuca 453
Bromo-4-chloro-3-indolyl
    phosphate (BCIP) 92

Burkholderia 26, 407
BYDV (Barley Yellow Dwarf Virus): 37, 80
   serotypes 80, 86

Ca²⁺/calmodulin dependent
   kinase 387
Cadmium 319
Caenorhabditis elegans 148, 222, 319
Calcinevrin 356
Callose 43, 62, 63, 217, 224, 299, 303,
   329, 401, 402, 432
Campherol 285
Canavanin 170
Capsid 31, 32
Capsomeres 31
Carbencillin 445
Carbon dioxide 7
Carotenoids 169
Carrot 23, 179, 273, 288, 289
Casben 195, 267, 454
Caspase 222, 255–261
Cauliflower 87
Ca-phosphatase 231
Capsidiol 189, 266, 267, 430, 431
Carbohydrates 6, 7, 20, 25, 28, 86, 98,
   111, 143, 149, 150, 157, 183, 185,
   194, 204, 221, 283, 288, 304, 329, 353
Catechol 171, 303
cDNA 64, 98, 99, 101, 184, 299, 451
Cecropine 459
Cecropine gene 459
Cell-to-cell movement 319, 320
Cellulase 7, 64, 65, 144, 147–149,
   189, 198
Cephalosporium 152
Ceramide 193, 354, 355
Ceratoulmin 151, 154, 155
Cercospora 151, 152
Cercospora nicotianae 153, 435, 455
Cereals 12, 14, 111, 161, 162, 182, 185,
   279, 366, 411, 413–416, 458
Chaconine 173
Chalcone 283–286, 305, 307, 330, 339, 401
Chalcone synthase 284, 305, 339, 401
Chemotrypsin 279–281
Chick pea 33, 271, 328
Chimaera 440
Chimeric gene 67
Chitin 111
Chitinase 149, 189, 210, 247, 273, 275,
   276, 292, 297, 299, 300, 309, 339, 389,
   401, 410, 426, 432, 455, 456, 464
Chitosan 111, 188–190, 193, 195, 234,
   282, 432, 433

Chloramfenicol 445
Chlorophyll-binding protein 445
Chloroplast 8, 19, 26, 32, 55, 56, 60, 61, 65, 69,
   250, 252, 254, 325, 337, 446
Chloroplast DNA 446
Chromatin 249, 256, 257, 259, 400
Chrysobactin 462
Cidial 428
Cinnamic acid 283, 284, 287
Citrus 87, 151, 267, 282, 434
Cladosporium 11, 88, 152, 227
Cladosporium cucumerinum 11, 289,
   331, 430
Cladosporium fulvum 11, 88, 173, 185,
   208–210, 212, 214, 219, 220, 223,
   224, 227, 239, 275, 328–330, 341, 358
Clavibacter 26, 87, 152, 366
Clavibacter michiganensis 92, 151
   subsp. *michiganensis* 93
Clavibacter michiganensis 104
   subsp. *sepedonicus* 107
Claviceps 411, 412
Clover 87, 219, 267, 273, 282, 367, 374–377,
   379, 383, 416, 417
CMV-2b suppressor 321
Coat protein 28, 95, 202, 214, 276, 316,
   429, 446–449, 451
Cochliobolus carbonum 130, 146, 213,
   334, 335, 338
Cochliobolus heterostrophus 142, 213, 333,
   335, 339, 340, 347
Cochliobolus victoriae 333–335, 338–340
Cold shock protein 434, 435
Colletotrichum 24, 100, 189, 427
Colletotrichum atramentarium 331
Colletotrichum gloeosporioides 455
Colletotrichum graminicola 141
Colletotrichum lindemutianum 88, 142,
   189, 199, 289
Commensalism 3–5
Comparable signalling molecules 315
Compatibility 113, 114, 134, 181, 183,
   203, 226, 328, 359
Competition 4–6, 10, 36, 83, 177, 339, 366,
   383, 398, 414, 418, 426, 450
Competitive ELISA 83, 86
Complementarity 78, 79, 183
Conidia 11, 21, 22, 24, 53, 57, 58, 65, 112,
   161, 210, 214, 410, 412, 425, 455
Conjugates 77, 78, 81–83, 85, 92, 93, 212,
   234, 265, 271, 285
Convulvolaceae 266
Coordinated breeding 417
Coronatin 151, 152
Corynebacterium 26, 87

Coriolus versicolor 354
Co-suppression 316
Co-suppressional gene silencing 315
Cowpea 87, 184, 267, 274
CP (Capsid Protein) 30, 35, 36, 158, 202,
    203, 207, 321, 356
Criphonectria 22
Criphonectria parasitica 425
Cross-protection 317, 423, 424, 446
Cross-resistance 446, 447
Cryptococcus 354
Cryptococcus neoformans 356
Cryptogenin (β-cryptogenin) 456
Cucumber 87, 168, 273, 278, 282, 285,
    289, 299, 300, 302, 303, 312, 425,
    430, 433, 434, 435
Cucumber mosaic virus (CMV) 32, 36,
    37, 278, 321
Cumarins 151, 283, 285
Cuticle 5, 6, 40, 45, 52, 59, 64, 66, 142,
    266, 408, 426, 460
Cutin 142
Cutinase 142, 143, 428
Cyanide 343, 344, 391, 409
Cyanide hydrolase 344
Cyanobacteria 366, 395–398
Cyclic peptides 151, 334, 354
Cycloadenylate 229, 230, 245, 425
Cyclooxygenase 310, 354
Cyclofillins 355, 356
Cyclosporin 355, 356
Cytochrome C 259, 260
Cytochrome-P-450 monooxygenase 64
Cytokines 7, 44, 222, 250, 356, 357,
    360, 363
Cytoskeleton 51, 69, 140, 157, 158, 256,
    260, 261, 387, 389

DAC-ELISA (Direct Antigen Coating-Enzyme
    Linked Immunosorbent Assay)
    see PTA-ELISA 86
Daidcein 269
DAS-ELISA (Double Antibody
    Sandwich-Enzyme Linked
    Immunosorbent Assay) 83–86
Datura stramonium 177
Defensin 169, 274
Defensive symbioses 409, 413, 414
Deleterious mobile genetic
    elements 318, 320
Demissine 173, 175
Dendritic cells 250
Derived gene 317
Determinate nodules 369, 374, 394

Desaturases 192
Diacylglycerine 310, 311
Dianthus caryophyllus 177
Dicer enzyme 318
Dicer-like RNase 321
Dichlorisonicotinic acid 299, 304, 435
Differential cultivars 75
Dihydrophenanthrene phytoalexins 455
Diisopropyl orthophosphate 428
Dipstick ELISA 92
Direct immunofluorescence 93
Direct ELISA format 85
Direct DAS-ELISA 85
Direct TBI 94
Diterpenes 168
Diterpenoid phytoalexin 454
DNA-dependent DNA ligase 109
DNA-dependent DNA
    polymerase 99, 101
Dominance 119, 121, 122, 131, 134
Dot-blot immunoassay (DBIA) 92, 96
Dot-ELISA 82, 92, 97
Drosophila 221, 222, 317, 318, 338,
    360, 361
dsRNA hairpin 318
dsRNA intermediate 317, 319
dsRNA size-reduction enzyme 315

E. coli 147, 201, 205, 206, 338, 348, 387,
    440, 442, 445, 456
Ectomycorrhiza 398, 399, 403, 404,
    407, 408
Eicosanoids 310–312, 354
Eicosopentaenic acid (EPA) 185, 192, 193
Electroporation 444
Elicitins 190, 456
Elicitor 186, 191–193
ELISA (Enzyme Linked Immunosorbent
    Assay) 77–93
Endogenous "target" gene 316
Endomycorrhiza 398, 408
Endonuclease 99, 104, 106, 149, 249,
    256, 291, 362, 397
Endophytes 366, 411–468
Endoplasmic reticulum 46, 51, 54, 60, 61,
    65, 72, 157, 369, 400, 451
Endopolygalacturonase 7, 195, 199
Epichloë 412, 413
Epicuticular wax 51
Epistasis 119–121
Epitope 79, 80, 451
Ergosterol 110–112, 114, 169
Erwinia 26, 87, 146, 147, 201, 206,
    301, 357

Erwinia amylovora 26, 102, 141, 206, 433, 460
Erwinia artroseptica 26, 357, 460
Erwinia carotovora 26, 195, 301, 357, 460, 462, 463
Erwinia chryzanthemi 26, 146, 147, 218, 348, 462
Erwinia stewartii 109
Erysiphe graminis 6, 12, 54, 88, 217, 271, 272, 289, 328, 453
Erysiphe orontii 51, 52, 54, 57, 58, 218
Ethidium bromide 99, 104
Ethylene 309
Evolution 5, 6, 11, 13, 14, 19, 42, 108, 154, 200, 222, 226, 227, 245, 255, 300, 310, 352, 360, 363, 365, 367, 374, 377, 378, 390, 403, 415, 465, 467, 469
Exobasidium 156
Exopolygalacturonase 7
Exopolysaccharides (EPS) 208, 390, 461
Extensive 51, 53, 75, 115, 182, 252, 273, 279, 322, 352
Extra-haustorial matrix (EHM) 8, 60–63
Extrahaustorial membrane (EM) 7, 8, 60, 61, 63

Fab 78, 85
FAME (Fatty Acid Methyl Ester) profile 115
Flavoprotein 259
Fc 43, 78, 83, 85
Fitness 212–215
Flagellins 25, 191, 357
Flavanoids 296
Fluorescein isothiocyanate 93
Fluorescence 93, 99, 103, 104, 324
Fluorescence microscope 77
Fluorescent:
    antibodies 76
    detector 103
    dye(s) 77, 93, 104, 106
    probes 97, 99, 103, 104
    substrate 82
Fomitopsis pinicola 22
Foreign phytoalexin gene 452
Formae speciales 12, 95
Formamide 343, 344
Fumonisin 238, 354, 362
Furanocumarins 185, 285, 286
Fusarium 22, 65, 93, 100, 111, 120, 154, 189, 409, 425, 428–430
Fusarium culmorum 111, 331
Fusarium graminearum 88, 111, 153, 154, 163
Fusarium oxysporum 88, 120, 127, 128, 141, 151, 152, 191, 223, 430

Fusarium sambucinum 111, 280, 348
Fusarium solani 88, 142, 143, 185, 191, 210, 280, 344, 345
Fusarium sporotrichoides 153, 154
Fusarium sulfureum 453
Fusicoccin 151, 157

G-proteins 222, 229, 230, 253, 260, 261, 308
Galactanase 7
Galacturonides 196
Galls 8, 27, 49, 156
Garlic 94,164
Gene analysis 321
Gene-for-gene 133, 134, 181, 183, 186, 202–204, 209, 213, 214, 218, 224, 225, 242, 294, 340, 374
Gene engineering 439, 444, 446, 450–455, 457, 461, 465
Gene expression 146, 290, 322, 436, 452
Gene of tachilepsin 460
Gene transfer 148, 315, 440, 441, 459
Genetic engineering 158, 449, 458–460, 462, 464
Genetic load 213
Genetic transformation 154, 208, 315
Germ tube 50–55, 62–64, 399, 411, 432
Geumannomyces graminis 89, 173, 343, 344
Gibberella fujikuroi 453
Gibberella pulcaris 347, 348
Gibberellins 7, 155, 195
Gloeosporium 24
Gloeosporium sorghi 344
Glomerella cingulata 146
Glomus 399, 401, 402, 407, 415
Glucanase 189, 192, 197, 275, 297, 309, 432, 464
Gluco-oxydase gene (GO) 463
Glucose 8, 62, 64, 142, 143, 147, 172, 173, 188, 189, 258, 283, 288, 344, 347, 353, 463
Glucosidase (β-glucosidase) 156, 344
Glucosinolates 174
Glutamate synthase 393, 394, 396
Glutamine synthetase 393, 394, 396, 461
Glyceollin 185, 188, 189, 191, 195, 266, 269, 271, 285, 329
Glycine 206, 288, 377
Glycine max 226, 293, 322, 331, 338, 355, 371, 376, 388, 417, 418
Glycine decarboxylase 338, 355
Glycogen 20, 54, 407
Glycolic acid 338

Glycoproteins 25, 45, 77, 86, 140, 143, 188,
    190, 191, 221, 265, 286–288, 290
Glycosides 171–174, 188, 296, 303, 435
Glycosyl hydrolase 60
Glyoxysomes 51
Golgi apparatus 61, 288, 289
Gomphrena globosa 203
Grafting assay 321, 325
Green fluorescent protein gene (GFP) 324
Green islands 277, 278
Guard model 225, 342
Guanin 13, 244, 261
Guignardia 427
Gunnera 366, 395, 396
GUS (β-Glucuronidase) 64, 67, 68, 290,
    291, 445, 447, 448
Gymnosporangium juniperi-virginiana 62

Hapten 81, 82
Harpins 206, 207, 433, 434
Hartig net 399, 404, 408
Haustorial apparatus 60, 62, 63
Haustorial mother cell 8, 50, 54–56, 59,
    61, 64
Haustorium 7, 8, 52, 55, 56, 60–63,
    124, 140, 141, 183, 209, 212,
    217, 328
HC-Pro protein 321
HC-toxin 334–336, 338
Head blight 88, 110, 114
HeLa cells 255
Hemibiotrophs 4, 22–24, 72, 339
Hemicellulose 143, 144, 147, 194,
    197, 286
Heterobasidium annosum 22, 425
Heterochromatin 249, 252
Heterocysts 395–398
Heterogeneous ELISA format 82
Histone 256, 257, 335, 338, 355
Homogeneous ELISA format 82
Homology-based recognition 315
Host specificity 375, 377–380, 384–386
Hot spots 202, 315
HPLC (High Performance Liquid
    Chromatography) 100, 111
HRPO (Horseradish peroxidase) 81, 82,
    97, 100
Hybridization 13, 26, 79, 96, 97, 99,
    100, 101, 104, 130, 131, 184, 200,
    204, 209, 285, 333, 340, 352, 380,
    439, 459
Hybridization as diagnostic
    methods 26
Hybridoma 79, 80, 93

Hydrogen peroxide 81, 152, 236–239, 262,
    287, 303, 304, 463
Hydrogen radicals 152, 236, 239
Hydrophobins 6, 140, 155, 405
Hypervariable sequence 78
Hypha 6, 12
Hypocrella 152
Hypovirulent strain 425
Hypersensitivity (HR) 247, 253

IgG(s) 78, 83, 85
Immobilization: 82
    antigen/antibodies on solid phase 83, 86
    on the surface of the optical cell 100
Immobilized DNA 96
Immunization 80, 81, 120, 311, 364,
    423, 426, 430, 432
Immunocapture PCR (IC-PCR) 102
Immunodiffusion 76, 77
Immunofluorescence 76, 77, 93
Immunoglobulin(s) 78–81, 85, 92–94,
    227, 357, 358, 361, 450, 451
Immunomodulators 353, 356, 362
Immunosuppressors 153, 198, 332, 341,
    342, 353–355
Impedins 327, 328, 342
Incompatibility 134, 181–183, 203–205, 222,
    226, 359, 361, 366, 425
Indeterminate nodules 367–369, 374,
    392–394
Indicator plants 75
Indirect DAS-ELISA 85
Indirect ELISA format 85
Indirect immunofluorescence 77
Indole-acetic acid 7, 167
Insertion 108, 200, 203, 204, 214, 215,
    273, 320, 439, 440, 445, 446
Integrin 139, 140
Intercalating agent
    in fluorescent probes 99
Interfering RNAs (siRNAs) 315, 317
Interferon 278, 363
Interleukin 221, 222, 357, 361, 363
Intermediate vectors 440
Iris 86
Iron-bound glycoprotein 460
Isoflavanone 296
Isoflavon oxidoreductase 454
Isoprene 168, 169

Jasmonic acid (JA) 152, 205, 233–235,
    241, 243, 244, 274, 300, 301, 305–308,
    354, 432, 435–437

Kanamycin 442, 445
Ketacin 428
Kievitone 266, 285

Lactoferrin 460, 465
Latent infection 87–89, 95
Lateral flow assay (LA) 92
LCR (Ligase Chain Reaction) 109
Lantinus edodes 354
Lectin 25,190, 194, 275, 288, 432
Leghemoglobin (Lb) 391, 395, 397, 402
Lespedezia buergeri 331
Lettuce 35, 37, 87, 267
Leucotriene 310, 312, 354
Ligase 109, 284, 285, 287, 439
Lignin 44, 62, 143, 167, 168, 185, 195, 196,
    250, 276, 283, 286, 287, 290, 299, 429
Linolenic acid 193, 233–235, 306,
    308, 311
Lipo-chito-oligosaccharides (LCOs)
    385, 386
Lipooxygenase 299, 233–235, 244, 245,
    283, 292, 295, 304, 305, 307, 310–312,
    339, 354, 357
Lipopolysaccharides (LPS) 244, 357, 361,
    390, 409, 410, 426
Local silencing signal 319
Lotus corniculatus 331
Lotus japonicus 370, 388 ,399, 402, 403
LRR domain 222, 224, 361, 389
LRR-region 221, 224, 225, 228, 294
Lubimin 185, 268, 272, 296, 330,
    347, 348
Luciferase 82, 445
Luciferin 82
Lily 86
Lymphocytes 79, 356, 363, 364
LysM domain 387
Lysozyme 460
Lytic peptides 459, 460
LZ-region 221, 222

Maackiain 271–273
Macrophage 238, 250, 254, 356, 360,
    361, 363
Magnoporthe grisea 6, 146, 212, 224
Maize 37, 38, 86, 141, 146, 153, 161, 172,
    252, 266, 274, 276, 312, 333, 335, 338,
    339, 347, 348, 354, 444
Malate 391, 393, 410
Malonyl-CoA 283, 453
Mannan 20, 353, 354
MAP-kinase 67, 229, 231, 244, 360

Marker genes 201, 441, 442, 445
Medicago sativa 331, 371, 388,
    417, 418
Medicago truncatula 370, 387,
    388, 403
Melampsora lini 132,135, 181, 219
Melanin 6, 140, 212, 427, 429
Melons 87, 89, 273, 299
Mesorhizobium 366, 377
Metalaxyl 428, 429, 431
Metalloprotease 255
Metatrexate 445
Methyl jasmonate (MeJA) 433, 455
Mevalonate 268
Microarray 67, 97, 98, 110, 241
Microbodies 51, 61
Micrococcus lysodeikticus 434
Microinjections 445
Micro-RNAs 317, 318
Microtitre plate(s) 82, 83, 93,
    100, 102
Milletia japonica 331
Mitochondria 9, 10, 20, 32, 44, 51, 54,
    60, 65, 69, 141, 250, 252, 254, 256,
    259–261, 311, 333, 337, 338,
    393, 406, 431, 446
Mitochondrial DNA 446
Molecular Beacons 104, 105
Momilacton 429
Monilia fructicola 190
Monolicollin 190, 193
Monoclonal antibodies 79, 80, 87, 93,
    451, 452, 462
Monoterpenes 168, 170
Monstera 248
Mosaics 33
Mucor circinelloides 317
Multiplex (multiprimer) PCR 102
Multipurpose vector 446
Mutations 35, 109, 115, 140, 146, 154,
    187, 200, 202–205, 210, 214, 220–222,
    231, 239–241, 251–253, 334, 338, 339,
    370–372, 380–382, 388, 390, 394,
    397, 402, 409, 433,447, 457,
    461, 465
Mutualism 3, 4, 411, 468
Mycobacterium tuberculosum 357
Mycobiont 20, 389, 399, 401, 403,
    405–408, 414
Mycolaminarin 341
Mycorrhiza 365, 366, 374, 388, 398–404,
    406–410, 414, 415, 468, 469
Mycosphaerella 24, 100, 143, 335
Mycosphaerella pinodes 89, 328–331
Myxomycetes 20, 21, 25

N₂ fixation 365–368, 371, 375, 377, 391, 395, 397, 398, 402, 410, 411, 418
NADPH-oxydase 229, 235, 236, 237, 268, 361, 362
NADPH-oxydative system 236, 254, 361
NBS-region 222, 223
Necrosis 4, 5, 31, 32, 35, 38, 41, 44, 49, 65, 67, 123–125, 152, 182, 190, 211, 217, 247–251, 253, 254, 262, 268, 287, 294, 299,302, 306, 329, 356, 362, 456
Necrotrophs 4, 6, 7, 10–12, 15, 16, 22, 23, 49, 50, 52, 64, 65, 68, 71, 72, 174, 241, 342
Nectria 22
Nectria haematococca 142, 146, 344
Nematodes 35, 39–42, 44–46, 59, 91, 100, 102, 147–150, 192, 225, 247, 290–298, 358, 389, 427, 431–433
Neomycin 445, 458
Neopinamine 428
Neotyphodium 412, 413
Nested PCR 101
Neurospora crassa 316, 344
NF-kB-complex 360
Nicotiana 202, 254, 258, 316
Nicotiana glutinosa 182
Nicotiana sylvestris 202, 214, 221
Nicotiana tabacum 64, 434
nif genes 365, 379, 391, 397, 407
Nitro blue tetrazolium (NBT) 92
Nitrocellulose membrane(s) 82, 92, 94, 96, 98
Nitrogenase 27, 365, 367, 369, 372, 379, 390–398, 407
Nitrogen-fixing symbiosis 365
p-Nitrophenol 81
p-Nitrophenyl phosphate 81
nod box 380, 381, 383
NodD protein 381–383
Nod factor 376, 383–390
Nodulation 27, 367, 371–375, 378–389, 402, 403, 417, 418, 468
Nodule bacteria 27, 139, 365, 367, 370, 382, 386, 387
Nodulins 369, 374, 389, 394, 402
NO-synthase 238
Nonadienals 233, 235
Noncompetitive ELISA 83
Nostoc 366, 395–397, 415
NSF (N-ethylmaleimide–sensitive factor) 60
Nucleus 9, 10, 43, 46, 50, 55, 60, 62, 69, 168, 172, 181, 182, 237, 248–250, 252, 256, 257, 259, 293, 312, 321, 361, 400

Oat 25, 35, 38, 152, 157, 173, 266, 267, 333, 335, 337–341, 343, 344
Ophiostoma ulmi 89, 141
Oligogalacturonides 145, 146, 195, 196, 234, 301, 308, 313
Oligonucleotide(s) 97, 99, 104, 278
Oligosaccharines 193–196, 198, 199, 290, 301
Olpidium brasscae 22, 35
Oncogenes 156, 441, 442
Onion 14, 21, 23, 87, 88, 164, 170, 171, 199, 266, 267, 357
Oomycetes 19–21, 24, 25, 59, 76, 77, 80, 86–90, 93, 96, 100, 101, 108, 109, 111, 114, 116, 130, 140, 188, 275, 428, 430
Opsin protein 464
Orchid mycorrhiza 366, 408
Ornithine carbomoyl transferase 461
Osmotin 276
Ovalbumin 460
Oxidase 81, 94, 149, 171, 196, 229, 235–237, 270, 276, 287, 288, 290, 292, 295, 299, 303, 304, 309, 310, 338, 361, 362, 401, 402, 408, 429, 432, 437, 455, 456, 463
Oxygen Paradox 391
Oxyproline 143, 189, 265, 286–289, 292, 309

Papain 30, 78, 210
Papaya 87, 424
Papilla 45, 62, 63, 163, 205, 401
Paranoid mutants 251–253
Parsley 285, 286, 306
Pathogenesis-related proteins (PR-proteins) 212, 273, 276
Pathotoxins 327, 332–336, 338–340, 342, 354
PAZ domains 318
PCR (Polymerase Chain Reaction) 96, 98, 100, 108, 455
p-cumarol-CoA 453
Pea 31, 33, 36, 37, 39, 87–89, 120, 189, 266, 271, 328, 330, 344, 367, 369, 370, 372, 373, 375–377, 379, 383, 386, 387, 402, 416–418
Peach 34, 42, 87, 157
Pear 14, 24, 26, 28, 39, 45, 88, 330, 334, 335, 337, 339, 340
Pectatelyase 145 ,147, 195, 196, 199, 206, 347
Pectolytic enzymes 26, 64, 65, 66, 69, 144, 145, 146, 147, 157, 308, 429, 462
Pepper 87, 176, 189, 192, 203, 204, 214, 266, 267, 294, 357, 430, 434

Peptoglycan 460
Peribacteroid membranes (PBM) 368
Periclinal walls 52
Peronospora parasitica 226, 252, 301
Peronospora tabacina 435
Peroxidase(s) 81, 94, 149, 196, 276, 287, 288, 290, 292, 295, 299, 303, 304, 309, 310, 338, 362, 401, 402, 408, 429, 432, 437, 455, 456
Pesticides 75, 235, 416, 426, 428
Phaseollin 189, 266, 269, 272, 345
Phaseolotoxin 151, 461
Phasolon 428
Phellinus igniarius 22
Phenylalanine 148, 202, 242, 283–286, 295, 305, 307, 309, 330, 401
Phenylalanine ammonium lyase (PAL) 284
Phenols 141, 165, 167, 170, 171, 187, 193, 286, 295, 354, 364, 429
o-Phenylene diamine 81
Phenylpropanoids 165, 283–286
Phoma exiqua 331
Phoma tracheophylla 151
Phosphatase 81, 94, 228, 231, 233, 253, 260, 261, 356, 389, 407
Phosphatidylserine 249
Phospholipase 65, 231–235, 253, 261, 311
Phospholipids 110, 170, 231, 232, 432, 463
Phyllaphora 427
Phylloplane 334, 433
Phthalophos 428
Phytoalexins 65, 164, 165, 184, 185, 189, 190, 193–195, 198, 247, 254, 265–269, 271–273, 285, 286, 295–297, 327, 330, 344, 348, 354, 364, 382, 390, 401, 405, 429, 430, 432, 452–456, 463, 465
Phytoanticipins 164, 165, 295, 305, 343, 430
Phytohormones 139, 155–157, 167, 195, 233, 235, 241, 366, 372, 405, 410, 411
Phytoncides 5, 164, 170
Phytophthora 11, 89, 93, 97, 100, 109, 111, 116, 118, 164, 190, 191, 237, 270, 286
Phytophthora cactorum 22
Phytophthora capsici 89, 190
Phytophthora cryptogea 89, 185, 190
Phytophthora gonapodyides 11
Phytophthora infestans 107, 109, 117, 164, 268, 272, 274, 281, 300, 354, 428
Phytophthora megasperma 271, 285, 289, 428, 453
Phytophthora nicotianae 430
Phytophthora sojae 14, 191

Phytoplasmas 28, 75, 77, 80, 87, 92, 94, 100, 102, 109, 452
Phytosterols 164, 168, 169
Phytotoxins 5, 150
Picnidia 56
Pigmentation 9, 316, 425
Pinosylvin monomethyltransferase 454
Pinosylvin synthetase (PSS) 454
Pisatin 185, 189, 190, 266, 271–273, 328, 344–347, 454
Pisum sativum 331, 370, 371, 373, 388, 402, 403, 417, 418
Plant growth-promoting rhizobacteria (PGPR) 366, 409, 411
Plant nutrition 410, 414
Plasmalemma 7, 8, 46, 51, 54–56, 59, 60, 62, 69, 183, 184, 192, 193, 211, 228, 330, 337, 367, 368
Plasmid 27, 108, 115, 156, 205, 324, 377, 379, 397, 440–446, 458
Plasmid pTi 440, 442, 445
Plasmodesmata 63, 158, 250, 311
Plasmodiophora brassicae 25, 174
Plasmopara viticola 430, 453
Pleurotus 425
Plum 34, 87, 104, 163
Plum pox polyvirus 104
Polyacrylamide 94, 99, 294
Polyamines 308
Polyclonal antibodies 79, 81, 87, 344
Polyclonal antisera 79, 81
Polystigma 427
Polyoxin 428
Polysaccharides 25, 80, 151, 187–190, 194, 195, 199, 200, 208, 236, 287, 303, 353, 354, 357, 390, 396, 401, 409, 426, 461
Polystyrene 82, 86
Polyvinylidene difluoride (PVDF) membranes 92
Poria coccos 354
Poria giganthea 425
Post-transcriptional gene silencing 316
Post-transcriptional termination 315
Potato 13, 26, 33, 87, 88, 164, 192, 193, 237, 267, 273, 282, 292, 294, 296, 306, 311, 323, 434, 463
Potato virus X (PVX) 32, 38, 321, 323
Powdery mildew 6, 49, 52–54, 57, 59, 60, 62–64, 88, 156, 162, 163
Primer(s) 99–104, 107–109
Primordium 367, 373, 374
pRi-plasmid 443–445
Probe(s) 96–100, 103, 104, 106, 200, 204, 209, 220, 344, 379, 380, 436
Probenazole 436

Programmed Cell Death 222, 236, 248, 250, 251, 258, 259, 295, 342, 459, 464

Promoter 64, 67, 68, 142, 210, 228, 274, 290–292, 322, 380, 381, 383, 394, 395, 440, 445–447, 451, 454, 456–458, 464

prp1-1 promoter 457, 458

35S-promoter of the cauliflower mosaic virus (CMV) 324, 445, 456–458

Prostacyclines 310, 354

Prostaglandines 310, 312

Proteases 29, 30, 65, 147, 152, 199, 210, 212, 255–258, 384, 459, 465

Protein A 85, 94, 226, 354

Protein kinase 140, 221–224, 226, 228, 230–233, 235, 237–239, 241, 244, 257, 260, 261, 360, 387–389

Protoplasts 10, 144, 176, 179, 184, 201, 224, 277, 278, 432, 444, 447

Peroxisomes 338

PR-proteins 194, 224, 243, 247, 252, 265, 273–276, 292, 295, 297, 299, 301, 302, 304, 309, 432, 435, 455

Pseudomonas 11, 14, 26, 27, 87, 116, 142, 206, 356, 409, 410, 425

Pseudomonas aeruginosae 68, 357

PA14 68–71

Pseudomonas fluorescens 11, 357, 409

Pseudomonas phaseolicola 27, 151, 152, 461, 462

Pseudomonas savastanoi 26, 27, 151, 152, 156, 223

Pseudomonas syringae 25, 104, 203, 224, 226, 238, 240, 286, 303, 341, 357, 429, 435, 459

Pseudomonas tabaci 26

Pseudoperonospora humuli 142

PTA-ELISA (Plate-Trapped Antigen Enzyme Linked Immunosorbent Assay) 86

Pterostilbene 454

Puccinia coronata 333, 340

Puccinia graminis 12, 50, 90, 111, 123, 182, 225, 328, 353, 432

Puccinia sorgi 120, 219

Puccinia striiformis 51

PVY-HC-Pro protein 321

Pyrasocarboxy acids 436

Pyrophosphotase 5, 22

Pythium ultimum 5, 22

Quercitin 285

Quinosan 428

Ralstonia 26, 27, 87

Ralstonia solanacearum 70, 93, 104, 206, 288, 357, 459

RAPD-PCR (Random Amplified Polymorphic DNA Polymerase Chain Reaction) 107, 108

Raspberry 31, 33, 35, 38, 42, 87, 162

Real-time PCR 103–107, 148

Recipient 201, 333, 379, 384, 439, 440, 444, 452, 464

Recombinant DNA 353, 440, 446, 462

Red algae 9

Regulatory elements 440, 445, 446

Replication 28–31, 36, 98, 99, 175–177, 202, 204, 214, 256, 276–278, 320, 439, 442, 446, 449

Resistance
    aquired 5, 178, 286, 298
    horizontal 117
    vertical 117

Respiration 59, 338, 391, 392, 430

Restrictases 98, 200, 201, 220, 439

Restriction endonuclease 99

Restriction sites 440

Resveratrol 267, 286, 453–455

Resveratrol methyltransferase 454

Retrotransposons 318, 320

Reverse transcriptase (RNA-dependent DNA polymerase) 101

RFLP (Restriction Fragment Length polymorphism) 96, 98, 99, 108

R-gene Xa-21 462

Rhinchosporium secalis 341

Rhizobiaceae 377, 378

Rhizobium 27, 366, 375–377, 379, 380, 382–384, 394, 417, 418

Rhizoctonia solani 5, 12, 22, 91, 107, 142, 199, 300

Rhizopus stoloniferum 195, 345

Rhizosphere 5, 366, 382, 409–411, 433, 469

Rhodamine isothiocyanate 93

Rhodostricta quercina 163

Rhodotorula rubra 354

Ribulosediphosphate-carboxylase 445

Rice blast 6, 90, 111, 112, 125, 219, 427, 434, 437

Rishitin 185, 188, 189, 267, 268, 272, 296, 306, 428

RNA-directed DNA methylation 317

RNA helicase 318

RNA-induced silencing complex (RISC) 318
RNA interference 315, 316
RNase III 318
RNA silencing 315–324
RNA transcripts 316, 317, 449
Root hairs 35, 367, 371, 380, 382, 387, 388, 405, 414
Root rot 13–15, 21, 22, 88–91, 111, 162, 173, 343, 425, 430, 468
ROS (Reactive Oxygen Species) 217, 231, 235–240, 248, 250, 252–254, 262, 268, 342, 430, 437, 463
RT-PCR (Reverse Transcription Polymerase Chain Reaction) 100–102
Rust 7–9, 12, 15, 24, 25, 49–57, 59, 61–64, 90, 91, 111–113, 120, 122–124, 126, 129, 130, 132–135, 156, 162, 172, 181, 182, 184, 214, 219, 220, 227, 249, 252, 329, 333, 340, 342, 432, 468

Saccharomyces cerevisiae 354
Salicylate hydroxylase 303
Salicylic acid (SA) 199, 223, 231, 234, 238, 239, 241, 242, 244, 251–254, 283, 285, 295, 299–305, 308, 362, 437, 464
Salmonella 206, 357
Saprotrophs 3, 4, 6, 10, 11, 13, 15, 28, 347, 405
SAR 243, 247, 261, 298–305, 308, 309, 312, 313, 410, 426, 433, 435, 437, 464, 468
Scab 4, 13, 14, 21, 24, 28, 88, 91, 111
Scizophyllum commune 354
Sclerotia 53, 412, 428
Sclerotinia sclerotiorum 146, 151, 428
Sclerotium 53, 54, 412
Scorpion™ 104, 105
Sensibilization 430, 431
Septoria 24, 91, 100, 107, 111, 113, 344, 430,434
Septoria lycopersici 344
Septoria nodorum 111, 434
Septoria tritici 107
Serological diagnostics 76
Serological detection 86
Serratia 409, 410
Sesquiterpene 164, 168, 171, 337
Sevin 428
Shigella flexneri 357
Shikimate 165, 166, 268
Shuttle vectors 446
Siderophore 410, 462

Signal exchange 370, 375, 384, 386, 405
Signaling 62, 140, 274, 275, 285, 299, 301–303, 310, 313, 375, 382, 386–389, 394, 469
Signalling systems 218, 227–229, 232, 233, 241, 243, 244, 245, 358, 359, 432
Silver:
   gel staining 99
Simbiotrophism 10
Single-chained Fv (scFv) 451
Single-stranded antisense-RNA 316
Singlet oxygen 152
Sinorhizobium 366, 375–377, 379, 380, 382–384, 418
Small antisense-RNA 317
Smut 7, 14, 15, 24, 25, 49, 54, 126, 162, 219, 281
Solanaceae 13, 26, 27, 173, 266, 279, 282, 288, 296
Solanine 173–175
Solanum demissum 173
Sophora japonica 273
Southern blotting 98
Soybean 253
Specialization 5, 10, 12–15, 26, 32, 35, 128, 137, 142, 146, 150
Spiroplasma 28, 452
Sporobolomyces 354
Squalene synthetase 164, 268, 270
Stem-loop precursor RNAs 318
Stemphylium loti 343
Stem rust 9, 12, 57, 90, 113, 123, 130, 162, 182, 214, 329
Sterols 7, 110, 111, 164, 168, 169, 173, 191, 268, 272, 291, 298
Stilbene 283–286, 453, 454
Stilbene synthase 284
Strawberry 14, 21, 87, 90, 334, 434
Streptavidin 82, 97
Suberin 44, 163, 276
Substomatal cavity 54, 55, 68, 69
Substomatal vesicle 50–52, 54, 55, 64, 126
Substrate (of enzymatic reaction) 82
Sugarcane 87, 333, 335
Superoxide 152, 229, 235–239, 244, 245, 276
Superoxide synthase 235, 237, 245
Suppressors 187, 243, 274, 318, 320, 321, 323, 324, 327–329, 332, 341, 342, 353–355, 390, 431
Sustainable agriculture 414, 415, 469
Symbiosis 467
Symbiosome 369, 371, 391, 396

Symptoms 1, 15, 32, 36, 49, 75, 83, 86,
  90, 147, 150, 152, 154, 251,
  323, 412
Synchytrium endobioticum 15
Syntaxin 60
Synthase 233, 235, 237–239, 241, 242, 244,
  245, 278, 284, 305, 339, 385, 393, 394,
  396, 401, 455
Systemic silencing signal 319
Systemin 282, 301, 307, 308

Taq polymerase 99, 101, 104, 106
Tabtoxin 151, 152
TaqMan® 104
Tangerine 162, 334, 335
Taphrina 7, 25, 156
Taphrinales 7, 25
Target gene 315–318, 324, 325, 379, 383,
  439, 440, 442–445, 457, 458, 464
Taumatins 275, 300
Taumatococcus daniellii 276
TBI (Tissue blot immunoassay) 94
T-DNA 156, 324, 440, 442–445, 458
Tentoxin 151, 152
Terpenoids 164, 168, 169, 172, 173, 266–268,
  270, 296, 348, 364
Thallomes 9
Thanatephorus cucmeris 22
Thionines 165, 169, 170, 276, 305,
  463, 465
Thielaviopsis basicola 91
Thresholds 109
Tilletia tritici 14, 91
TIR-region 221, 222
Tobacco 14, 22, 34, 152, 178,
  182, 202, 237, 267, 274, 276,
  282, 285, 435
Tobacco rattle virus (TRV) 32, 35, 38,
  42, 323
Tolypocladium 356
Toll-protein 222, 360
Tomatinase 344
Tomatine 173, 174
Tomato 27, 31, 87, 282, 289, 312
Tonoplast 51, 54, 171, 174, 232, 250, 251,
  257, 285
Topoisomerase 256
Tracheomycosis 22
Transcriptional regulators 387
Transferases 184, 240
Transferrin-encoding genes 462
Transgenosis 439, 445, 447
Transgenic dihaploid 444
Transgenic petunia 316

Transgenic plant 322, 440, 445, 448–450,
  453, 464
Transient expression systems 322, 323
Transient silencing suppressor 324
Transitive RNA silencing 317, 319
Translation 28–31, 36, 140, 179, 209,
  269, 279, 318, 449
Transport protein 30, 63, 64, 157, 158,
  169, 202, 203, 312, 449, 450
Transposons 200, 222, 379
Transposon mutagenesis 154, 200, 220,
  227, 251, 252, 344
Tricarboxylic acid (TCA) cycle 393
Trichoderma 91, 210, 409, 426
Trichothecene 107, 153, 154
Tricyclasol 427, 428
Trifluraline 430
Trifolium pratense 331, 371
Trifolium repens 331
Triggers of RNA silencing 316
Triterpenes 168
Thrombaxanes 354
Trypsin 276, 279–281, 297, 305
Tth polymerase 101
T-toxin 335, 338
Tulip 14, 86, 172, 343, 435
Tumors 8, 26, 27, 156
Turgor pressure 6, 19, 140, 356
Turnip 29, 32, 34, 38, 87, 267,
  321, 447
Turnip mosaic virus 447
Type III secretion 206–208, 243

Umbelliferae 266
Untranslatable coat protein 316
Uredospore(s) 112
Uromyces phaseoli 328, 429
Uromyces striatus 62
Uromyces vignae 184
Ustilago maydis 25

Vaccination 351, 423
Valsa 22
Vascular transport of viruses 319
Venturia ineaqualis 14
Verticillium 7, 14, 22, 91, 100, 120, 126,
  157, 162, 163, 195, 199, 409, 427, 430
Verticillium dahliae 14, 91, 126, 163,
  195, 199
Vicia faba 331, 371
Victorin 333–335, 337, 338, 340, 341, 354
Vigna sinensis 331
Vir-genes 141, 187, 327, 341, 440, 443

Vir-region 442, 443
Viroids 31, 59, 75, 95, 100–102, 108, 109
Virus-induced gene silencing
 (VIGS) 316, 323
Vitis vinifera 453
Vivotoxins 139, 150, 151, 155, 157, 332

Watermelon 299
Western blotting 93, 94
Wheat 7, 12–14, 35, 38, 51, 53–56, 61, 76,
 86, 89, 90, 91, 94, 111–114, 123, 124,
 126, 129–132, 134, 152, 153, 157, 161,
 162, 172, 179, 182, 198, 219, 220, 225,
 266, 280, 281, 289, 294, 317, 319, 323,
 328, 329, 344, 410, 411, 430, 432, 434,
 435, 437, 440, 444

Xanthomonas 26, 27, 87,116, 142, 206
Xanthomonas axonopodis 27
Xanthomonas campestris 151, 461
Xanthomonas malvacearum 26
Xanthomonas oryzae 124, 219
Xanthomonas vesicatoria 204, 224, 357
Xenobiotics 411, 415, 439
Xylanase 7, 144, 146, 147, 191

Yeast 20,140, 157, 223–225, 231, 338,
 342, 362
Yersinia pseudotuberculosum 206, 357

Zoospores 5, 21, 24, 35, 80, 141, 142, 170, 274
Zea mays 252

and (Lotan 1970; Apostol et al. 1976) . . . . . . the diet
efficient plant nutrients . . . . . (Lotan et al. 1976; Mattson
important plant nutrients decreasing . . . growth (Herms and . . . 1992) (Lou
Despommier, Beille, 2002; . . . . . . browser . . . . puzzlement appears . . .
also only present . . . in three inducement resistance of plant those (Herms
Douglas, . . . (1995) (Chapman 1974) . . . .
. . . between interactions . . . . . (Chapman 1974)
. . . .
Brough that would be to more effective planting (Lee 1980 should also . . .
. . . effect to post . . . population control . . . . . . decline . . . may single